Phytoplankton and Trophic Gradients

Developments in Hydrobiology 129

Series editor
H. J. Dumont

Phytoplankton and Trophic Gradients

Proceedings of the 10th Workshop of the International Association of Phytoplankton Taxonomy & Ecology (IAP), held in Granada, Spain, 21–29 June 1996

Edited by

M. Alvarez-Cobelas, C.S. Reynolds, P. Sanchez-Castillo & J. Kristiansen

Reprinted from Hydrobiologia, volumes 369/370 (1998)

Springer-Science+Business Media, B.V.

Library of Congress Cataloging-in-Publication Data

A C.I.P. Catalogue record for this book is available from the Library of Congress.

ISBN 978-90-481-5067-0 ISBN 978-94-017-2668-9 (eBook)
DOI 10.1007/978-94-017-2668-9

Printed on acid-free paper

Hydrobiologia **369/370**: v–vii, 1998.
M. Alvarez-Cobelas, C. S. Reynolds, P. Sanchez-Castillo & J. Kristiansen (eds), Phytoplankton and Trophic Gradients.

Contents

Seasonal and Long-Term Dynamics

Taxonomy and Ecology of Photosynthetic Flagellates

Foreword

The International Association of Phytoplankton Taxonomy and Ecology exists to share understanding and knowledge among students of the plankton. Particular regard is paid to the composite question: what lives where and why? The Association's 10th Workshop was held in Granada, Spain, in June 1996. Fifty-three people from nineteen countries attended the meeting which, following precedent, comprised two main parts. The first reviewed the progress in ecology: the particular emphasis this time was laid on the relation between species composition and the trophic gradient. The second, directed towards taxonomical issues, concentrated on problems concerning planktonic Chrysophytes, Volvocales and Euglenophytes living at the extremes of the trophic spectrum. The taxonomy of these divisions have remained relatively less well-studied than that of other phytoplankton groups.

The contributions to these proceedings have been arranged in sections. In the first of these, two overview papers set the scene. Rojo provides an analysis of the literature to reveal the extent of, and the significant gaps in our current knowledge. Reynolds reviews some past approaches and suggests that no single set of factors (e.g. nutrients) satisfactorily explains the selection of phytoplankton across trophic gradients. Instead, the trophic spectrum appears to be a probabilistic outcome of several dimensions of variability that impinge upon phytoplankton species selection.

Twelve papers comprise the section on community structure. Some of these, e.g. the papers by Dasí et al., Díaz et al., Huszar et al., Izaguirre et al., Lepistö & Rosenström and Trifonova, are regional studies. Collectively, they span all latitudes from Antarctica to equatorial Brazil. Spijkerman & Coesel present an ecophysiological study distinguishing the preferences of desmids characterizing different parts of the trophic gradient. Another paper (Vörös et al.) considers picoplankton distribution across light- and nutrient-gradients, whereas two papers (Reynolds et al., Romo) describe the responses of phytoplankton communities to deliberate attempts to raise the pH by artificial enrichment. Komárková reviews her work on top-down effects on phytoplankton in fish ponds, lakes, and reservoirs, subject to variable fish-stocks.

Temporal dynamics, viewed at a variety of scales, are treated in the next nine contributions. Seasonal studies carried out in lakes (Büsing, Naselli-Flores & Barone, Padisák et al.) and rivers (Gosselain et al., Ha et al.) are presented. The paper by Wiedner & Nixdorf concentrates on phytoplankton dynamics under ice in temperate lakes of differing trophic status. Three long-term studies are analysed. Juhos & Vörös adopt rank-frequency diagrams to reveal changes in the community responses to contrasted levels of eutrophication in the two basins of Lake Balaton. Ruggiu et al. report responses of phytoplankton to a lowering of the nutrient load into Lake Maggiore. Stoyneva presents some interesting effects on the phytoplankton of a lower-Danube floodplain lake, which arises through changes in the management of the catchment area and the variable connection of the lake to the river.

The section on taxonomy and autoecology includes papers on Volvocales (Angeler, Schlegel et al.), Chrysophytes (De Hoyos et al., Olrik, Peterfi et al.), Euglenophytes (Conforti, Tell) and mixed assemblages (Domingos & Menezes, Nixdorf et al.).

The editors are grateful to all the contributors for their cooperation in the preparation of the manuscripts. We are grateful to the many referees who kindly reviewed manuscripts for us. The workshop was supported by the Spanish Higher Council of Scientific Research (CSIC), the County of Andalucia (Junta de Andalucía) and the University of Granada. On behalf of all the workshop participants, we are pleased to express our gratitude. The technical advice from the editorial team at HYDROBIOLOGIA (Mr René Mijs, Prof. Henri Dumont and Mrs Simonne Wellekens) has been valued and is also gratefully acknowledged.

THE EDITORS

List of participants

ÁCS, Eva. Eötvös Lovánd University, Fac. Microbiology, Muzeum Kkt 4A, H-1088 Budapest, Hungary.
acs@ludens.elte.hu

ALVAREZ-COBELAS, Miguel. Centro de Ciencias Medioambientales (CSIC), Serrano 115 dpdo., E-28006 Madrid, Spain.
malvarez@fresno.csic.es

ANGELER, David. Institut für Pflanzenphysiologie, Hydrobotanische Abteilung, Althanstrasse 14, A-1090 Vienna, Austria.
dangel@pflaphy.pph.univie.ac.at

ANSOTEGUI, Arantxa. Lab. Ecología, Fac. Ciencias, Univ. País Vasco, Apdo. 644, E-48080 Bilbao, Spain.

BARONE, Rosella. Dip. Scienze Botaniche, Via Archirafi 38, I-90123 Palermo, Italy.

BARRERA José. Rhone-Poulenc Química S.A., Capitán Haya 1, Pl. 11, E-28020 Madrid, Spain.

BÜSING, Nanna. Institut für Botanik, Sternwartestrasse 15, A-6020 Innsbruck, Austria.
nanna.buesing@uibk.ac.at

CALLIERI, Cristiana. CNR-Istituto Italiano di Idrobiologia, Largo Tonolli 50/52, I-28048 Pallanza-Verbania, Italy.
callieri@iii.to.cnr.it

CHORUS, Ingrid. Institut für Wasser, Boden und Luft B112, Corresplatz 1, D-1000 Berlin 33, Germany.

COMÍN, Francisco A. Dept. Ecología, Fac. Biología, Univ. Barcelona, Avgda. Diagonal 645, E-08028 Barcelona, Spain.
comin@porthos.bio.ub.es

CONFORTI, Visitación. Dept. Biología, Fac. Ciencias Exactas y Naturales, 1428 Ciudad Universitaria, Núñez, Buenos Aires, Argentina.
conforti@bg.fcen.uba.ar

CRUZ-PIZARRO, Luis. Dept. Ecología, Fac. Ciencias, Univ. Granada, Avda. Fuentenueva s/n., E-18001 Granada, Spain.
lcruz@goliat.ugr.es

DASÍ, María José. Dept. Microbiología y Ecología, Fac. Biología, Univ. Valencia, E-46100 Burjassot, Valencia, Spain.

DE HOYOS, Caridad de. Dept. Ecología, Fac. Biología, Univ. Salamanca, E-37008 Salamanca, Spain.
dehoyos@gugu.usal.es

DÍAZ, Mónica M. Centro Regional Universitario de Bariloche, Unidad Postal Universidad del Comahue, 8400 Bariloche, Argentina.
mdiaz@uncmai.edu.ar

DOKULIL, Martin. Institut für Limnologie, Österreichische Akademie der Wissenschaften, Gaisberg 116, A-5310 Mondsee, Austria.
martin.dokulil@oeaw.ac.at

GOSSELAIN, Veronique. Unité d'Ecologie des Eaux Douces, FUNDP, Rue de Bruxelles 61, B-5000 Namur, Belgium.
veronique.gosselain@fundp.ac.be

GRIGORSZKY, István. Dept. Ecology, L. Kossuth Univ., P.O. Box 71, H-4010 Debrecen, Hungary.
gege@tigris.klte.hu

HA, Kyong. Dept. Biology, Pusan National University, Pusan, 609-735, South Korea.
kha@hyowon.cc.pusan.ac.kr

HERNÁNDEZ, Pilar. Dept. Ciencias Ambientales, Univ. Alicante, Aptdo. 99, E-03080 Alicante, Spain.
pilar.hernandez@aitana.cpd.ua.es

HÖRNSTRÖM, Einar. Stockholm University, Institutet for Tillampad, Miljoforskning, S-10691 Stockholm, Sweden.

HUSZAR, Vera L. M. Lab. Ficologia, Dept. Botânica, Museu Nacional, Univ. Federal Rio de Janeiro, Quinta da Boa Vista, São Cristovão, 20940-040 Rio de Janeiro, Brazil.

IZAGUIRRE, Irina. Dept. Ciencias Biológicas, Fac. Ciencias Exactas y Naturales, 1428 Ciudad Universitaria, Núñez, Buenos Aires, Argentina.
iri@bg.fcen.uba.ar

JOO, Gea-Jae. Dept. Biology, Pusan National University, Pusan, 609-735, South Korea.
gjjoo@hyowon.cc.pusan.ac.kr

JUHOS, Szilveszter. Balaton Limnological Research Institute, Hungarian Academy of Sciences, H-8237 Tihany, Hungary.
ndgy@tres.blki.hu

KIM, Hyun-Woo. Dept. Biology, Pusan National University, Pusan, 609-735, South Korea.
hwkim7@hyowon.cc.pusan.ac.kr

KISS, Keve T. Hungarian Danube Research Station, Hungarian Academy of Sciences, Javorká S.u. 14, H-2131 Göd, Hungary.

KLING, Hedy. Dept. Fisheries and Oceans, Freshwater Institute, Winnipeg, Manitoba, R3T 2N6, Canada.
hedy-kling@fwi.dfo.ca

KOMÁRKOVÁ, Jaroslava. Hydrobiological Institute ASCR, Na Sádkách 7, CZ-37005 České Budějovice, Czech Republic.
jarkakom@hbu.cas.cz

KRISTIANSEN, Jørgen. Institut for Spøreplanter, Univ. Køvenhavn, Øster Farimagsgade 2D, DK-1353 Køvenhavn, Denmark.
joergenk@bot.ku.dk

LEITAO, Maria. Bi-Eau, 14 rue Volney, F-49000 Angers, France.

LEPISTÖ, Liisa. National Board of Waters and Environment, Pohj. Rautatiekatu 21 B, P.O.B. 250, SF-00101 Helsinki 10, Finland.
liisa.lepisto@ryh.fi

LINARES, Eduardo. Dept. Biología Vegetal, Fac. Ciencias, Univ. Granada, Avda. Fuentenueva s/n., E-18001 Granada, Spain.

MENEZES, Mariângela. Lab. Ficologia, Dept. Botânica, Museu Nacional, Univ. Federal Rio de Janeiro, Quinta da Boa Vista, São Cristovão, 20940-040 Rio de Janeiro, Brazil.

MIRACLE, Rosa. Dept. Microbiología y Ecología, Fac. Biología, Univ. Valencia, E-46100 Burjassot, Valencia, Spain.

MISCHKE, Ute. Technische Universität Cottbus, Lehrstuhl Gewässerschutz, Seestrasse 45, D-15526 Bad Saarow, Germany.

MORABITO, Giuseppe. CNR-Istituto Italiano di Idrobiologia, Largo Tonolli 50/52, I-28048 Pallanza-Verbania, Italy.

NASELLI-FLORES, Luigi. Dip. Scienze Botaniche, Via Archirafi 38, I-90123 Palermo, Italy.
nase001@it.net

NAVARRO, María José. Dept. Ingeniería Química, Univ. Alicante, Aptdo. 99, E-03080 Alicante, Spain.

NIXDORF, Brigitte. Technische Universität Cottbus, Lehrstuhl Gewässerschutz, Seestrasse 45, D-15526 Bad Saarow, Germany.

OLRIK, Kirsten. Miljøbiologisk Laboratorium Aps, Baunebjergvej 5, DK-3050 Humlebæk, Denmark.

PADISÁK, Judit. Balaton Limnological Research Institute, Hungarian Academy of Sciences, H-8237 Tihany, Hungary.
padisak@tres.blki.hu

PANZANI, Pierisa. CNR-Istituto Italiano di Idrobiologia, Largo Tonolli 50/52, I-28048 Pallanza-Verbania, Italy.

PÉREZ-MARTÍNEZ, Carmen. Dept. Ecología, Fac. Ciencias, Univ. Granada, Avda. Fuentenueva s/n., E-18001 Granada, Spain.
cperez@goliat.ugr.es

RAFN, Pia Dorthea. Miljøbiologisk Laboratorium Aps, Baunebjergvej 5, DK-3050 Humlebæk, Denmark.

REYNOLDS, Colin S. NERC Institute of Freshwater Ecology, The Ferry House, Ambleside, GB-LA 22 0LP, Cumbria, United Kingdom.
csr@wpo.ife.ac.uk

ROJO, Carmen. Dept. Microbiología y Ecología, Fac. Biología, Univ. Valencia, E-46100 Burjassot, Valencia, Spain.
carmen.rojo@uv.es

ROMO, Susana. Dept. Microbiología y Ecología, Fac. Biología, Univ. Valencia, E-46100 Burjassot, Valencia, Spain.
susana.romo@uv.es

ROTT, Eugen. Institut für Botanik, Sternwartestrasse 15, A-6020 Innsbruck, Austria.
eugen.rott@uibk.ac.at

RUGGIU, Delio. CNR-Istituto Italiano di Idrobiologia, Largo Tonolli 50/52, I-28048 Pallanza-Verbania, Italy.
ruggiu@iii.to.cur.it

SÁNCHEZ-CASTILLO, Pedro. Dept. Biología Vegetal, Fac. Ciencias, Univ. Granada, Avda. Fuentenueva s/n., E-18001 Granada, Spain.
psanchez@goliat.ugr.es

SCHLEGEL, Ines. Institut für Gewässerökologie und Binnenfischerei, Alte Fischerhütte 2, D-16775 Neuglobsow, Germany.
schlegel@igb-berlin.de

SCHMIDT, Mechthild. Technische Universität Cottbus, Lehrstuhl Gewässerschutz, Seestrasse 45, D-15526 Bad Saarow, Germany.

SENDRA, María Dolores. Dept. Microbiología y Ecología, Fac. Biología, Univ. Valencia, E-46100 Burjassot, Valencia, Spain.

SPIJKERMAN, Elly. Dept. Aquatic Ecology, Univ. Amsterdam, Kruislaan 320, NL-1098 Amsterdam, The Netherlands.
postma@bio.uva.nl

STOYNEVA, Maya P. Sofia University 'St.Kl.Ohridski', Fac. Biology, Dept. Botany, Bld. Dr Zankov 8, Sofia 1000, Bulgaria.
subf@sf.cit.bg

TELL, Guillermo. Dept. Biología, Fac. Ciencias Exactas y Naturales, 1428 Ciudad Universitaria, Núñez, Buenos Aires, Argentina.
tell@bg.fcen.uba.ar

TRIFONOVA, Irina. Institute of Lake Research RAS, Sevastijanova 9,196199 St.-Petersburg, Russia.
trif@lake.spb.su

VICENTE, Eduardo. Dept. Microbiología y Ecología, Fac. Biología, Univ. Valencia, E-46100 Burjassot, Valencia, Spain.
eduardo.vicente@uv.es

WIEDNER, Claudia, Technische Universität Cottbus, Lehrstuhl Gewässerschutz, Seestrasse 45, D-15526 Bad Saarow, Germany.

WILLÉN, Eva. Center for Environmental Monitoring, Swedish University of Agricultural Sciences, P.O. Box 7050, S-75007 Uppsala, Sweden.
eva.willen@ma.slu.se

Hydrobiologia **369/370:** 1–9, 1998.
M. Alvarez-Cobelas, C. S. Reynolds, P. Sanchez-Castillo & J. Kristiansen (eds), Phytoplankton and Trophic Gradients.

Differential attributes of phytoplankton across the trophic gradient: a conceptual landscape with gaps

Carmen Rojo
Department of Ecology, Faculty of Biology, University of Valencia, c/ Dr Moliner s/n, E-46100 Burjassot, Valencia, Spain

Key words: phytoplankton, ecological variables, trophic state, eutrophication, freshwater

Abstract

I provide a conceptual landscape of attributes of phytoplankton related to trophic gradients, by reviewing 69 papers and books on eutrophication, with special reference to the ecology of the phytoplankton. Forty-eight ecological variables such as total phosphorus and Chl-*a*, are used, each related to the trophic gradient measured. They are subdivided in structural features of the phytoplankton, functional or dynamic features, and other relevant properties of the plankton community. Only twelve of the forty-eight variables are statistically related to trophic gradient. In most cases, variability across a trophic gradient is only nominally described. Less attention is given to functional, structural or dynamic variables. Some associations occur because the variables are interrelated, for example, photosynthesis and Chl-*a*, or because the mathematical comparison is between a whole factor and a part of it. Many important environmental features which trigger or affect phytoplankton behaviour are independent of trophic state. Therefore, trophic states might be viewed as a multi-dimensional space, the axes of which are defined by the various independent variables, and not merely a simple gradient of fertility. The alternative approaches to the search for pattern are either the single *global* pattern (which seeks to explain the ecological behaviour of phytoplankton in all freshwater systems) or multiple *local* patterns (where only a small set of specific variables is solved for each question and system). If the ultimate goal is a paradigm of phytoplankton dynamics, biomass models, and tools for the prediction of species composition, phytoplankton ecologists should adopt the theories of nonlinear dynamic complex systems.

Introduction

Many concepts in ecology are only loosely defined. This vagueness often causes misunderstanding, through interpretational semantics. Eutrophication, trophic gradient and trophic states provide good examples of this. The term and the basic understanding are qualitatively familiar, yet debate continues about how the classical trophic states, from oligotrophic to hypertrophic, are to be categorised and assessed on gradients of concentration of major nutrients (nitrogen, phosphorus and others), or the capacity to sustain the growth of primary producers (Moss, 1980; Reynolds, 1984a; Harris, 1986; Harper, 1992). Similarly, the relationship between the concentration of total phosphorus and of chlorophyll *a*, recognised for almost a century, has been used as a surrogate for grades of eutrophication (OECD, 1982). Yet the effects of morphometry, cascading interactions and the recently recognized principle of alternative states (Scheffer et al., 1993) complicate even this simple picture.

Plankters are amenable organisms for ecological study because of their fast growth and well-researched intra-interspecific and inter-specific relationships (for example, Tilman, 1982, on competition; Hutchinson, 1961 on diversity; Reynolds, 1980 on succession; Carpenter & Kitchell, 1993 on food-web dynamics). Moreover, researchers have different objectives (ecological answers, management solutions, etc.) and different research specialisms (from molecular to community biology). Consequently, there is a vast amount of information available about the many attributes of phytoplankton. However, there is no established high-order pattern, no falsifiable paradigm, about the behaviour of phytoplankton across a trophic gradient. Progress in both understanding and predicting events in nature has been hampered for several reasons, including the difficulties arising in comparisons of data assembled using different techniques, collected from different lake types, or confounded by weak experimental approaches. Most serious is the lack of a conceptual overview that allows local or single results to be fitted to a whole pattern.

The aim of this paper is to outline a conceptual landscape of attributes of phytoplankton form and function which incorporates demonstrable sensitivities to variation in trophic state. In particular, the following questions are addressed: Do we know enough to understand the behaviour of phytoplankton across the trophic gradient? Where are the gaps in this landscape? Should oligo-meso-eutrophic states be described as a productivity gradient? Would not relating eutrophication with ecology of primary producers give trivial results? Kalff & Knoechel (1978) wrote that *'advances in understanding and predicting the dynamics of phytoplankton, in both oligotrophic and eutrophic lakes, would require a more holistic view of phytoplankton as a dynamic component of lake ecosystem'*. Almost twenty years later, do we know how extensive a holistic view of these problems should be?

I think that the challenge is to achieve the best and most complete analytical research in order to understand the mechanisms and, at the same time, to attempt to predict the dynamics of phytoplankton, treating the issue as a problem of emergent properties of a complex system. With this challenge in mind, we should bear in mind that *'It is very easy to be so busily occupied with trivialities that the greater thoughts are excluded. We must therefore beware lest technological progress becomes a bar to creative though'* (Lund, 1964).

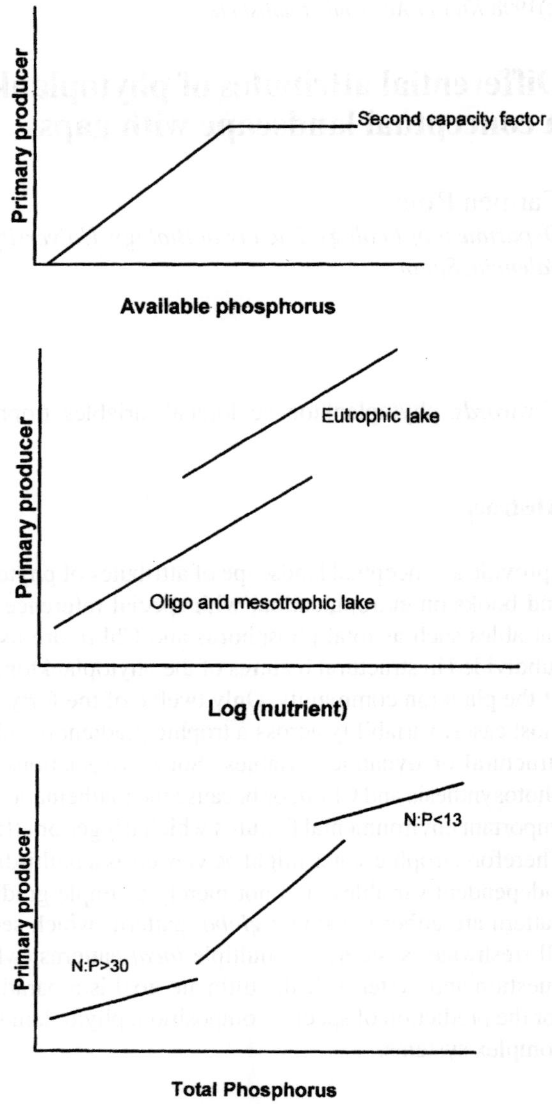

Figure 1. The figure shows different nonlinear relationships between primary producers and eutrophication (modified from Reynolds 1992, White et al. 1988 and Prairie et al. 1989, respectively).

Material and methods

The approach adopted has been to review a selection of papers and books dealing with eutrophication, in the context of phytoplankton ecology. Mostly, works with data on different levels of eutrophication and general papers have been used; few other papers relevant to the issue, for example on the oligotrophication in Lake Constance, have been extensively consulted. The ecological variables sought followed the criteria set by Odum (1985), Reynolds (1988), Schindler (1990), Pimm (1991) and Seip & Reynolds (1995). Both quantitative and qualitative relationships were used. Quantified variables were related to trophic gradient either positively, negatively or zero, and their explained variances were noted whenever available. Qualitative variables were YES or NO related to the trophic gradient (i.e. dominant taxa). When a trend of change in relationships across the trophic gradient was detected, it is shown by + or −: From oligotrophic to hypertrophic, eutrophication may first drive a positive relationship,

Table 1. Ecological variables of phytoplankton (column Y) as possible dependent variables and TP or Chlorophyll *a* (X columns) as independent descriptors of trophic gradient. Qualitative variables are (Y) or not (N) related to trophic gradient, metric variables were related positively, negatively or zero (+,−,0) to trophic gradient and its variance explained is shown when data exist. See more explanation in the material and methods section. R^2: variance explained by regression. n: number of samples. S: source, this number corresponding to:1. Harper (1992). 2. Moss (1980). 3. Smith (1979). 4. Alvarez-Cobelas & Rojo (1994). 5. Stockner (1991). 6. Takamura & Nojiri (1994). 7. Alvarez-Cobelas & Jacobsen (1992). 8. Blomqvist et al. (1994). 9. Harris (1986). 10. Harris (1994). 11. Reynolds (1988). 12. Watson et al. (1992). 13. Tsuda et al. (1992). 14. Gaedke & Schweizer (1993). 15. Sommer et al. (1993). 16. Alvarez-Cobelas & Rojo (1994). 17. Baines et al. (1994). 18. Sprules & Munawar (1986). 19. Reynolds & Bellinger (1992). 20. Makulla & Sommer (1993). 21. Reynolds (1992). 22. Carney & Elser (1990). 23. Elser et al. (1990). 24. Prairie et al. (1989). 25. Carpenter et al. (1992). 26. Elser & George (1993). 27. Carrillo et al. (1995). 28. Marshall & Peters (1989). 29. De Angelis (1992). 30. Rawson (1956). 31. George & Jones (1987). 32. Carlson (1977). 33. Sas (1989). 34. Bloesch & Uehlinger (1990). 35. Golterman (1988). 36. Trifonova (1988). 37. Munawar & Munawar (1986). 38. Kalff. & Knoechel (1978). 39. Dillon & Rigler (1974). 40. Steinberg & Hartmann (1988). 41. Schindler & Fee (1974). 42. von Pavoni (1963). 43. Gliwicz (1975). 44. Smith (1990). 45. OECD (1982). 46. White et al. (1988). 47. Reynolds (1984b). 48. Dillon et al. (1988). 49. Sommer (1986). 50. Kalff & Watson (1986). 51. Carpenter & Kitchell (1993). 52. Chow-Fraser et al. (1994). 53. Canfield et al. (1989). 54. Baines & Pace (1994). 55. Trifonova (1989). 56. Smith et al. (1987). 57. Kalff (1991). 58. Seip et al. (1992). 59. Moss (1973). 60. Findlay & Kasian (1987). 61. Smith (1986). 62. Seip & Reynolds (1995). 63. Schindler (1990). 64. Smith (1985). 65. Anon (1990). 66. Reynolds (1984a). 67. Sprules al. (1983). 68. McCauley & Kalff (1981). 69. Pahl-Wostl (1995)

(Y) Phytoplankton	(X) TP R^2	n	S	Chl-*a* R^2	n	S
Overall features						
Chlorophyll *a*	0.71+	875	12			
	0.95+	55	39			
	0.69+	133	24			
	0.87+	61	28			
	0.70+	31	52			
	0.87+	18	55			
	0.46+	18	35			
	0.73+	43	32			
	+		41			
	+,0		21			
Phytoplankton biomass	0.60+	451	12			
	0.70+	28	38			
	0.62+		48			
	0.62+	31	52			
	0.88+	16	36	0.85	31	52
Yearly Primary production	0.50+	49	45	0.89+	83	17
	0.49+	16	55	0.57+	49	45
	0.28+	96	57			
	Y		63			
Volumetric rates of photosynthesis	0.95+	38	3	0.81+	49	3
	0.24+	98	16	0.82+	98	16
Areal phytoplankton photosynthesis	0.15+	98	16	0.34+	98	16
Photosynthetic capacity (P/B)	ns, −	98	16			
Corrected phytoplankton biomass (without diatoms)	0.82+		48			
Structural features						
% taxonomic groups	*Y*		49, 55, 52, 44, 37, 36			
Species composition	*Y*		66, 30, 7, 49			
Dominance	*Y*		4, 30, 63			
Diversity	−		38, 63			

Table 1. (continued)

(Y) Phytoplankton	(X) TP R^2	n	S	Chl-*a* R^2	n	S
% picoplankton	0.17–	42	6	0.27–	42	6
	–		5			
Community size	+		9	+		13
	+		63			
Cyanobacterial biomass	0.30+	165	53			
	Y		56			
Richness	–		38			
% Centrales: Pennales	*Y*		4			
Relative diatom biomass	*N*		4			
Functional features						
Edible biomass	0.22–	101	12			
	+, –, +		42,43			
	+		60			
	0		38,18			
Inedible biomass	0.44+	101	12			
	+		38, 63			
Specific growth rate	–		9			
Growth	+		9			
Loss by grazing	–, +, –		23, 22			
Functional class	*Y*		19, 51			
Ecological strategies(r-k, C-S-R)	*Y*		11			
Cell volume	*Y*		63			
Nutrient turnover rate	–, +, –		22			
	–		9			
	–		29			
Pigments, N, & P sinking flux				*Y*		54
C sinking flux				0.86+	15	17
				0.91+	15	54
Dynamic pattern						
Seasonality of biomass	*Y*		14, 44, 49, 55			
			38, 28, 37, 36			
Variance of biomass	+		49, 37, 55			
	+		38, 36			
Succession rate	+, 0		44, 49			
Variance of Chl-*a*	*Y*		28	*Y*		28
Cyanobacterial blooms	*N*		38			
Diatom dynamics	*Y*		4			
Chl-*a* phenology	*Y*		28	*Y*		28
Predictability	*Y*		10			
Plankton community						
Phytoplankton/zooplankton	+		67, 68			
Nannoplankton/zooplankton	0		68			
Netplankton/zooplankton	+		68			
Connectivity and food chain levels	–, +, –		23			
Resistance	–		29			
Resilience	+		25, 29			
Top-down: bottom-up flux	+		9, 10, 26, 27, 34, 25			

but later a negative one. Sources of information are noted in the legend of Figure 1.

Results and discussion

Forty-eight attributes of phytoplankton were related to trophic subdivisions, including total phosphorus and Chl-*a*, and these are summarised in Table 1. The entries are grouped into categories embracing overall features of community of phytoplankton (7 variables), structural features of the community (10 variables), functional features (17), dynamic features of specific populations (7), and interactive features of the community (7).

Heterogeneous characteristic of patterns

Table 1 includes quantitative and qualitative variables. Although Kalff & Knoechel (1978) thought that *'phytoplankton ecologists make vast numbers of measurements, frequently subjected to correlation analysis whose relationships may be spurious'*, only twelve of the variables in Table 1 have been related to trophic gradient, for example, yearly primary production. This implies that reviews of this topic will always be subjective or anecdotal. A quantitative meta-analysis of a batch of studies still seems a remote prospect (Arnqvist & Wooster, 1995). In many cases, the difficulty is that comparable measurements of quantitative variables, e.g. nutrient turnover rate or community size, are not available. Thus, the variability studied in phytoplankton ecology across the trophic gradient has been a qualitative exercise. In other cases, the relationships themselves are wholly descriptive, as, for example, in the case of species succession in lakes of differing trophic state (Reynolds, 1984a).

Evidence of gaps

The relationship of each of the variables in Table 1 to the trophic gradient together build a landscape, but one which has substantial gaps, representing poorly investigated areas. Some gaps appear among the functional variables (intensive features: minimum sinking rate, loss rate, affinity for P, N:P quota, temperature optimum for growth, light optimum for growth, resource ratio, competition, photosynthetic efficiency, SA/V). In many cases, attributes were measured in controlled laboratory experiments and only a small number of species were involved. Consequently, it is difficult to distinguish which features suit algae to one trophic state or another, and which determine how they will perform in communities of different lakes and trophic states. For instance, we suspect a change in the size-structure of communities across a trophic gradient (Sprules & Munawar, 1986) but we do not know enough about species-specific efficiencies and growth rates within the community to explain the expectation plausibly. Moreover, to standardise data for comparative purposes remains an outstanding problem; some variables are technically easier to quantify and are in more widespread use than others (such as Chl-*a*, primary production) but it is impossible for different researchers to measure the biovolume of phytoplankton using the same criteria and with the same precision (for sources of error, see Rott, 1981). The knowledge of gaps about structural variables provide another example: such quantitative variables as diversity or species richness depend on the diligence and experience of the researcher. As these variables are often used in local studies, comparisons among data sets can be very unreliable.

The temporal dynamics of variables should provide the most important information for the prediction of community responses. Many time series exist but have not been compared as a function of trophic gradient. For example, we know that, in a given year, the variance of total biomass is higher in eutrophic lakes (Trifonova 1988, 1989), but we do not know how the variance of biomass of either Cyanobacteria or Chrysophytes responds to differences in trophic state.

These important problems hinder progress towards formalisation and generalisation, and hamper the quest for general patterns. In short, despite the prodigious efforts of numerous individuals, few data can yet be rigourosly compared or sited firmly in the trophic landscape.

Some relationships are naïve

If our interest is, for example, the variation of fish biomass across a trophic gradient, the latter might be characterized by the concentration of primary producers. The basic cause (fertilization) is thus quantified by its direct effect (primary production), assuming a robust relation between the two. However, when our interest is focused on the ecology of primary producers at different trophic states, it would be quite naïve to take the concentration of Chl-*a* as a measure of eutrophication. In Table 1, the ecological attributes of the phytoplankton (dependent variables) are related not only to concentration of phosphorus (independent variable), but also to Chl-*a*. In some cases, we have variables whose definitions are related as well, for example, photosynthesis and Chl-*a*. In yet others, the mathematical problem of part *vs.* whole appears, for instance, the percentage of picoplankton (measured as Chl-*a*) *vs.* total Chl-*a* (Pavoni, 1963; Takamura & Nojiri, 1994), or biomass of cyanobacteria vs. total biomass (Canfield et al., 1989). Trivial relationships (may be tautologic) are a handicap to progress. Moreover, the total biomass of phytoplankton, Chl-*a* and total phosphorus are not linearly interrelated (Watson et al., 1992). Even Chl-*a* vs. total phosphorus shows different slopes at differing trophic states (White et al., 1988): there is not even a consistent Chl-*a* to total phytoplankton that holds for both oligotrophic and eutrophic lakes (Chow-Fraser et al., 1994). Thus, if we use its biological effects to describe eutrophication, we should be surprised if some features of the phytoplankton appear insensitive.

Multivariate relationships and non-linearity

Table 2 lists environmental features which may trigger or affect particular traits of phytoplankton behaviour. These independent of trophic state include light availability (Carlson, 1977), lake depth (Marshall & Peters, 1989; Smith, 1990), and the ratios between resources such as Si:P or Si:Light (Tilman, 1982; Makulla & Sommer, 1993). Qualitative descriptions of phytoplankton (dominant species, percentage of taxonomic groups, ecological strategies, etc.) linking trophic gradient to another independent variable may also reveal important information. Patterns of distribution of species against the second independent variable sometimes coincide with trophic-state ordination. Examples include nutrient fertility and carbon dioxide (Moss, 1973), trophic states and seasonality (Reynolds, 1984a) and classification of primary life-history strategies (Grime, 1979; Reynolds, 1988).

Statistically, multiple regressions relating attributes of the phytoplankton to two or more independent variables and tested by stepwise regression may be used to partition the variance in the dependent variable (Prairie et al., 1989; Smith, 1986; Seip & Reynolds, 1995). Therefore, trophic states should perhaps be conceptualised as multi-dimensional spaces, defined by key independent variables, rather than as a simple gradient of fertility. A good knowledge of how each of these key variables influences the performance of phytoplankton in each given system might then permit us to formulate and, ultimately compare, multiple-regression coefficients that avoid spurious correlations. Such a regression would both define the pattern of behaviour in a given system and contribute coefficients to the wider, general relationships.

Multivariate relationships are often nonlinear (Figure 1). In some cases, functional relationships between variables are complex (i.e. sigmoidal), or a threshold effect is observed, e.g., different trends of the same relationship at different stages (Figure 1a). For example, Chl-*a* vs. TP in 133 lakes shows a sudden change of slope above 100 μg l^{-1} of TP (Prairie et al., 1989). We may note other cases (Table 1): the relationship between edible biomass and trophic gradient (TP) is positive in oligotrophic lakes, negative in mesotrophic cases and positive again in eutrophic sytems (Pavoni, 1963; Gliwicz, 1975). Losses to grazing (Carney & Elser, 1990; Elser et al., 1990) and nutrient turnover (Carney & Elser, 1990) respond positively in mesotrophic lakes but negatively at either end of the trophic spectrum.

Interactions among independent variables also show nonlinear co-variation and responsive qualitative changes (Figure 1b,c). Examples include: (i) the average ratios of Chl-*a* to particulate nutrients are similar for oligotrophic and mesotrophic lakes, but are about two times greater in eutrophic lakes. The higher ratio in eutrophic waters is likely a Chl-*a* response to reduced light availability (White et al., 1988) and, thus to the specific concentration of Chl a; (ii) coefficients and precision of Chl-*a* vs. TP regression equations vary systematically and concommitantly with TN:TP ratios (Prairie et al., 1989). Moreover, any threshold effect is often due to an interacting variable which is not used in the equation; for example, the yield of Chl-*a* for TP is slower at higher concentrations of phosphorus. This is explained by the critical intervention of other limiting resources, such as light, nutrients, etc. (Reynolds, 1992).

Understanding of phytoplankton behaviour

Taking all the above into account, it is reasonable to deduce that responses to eutrophication are subject to a complex web of relationships among ecological variables and the attributes of phytoplankton. These are influenced by trophic gradient, but are not their simple, direct consequence. The degree of formalisation of the component relationships is, at present, inconsistent and there remain large conceptual gaps in many areas. The search for the nature of pattern (the struc-

Table 2. Main ecological variables of a limnological system out of a trophic framework that are related to phytoplankton

Abiotic			Biotic
physical	edaphic	morphological	community
Transparency	Conductivity	Z_{eu}/Z_m	Zooplankton
Light availability	TN, DIN, SRP	Mixing regime	Biomass of fish
PAR attenuation	CO_2	Latitude	Biomass of bacteria
Temperature	pH	Turbulence	
N^2	TN:TP	Depth and shape of lake	
Seasonality	DOM	Hydraulic disturbance	
	Acidification		
	Carbonaceous		

Table 3. Some concepts from local to global approach involved in ecological studies

Local		*Global*
Geographic		
microcosms		
Lake	Region	World
Biotic		
species		
genera	assemblages	overall measurements
group		
Trophic sector		
oligotrophic		
mesotrophic		oligotrophic-hypertrophic
eutrophic		
hypertrophic		
Independent variables		
X		
X_1, X_2		$X_1 \cdot X_2, \{X_1, X_2, X_3\}$
	X, X^2	

ture of the web), and for establishing its general validity (assessing the strength of its main components) has several facets. We need to syntesise the single *global* pattern from a multiplicity of well-quantified *local* patterns (Table 3). This requires us: (i) to close the gaps in the information, by selecting the key overall state variables, by using multivariate approaches and by designing more complex statistical analyses of correlation, which describe the ecology of phytoplankton in all freshwater systems; and (ii) to model simple patterns (statistical or otherwise) based on small numbers of specific variables drawn from large numbers of contrasting systems. This dual approach combines the respective advantages of either. The focus of future research should bridge the dimensions of biological knowledge (naturalistic, mechanistic and conceptual), of spatio-temporal scale (local, regional, planetary) and, also, of the trade-offs among precision, generality and realism (Levins, 1984).

Finally, if our ultimate goal is a paradigm for the dynamics of phytoplankton, biomass models and the tools for prediction of species composition, I would encourage phytoplankton ecologists to adopt the theory of nonlinear, dynamic, complex systems. Ecosystem properties may well arise from some simple basic rules played out with intricate complexity. Such a conceptual advance explains why high level predictions about biomass work, but also suggest the possibility of low level predictions about such things as the presence or absence of species. These concepts are closely related to self-organisation and system autonomy (independence from external inputs of energy and material, following Pahl-Wostl, 1995). Moreover, the predictibility of system behaviour and trophic stages are related (Harris, 1994).

Acknowledgements

I thank Dr Colin Reynolds (NERC Institute of Freshwater Ecology, Ambleside, UK) very much for his critical reading of the manuscript and linguistic help.

References

Alvarez-Cobelas, M. & B. Jacobsen, 1992. Hypertrophic phytoplankton: an overview. Freshwat. Forum 2: 184–199.

Alvarez-Cobelas, M. & C. Rojo, 1994a. Spatial, seasonal and long-term variability of phytoplankton photosynthesis in lakes. J. Plankton Res. 16: 1691–1716.

Alvarez-Cobelas, M. & C. Rojo, 1994b. Factors influencing the share of planktic diatoms in lakes. Arch. Hydrobiol. Algol. Stud. 74: 73–104.

Anon, 1990. Toxic blue green algae. National Rivers Authority, Peterborough, England, 128 pp.

Arnqvist, G. & D. Wooster, 1995. Meta-analysis: synthesizing research findings in ecology and evolution. TREE 10: 236–240.

Baines, S & M. L. Pace, 1994. Relationships between suspended particulate matter and sinking flux along a trophic gradient and implications for the fate of planktonic primary production. Can. J. Fish. Aquat. Sci. 51: 25–36.

Baines, S. B., M. L. Pace & D. M. Karl, 1994. Why does the relationship between sinking flux and planktonic primary production differ between lakes and oceans? Limnol. Oceanogr. 39: 213–226.

Bloesch, J. & U. Uehlinger, 1990. Epilimnetic carbon flux turnover of different particle size classes in oligo-mesotrophic Lake Lucerne, Switzerland. Arch. Hydrobiol. 118: 403–419.

Blomqvist, P., A. Pettersson & P. Hyenstrand, 1994. Ammonium-nitrogen: A key regulatory factor causing dominance of non-nitrogen-fixing cyanobacteria in aquatic systems. Arch. Hydrobiol. 132: 141–164.

Canfield, D. E., E. Philips & C. M. Duarte, 1989. Factors influencing the abundance of blue-green algae in Florida Lakes. Can. J. Fish. Aquat. Sci. 46: 1232–1237.

Carlson, R. E., 1977. A trophic state index for lakes. Limnol. Oceanogr. 22: 361–369.

Carney, H. J. & J. J. Elser, 1990. Strength of zooplankton-phytoplankton coupling in relation to lake trophic state. In M. M. Tilzer & C. Serruya (eds), Large Lakes. Springer, Berlin: 615–631.

Carpenter, S. R., K. L. Cottingham & D. E. Schindler, 1992. Biotic feedbacks in lake phosphorus cycles. TREE 7: 332–335.

Carpenter, S. R. & J. F. Kitchell (eds), 1993. The trophic cascade in lakes. Cambridge University Press, Cambridge, 383 pp.

Carrillo, P., I. Reche, P. Sanchez-Castillo & L. Cruz-Pizarro, 1995. Direct and indirect effects of grazing on the phytoplankton seasonal succession in an oligotrophic lake. J. Plankton Res. 17: 1363–1379.

Chow-Fraser, P., D. O. Trew, D. Findlay & M. Stainton, 1994. A test of hypotheses to explain the sigmoidal relationship between total phosphorus and Chlorophyll *a* concentrations in Canadian lakes. Can. J. Fish. Aquat. Sci. 51: 2052–2065.

De Angelis, D. L., 1992. Dynamics of nutrient cycling and food webs. Chapman & Hall, 270 pp.

Dillon P. J. & F. H. Rigler, 1974. The phosphorus-chlorophyll relationship in lakes. Limnol. Oceanogr. 19: 767–773.

Dillon, P. J., K. H. Nicholls, B. A. Locke, E. de Grosbois & N. D. Yan, 1988. Phosphorus-phytoplankton relationships in nutrient-poor soft-water lakes in Canada. Verh. int. Ver. Limnol. 23: 258–264.

Elser, J. J., H. J. Carney & C. R. Goldman, 1990. The zooplankton-phytoplankton interface in lakes of contrasting trophic status: an experimental comparison. Hydrobiologia 200/201: 69–82.

Elser, J. J. & N. B. George, 1993. The stoichiometry of N and P in the pelagic zone of Castle Lake, California. J. Plankton Res. 15: 977–992.

Findlay, D. L. & S. E. M. Kasian, 1987. Phytoplankton community responses to nutrient addition in Lake 226, Experimental Lakes Area, Northwestern Ontario. Can. J. Fish. Aquat. Sci. 44: 35–46.

Gaedke, U. & A. Schweizer, 1993. The first decade of oligotrophication in Lake Constance. I. The response of phytoplankton biomass and cell size. Oecologia 93: 268–275.

George, D. G. & D. H. Jones, 1987. Catchment effects on the horizontal distribution of phytoplankton in five of Scotland's largest freshwater lochs. J. Ecol. 75: 43–59.

Gliwicz, Z. M., 1975. Effect of zooplankton grazing on photosynthetic activity and composition of phytoplankton. Verh. int. Ver. Limnol. 19: 1490–1497.

Golterman, H. L., 1988. Chlorophyll-Phosphate relationships, a tool for water management. In F. E. Round (ed.), Algae and the aquatic environment. Biopress Ltd., Bristol: 205–224.

Harper, D., 1992. Eutrophication of freshwaters. Chapman & Hall. London, 327 pp.

Harris, G. P., 1986. Phytoplankton ecology. Structure, function and fluctuation. Chapman and Hall, London, 384 pp.

Harris, G. P., 1994. Pattern, process and prediction in aquatic ecology. A limnological view of some general ecological problems. Freshwat. Biol. 32: 143–160.

Hutchinson, G. E., 1961. The paradox of the plankton. Am. Nat. 95: 137–145.

Kalff, J., 1991. The utility of latitude and other environmental factors as predictors of nutrients, biomass and production in lakes worldwide: problems and alternatives. Verh. int. Ver. Limnol. 24: 1235–1239.

Kalff, J. & R. Knoechel, 1978. Phytoplankton and their dynamics in oligotrophic and eutrophic lakes. Ann. Rev. Ecol. Syst. 9: 475–495.

Kalff, J. & S. Watson, 1986. Phytoplankton and its dynamics in two tropical lakes: a tropical and temperate zone comparison. Hydrobiologia 138: 161–176.

Levins, R., 1984. The strategy of model building in population biology. In E. Sobe (ed.), Conceptual Issues in Evolutionary Biology. The MIT Press, Cambridge: 18–27.

Lund, J. W. G., 1964. Primary production and periodicity of phytoplankton. Verh. int. Ver. Limnol. 15: 37–56.

Makulla, A. & U. Sommer, 1993. Relationships between resource ratios and phytoplankton species composition during spring in five north German lakes. Limnol. Oceanogr. 38: 846–856.

Marshall, C. T. & R. H. Peters, 1989. General patterns in the seasonal development of chlorophyll a for temperate lakes. Limnol. Oceanogr. 34: 856–867.

McCauley, E. & J. Kalff, 1981. Empirical relationships between phytoplankton and zooplankton biomass in lakes. Can. J. Fish. Aquat. Sci. 38: 458–463.

Moss, B., 1973. The influence of environmental factors on the distribution of freshwater algae: an experimental study. IV. Growth of test species in natural lake waters and conclusions. J. Ecol. 61: 193–211.

Moss, B., 1980. Ecology of fresh waters. Blackwell Scientific Publications. Oxford, 332 pp.

Munawar, M. & I. F. Munawar, 1986. The seasonality of phytoplankton in the North American Great Lakes, a comparative synthesis. Hydrobiologia 138: 85–115.

Odum, E. P., 1985. Trends expected in stressed ecosystems. BioScience 35: 419–422.

OECD, 1982. Eutrophication of waters. Monitoring, assessment and control. OECD, Paris, 153 pp.

Pahl-Wostl, C., 1995. The dynamic nature of ecosystems. Chaos and order entwined. John Wiley & Sons, Chichester, 267 pp.

Pimm, S. L., 1991. The balance of nature? University of Chicago Press, Chicago, Illinois, 434 pp.

Prairie, Y. T., C. M. Duarte & J. Kalff, 1989. Unifying nutrient-chlorophyll relationships in lakes. Can. J. Fish. Aquat. Sci. 46: 1176–1182.

Rawson, D. S., 1956. Algal indicators of trophic lake types. Limnol. Oceanogr. 1: 18–25.

Reynolds, C. S., 1980. Phytoplankton assemblages and their periodicity in stratifying lake systems. Holarctic Ecol. 3: 141–159.

Reynolds, C. S., 1984a. Phytoplankton periodicity: the interactions of form, function and environmental variability. Freshwat. Biol. 14: 111–142.

Reynolds, C. S., 1984b. The ecology of freshwater phytoplankton. Cambridge Univ. Press, New York, 384 pp.

Reynolds, C. S., 1988. Functional morphology and the adaptative strategies of freshwater phytoplankton. In C. D. Sandgren (ed.), Growth and reproductive strategies of freshwater phytoplankton. Cambridge University Press, Cambridge: 388–433.

Reynolds, C. S., 1992. Eutrophication and the management of planktonic algae: what Vollenweider couldn't tell us. In D. W. Sutcliffe & J. G. Jones (eds), Eutrophication: research and application to water supply. Freshwater Biological Association, Ambleside: 4–29.

Reynolds, C. S. & E. G. Bellinger, 1992. Patterns of abundance and dominance of the phytoplanton of Rostherne Mere, England: evidence from an 18-year data set. Aquat. Sci. 54: 1015–1621.

Rott, E., 1981. Some results from phytoplankton counting intercalibrations. Schweiz. Z. Hydrol. 43: 34–62.

Sas, H. (ed.), 1989. Lake restoration by reduction of nutrient loading. Academia Verlag, Sankt Agustin, 497 pp.

Scheffer, M., S. M. Hosper, M. L. Meijer, B. Moss & E. Jeppesen, 1993. Alternative equilibria in shallow lakes. TREE 8: 275–278.

Schindler, D. W., 1990. Experimental perturbations of whole lakes as test of hypotheses concerning ecosystem structure and function. Oikos 57: 25–41.

Schindler, D. W. & E. J. Fee, 1974. Experimental lakes area: whole-lake experiments in eutrophication. J. Fish. Res. Bd Can. 31: 937–953.

Seip, K. L., H. Sas & S. Vermij, 1992. Changes in Secchi disk depth with eutrophication. Arch. Hydrobiol. 124: 149–165.

Seip, K. L. & C. S. Reynolds, 1995. Phytoplankton functional attributes along trophic gradient and season. Limnol. Oceanogr. 40: 589–597.

Smith, V. H., 1979. Nutrient dependence of primary productivity in lakes. Limnol.Oceanogr. 24: 1051–1064.

Smith, V. H., 1985. Predictive models for the biomass of blue-green algae in lakes. Water resour. Bull. 21: 433–439.

Smith, V. H., 1986. Light and nutrient effects on the relative biomass of blue-green algae in lake phytoplankton. Can. J. Fish. Aquat. Sci. 43: 148–153.

Smith, V. H., 1990. Phytoplankton responses to eutrophication in inland waters. In I. Akatsuka (ed.), Introduction to Applied Phycology. Academic Publishing, The Hague: 231–249.

Smith, V. H., E. Willén & B. Karlsson, 1987. Predicting the summer peak biomass of four species of blue-green algae (Cyanophyta/Cyanobacteria) in Swedish lakes. Water resour. Bull. 23: 397–402.

Sommer, U., 1986. The periodicity of phytoplankton in Lake Constance (Bodensee) in comparison to other deep lakes of central Europe. Hydrobiologia 138: 1–7.

Sommer, U, U. Gaedke & A. Schweizer, 1993. The first decade of oligotrophication of Lake Constance. II. The response of phytoplankton taxonomic composition. Oecologia 93: 276–284.

Sprules, W. C., J. M. Casselman & B. J. Shuter, 1983. Size distribution of pelagic particles in lakes. Can. J. Fish. Aquat. Sci. 40: 1761–1769.

Sprules, W. G. & M. Munawar, 1986. Plankton size spectra in relation to ecosystem productivity, size, and perturbation. Can. J. Fish. Aquat. Sci. 43: 1789–1794.

Steinberg, C. E. W. & H. M. Hartmann, 1988. Planktonic bloom-forming Cyanobacteria and the eutrophication of lakes and rivers. Freshwat. Biol. 20: 279–287.

Stockner, J. G., 1991. Autotrophic picoplankton in freshwater ecosystems: the view from the summit. Int. Revue ges. Hydrobiol. 76: 483–492.

Takamura, N. & Y. Nojiri, 1994. Picoplankton biomass in relation to lake trophic state and the TN:TP ratio of lake water in Japan. J. Phycol. 30: 439–444.

Tilman, D., 1982. Resource competition and community structure. Princeton Univ. Press, Princeton, 296 pp.

Trifonova, I. S., 1988. Oligotrophic-eutrophic succession of lake phytoplankton. In F. E. Round (ed.), Algae and the aquatic environment. Biopress Ltd., Bristol: 107–124.

Trifonova, I. S., 1989. Changes in community structure and productivity of phytoplankton as indicators of lake and reservoir eutrophication. Arch. Hydrobiol. Beih. 33: 363–371.

Tsuda, R., M. Kumagai & Y. Kakui, 1992. Spatial changes of phytoplanktonic size spectra in Lake Biwa. Hydrobiologia 243/244: 137–140.

von Pavoni, M., 1963. Die Bedeutung des Nannoplanktons im Vergleich zum Netzplankton. Schweiz. Z. Hydrol. 25: 219–341.

Watson, S., E. McCauley & J. A. Downing, 1992. Sigmoid relatioships between Phophorus, Algal Biomass, and Algal Community Structure. Can. J. Fish. Aquat. Sci. 49: 2605–2610.

White, E., G. Payne & S. Pickmere, 1988. A limitation to the usefulness of chlorophyll as a biomass indicator in eutrophication studies. Verh. int. Ver. Limnol. 23: 598–601.

Hydrobiologia **369/370:** 11–26, 1998.
M. Alvarez-Cobelas, C. S. Reynolds, P. Sanchez-Castillo & J. Kristiansen (eds), Phytoplankton and Trophic Gradients.

What factors influence the species composition of phytoplankton in lakes of different trophic status?

C. S. Reynolds
NERC Institute of Freshwater Ecology, Windermere Laboratory, GB-LA22 0LP Ambleside, UK

Key words: phytoplankton, lake typology, trophic status

Abstract

The paper articulates some present concepts relating to the selection of phytoplankton along trophic gradients. Concerns over lake eutrophication have heightened the importance of nutrients but it is not obvious that interspecific differences in the nutrient requirements of algae genuinely segregate species except under chronic deficiencies. The selectivity supposedly generated by altered resource ratios is re-examined. It is argued that ratios explain very little of the distribution of species with respect to trophy. However, changing nutrient loading does have consequential impacts on the availability of other requirements including light and carbon dioxide. It is argued that the trophic spectrum is not a single dimension of a single factor but, rather, a template of factors covarying in consequence of the larger levels of biomass that are supported, and which weight in favour of the growth and survival prospects of particular kinds of planktonic algae. The trophic spectrum is a probabalistic outcome of several dimensions of variability.

Introduction

Over the past seventy or eighty years, the study of freshwater phytoplankton has evolved in a quite idiosyncratic manner. From the initial fascination with the organisms found in open water and the detection of where and when particular populations might be prevalent, the dominating trends have been to gauge the collective productivity of phytoplankton and to assess its abundance in relation to nutrients and to the consumers of its production. In the most recent past, the greatest interest has been driven by the quest to manage water quality in lakes and reservoirs and to find ways to overcome the overproduction of acknowledged nuisance species. Thus, the circle has turned fully and the questions asked today are the same ones which engaged the pioneers of phytoplankton ecology: 'what lives where and why?'.

This curious revolution underpins the discussions of the 10th Workshop of the International Association of Phytoplankton Taxonomy and Ecology, convened at Granada, Spain, in June, 1996. On the one hand, phycologists have little difficulty in distinguishing what they classify intuitively as (say) 'oligotrophic species', 'eutrophic indicators' or 'mesotrophic assemblages'; they can propose schemes like Table 1 to summarise their current knowledge, without fear of more than minor quibbles of detail. On the other hand, there has been in the last two or three decades a vast expansion in knowledge about the physiology of carbon fixation and the consumption of nutrients by phytoplankton, together with some remarkable advances in molecular biology and environmental physics. Surely, we are by now much better equipped to be able to isolate the factors which define the spatio-temporal ranges of phytoplankton species?

There is, of course, already a large literature devoted to selective processes in the phytoplankton. Much is apparently or actually contradictory. Much is viewed in terms of single dimensions, such as resource ratios or trophic cascading, and much is viewed in the context of rigid equilibrium states. What is needed is a basis for relating the differing arrays of adaptive attributes and trade-offs of individual species of phytoplankton and how these might interact with the totality of processes

Table 1. A provisional trophic spectrum of major genera of phytoplankton

Trophic status	ULTRAOLIGOTROPHIC..HYPER-EUTROPHIC
nutrient supply	strongly deficient.................................adequate.................................saturating
alkalinity	acid...alkaline................calcareous
clarity	clear...turbid
Diatoms	*C.glomerata/C.comensis....C.meneghiniana...S.minutulus....S.neoastraea....S.rotula....S.hantzschii* *....Urosolenia.......* *....Tabellaria........................Asterionella...................Fragilaria...........Diatoma.....* *..Aulacoseira distans....A.subarctica..........A.ambigua...A.granulata.......* *.............Melosira varians........*
Chlorophytes	*.......................................Chlorella spp...* *.......................Chlamydomonas...* *...........................Scenedesmus...........................* *Gonium...Eudorina...Pandorina* *...........Coelastrum, Pediastrum* *.......Sphaerocystis, Gemellicystis...........* *..Staurodesmus.......Cosmarium........Staurastrum.......Closterium.....*
Cyanobacteria	*..Merismopedia..Gloeotrichia..Coelosphaerium...Planktothrix..Limnothrix/Pseudanabaena.* *..A.solitaria.......Gomphosphaeria..........................Microcystis.......* *...A.lemmermanni.......A.flos-aquae/A.circinalis.....* *...Aphanizomenon...*
Dinoflagellates	*...............................Peridinium, Ceratium...............*
Cryptophytes	*...................................Rhodomonas...........Cryptomonas..*
Chrysophyceae	*Dinobryon...* *Uroglena.......* *Mallomonas............* *Synura.....* *Chrysosphaerella...*
Euglenoids	*Euglena* *Phacus* *Lepocinclis*

which together determine the dynamic variability of pelagic environments.

The objective cannot be attained in a single workshop, much less in a single paper. The intention of this contribution is to rationalise our intuitions about the assembly of broad but distinctive associations of phytoplankton species. It begins with an overview of current perceptions of the role and selectivity of competition. It then attempts to draw together the factors which, simultaneously or sequentially, are most relevant to the intuitive foundation of the 'trophic spectrum'. Then, following the habitat-template approach, it is proposed that algal attributes are matched to the opportunities provided by environmental and that there is an encouraging fit of phytoplankton species to the range of habitats described by the trophic spectrum. Finally, some well-documented transitions in trophic state from the English Lake District serve to emphasise the generalist nature of the floristic responses.

Current perceptions of phytoplankton distribution in relation to trophic status

The first impediment to a crisp understanding of the issues is one of semantics. The terms oligotrophy and eutrophy were introduced in limnology's founding years to distinguish the classical differences between deep, clear, Caledonian types of mountain lake and the shallow, productive, Baltic type of lowland lake (Thienemann, 1918; Naumann, 1919). Presented with samples of water from either kind of lake, the experienced plankton biologist would have no difficulty in distinguishing which was from which, merely on the basis of the assemblage of phytoplankton present. Desmids, Chrysophyceae and diatoms of the genera *Tabellaria* and *Cyclotella* characterise the sparse but diverse plankton of the oligotrophic Caledonian-type lakes; Cyanobacteria (*Anabaena, Aphanizomenon, Microcystis*) and the diatoms *Asterionella, Aulacoseira, Fragilaria* and *Stephanodiscus* are indicative of more eutrophic conditions (Rawson, 1956). This classification may now seem a little facile, both in terms of the classification of lakes and the ecologically indicative groupings (see, for instance, Olrik, 1994), but it is important to recognise that the understanding of lake evolution was that, through progressive siltation and accumulation of catchment exports, oligotrophic lakes slowly turned into eutrophic ones (Pearsall, 1921). This is presumably the origin of the noun 'eutrophication'. When, mostly in the latter half of the present century, the enrichment of rivers and lakes with nitrogen and phosphorus led to enhanced plankton biomass and often biassed in favour of the Baltic-type assemblages, it was quite reasonable to regard this as accelerated eutrophication. This notion is also discredited for its naïvety but the understanding that nutrient enrichment leads to changes in algal quantity and quality is firmly entrenched. The distribution of species against some measure of nutrient (usually phosphorus) availability is for most of us, I suspect, the 'trophic spectrum'. In short, the pattern shown in Table 1 relates to the nutrient axis and not to the other correlatives of trophic state used by Thienemann and Naumann.

A second general point also needs to be re-stated at the outset, to the effect that it is not valid to ascribe phyla or classes exclusively to one part of the trophic range. Diatoms, Chlorophytes and Cyanobacteria occur right across the width of the spectrum, embracing from extreme (ultra-) oligotrophy to extreme (hyper-) eutrophy. Desmids, centric diatoms, even the Chrysophyceae, occupy substantial horizontal blocks. Even within single orders, such as the Oscillatoriales, and the same ostensible genera e.g. *Planktothrix*, representation stretches from the occurrence of *P. rubescens* (D.C. ex Gom.) Anagn. et Kom, in deep, mesotrophic alpine lakes to the dominance of *P. agardhii* (Gom.) Anagn. et Kom. in exposed shallow basins of high nutrient content. Indeed, the diagnostically more helpful species-associations run vertically through Table 1, allying multiphyletic floral components of (say) nutrient-rich ponds (*Chlorella, Scenedesmus, Ankistrodesmus* and Euglenoids), of eutrophic lakes at the time of their summer maxima (*Ceratium, Microcystis, Aulacoseira granulata* (Ehr.) Simonsen), of mesotrophic temperate lakes in spring (with *Cyclotella* spp. and *Aulacoseira subarctica* (O.Müller) Haworth), or oligotrophic lakes, with *Dinobryon, Uroglena, Gemellicystis* and *Sphaerocystis*. Olrik (1994) and Reynolds (1996) provide further examples of recognisable associations of phytoplankton that are indicative of particular limnetic habitats.

The abiding challenge is to explain *why* the algae should be distributed in this way. What is about the trophic status of a habitat that should select for certain groups of species when an alternative status favours others? After all, algal protoplasm is built from largely identical elements, combined in consistent mutual general proportions.

Current perceptions of phytoplankton species selection by trophic status

Possible answers to this challenge are strongly polarised between two paradigms. One set of explanations is assembled around an equilibrium view of the world, where the biological demand is constrained by resources to the extent that survival and success are the outcome of fierce interspecific competition. The other invokes a less deterministic role for a dynamic resource base that offers fluctuating opportunities for exploitation and changing probabilities for success to species which happen to be better adapted to that part of the conceptual 'habitat template'. While the respective adherents to either paradigm are not necessarily in open dispute and their theories are not altogether mutually exclusive, they nevertheless differ quite fundamentally and, really, the hypotheses are overdue for rigourous scrutiny.

Resource competition

Largely because of the key role they play in lake eutrophication (*sensu* that based upon accelerated anthropogenic enrichment), nutrients have been properly judged to play a similar role in species selection. An attractive deduction might be that there are species which show some higher requirement for a given nutrient and which would be disadvantaged at low availability, whereas their extra ability to exploit enrichment would help them to move to dominance when the opportunity arose. This is clearly understandable in the case of the special requirement of many diatoms for large quantities of skeletal silica. Conversely, the ability of nitrogen-fixing Cyanobacteria to become independent from the impediment of dwindling supplies of combined inorganic nitrogen experienced by other algae (Riddolls, 1985).

There are demonstrable interspecific differences in the cell-specific rates of nutrient-uptake and in the external concentrations required to saturate them (Nalewajko & Lean, 1978, *a.o.*): some species take up nutrient faster than others (i.e., they are rate-adapted); others can satisfy their needs at lower external concentrations than others (i.e., they are affinity-adapted). Moreover, the performance characteristics of nutrient uptake differ not only among algal species but among the various nutrients too. This means that a given alga can function relatively better than another at a low concentration of one nutrient (say, phosphorus) but to be, perhaps, less efficient in taking up another (say, silicon).

This was the hypothesis tested by Tilman and Kilham (1976) in their classic experiments using cultures of *Asterionella formosa* Hass. and *Cyclotella meneghiniana* Kütz. These diatoms grow at comparable rates and, weight-for-weight, consume similar amounts of silicon and phosphorus. However, their affinites for small amounts of the nutrient, as indicated by the half-saturation constants for uptake (K_U), differ significantly. *Asterionella* has a lower K_U for phosphorus uptake than *Cyclotella* (0.02–0.04 against 0.25 μmol l^{-1}) but a higher K_U than *Cyclotella* for silicon uptake (3.9 as against 1.4 μmol l^{-1}). This means that, other factors notwithstanding, *Asterionella* will grow further down a gradient of declining phosphorus than will *Cyclotella* but *Cyclotella* should perform better than *Asterionella* when silicon falls to low concentrations. From these observations, Tilman & Kilham correctly predicted the outcome of competition experiments in continuous cultures (which one or other 'wins') and the consequence of supplying resources in concentrations likely to limit the phosphorus-uptake kinetics of the one and the silicon-uptake kinetics of the other (they co-exist). A further corollary is that when the algae grow in natural waters where the silicon or the phosphorus concentration is consistently or frequently at such low levels, then one species has a consistent or frequent growth advantage over the other and, in time, is more likely to dominate over the other. Later work (Tilman et al., 1982) extended the number of diatoms to which this logic might be applied, yielding a spectrum of ratios of limiting Si to limiting P at which different species were considered to be superior competitors.

At about the same time, G.-Y. Rhee's group were making analogous discoveries about the comparative algal uptake rates of nitrogen and phosphorus. Taking the logic one stage further, it was supposed that the ratio in which the nutrients were taken up under these circumstances might predict their relative performances along gradients of nutrient limitation. Rhee & Gotham (1980) found N:P ratios (by weight) of < 10 supported *Microcystis* and a *Melosira* sp., while three species of Chlorococcales survived better when the ratio of limiting concentrations was pitched nearer 25. The work of these two schools in particular have established a powerful case for the resource-based competition theory. The idea that there could be as many co-existent species as there are limiting resources quickly followed (Petersen, 1975) and the inclusion of light among the quantifiable resources (Huisman & Weissing, 1995) provided an added dimension to the applicability of the theory.

The most familiar adoption of the resource-ratio theory relates to the species dominance of phytoplankton, most particularly to the alleged dependence of dominance by cyanobacteria upon total nitrogen to total phosphorus. From Smith (1983) to Levich (1996), the role of 'N:P ratios in selecting for algal dominance' and 'the appearance of cyanobacterial blooms owes to falling N:P ratios' have become traditions in phytoplankton ecology.

The fallacy of these abstractions is that they confuse resource capacity with the consumptive processes (Reynolds, 1992). It is relatively simple to determine that some lakes receive modest external loads of nitrogen and phosphorus and in molecular ratios which differ significantly from the approximate stoichiometric parity of cytoplasm (16:1). Accordingly, it can be anticipated that uptake by growing plankton (supposing temperature, light, the supply of carbon and seventeen or so other elements to be adequate) should lead developing populations to a situation where one or other of these resources is exhausted. Because the starting ratio indicates whether N or P is likely to run out first, the initial ratio of TN:TP might well enable us to predict, for example, whether nitrogen-fixing cyanobacteria might gain a selective advantage. Even this approach has its limitations (see Pick & Lean, 1987). Any other argument for a selective role of the nutrient resource ratio is, I believe, difficult to justify.

Three reasons for this assertion are re-advanced. One is purely interpretational. The ratio between any pair of nutrients is a fortuitous consequence of the absolute availabilities of each. If the growth requirements of both *Asterionella* and *Cyclotella* for phosphorus and silicon are simultaneously exceeded, then neither is 'limited' either by the ratio of the resources or by its performance in relation to the other. They are not even competing, in the sense that the growth of one necessarily impinges upon the growth of the other. The data-set investigated by Reynolds (1987a) suggested that temperate lakes failing to produce significant summer crops of bloom-forming Cyanobacteria (*Anabaena, Aphanizomenon* or *Microcystis*; *Planktothrix* associations are quite another matter) were simply low in phosphorus (thus, with high TN:TP ratios). Increasing its concentration sufficiently overcomes the phosphorus deficiency, to the extent that more of the nitrogen is consumed to meet the demand of the enhanced biomass. In mesocosm experiments, Reynolds (1986) found that dominance of nitrogen-fixing cyanobacteria could be engineered with external loads of < 10 g N m^{-2} and up to 1.2 g P m^{-2} (molecular ratio 18:1) but at greater loads of either, other algae thrived to the exclusion of nitrogen-fixers, *even when the ratio between them was preserved.*

The selective relevance of the ratio between saturating concentrations of nutrients is drawn into question by the results of Sommer's (1993) investigation of the relative biomass of each of 16 main species produced in Plußsee, Germany, against a background of changing resource ratios. The predicted responses were detected only at minimal or maximal ratios and that these could be 'lagged by up to six weeks'. Optimal mid-range ratios 'were rare'. Changes in ratio are symptomatic of the depletion and not the driver of the algal response.

The second point is that the uptake capabilities of algae far exceed the maximum requirements of the maximum sustainable growth rates (Reynolds, 1990, 1992). This may be an adaptation to life in chronically low-nutrient environments but it is important to recognise that all investigated phytoplankton are able to sustain maximum replication rates (at 20 °C) concentrations of $< 10^{-7}$ mol P l^{-1} (< 3 μg SRP l^{-1}), even though this is usually far below the level required to saturate maximal uptake rates. The corollary may be suggested that if the external concentration of soluble phosphorus exceeds 3 μg P l^{-1}, algae are scarcely 'looking for' phosphorus, much less 'competing' for it. Manifestly, it is *not limiting* algal growth. The first condition of the Tilman-Rhee model is violated. Moreover, external concentrations should be significantly and consistently below growth-rate saturating levels *before* selection favours species with high affinities for nutrient uptake.

The third argument is drawn from contemporary advances in the molecular biology of nutrient uptake (for an excellent review, readily assimilable by ecologists and non-biochemists, see Mann, 1995) . The receptor/kinase uptake and internal transport pathways employed by live cells to deliver scarce resources to internal sites of assimilation are each dedicated to the passage of a single type of molecule. Each can be so replete that the receptor sites cannot accept further nutrient molecules or the latter are so relatively sparse that they are processed at once (a pattern of activity which fits and explains the classic Monod uptake model). Each signals its activity to the genome through proteins which suppress the expression of the genes which initiate cell-closedown reactions to resource limitation. The undersupplied pathway system informs the cell that the resource is limiting, therefore it is the shortage of that resource which determines the cell response. No molecular means whereby the cells 'sample' ratios has yet been discovered.

Other authors have expressed difficulties in interpreting surveys of the relative abundance of cyanobacteria in terms of N:P ratios (Trimbee & Prepas, 1987; Jensen et al., 1994). It is clear that many other factors are involved (*inter alia*,the sources of carbon and nitrogen: see Shapiro, 1990; Blomqvist et al., 1994) and the match and mis-match of events during the potential growth period in any given year are crucial (George et al., 1990; Reynolds & Bellinger, 1992).

None of this is to say that nutrients are unimportant. Clearly, we can cite concentrations of given resources below which we might anticpate the rates of cell replication to be governed by the rate of resource supply. In lakes where the nutrient concentrations continuously fail to saturate the growth-sustaining requirement, there will be, by definition, an absolute resource limitation and this will surely operate in favour of algae able to maintain the best net performance against the deficiency. Success in one year may well mean a greater carry over of inoculable cells to the following year, so biassing the species composition further by a kind of compositional inertia. Thus, in lakes where the phosphorus concentration is always substantially below 1–2 μg P l^{-1}, the phytoplankton is likely to tend quickly towards dominance by high-affinity species. An analogous argument will apply to lakes where the nitrogen concentrations are substantially below 50–100 μg N l^{-1}. In both cases, outcome is independent of nutrient ratios.

In richer lakes, where converse argument insists that the algal replication is NOT nutrient limited intitially, algal growth might well proceed until such growth-rate limiting concentrations were encountered. Can we deduce the impact of a critically declining resource, save that it is about to be denied to any of the contesting species? Suppose a typical scenario in which populations of *Asterionella* and *Cyclotella* are growing simultaneously in a natural lake, building stocks towards a vernal bloom. They have phosphorus and silicon available to them, at least initially, in concentrations adequate to saturate the requirements of the fastest rates of growth that the temperature and underwater light climate will sustain. Let us further suppose that the *Asterionella* began its growth earlier in the year (it is, demonstrably, a better light antenna than *Cyclotella*) and from a larger inoculum: this helps us to understand why *Asterionella* cells already outnumber those of *Cyclotella* (let us say, by 4000 to 800 cells ml^{-1}). The next day, the concentration of silicon is depleted to below 3.9 μmol Si l^{-1} (i.e., close to the 0.5 mg SiO_2 l^{-1}, famously observed in nature by Lund, 1950), at which point the growth rate of the *Asterionella* falls to that determined by the rate of silicon uptake and deployment (it *is* silicon-limited) but that of the *Cyclotella* continues to be determined by light and temperature alone (it is *not* silicon-limited). What does this mean practically? At the end of that day, *Cyclotella* numbers have continued to increase (by, say, 15%, to 920) but now more strongly than those of the *Asterionella* (by, say, only 5%, to 4200). By the third day, the demand for silicon by both has effectively exhausted the remaining silicon, so neither alga can increase at all. The critical concentrations and the critical ratios have been passed, the performances have responded as predicted. The brief episode of resource competition has not broken the dominance of *Asterionella.*

In consequence of these considerations, it is prudent to advise against too literal an application of resource competition theory to the selection of phytoplankton across the trophic spectrum.

Weighted opportunities; the habitat-template approach

The alternative paradigm starts with a recognition that there are estimated to be upwards of 4000 described species of freshwater phytoplankton, that endemism is rare and many are highly cosmopolitan in distribution. All grow as well as they can wherever and whenever they can. It is abundantly obvious that not all species occur everywhere, certainly not in equal numbers, and it is no less clear that, at given times and under given conditions, some species or groups of species tend to increase in biomass rather more strongly than others. Thus, they become, at least temporarily, better represented. It may be deduced that recruitment and dominance are not wholly stochastic and that certain preferences ('weightings') must operate in favour of some of the species. These therefore perform relatively better than others when the relevant conditions select for their particular attributes. These attributes are generally in the category of *pre-adaptations* and therefore tend to be permanent, quantifiable features of the organisms. Their performances have sufficient fidelity and consistency to be invoked experimentally and out of natural sequence (Reynolds et al., 1984; Reynolds & Reynolds, 1985) and, thus, to conform to an ecological *pattern.* Together, pre-adaptations and discernable patterns allow us to determine what have become known as the growth *strategies* of algae and, in turn, these assist us to interpret, understand and even predict *processes* of community assembly. In this way, an array of similarly-adapted species (in another terminolgy, species with broad niche-overlap) are able simultaneously to increase and, between them, to constitute a species-assemblage, arguably characteristic for the conditions obtaining. This logic does not invoke competition as a continuously deterministic component. Indeed, it does not predict a dominant species, beyond that it will be one of a cluster of candidates. The only element of trade-off is that an alteration in environmental conditions might then 'weight' other attributes and adaptations, of alternative species with alternative strategies; alternative patterns of assembly are assumed, leading to the assembly of a different kind of community.

This model anticipates the shift in dominance, for instance, from the diatom-dominated communities of cold, well-mixed water columns to the resource-

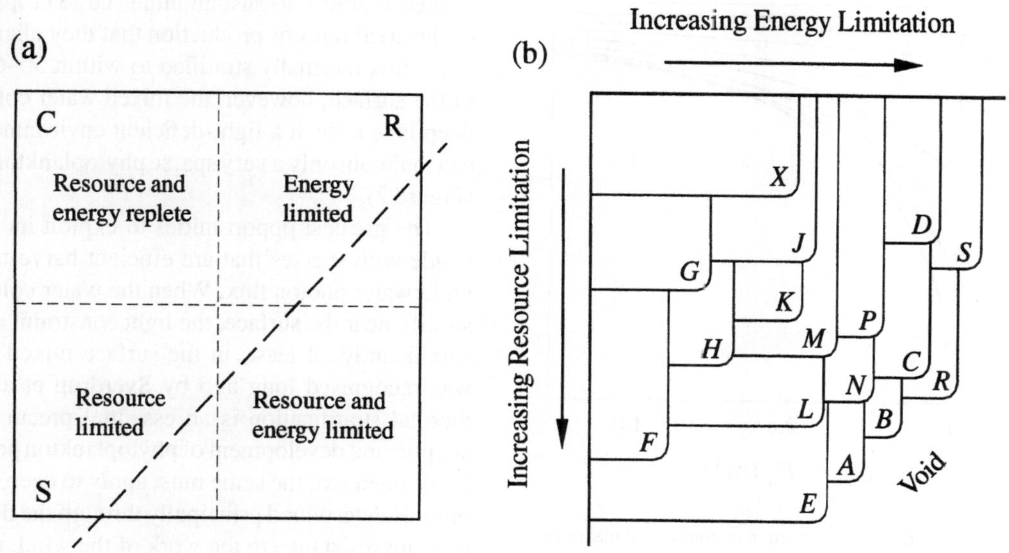

Figure 1. A simple habitat template involving (a) a 2 × 2 contingency table to give combinations of resource- and energy- limitation and repleteness (together with the type of plant-strategy – C, S or R-favoured; habitats simultaneously deprived of energy and resources are untenable); in (b), the distributions of some distinct associations of phytoplankton (Reynolds, 1987a, b, 1996) in relation to the template are represented. Redrawn from various figures in Reynolds (1996).

partitioned, spatially-structured assemblages of the shallow mixed depth of a thermally stratified lake in summer, wherein (say) motile dinoflagellates or buoyancy-regulating colonial cyanobacteria are the probable dominants (Reynolds, 1996). This model makes no more detailed a prediction (which species, how much and on what date?). It invokes no precise mechanisms and makes no prediction of any outcome, beyond that a certain kind of community is more likely to be assembled than another, simply because particular environments influence the relative success of the individual organisms present. I do not consider this imprecision to be either inadequate or disappointing. It is a fair statement of the level of predictability that it is realistic to expect. To do more is as elusive as predicting to-morrow's rainfall.

We may well have the knowledge, however, to calculate the less precise (but more accurate) probability about whether or not it will rain to-morrow. The key question persists, 'what kinds of algae live better than others under proscribable trophic conditions?'. This may not seem to differ greatly from the resource-ratio problem, save that the challenge is now to be able to discern and to match the attributes and performances of given species, or given groups of species with similar preadaptations, to a 'habitat template' (Southwood, 1977). In short, which habitats offer what opportunities to which species? The template approach identifies circumstances where given species perform better or less well than others. The outcome differs in being one of probabilities and not preordination.

Through previous work (Reynolds, 1987a, b, 1988, 1993), I have been able to demonstrate broad distributions of common species in the phytoplankton of lakes, against gradients of mean underwater insolation (I^*) and phosphorus availability ([P]). They are well-matched, on the one hand, to the reported adaptabilities of photosynthetic efficiencies (α) and kinetics of phosphorus uptake of the same algae in culture and, on the other, to the scatter of species-specific morphologies (maximum dimension, surface area and volume). Following Grime (1979), these broad strategies, can be represented in 2 × 2 contingency tables, which feature high- or low- light- energy and nutrient resources (Figure 1a). The low energy-low resource eventuality is consistently unoccupied – all the species take up positions defined by the other three apices, hence the triangular representation (Figure 1b). Again, following Grime (1979), the apices were labelled C, S and R, although the original interpretations of these symbols have been revised (Reynolds, 1995). This has not altered the validity of Reynolds' (1987a) proposal that the associations represented by dominant species fit to the broad template defined by these two axes.

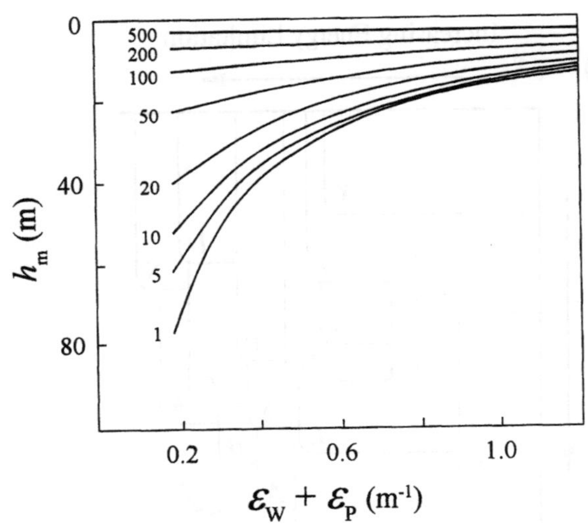

Figure 2. A matrix to show the distribution of contours of maximum supportable phytoplankton chlorophyll content (in μg l^{-1}) in terms of mixed water-column depth (h_m) and the combined coefficient of vertical light attenuation due to pure water (ϵ_W) and any colour and particulate material (ϵ_P). From relationships discussed in Reynolds (1992).

How might this concept be applied to the question of gradients in the trophic states of water bodies? What alternative explanation can it offer for (say) the continuing problem of the consistent linkage of the appearance of bloom-forming cyanobacteria in lakes with their eutrophication by enhanced phosphorus loading? The following sections show how further template axes distinguish other measurable habitat gradients to which the attributes of algae might respond, yet that many of these are additive to or consequential upon, or at least co-vary with, habitat changes across the trophic spectrum.

Factor covariability with trophic state

The physical environment

The classical differentiation of Caledonian and Baltic type lakes separated not merely their nutrient contents so much as their metabolic properties: the area-specific levels of biological production, maintained biomass and hypolimnetic oxygen consumption are greater among the Baltic-types, which Naumann (1919) categorised 'eutrophic'. The low concentrations of phytoplankton and the high clarity that are conversely characteristic of the oligotrophic Caledonian-types attest to their inability to sustain either large crops of biota or the areal rates of production that they elicit. Except when it is thermally stratified to within 50–60 meters of the surface, however, the mixed water column of a deep lake is such a light-deficient environment that it can maintain only a very sparse phytoplankton biomass (Figure 2).

The greatest opportunities to exploit its resources reside with species that are efficient harvesters of the underwater photon flux. When the water column does stratify near the surface, the light constraint is reduced significantly, at least, in the surface mixed layer. As was recognised long ago by Sverdrup et al. (1942), thermal stratification is an essential precursor to the support and development of phytoplankton populations in the open sea; the same must apply to deep lakes.The onset is determined principally through the developing buoyant resistance to the work of the wind, usually as a result of heating at the surface. This in turn leads to the shrinkage of the mixed layer, wherein the average insolation increases sharply. We might deduce a certain factor co-variance among temperature (θ), mixed depth (h_m) and insolation (I^*).

Resources and energy

However, growing populations exploiting the opportunities provided at the onset of stratification must be expected, sooner or later, to encounter the resource-dependent capacity limit. Among the species which are present, the opportunities for assembling dominant populations are weighted towards those which are quickest to respond, or which field the largest inoculum. Persistent stratification should be expected to favour species which are best adapted to exploit the vertical structure, in order to satisfy at least their maintenance requirements for energy and resources. Conserving existing biomass is an essential component of this strategy.

In the shallow lake, certainly one less than 5 m or so, the light-determined carrying capacity often saturates the supportive capacity of the nutrient resources. As nutrients are depleted, the opportunity for self-regulating, biomass-conserving larger species to take over from the invasive colonists is progressively established. Where nutrient loading is high, however, the onset of nutrient limitation might not intervene until after a large biomass has been established. Shallow systems, by definition, cannot export products to depth. There remains an on-going capacity to recycle the same nutrient over and over again (Reynolds, 1992;

Moss et al., 1996); provided the retention time is also long, enriched shallow lakes can maintain rather higher average crops than the external load appears to permit. Thus, for substantial periods, shallow lakes often have the energy- and nutrient- potential to maintain substantial levels of plant biomass. This may be either planktonic or macrophytic (Scheffer et al., 1993).

The carbon problem

Under such circumstances, assembling a substantial area-specific standing crop introduces new resource problems. Apart from the self-imposition of an energy limitation in the mixed layer, photosynthetic productivity is liable to become constrained by the carbon supply. The carbon dioxide content of lake water owes partly to that which was dissolved in the inflowing water, but this is modified by subsequent exchanges through uptake, respiration and atmospheric invasion, all at well-investigated rates (Maberly, 1996). In unbuffered waters (that is, relatively lacking in bicarbonate), rapid CO_2 withdrawal by otherwise nutrient-replete algal growth leads to raised pH levels and to selective weighting in favour of those species with efficient carbon-concentrating mechanisms, including the bloom-forming cyanobacteria (see especially, Shapiro, 1990). On the other hand, CO_2 levels not exploited thus are maintained close to the air-equilibrated concentration (0.5–1.0 mg l^{-1}; < 0.1 mmol l^{-1}). This is sufficient to keep the pH on the acid side of neutrality, where the bloom-forming cyanobacteria perform relatively poorly. While it is present, bicarbonate not only regulates the pH but its dissociation yields assimilable carbon dioxide. Many algae exploit this reaction through bicarbonate-concentrating mechanisms and the production of carbonic anhydrase. There is growing evidence (Saxby, 1990; Saxby-Rouen et al., 1996) that many common chrysophytes lack this ability, which thus weighs against their opportunities to flourish in either alkaline lakes or in eutrophied soft-water lakes. It may be a shortage of carbon and not a surfeit of phosphorus that precludes these algae from the eutrophic part of the spectrum (cf. Rodhe, 1948). Such effects are never easy to tease out in general survey data, because the concentrations of nutrients across the Baltic-Caledonian spectrum roughly co-vary with the hardness component, dominated by the calcium bicarbonate content. In the case of the pH-CO_2 conundrum, even good experimental design failed to separate their close co-variability until relatively recently (Saxby, 1990).

Herbivory

A third consequence of phytoplankton community assembly is the provision of an increasingly abundant, high-quality food resource for zooplankton. Moreover, above definable concentration thresholds (some 0.1–0.5 mg C l^{-1} in the case of *Daphnia* spp.: Lampert, 1977), filter-feeding represents a superior foraging strategy to selective feeding. It is well-known that, in most mesotrophic and eutrophic systems, Daphnids tend to the most prominent planktonic herbivores, rather than the calanoid copepods that generally dominate in oligotrophic lakes. The population dynamics of well-nourished *Daphnia* spp also differ from those of calanoids in being very responsive to resource sufficiency (Ferguson et al., 1982). Growth, fecundity and egg development times are sensitive to the food supply: under good conditions, Thompson et al. (1982) found that successive cohorts (generations) were recruited at two-week intervals, each generation increasing the aggregate filtration capacity twelve- to thirteen-fold. Given the size-dependent filtration capacities of individual animals of 5–60 ml of water per day (Burns, 1969), it is not long before a population is achieved that is capable of filtering the entire water volume each day. Removal at this rate is beyond the capacity of most algae to recoup by growth under normal conditions; the grazing down of eutrophic phytoplankton to create conditions of high water clarity is also a familiar phenomenon. The supposition that some species of phytoplankton attain sizes beyond the filtration capacity of extant *Daphnia* populations and hence gain a dynamic advantage over smaller, faster-growing algae is also well-supported (Reynolds et al., 1982; Gliwicz, 1990).

Factor interaction

In these ways, judgements as to the trophic condition of lakes and the plankton that they support should not be confined to the nutrients available but should relate to a package of factors, including basin morphometry, mixing dynamics (thus, local climate and, hence, latitude, altitude and relative exposure), water clarity and alkalinity in addition to the size and nature of the nutrient-resource base. That they are determined independently does not prevent some of them from being frequently and mutually co-variable. Thus, most eutrophic lakes are in lowlands, are relatively small, shallow, are at least mildly calcareous and are, indeed, relatively well-supplied with the nutrients critical to the support of plant biomass. Most deeper lakes and smaller ones of mountainous areas are judged to be oligotrophic and resource-poor. The sediments of deep lakes do not recycle nutrients as efficiently as those of shallow sediments or, by implication, those of shallow lakes. Poverty of bases allows pH to drift into acidity, although, without bicarbonate buffering, carbon dioxide demand of an expanding phytoplankton can soon raise the pH to selective levels ($\geq$pH 8.3–8.6: see Moss, 1973a; Talling, 1976). When it comes to regulating the wide differences in species composition associated with the trophic spectrum, several factors require to be satisfied. Even though some of these may co-vary in a way which makes their positive effects hard to distinguish, the failure of just one is likely to be decisive. It is thus quite easy to misinterpret critical environmental transitions leading to changes in the species composition of planktonic assemblages.

The latter is not quite a re-statement of Liebig's Law of the Minimum, though the idea that growth may proceed as far as organisms can manage it is common to both ideas. The present view is that not all species have the same 'minimal level' and that the species-specific reaction is a decline rather than a cut-off, the slope of which is generally well predicted from the rising portion of a Monod curve.

Studies on the structure of the phytoplankton in atypical lake systems or extreme conditions are often invaluable in the recognition of the niche boundaries of given species. The importance of interactions of factors influencing the dynamics of mixing has sometimes been shown by very large interannual differences in total phytoplankton and relative representation of key species. Early onset of near-surface stratification, as opposed to persistence of prolonged or frequent mixing events, promotes quite striking differences in the composition of the phytoplankton but it can be demonstrated that the species-specific growth responses are wholly consistent (Reynolds & Reynolds, 1985; Reynolds & Bellinger, 1992). Note is also taken of the descriptions of phytoplankton in nutrient-poor marl lakes (which carry a sparse biomass, qualitatively that of a eutrophic lake: Lund, 1961) and of phosphorus-rich acid lakes (abundant biomass of acidophils: Swale, 1968): these instances are extremely helpful to our appreciation of the selection of 'oligotrophic' and 'eutrophic' species.

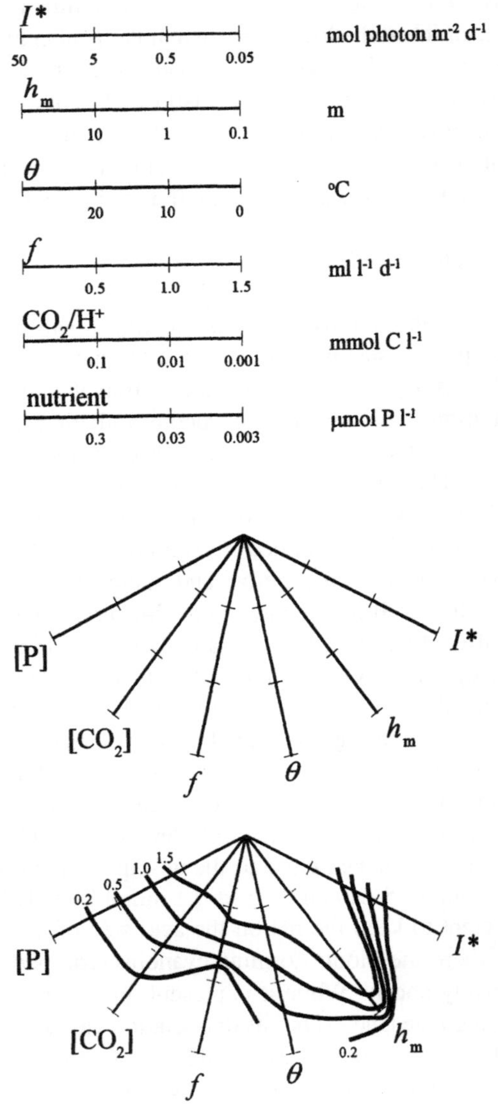

Figure 3. The design of a more advanced habitat template, involving a selection of axes to represent gradients of habitat factors, involving mean underwater irradiance (I^*), mixed-layer depth (h_m), water temperature (θ), the aggregate fitration rate of the zooplankton biomass (f), the concentation of carbon dioxide as a surrogate of acidity/alkalinity, zooplankton filtration and the concentration of biologically available phosphorus ([P], soluble and intracellular fractions). The axes are assembled into a hexacle (middle diagram) to which contours of the replication rate of *Chlorella* (data of Reynolds, 1988, 1993) are added in the bottom diagram.

Matching the adaptive strategies of phytoplankton to habitat templates

Habitat templates may be developed to embody any number of supposedly diagnostic axes of environmen-

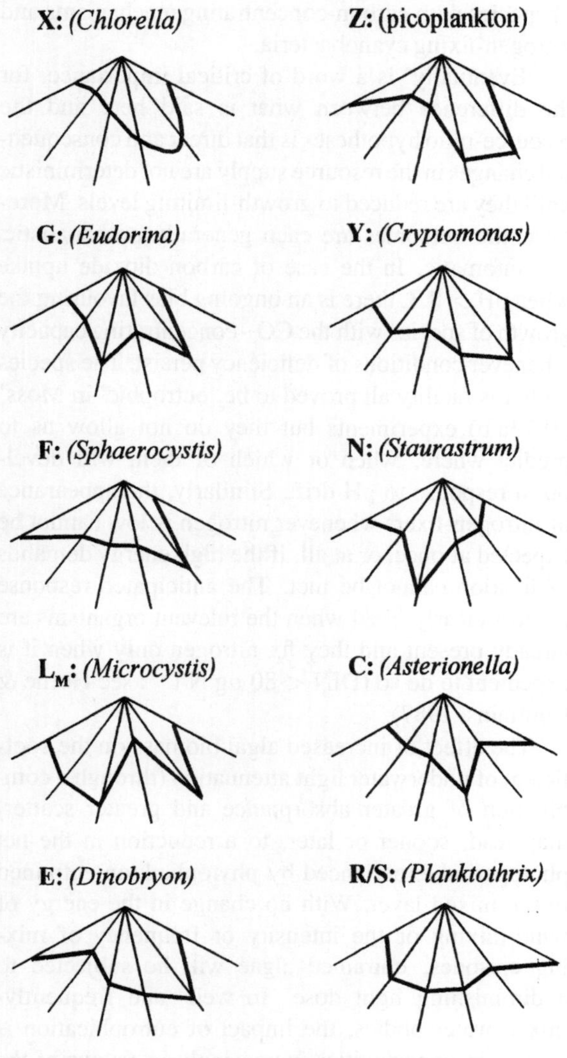

Figure 4. Suggested qualitative hexacle shapes for some common genera of phytoplankton: the letters refer to the phytoplankton associations of which they are usually representative (Reynolds, 1996), and as marked in Figure 1b.

tal variability, to correspond to differential attributes and adaptations of the various species of phytoplankton. The example developed in Figures 3 and 4 envisages a hexacle, with arms representing the specific requirements or tolerances of named algae to separate variable characters of the environment. Specifically, those chosen describe critical ranges of I^*, the integral of underwater irradiance; h_m, the mixed depth of the water column, especially as it is relevant to suspension; θ, the water temperature; f, the feeding rate or, more precisely, the filtration capacity of the herbivores present; acidity or alkalinity (represented here as carbon dioxide in mmol l^{-1}); and nutrient, being the external concentration required to saturate the maximum growth rate of the alga (phosphorus is considered critical in the example, scaled in μmol l^{-1}). Against each of these axes, it is possible to plot the growth-rate responses of given algae to the given variable, say by plotting r' as a function of θ, and the net replication rate, r_{20}, as dependent variable in each case. Differences in growth-rate responses of specific algae to temperature change are also plottable and approximations to the effects of mixed depth on sinking loss and filtration rates on grazing loss rates are available for some species (Reynolds, 1988, 1993). Discriminatory data on the tolerances and requirements of algae with respect to the carbon supply have been accorded less attention from ecologists than they merit but important experimental data sets assembled by (*inter alia*) by Moss (1973b), Shapiro (1990) and Maberly (1996) are available.

The diagnostic axes may then be arranged to provide a compound figure for each species, paying careful attention to orientation (the best net performance is shown to the left and, as assembled, towards the focal root of the hexacle). Then, contours may be constructed, to link intersections of the same growth rate on the respective axes. The completed example in Figure 3 is for *Chlorella*. The concept is that the 'rose' now specifies the kinds of habitat in which *Chlorella* should be able to thrive – relatively rich in phosphorus and inorganic carbon, over a wide range of temperatures with tolerance of mild grazing; it is largely insensitive to mixed depths > 0.2 m, provided its I^* demands are met. Conversely, it suggests the habitats in which the alga would fare poorly.

At the present time, too few systematic investigations of specific algae have been made to construct quantitative templates for very many species. Sufficient data from field experiments in limnetic enclosures have been obtained, however, for it to be possible to propose tentative shapes to fit the simple habitat template (Figure 4). Compared to the sketch representing *Chlorella*, itself based upon the fitted contour map in Figure 3, the various silhouettes suggest alternative habitat preferences of other common genera. The algae are assigned to the respective floristic associations of which they are alleged to be representative (Reynolds, 1996), in the expectation that the behaviour denoted is typical of the other members of the relevant association. Moreover, compounding species-specific information with a representation of the opportunities that a given habitat provides helps us to predict for or against which species and attributes the habitat will weight. Thus, the axes chosen and the sketches thus assembled are able to discern such key sensitivities as those of *Eudorina* (Association G of Reynolds, 1996), *Sphaerocystis* (F) and *Microcystis* (L_M,M) to insolation; those of non-motile diatoms and desmids (C,N) to mixed depth; those of larger algae (L,M) to low water temperatures; those of small ones (X,Y,Z) to heavy grazing; and those of chrysophytes (E) to free carbon dioxide levels. Conversely, the tolerance of poor insolation and photadaptability of *Asterionella* (C) and *Planktothrix* (R,S) are acknowledged to be positively selective in light-deficient deep mixed layers.

In spite of its present reliance upon too many pseudoquantities, themselves often deduced qualitatively from observations upon natural distributions (they can scarcely then be invoked to 'explain' natural distributions), the approach does constitute a testable hypothesis. The collection of hard laboratory data about the attributes of type species is required to support or negate the hypothesis. Besides, some of the axes will always be complementary, if not directly additive. For the present, the preference is to concentrate upon the [P] and I^* axes, for which many more data are available, and for which approach, consistent quantitative data exist (see e.g., Reynolds 1987a,b, 1988, 1993).

Eutrophication and the trophic spectrum

Supposing the habitat-template hypothesis to be substantially valid, what factors are decisive in altering the species composition of phytoplankton in lakes subject to eutrophication? Taking the most general definition of this term, which refers to the excessive algal growth consequential upon an enhanced supply of nutrients (Harper, 1992), we may also take it that increased chemical fertility and the capacity to support additional biomass are not at issue. To understand the changes in floristic dominance, we need to trace the impacts of the secondary relationships.

Among the many such secondary effects, we may enumerate altered demands on elements whose rates of supply are either conservative (such as silicon), or are governed by chemical equilibria (such as carbon dioxide), or are varied but in other proportions (classically nitrogen). In each instance, there is a predicable shift in weighting towards particular groups of species. In response to the effects cited, we can expect eventually weighting in favour, respectively, of non-diatoms, of species with carbon-concentrating mechanisms and nitrogen-fixing cyanobacteria.

'Eventually' is a word of critical importance, for the difference between what is said here and the resource-ratio hypotheses is that direct and consequential changes in the resource supply are not deterministic until they are reduced to growth-limiting levels. Moreover, the outcomes are each generally probabalistic, not automatic. In the case of carbon dioxide uptake when $pH > 8.3$, there is an ongoing bias favouring the growth of species with the CO_2-concentrating capacity whenever conditions of deficiency persist. The species with this facility all proved to be 'eutrophic' in Moss' (1973a,b) experiments but they do not allow us to predict where, when or which of them will develop in response to pH drift. Similarly, the appearance of nitrogen-fixers whenever nitrogen is low cannot be expected at once, or at all, if the high energy demands of fixation cannot be met. The anticipated response is most clearly filled when the relevant organisms are already present and they fix nitrogen only when it is expedient to do so (DIN $< 80\ \mu g\ N\ l^{-1}$: see Horne & Commins, 1987).

The effect of increased algal biomass on the coefficient of underwater light attenuation (through a combination of greater absorptance and greater scatter) may lead, sooner or later, to a reduction in the net photoperiod experienced by phytoplankton entrained in the mixed layer. With no change in the energy of wind-mixing or the intensity or frequency of mixing episodes, entrained algae will be subjected to a diminishing light dose. In well- and frequently-mixed water bodies, the impact of eutrophication is to weight opportunities increasingly in favour of the more photosynthetically-efficient species and the more photoadaptable ones (slender and filamentous species, with a demonstrable capacity to enhance their content of chlorophyll and accessory pigments). Another interesting consequence of eutrophication of physically-variable systems, is that the change in the light dose with the switch from mixed to stagnating conditions is made considerably more severe. If such switches are not to be fatal, pre-adaptations permitting algae to self-regulate, founded on rapid rates of controlled movement, gain a selective premium.

Increased algal biomass is ultimately an increased resource to consumers. A greater zooplankton biomass is presumed to exercise a feedback on the algal production. If literally true, two immediate effects are observable. One is an increasing selection in favour of the algae which, for reasons of size, palatability or accessibility, are avoided and against the smaller, more attractive and immotile ones. The other is that because algae grow faster, they also grow further before grazers 'catch up', by which time, the efficiency of the collective feeding strips out the resource altogether. The effect is one of much greater oscillations in the biomass of algae and their herbivorous dependents. There are further subtle effects, including the consequences on the composition of the zooplankton, the scale and operation of the microbial loop and recycle pathways and the kinds of planktivore that are most advantaged. The promotion of the decomposition cycle, the impact upon the hypolimnetic oxidative capacity and the faunal tolerances of low redox and its consequences may all ensue from a modicum of phosphorus enrichment. Each may feed back another factor in the selection of the biota.

To provide just one illustration of how this paradigm might be applied to the interpretation to species selection across the trophic spectrum, let us revert to the long-standing conundrum of why the Cyanobacteria are so commonly associated with eutrophic lakes. Even if the statement were literally true and it did not ignore either the distribution of picocyanobacteria or that of the *Merismopedia-Peridinium inconspicuum* assemblage (L_O), or even those of the assemblages dominated by members of the Oscillatoriales (R,S) which are anathemous to the bloom-forming genera, the explanations for the dominance of *Microcystis*, *Aphanizomenon* and *Anabaena flos-aquae* continue to be contested vigourously (Jensen et al., 1994). The habitat-template approach helps us to appreciate that the problem these genera have is that most temperate lakes are *too cold* for them through a major part of the year. Then, when they are able to grow, they encounter a nutrient deficiency because something else (the spring diatom bloom) has already stripped out the available phosphorus and transported it, intracellularly, to the bottom sediment. Once a lake is sufficently enriched for the ability of the spring-bloom to act as a nutrient-sponge to have become saturated, then the residue that persists into the clarity and warmth of a temperate summer is quite suddenly able to support these pervasive cyanobacteria. In tropical lakes, where the temperatures are adequate for cyanobacterial growth the year round, this problem does not arise: populations may remain vegetative through the year, their superior survivorship often assisting them to dominance.

None of this is to negate the advantage that these Cyanobacteria derive from low half-saturation coefficients for carbon uptake, or (where appropriate) from being able to fix atmosperic nitrogen, or from being the least favourite food of everything except ciliates. It is undeniable that other indirect effects of nutrient enrichment (like the formation of soft, organic-rich, low-redox sediments) aid their perennation. Neither are Cyanobacteria protected from vulnerability to deep mixing or (especially) rapid flushing. However, it does allow us to form a paradigm about the kinds of lakes where these cyanobacteria are likely to be prevalent and the conditions likely to promote their abundance in any one year. It is, indeed, a prime illustration of the application of the template and the assessment of the prospects of fulfilling the opportunity for population growth.

Trophic-state change – Temporal aspects

The closing section focuses on the fact that the template-predicted effects of a change in the chemical or physical status of a lake are not immediate, nor need they occur at all if the favoured species are not already present in the system, as rarities, or their propagules are not so 'mobile' (*sensu* between sites) that they can be expected to arrive imminently to respond to the new conditions.

I can point to several instances of progressive eutrophication of lakes in the English Lake District where there have been qualitatively similar responses but where the timing is really quite stochastic.

All the lakes of the series lie in glacially-excavated basins with upland, thin-soiled catchments. They differ in size, depth and flushing rates and they do, as a series, cover a range of 'trophic states' within the understanding of this paper (see Jones, 1972; Kadiri & Reynolds, 1993). They are all located in an area of relatively high annual rainfall, are poor to very poor in bicarbonate buffering capacity and they are mostly naturally deficient in phosphorus.

Blelham Tarn (area: 0.1 km^2; mean depth: 6.8 m; mean hydraulic retention time: 50 d) has a small catchment, mainly devoted to agriculture, with some montane pasture and woodland. In 1962, a small sewage treatment plant was commissioned at the village of Outgate, final effluent from which was discharged to one of the afferent streams feeding the lake. Lund (1978) has recorded the changes in the nutrient loads and the phytoplankton populations as the lake rapidly moved to being 'eutrophic' with an anoxic hypolimnion. From having been always < 4 μg l^{-1}

prior to 1962, the winter concentration of soluble reactive phosphorus rose steeply to > 10 μg P l^{-1} by 1966, where it has since remained. Algal species indicative of eutrophic conditions now form significant components of the annual planktonic crop, although they were rare (*Aphanizomenon flos-aquae*, first recorded in 1955) or appeared in the lake since the commissioning of the works: they include *Fragilaria crotonensis* (from 1962), *Closterium limneticum* (1965), *Anabaena solitaria* (1966), a small *Stephanodiscus*, accorded by Lund to *S.astraea* (Ehr.) Grun. var. *minutula* (Kütz.) Grun. (1973) and *Microcystis aeruginosa* (1974). In nearby Grasmere (area: 0.65 km^2; mean depth: 7.7 m; mean hydraulic retention time: 25 d), where a sewage works was commissioned in 1971, a similar rise in winter SRP (from < 4 μg l^{-1} to 10 –12 μg l^{-1} by 1980) was not accompanied by any immediate change in its previously oligo-mesotrophic plankton flora, ostensibly because the frequency of efficient flushing prevented inocula, necessarily imported from outwith its montane catchment, to establishing in the lake (Reynolds & Lund, 1988). However, in the wake of two dry summers (1989, 1990), many eutrophic species took hold: *Aphanizomenon flos-aquae*, *Anabaena flos-aquae* and *Fragilaria crotonensis* have been conspicuous since 1990 and what is believed to be the same species of *Stephanodiscus* from 1991. *Ceratium hirundinella* (1990) and *Microcystis aeruginosa* (1991) were also recorded in the lake for the first time but, to date, these have failed to establish themselves as regular dominants. Meanwhile, Grasmere has attained a total phosphorus content (maximum 70 μg P l^{-1}) and a reducing hypolimnion to qualify, within the modern definition, as a eutrophic lake.

From 1945 to 1965, the winter SRP in Windermere was generally < 2 μg P l^{-1} but, here too, revised sewage disposal arrangements came into commission at Ambleside (on the lake's north basin) and Tower Wood (on the south basin). A steady rise in the winter SRP followed, especially in the south basin where 28 μg P l^{-1} was reached in 1991. In the north basin, the levels reached have been 8–10 μg l^{-1}. Removal of phosphorus from the effluents at Ambleside and Tower Wood since 1992 brought immediate reductions in the winter SRP-levels in the lake (north basin, 1995: 5 μg P l^{-1}; south basin < 11 μg l^{-1}). Considering only the north-basin populations, *Aphanizomenon* was recorded at > 10 cells ml^{-1} for the first time in 1981 and reached > 2 500 cells ml^{-1} in every summer from 1985–1992. Similarly, *Fragilaria crotonensis* was very rare before 1978 but achieved > 100 ml^{-1} in 1980 and > 2 500 ml^{-1} since 1989. It has remained similarly abundant today. However, the dominant organism during most summers of the 1980s, *Tychonema bourrelleyi* (see Heaney et al., 1996), has completely disappeared since 1993.

These cases illustrate qualitatively similar responses to similar chemical stimuli but in contrasted lake basins. It appears that no change is obligate and, even if it remains a strong probability, it may require some other event to secure it. The section may be concluded by drawing attention to the case, reported elsewhere in this volume (Reynolds et al., 1998), where the SRP content of another lake in the English Lake District, Seathwaite Tarn, was raised artificially from < 1 to > 20 μg P l^{-1} but without any new addition to the plankton flora in two years, much less any recognised 'eutrophic species'.

Conclusions

The composition of phytoplankton from a given water is very often an excellent indication of the trophic state of the water body. It is simple to deduce that an alteration in species composition is a consequence of increased nutrient loading but the mechanism for the change is rarely directly due to the nutrient availability. Certainly and within defineable limits, the level of supportable biomass varies with the amounts of nutrient supplied but this often places constraints on the supply of other resources. For instance, diminution in the availability of carbon dioxide and in the underwater irradiance will weight more strongly in favour of species with active carbon-dioxide concentrating mechanisms places and which are better or more photo-adaptable light antennae. This conclusion endorses that of Moss (1973b): with the advent of 'eutrophication' as an issue, perceptions of the oligotrophic-eutrophic spectrum became dominated by the role of nitrogen and, especially, phosphorus. In reality, the trophic spectrum should not be regarded as a single dimension of a single factor but, rather, as a swathe of interelated factors co-varying in response to productivity demands on the totality of resources. Stresses placed on the supply of carbon or light bias selection in favour of species known for their tolerance or adaptability to these conditions. In the wake of nutrient enrichment, however, compositional change is neither immediate nor inevitable. Rather, change is probabalistic. The outcomes nourish our intuitions about the distribution of phytoplankton across the trophic spectrum.

Acknowledgements

I am grateful to Dr Miguel Alvarez Cobelas, for suggesting the topic of this discussion, and to Dr Judit Padisák for the benefit of her perceptive comments. The manuscript benefitted from the encouraging and generous remarks of two anonymous reviews, for which I remain grateful.

References

Blomqvist, P., A. Petterson & P. Hyenstrand, 1994. Ammonium-nitrogen: a key regulatory factor causing dominance of non-nitrogen fixing cyanobacteria in aquatic systems. Arch. Hydrobiol. 132: 141–164.

Burns, C. W., 1969. Relation between filtering rate, temperature and body size in four species of *Daphnia*. Limnol. Oceanogr. 14: 693–700.

Ferguson, A. J. D., J. M. Thompson & C. S. Reynolds, 1982. Structure and dynamics of zooplankton communities maintained in closed systems, with special reference to the food supply. J. Plankton Res. 4: 523–543.

George, D. G., D. P. Hewitt, J. W. G. Lund & W. J. P. Smyly, 1990. The relative effects of enrichment and climatic change on the long-term dynamics of *Daphnia* in Esthwaite Water, Cumbria. Freshwat. Biol. 23: 55–70.

Gliwicz, Z. M., 1990. Why do cladocerans fail to control algal blooms? Hydrobiologia, 200/201: 83–97.

Grime, J. P., 1979. Plant strategies and vegetation processes. Wiley, Chichester.

Harper, D. M., 1992. Eutrophication of freshwaters. Chapman and Hall, London.

Heaney, S. I., J. E. Parker, C. Butterwick & K. J. Clarke, 1996. Interannual variability of algal populations and their influence on lake metabolism. Freshwat. Biol. 35: 561–577.

Horne, A. J. & M. L. Commins, 1987. Macronutrient controls on nitrogen fixation in planktonic cyanobacterial populations. New Zealand J. mar. freshwat. Res. 21: 413–423.

Huisman, J. & F. J. Weissing, 1995. Competition for nutrients and light in a mixed water column: a theoretical analysis. Am. Nat. 146: 536–564.

Jensen, J. P., E. Jeppesen, K. Olrik & P. Kristensen, 1994. Impact of nutrients and physical factors on the shift from Cyanobacteria to Chlorophyte in shallow Danish lakes. Can. J. Fish. aquat. Sci. 51: 1692–1699.

Jones, J. G., 1972. Studies on freshwater micro-organisms. Phosphatase activity in lakes of differing degrees of eutrophication. J. Ecol. 60: 777–791.

Kadiri, M. O. & C. S. Reynolds, 1993. Long-term monitoring of the conditions of lakes: the example of the English Lake District. Arch. Hydrobiol. 129: 157–178.

Lampert, W., 1977. Studies on the carbon balance of *Daphnia pulex* De Geer as related to environmental conditions. IV. Determination of the 'threshold' concentration as a factor controlling the abundance of zooplankton species. Arch. Hydrobiol. (Suppl.) 48: 361–368.

Levich, A. P., 1996. The role of nitrogen-phosphorus ratio in selecting for dominance of phytoplankton by cyanobacteria or green algae and its application to reservoir management. J. aquat. Ecosyst. Hlth. 5: 55–61.

Lund, J. W. G., 1950. Studies on *Asterionella formosa* Hass. II. Nutrient depletion and the spring maximum. J. Ecol. 38: 1–35.

Lund, J. W. G., 1961. The algae of the Malham Tarn district. Field Stud. 1: 85–115.

Lund, J. W. G., 1978. Changes in the phytoplankton of an English Lake, 1945–1977. Hydrobiol. J. 14: 6–21.

Maberly, S. C., 1996. Diel, episodic and seasonal changes in pH and concentrations of inorganic carbon in a productive lake. Freshwat. Biol. 35: 579–598.

Mann, N. H., 1995. How do cells express nutrient limitation at the molecular level? In I. Joint (ed.), Molecular ecology of aquatic microbes. Springer, Berlin: 171–190.

Moss, B., 1973a. The influence of environmental factors on the distribution of freshwater algae: an experimental study. II. The role of pH and the carbon dioxide-bicarbonate system. J. Ecol. 61: 157–177.

Moss, B., 1973b. The influence of environmental factors on the distribution of freshwater algae: an experimental study. IV. Growth of test species in natural lake waters and conclusions. J. Ecol. 61: 193–211.

Moss, B., P. Johnes & G. Phillips, 1996. The monitoring of ecological quality and the classification of standing waters in temperate regions: a review and proposal based on a worked scheme for British waters. Biol. Rev. 71: 301–339.

Nalewajko, C. & D. R. S. Lean, 1978. Phosphorus kinetics-algal growth relationships in batch cultures. Mitt. int. Ver. theor. angew. Limnol. 21: 184–192.

Naumann, E., 1919. Några synpunkter angående limnoplanktons ökologi med särskild hänsyn till fytoplankton. Svensk bot. Tidskr., 13: 129–163.

Olrik, K., 1994. Phytoplankton – Ecology. Miljøsministiert, København.

Pearsall, W. H., 1921. The development of vegetation in English lakes, considered in relation to the general evolution of glacial lakes and rock basins. Proc. r. Soc. Lond. B 92: 259–284.

Petersen, R., 1975. The paradox of the plankton: an equilibrium hypothesis. Am. Nat. 109: 35–49.

Pick, F. R. & D. R. S. Lean, 1987. The role of macronutrients (C,N,P) in controlling cyanobacterial dominance in temperate lakes. New Zealand J. mar. freshwat. Res. 21: 425–434.

Rawson, D. S., 1956. Algal indicators of trophic lake types. Limnol. Oceanogr. 1: 18–25.

Reynolds, C. S., 1986. Experimental manipulations of the phytoplankton periodicity in large limnetic enclosures in Blelham Tarn, English Lake District. Hydrobiologia 138: 43–64.

Reynolds, C. S., 1987a. Community organization in the freshwater plankton. In J. H. R. Gee & P. S. Giller (eds), Organization of communities, past and present. Blackwell Scientific Publications, Oxford: 297–325.

Reynolds, C. S., 1987b. The response of phytoplankton communities to changing lake environments. Schweiz. Z. Hydrol. 49: 220–236.

Reynolds, C. S., 1988. Functional morphology and the adaptive strategies of freshwater phytoplankton. In C.D.Sandgren (ed.), Growth and reproductive strategies of freshwater phytoplankton. Cambridge University Press, New York: 388–433.

Reynolds, C. S., 1990. Temporal scales of variability in pelagic environments and the responses of phytoplankton. Freshwat. Biol. 23: 25–53.

Reynolds, C. S., 1992. Eutrophication and the management of planktonic algae: what Vollenweider couldn't tell us. In D. W. Sutcliffe & J. G. Jones (eds), Eutrophication: research and application to water supply. Freshwater Biological Association, Ambleside: 4–29.

Reynolds, C. S., 1993. Swings and roundabouts: engineering the environment of algal growth. In K. H. White, E. G. Bellinger, A. J. Saul, M. Symes & K. Hendry (eds), Urban waterside regeneration, problems and prospects. Ellis Horwood, Chichester: 330–349.

Reynolds, C. S., 1995. Successional change in the planktonic vegetation: species, structues, scales. In I. Joint (ed.), The molecular ecology of aquatic microbes. Springer-Verlag, Berlin: 115–132.

Reynolds, C. S., 1996. Plant life of the pelagic. Verh. int. Ver. theor. angew. Limnol. 26: 97–113.

Reynolds, C. S. & E. D. Bellinger, 1992. Patterns of abundance and dominance of the phytoplankton of Rostherne Mere, England: evidence from an 18-year data set. Aquat. Sci. 54: 10–36.

Reynolds, C. S., G. H. M. Jaworski, J. V. Roscoe, D. P. Hewitt & D. G. George, 1998. Responses of the phytoplankton to a deliberate attempt to raise the trophic status of an acidic, oligotrophic mountain lake. Hydrobiologia 369/370: 127–131.

Reynolds, C. S. & J. W. G. Lund, 1988. The phytoplankton of an enriched, soft-water lake, subject to intermittent hydraulic flushing (Grasmere, English Lake District). Freshwat. Biol. 19: 379–404.

Reynolds, C. S. & J. B. Reynolds, 1985. The atypical seasonality of phytoplankton in Crose Mere, 1972: an independent test of the hypothesis that variability in the physical environment regulates community dynamics and structure. Br. phycol. J. 20: 227–242.

Reynolds, C. S., J. M. Thompson, A. J. D. Ferguson & S. W. Wiseman, 1982. Loss processes in the population dynamics of phytoplankton maintained in closed systems. J. Plankton Res. 4: 561–600.

Reynolds, C. S., S. W. Wiseman & M. J. O. Clarke. 1984. Growth- and loss-rate responses of phytoplankton to intermittent artificial mixing and their potential application to the control of planktonic algal biomass. J. appl. Ecol. 21: 11–39.

Rhee, G.-Y. and I. J. Gotham, 1980. Optimum N:P ratios and coexistence of planktonic algae. J. Phycol. 16: 486–489.

Riddolls, A., 1985. Aspects of nitrogen fixation in Lough Neagh. II. Competition between *Aphanizomenon flos-aquae, Oscillatoria redekei* and *Oscillatoria agardhii.* Freshwat. Biol. 15: 299–306.

Rodhe, W., 1948. Environmental requirements of freshwater plankton algae: experimental studies in the ecology of phytoplankton. Symbol. bot. Upsal. 10: 5–149.

Saxby, K. J., 1990. The physiological ecology of freshwater chrysophytes with special reference to *Synura petersenii.* Ph.D. Thesis, University of Birmingham.

Saxby-Rouen, K. J., B. S. C. Leadbeater & C. S. Reynolds, 1996. Ecophysiological studies on *Synura petersenii* (Synurophyceae). Beih. Nova Hedwigia 114: 111–123.

Scheffer, M., S. H. Hosper, M.-L. Meijer, B. Moss & E. Jeppesen, 1993. Alternative equilibria in shallow lakes. TREE 8: 275–279.

Smith, V. H., 1983. Low nitrogen to phosphorus favour dominance by blue-green algae in lake phytoplankton. Science 225: 669–671.

Shapiro, J., 1990. Current beliefs regarding dominance by bluegreens: the case for the importance of CO_2 and pH. Verh. int. Ver. theor. angew. Limnol. 24: 38–54.

Sommer, U., 1993. Phytoplankton competition in Plußsee: a field test of the resource-ratio hypothesis. Limnol. Oceanogr. 38: 838–845.

Southwood, T. R. E., 1977. Habitat, the templet for ecological stategies? J. anim. Ecol. 46: 337–365.

Sverdrup, H. U., M. W. Johnson & R. H. Fleming, 1942. The oceans. Their physics, chemistry and general biology. Prentice Hall, New York.

Swale, E. M. F., 1968. The phytoplankton of Oak Mere, Cheshire, 1963–1966. Br. phycol. Bull. 3: 441–449.

Talling, J. F., 1976. The depletion of carbon dioxide from lake water by phytoplankton. J. Ecol. 64: 79–121.

Thienemann, A., 1918. Untersuchungen über die Beziehungen zwischen dem Sauerstoffgehalt des Wassers und der Zusammensetzung der Fauna in norddeutschen Seen. Arch. Hydrobiol. 12: 1–65.

Thompson, J. M., A. J. D. Ferguson & C. S. Reynolds, 1982. Natural filtration rates of zooplankton in a closed system: the derivation of a community grazing index. J. Plankton Res. 4: 545–560.

Tilman, D. & S. S. Kilham, 1976. Phosphate and silicate uptake and growth kinetics of the diatoms *Asterionella formosa* and *Cyclotella meneghiniana* in batch and semi-continuous culture. J. Phycol. 12: 375–383.

Tilman, D., S. S. Kilham & P. Kilham, 1982. Phytoplankton communiy ecology: the role of limiting nutrients. Annu. Rev. Ecol. Syst. 13: 349–372.

Trimbee, A. M. & E. E. Prepas, 1987. Evaluation of total phophorus as a predictor of the relative biovolume of blue-green algae with emphasis on Alberta lakes. Can. J. Fish. aquat. Sci. 44: 1337–1342.

Hydrobiologia **369/370:** 27–43, 1998.
M. Alvarez-Cobelas, C. S. Reynolds, P. Sanchez-Castillo & J. Kristiansen (eds), Phytoplankton and Trophic Gradients.

Summer phytoplankton assemblages across trophic gradients in hard-water reservoirs

M. J. Dasí, M. R. Miracle, A. Camacho, J. M. Soria & E. Vicente
Departament de Microbiologia i Ecologia, Facultat de Biologia, Universitat de València, E-46100 Burjassot, Spain

Key words: phytoplankton, trophic level, reservoirs, hypertrophic ecosystems, nutrients, diversity

Abstract

Summer phytoplankton assemblages are described and characterised according to their prevalence in a series of hard-water reservoirs of eastern Spain that had been classified in trophic categories on OECD criteria. Distribution patterns of phytoplankton species were ordinated statistically by principal components analysis (PCA). The first component was strongly related to trophic gradient and it particularly discriminated the eutrophic and hypertrophic reservoirs. The second component segregated life-forms, so that (1), on the oligo-mesotrophic side, large dinoflagellates were separated from small centric diatoms, unicellular chrysophytes and filamentous ullotrichales and, on the eu-hypertrophic side (2), colonial greens and large desmids were separated from unicellular volvocales and small centric diatoms.

The large differences between eutrophic and hypertrophic reservoirs were also clearly identified in a second PCA, in which physical and chemical factors were used with the principal components solved from the phytoplankton data. From these results, a new trophic category was discerned, for which we propose the name 'holotrophic'. This category applies to water bodies having the following main features: (1) concentrations of chorophyll, total P and total N in the range of the hypertrophic systems, but with much higher concentrations of dissolved phosphorus and ammonia and (2) phytoplankton predominantly composed by unicellular green flagellates (*Pteromonas*, *Chlamydomonas*) and chlorococcales (*Scenedesmus*), without cyanobacterial blooms.

Introduction

Ortiz-Casas & Peña (1984) showed that OECD criteria (OECD, 1982), relating nutrient load to eutrophication responses, are valid for Spanish reservoirs. When broad-scale patterns of phytoplankton distribution were explored in a set of more than 100 Spanish reservoirs, trophic level was also an important explanatory factor (Margalef et al., 1982; Sabater & Nolla, 1991; Riera et al., 1992). However the resulting ordinations of the reservoirs shown in these works, based on phytoplankton species composition, consistently matched ordinations based on physical and chemical varables, in which two types of reservoirs, according to siliceous and calcareous rock dominance, were first established (Margalef et al., 1976). In the western part of the Iberian Peninsula, watersheds lie on igneous rocks or slates, which are responsible for the low alkalinities of reservoir waters in the region (less than 1 meq l^{-1}). In contrast, watersheds in the eastern part, lie on limestone and sedimentary soils and reservoirs generally have high alkalinities. Subgroups within these two sets of reservoirs were then distinguished by trophic level (indicated mainly by Chlorophyll-*a*), which became a clearer descriptor when only summer data was used (Estrada, 1975; Riera et al., 1992).

Phytoplankton composition has been shown to be influenced by two composite factors: water hardness and trophic status. The outcome of their interaction was recognised long ago (Moss, 1973). The important role of these two main variables (essentially, pH-inorganic C availability, on the one hand; the nutrients, N and P, on the other) in regulating phytoplankton composition has also been revealed in field experiments (e.g.

Reynolds, 1986; Shapiro, 1989; Sommer, 1989). The first composite factor is more related to geology and climate; the second to anthropogenic influences. Thus, in this exploration of the validity of the trophic spectrum, illustrated by data from fifteen hard-water reservoirs, it was necessary to suppress all other sources of variation as far as possible, especially the alkalinity-hardness factor. The reservoirs are located within an area of quite uniform geography, hydrography and mesomediterranean-dry climate. Moreover, the study was confined to a summer period, because between-site differences in trophic state are most apparent due to rainfall is least and, so sewage-enriched inflows are poorly diluted. The aim of this paper is to explore statistically the relationships between the phytoplankton communities of these high-alkalinity reservoirs and their various nutrient inputs. We then seek to compare these patterns with those detected in other regional studies, in Spain and elsewhere in the Mediterranean (Naselli-Flores & Barone, 1994) as well as those observed in other types of lake (Reynolds, 1982, 1984a), especially at the hypertrophic end of the spectrum (Alvarez-Cobelas & Jacobsen, 1992).

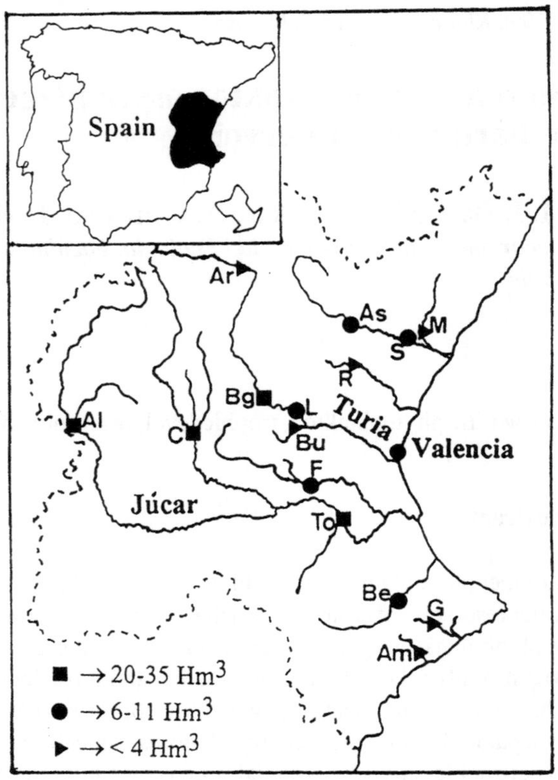

Figure 1. Geographic location of the studied reservoirs and their corresponding main river courses. The hydrographic zone of river Júcar is indicated in black. Abbreviations of reservoirs as in table 1.

Material and methods

Study sites

Spanish Mediterranean climate is characterised by the irregularity of rains, resulting in a scarcity of water resources. Because of this, a great number of reservoirs has been built, mostly during this century, to meet water needs. There are now more than one thousand large dams in Spain, in contrast to the small number of lakes in the country. Water supply, irrigation and flood control are the main uses of these reservoirs.

We studied the main reservoirs of the central Mediterranean hydrographic zone of Spain, named after its main river drainage area, the basin of the river Júcar (Figure 1). This river and its main tributary, Cabriel, feed the largest reservoirs of the zone, viz., Alarcón, Contreras and Tous. Their water volumes at the time of the study, summer 1995, represented only a very small percentage (3–5%) of their capacities. The smaller reservoirs then stored 10–30% of their capacities. The smaller Forata Reservoir is located on the second main tributary of the Júcar, the Magro. The large reservoir, Benageber, is on the second most important tributary, the Turia, which also has the small headwater reservoir, Arquillo. Three medium-sized reservoirs (Loriguilla, Arenós and Sichar) on the third most important tributary (the Millares) were included. The other resevoirs are relatively small, on short streams draining lesser subcatchments. Beniarrés and Regajo both receive urban wastewater and neither were as below capacity as the others (50% full by volume).

Most of the reservoirs studied were constructed between 1955 and 1967, except for Contreras, Arenós and Tous, which are more recent. They are hard-water reservoirs with alkalinities varying between 2 and 4 meq l^{-1} at the end of summer. Calcite precipitation, which usually takes place during summer, is more intense in the more eutrophic reservoirs. Then, in María Cristina Reservoir, alkalinity was lowered down to a seasonal minimum of 1.7 meq l^{-1}. In this case, pH is exceptionally high, reaching 10, whereas all the other measured values were in the range pH 8–9. Conductivities in these reservoirs varied from 0.5 to 1 mS cm^{-1} and, in most of them, sulphate concentrations (meq l^{-1}) exceeded bicarbonate alkalinity (Group III waters in the typology of Armengol et al., 1991).

The anthropogenic influence is very different between reservoirs. Some of them receive waters with only agricultural contamination (Tous, Contreras, Guadalest, Arenós and Arquillo), while others, such as Sichar, Amadorio, Benageber and Loriguilla, receive diluted wastewater from villages. The reservoirs, Forata, Beniarrés and María Cristina, all receive important sewage loads of both industrial and domestic wastewater with a high content in organic matter and inorganic nutrients. Buseo reservoir is still heavily affected by the massive input of ash and woody debris received as a consequence of the forest fire which extended throughout the catchment in 1994.

Methods

The reservoirs were sampled on four occasions, separated approximately by 20-day intervals, between early July and late September, 1995. Two points were sampled in each case, one near the dam and at approximately the deepest point, the other near to the river inflow. Water samples for chemical and phytoplankton analysis were collected from selected depths using a Ruttner bottle. Only epilimnetic samples, within the photic zone, have been used in this work (usually two from near the dam and one at the inflow).

Dissolved oxygen, temperature, pH, Eh, conductivity and light penetration were determined *in situ* along vertical profile with the appropriate sensors or electrodes. Secchi disk was also measured at every sampling site. Chemical analyses were performed according to Standard Methods (APHA, AWWA, WEF, 1992) and Golterman et al. (1978). The main dissolved inorganic nutrients – nitrate, nitrite, ammonia, soluble phosphorus and silica – were determined in filtered samples. Measurements were also made of total phosphorus, BOD and total iron. Photosynthetic pigments were determined spectrophotometrically in a 90% acetone-DMSO (1:1) extract after filtration of the sample through a GF/F glass fibre filter.

Water samples for phytoplankton were fixed *in situ* with Lugol's solution, and counted in the laboratory on an Olympus inverted microscope at 1000X, using the Utermöhl's method (1958). Scanning electron microscopy was used to determine the centric diatom species. At least twenty individuals from each taxa were measured to obtain their cell volume. Biovolumes were calculated using average dimensions and geometric formulae suggested by Rott (1981). Volumes of coenobial and some colonial forms (*Scenedesmus, Actinastrum, Dictyosphaerium*) were considered as the aggregate of the volumes of the individual cells, whereas compact colonial forms (*Pediastrum, Coelastrum*) were considered as 'biomass units', using the dimensions of the colony to calculate the biovolumes. The volume of blue-green filaments has been calculated from average lengths and widths of filaments.

Biovolume was used as a measure of biomass. Biomass diversity of each sample was based on Shannon-Wiener index: $H' = -\Sigma p_i \times \log_2 p_i$, where p_i is the proportion of biovolume of species i in the sample. Species with proportions below 0.5% of total biomass were not considered. The equitability of each sample was also calculated from $E = H'/\log_2 s$, where s is the number of species used to calculate H'. Two principal component analysis without rotation (PCA-1 and PCA-2) were made using SPSS software (Norusis, 1990). PCA-1 was performed on a selected phytoplankton group of 63 species, using log-transformed densities (cells/ml), and PCA-2 was done by considering as variables the first two components of PCA-1 and log-transformed physico-chemical variables (except temperature and pH). All variables and factors have been standardised.

Results

The reservoirs are arranged in Table 1 in order of their chlorophyll-a content. They were assigned to trophic categories according to chlorophyll, TP and transparency ranges (OECD, 1982). Since the trophic spectrum is a continuum, the allocation of the reservoirs to the various categories is not always easy. In the cases of those reservoirs with transitional characters, we found it to be more clear to use compound categories. Other main physico-chemical variables related with trophic level, measured during summer 1995, have been summarised in Table 1. At the end of the sequence, we have included two reservoirs which seem not to be aligned with the trophic series. The Beniarrés Reservoir (1) is distinguished for its high concentrations of soluble phosphorus and ammonia, low subsurface oxygen concentrations and reduced photic depth; we propose a new category, 'holotrophic', to cover this. The highly stained water of the Buseo Reservoir (2), with abundant humic material, is considered to be dystrophic.

The gradient becomes still more evident when the relationships of chlorophyll-*a* to total phosphorus and to phytoplankton biomass are considered (Figure 2A and B). Both regressions are highly significant: coefficients of determination were $r^2 = 0.63$ for log-

Table 1. Physical and chemical features of studied reservoirs, ordered according to trophic status. Concentrations (means and ranges) are only given for the photic epilimnetic zone. The two last reservoirs are differentiated from the main sequence ordered by mean chlorophyll. Abbreviations used for reservoirs in the graphs are indicated in brackets, next to the corresponding names. Thickness of the anoxic layer correspond to its heigth from the bottom near the dam, at the time of its measured maximum and the total depth at this point and time is just by indicated.

Reservoir		Epilimnetic Chlorophyll *a* ($\mu g\ l^{-1}$)		Total phosphorus (μM)		Dissolved oxygen (% saturation)		Photic depth (1% light) (m)		Maximum anoxic hypolimnion (m)	
		Mean	Min-Max.	Mean	Min-Max.	at surface	at 2 m depth	Minimum	Maximum	Thickness of anoxia	Depth at dam
Oligo-mesotrophic											
Tous	(To)	1.9	1.1–2.2	0.49	0.29–0.68	100 ± 9	98 ± 8	8	12.5	3.5	15
Contreras	(C)	2.0	0.7–3.7	0.61	0.35–1.00	96 ± 13	94 ± 13	10	15	5	28.5
Guadalest	(G)	2.5	1.4–3.4	0.81	0.68–1.10	100 ± 15	100 ± 12	8	12	0	24
Arenós	(As)	5.6	1.2–8.8	0.75	0.51–0.96	101 ± 8	101 ± 7	6.8	10.2	0	20
Arquillo	(Ar)	6.9	2.9–10.6	1.10	0.61–1.54	84 ± 14	81 ± 16	4.6	6.5	2.5	9
Meso-eutrophic											
Sichar	(S)	7.9	4.6–14.5	1.11	0.71–1.90	102 ± 9	100 ± 10	4.5	5.7	7	12
Amadorio	(Am)	10.4	6.3–14.8	1.35	0.80–2.41	105 ± 15	104 ± 20	3.5	7.2	9	14.2
Benageber	(Bg)	18.7	8.5–34.3	1.39	0.90–2.26	119 ± 26	111 ± 17	4	10	6.5	17.5
Loriguilla	(L)	20.2	12.3–28.6	1.29	0.74–1.74	144 ± 29	132 ± 32	4.5	7.5	3.5	15.5
Eutrophic											
Alarcón	(Al)	19.9	13.9–39.0	2.28	0.77–4.10	112 ± 18	112 ± 17	4.2	5.0	0	12
Regajo	(R)	40.9	9.6–50.2	2.22	1.41–4.84	136 ± 28	120 ± 18	3.3	6.5	5	12
Hypertrophic											
Forata	(F)	42.9	7.1–190.7	7.41	1.42–12.62	131 ± 52	106 ± 23	2.5	6.5	3.5	14.5
M. Cristina	(M)	250.0	57.7–439.6	11.61	8.52–16.23	161 ± 31	153 ± 37*	Bottom	Bottom	0	1.5
Holotrophic											
Beniarrés	(Be)	122.0	20.7–232.8	11.28	7.94–18.00	200 ± 88	83 ± 42	1.3	2	6.5	13.6
Dystrophic											
Buseo	(Bu)	25.8	12.0–46.4	4.66	1.58–11.03	108 ± 26	78 ± 20	1.3	2.2	3	8.4

Reservoir		Soluble Phosphate		Soluble Silica		Total inorganic		Nitrate (μM)		Nitrite (μM)		Ammonia (μM)	
		Mean	Min-Max.	Minimum	Maximum	Minimum	Maximum	Minimum	Maximum	Minimum	Maximum	Minimum	Maximum
Oligo-mesotrophic													
Tous	(To)	0.08	<0.03–0.14	64.3	116.0	31.0	116.6	29.1	115.8	0.23	0.58	0.2	1.6
Contreras	(C)	0.04	<0.03–0.09	25.8	77.6	41.2	53.8	34.4	53.1	0.02	2.68	<0.1	4.8
Guadalest	(G)	0.08	<0.03–0.18	84.7	127.1	39.6	57.8	31.9	56.3	0.02	0.58	<0.1	7.1
Arenós	(As)	0.05	<0.03–0.08	107.9	135.8	69.2	79.4	62.1	73.7	0.86	1.10	4.7	7.4
Arquillo	(Ar)	0.06	<0.03–0.53	52.0	105.2	7.8	15.6	0.3	7.7	<0.01	0.14	3.5	14.4
Meso-eutrophic													
Sichar	(S)	0.11	<0.03–0.28	98.8	127.5	1.9	7.7	<0.03	4.0	0.06	0.84	1.6	5.2
Amadorio	(Am)	0.06	<0.03–0.24	53.4	130.7	1.3	26.3	0.04	23.5	<0.01	0.15	0.8	22.0
Benageber	(Bg)	0.08	<0.03–0.14	60.9	91.7	26.6	80.5	25.8	56.1	0.24	2.16	0.1	23.0
Loriguilla	(L)	0.09	<0.03–0.21	51.2	85.7	46.6	76.1	37.0	71.8	0.07	0.82	0.7	13.6
Eutrophic													
Alarcón	(Al)	0.08	<0.03–0.21	8.67	9.6	<0.2	30.8	0.05	22.2	<0.01	0.11	<0.1	15.0
Regajo	(R)	0.11	<0.03–0.24	27.6	79.1	1.00	21.5	0.1	6.0	0.01	0.10	0.4	21.2
Hypertrophic													
Forata	(F)	0.17	<0.03–0.47	0.35	2.2	<0.2	56.1	<0.03	39.0	0.02	4.27	<0.1	15.6
M. Cristina	(M)	0.15	<0.03–0.24	0.19	4.8	3.6	9.5	0.1	6.1	0.07	1.36	2.0	3.8
Holotrophic													
Beniarrés	(Be)	3.00	0.51–11.91	47.9	148.0	1.0	106.7	<0.03	31.2	0.02	28.39	0.8	65.4
Dystrophic													
Buseo	(Bu)	0.22	<0.03–0.47	43.7	108.2	17.4	43.5	8.3	40.3	0.02	0.35	2.6	19.6

* At 1 meter depth.

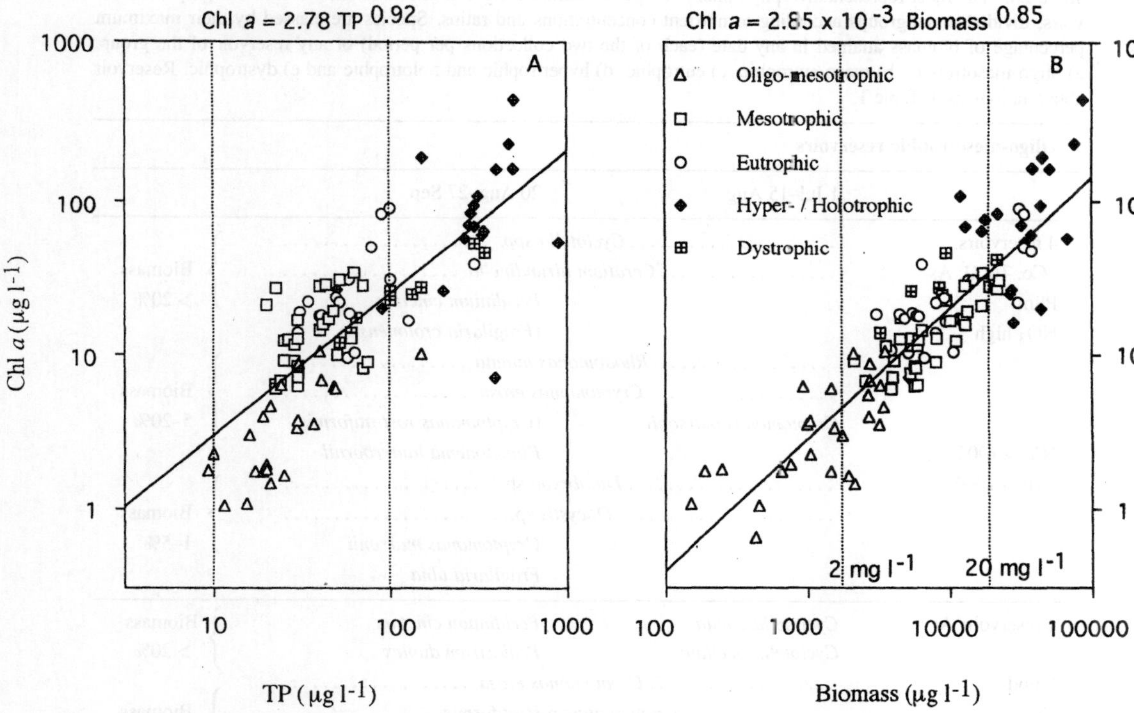

Figure 2. Linear regressions between trophic indicators in the samples from the photic zone of each reservoir. Horizontal lines indicate the most frequently given boundary values (see Reynolds, 1980) for Chlorophyll-*a* and biomass to separate oligotrophic, in one end, and hypertrophic reservoirs, in the opposite end. a) Chl *a*:TP b) Chl *a*: Biomass.

transformed Chlorophyll-*a* and TP, and $r^2 = 0.76$ for log-transformed Chlorophyll-*a* and biomass). Some overlap appears between the reservoirs allocated in the meso-eutrophic and eutrophic categories, however the delimiting values given by Reynolds (1980) for total cell volume to define trophic categories ($< 2\ \mu g\ l^{-1}$ for oligotrophic and $> 20\ \mu g\ l^{-1}$ for hypertrophic) are approximately met.

Transparency (Table 1) is less well correlated, in many cases due to the interference from calcite precipitation. Tables 2(a–e) show the various assemblages of phytoplankton present in each trophic category, by reference to the most representative species on the basis of total biomass. Oligo-mesotrophic reservoirs (Table 2a) must be strongly limited by phosphorus, whereas nitrate, the dominant form of inorganic nitrogen, shows high concentrations. They have short residence times and their sediments become anoxic only during short periods of high thermal stability. In these reservoirs, species achieving > 20% biomass in early summer are *Cyclotella* spp. and *Ceratium hirundinella*. At the end of summer the dinoflagellate *Peridinium cinctum* is also important. In one reservoir (Tous), probably due to earlier mixing, *Fragillaria crotonensis* was abundant, but then *C. hirundinella* was not found. Cryptophyte species were important during all summer; *Dinobryon* was also a characteristic of these reservoirs. We included Arquillo in a subgroup of these reservoirs: its position is to the mesotrophic side, since it has only slightly higher Chlorophyll-*a* concentrations and experiences oxygen depletion in the bottom waters. It has occasionally moderate phosphorus concentrations, but usually this nutrient was also limiting. Inorganic N did not reach such high concentrations as in the other four reservoirs and ammonia is often the inorganic N compound with higher concentrations. Phytoplankton in Arquillo differed, because *Ceratium* was not present and *Cyclotella spp.* diminished at the end of summer, when there was an increasing percentage of *Fragillaria ulna* var. *acus*. The abundance of chrysophytes is the main characteristic of this reservoir, although, as a result of their small cell size their biomass percentage is not high.

The group of meso-eutrophic reservoirs (Table 2b) have low phosphate concentrations but they are higher than those of the first group. Two subgroups could

Table 2a. The most representative phytoplankton species during early and late summer, for each category of reservoirs, divided in subgroups according to nutrient concentrations and ratios. Species are classed by their maximum percentage of biomass attained in any date (each of the two collections per period) or any reservoir of the group. a) oligo-mesotrophic, b) meso-eutrophic, c) eutrophic, d) hypertrophic and holotrophic and e) dystrophic. Reservoir abbreviations as in Table 1.

Oligo-mesotrophic reservoirs

	1 Jul–15 Aug	20 Aug–27 Sep	
4 reservoirs:	.... *Cyclotella spp.*	....	Biomass > 20%
Co, To, G, As	.... *Ceratium hirundinella*	....	
Plim		*Peridinium cinctum*	
NO_3 high		(*Fragilaria crotonensis*)	
	.... *Rhodomonas minuta*	....	Biomass 5–20%
	.... *Cryptomonas erosa*	....	
	Cryptomonas marsonii	(*Cryptomonas rostratiformis*)	
N/P > 1000		*Planctonema lauterbornii*	
Si/P > 1000	.... *Dinobryon* sp.	....	Biomass 1–5%
	.... *Oocystis* sp.	....	
		Cryptomonas marsonii	
		Fragilaria ulna	
1 reservoir: Ar	*Cyclotella comta*	*Peridinium cinctum*	Biomass > 20%
	Cyclotella ocellata	*Pediastrum duplex*	
Pmod	.... *Cryptomonas erosa*	....	
NO_3	.... *Cryptomonas rostratiformis*	....	Biomass 5–20%
	.... *Chrysococcus spp.*	....	
	Cryptomonas marsonii	*Rhodomonas minuta*	
	Trachelomonas sp.	*Lepocinclis sp.*	
N/P 20–500	.... *Kephyrion globosum*	....	
Si/P > 1000	.... *Planctonema lauterbornii*	....	Biomass 1–5%
	Pediastrum duplex	*Cryptomonas marsonii*	
	Rhodomonas minuta	*Peridinium umbonatum*	
	Fragilaria acus		

Table 2b.

Meso-eutrophic reservoirs

	1 Jul–15 Aug	20 Aug–27 Sep	
P Low	.... *Peridinium cinctum*	....	Biomass > 20%
Si/P 500–(> 1000	.... *Planctonema lauterbornii*	....	
		(*Microcystis* sp.)	
		(*Carteria* sp.)	
2 reservoirs: L, Bg	.... *Oocystis sp*	....	
NO_3 high	.... *Cryptomonas erosa*	....	
N/P 500–(> 1000	.... *Cryptomonas rostratiformis*	....	Biomass 5–20%
	.... *Ceratium hirundinella*	....	
	.... *Peridinium umbonatum*	....	
2 reservoirs: Am, S	.... *Pediastrum simplex – Pediastrum duplex*	....	
NO_3 low	(*Carteria* sp.)		
N/P 50–300	.... *Scenedesmus linearis*	....	
	.... *Pediastrum boryanum*	....	Biomass 1–5%
		Nitzschia palea	
		Lepocinclis sp.	
		Staurastrum sp.	
		Phacus longicauda	

() Only found in one of the reservoirs.

Table 2c.

<table>
<tr><th colspan="4">Eutrophic reservoirs</th></tr>
<tr><th></th><th>1 Jul–15 Aug</th><th>20 Aug–27 Sep</th><th></th></tr>
<tr><td>1 reservoir: Al</td><td>Microcystis sp.</td><td>Cryptomonas erosa</td><td rowspan="2">Biomass > 20%</td></tr>
<tr><td></td><td></td><td>Pediastrum duplex</td></tr>
<tr><td>Pmod</td><td colspan="2">Peridinium cinctum</td><td></td></tr>
<tr><td>$NO_3 > NH_4$</td><td colspan="2">Cryptomonas rostratiformis</td><td></td></tr>
<tr><td></td><td>Cryptomonas erosa</td><td>Ceratium hirundinella</td><td rowspan="3">Biomass 5–20%</td></tr>
<tr><td></td><td>Oocystis sp.</td><td>Pediastrum simplex</td></tr>
<tr><td></td><td>Cyclotella atomus</td><td>Staurastrum sp.</td></tr>
<tr><td></td><td>Scenedesmus acuminatus</td><td></td><td></td></tr>
<tr><td></td><td>Coelastrum astroideum</td><td></td><td></td></tr>
<tr><td>N/P 300–(> 1000)</td><td colspan="2">Cryptomonas marsonii</td><td></td></tr>
<tr><td>Si/P 200–(>1000)</td><td colspan="2">Pediastrum boryanum</td><td></td></tr>
<tr><td></td><td>Pediastrum sp.</td><td>Oocystis sp.</td><td rowspan="5">Biomass 1–5%</td></tr>
<tr><td></td><td>Staurastrum sp.</td><td>Rhodomonas minuta</td></tr>
<tr><td></td><td>Aulacoseira sp.</td><td>Scenedesmus acuminatus</td></tr>
<tr><td></td><td>Gomphosphaeria sp.</td><td>Stephanodiscus minutulus</td></tr>
<tr><td></td><td>Tetraedron caudatum</td><td>Carteria sp.</td></tr>
<tr><td>1 reservoir: R</td><td>Carteria sp.</td><td>Pediastrum simplex</td><td rowspan="2">Biomass > 20%</td></tr>
<tr><td></td><td>Merismopedia sp.</td><td>(simplex + echinulatum)</td></tr>
<tr><td>Pmod</td><td colspan="2">Cryptomonas erosa</td><td></td></tr>
<tr><td>$NO_3 < NH_4$</td><td colspan="2">Peridinium palatinum</td><td></td></tr>
<tr><td></td><td colspan="2">Staurastrum sp.</td><td rowspan="3">Biomass 5–20%</td></tr>
<tr><td></td><td colspan="2">Cosmarium spp.</td></tr>
<tr><td></td><td>Oocystis sp.</td><td>Ceratium hirundinella</td></tr>
<tr><td></td><td></td><td>Merismopedia sp.</td><td></td></tr>
<tr><td></td><td colspan="2">Pediastrum boryanum</td><td></td></tr>
<tr><td>N/P 30–200</td><td colspan="2">Trachelomonas sp.</td><td rowspan="4">Biomass 1–5%</td></tr>
<tr><td>Si/P 200–500</td><td colspan="2">Scenedesmus acuminatus</td></tr>
<tr><td></td><td colspan="2">Coelastrum astroideum</td></tr>
<tr><td></td><td>Cyclotella ocellata</td><td>Peridinium umbonatum</td></tr>
<tr><td></td><td>Coelastrum reticulatum</td><td>Oocystis sp.</td><td></td></tr>
</table>

Table 2d.

<table>
<tr><th colspan="4">Hypertrophic reservoirs</th></tr>
<tr><th></th><th>1 Jul–15 Aug</th><th>20 Aug–27 Sep</th><th></th></tr>
<tr><td>1 reservoir: F</td><td colspan="2">Carteria sp.</td><td></td></tr>
<tr><td></td><td>Oocystis sp.</td><td>Cryptomonas erosa</td><td rowspan="3">Biomass > 20%</td></tr>
<tr><td>Pmod</td><td>Pediastrum boryanum</td><td>Pediastrum simplex</td></tr>
<tr><td>N high</td><td></td><td>(simplex + sturnii)</td></tr>
<tr><td></td><td>Pediastrum simplex</td><td>Stephanodiscus cf. hantzschii</td><td rowspan="3">Biomass 5–20%</td></tr>
<tr><td></td><td>Staurastrum sp.</td><td></td></tr>
<tr><td></td><td>Phormidium sp.</td><td></td></tr>
<tr><td>N/P 3–(> 400)</td><td>Ceratium hirundinella</td><td>Rhodomonas minuta</td><td rowspan="6">Biomass 1–5%</td></tr>
<tr><td>Si/P 3–1000</td><td>Microcystis sp.</td><td>Pediastrum duplex</td></tr>
<tr><td></td><td>Euglena sp.</td><td>Pediastrum boryanum</td></tr>
<tr><td></td><td>Merismopedia sp.</td><td>Coelastrum astroideum</td></tr>
<tr><td></td><td></td><td>Staurastrum sp.</td></tr>
<tr><td></td><td></td><td>Phacotus lenticularis</td></tr>
</table>

Table 2d. Continued.

1 reservoir: M	*Microcystis* sp.	*Synechocystis* sp.	Biomass > 20%
P low		*Oscillatoria* sp.	
N low	*Cyclotella meneghiniana*	*Anabaenopsis circularis*	Biomass 5–20%
	Pediastrum simplex var. simplex		
	Trachelomonas cf. volvocina		
	.... *Scenedesmus acuminatus*		
	Coelastrum astroideum	*Cryptomonas erosa*	Biomass
N/P 20–100	*Oscillatoria* sp.		
Si/P 4–400	*Oocystis* sp.		
	Cyclotella atomus		
	Staurastrum sp.		
	Cryptomonas rostratiformis		
Holotrophic reservoir: Be			
	.... *Scenedesmus acuminatus*		Biomass > 20%
P high	.... *Cryptomonas erosa*		
N high	.... *Stephanodiscus minutulus*		Biomass 5–20%
	Merismopedia sp.	*Peridinium cinctum*	
		Oocystis sp.	
		Pteromonas angulosa	
	.... *Chlamydomonas sp2.*		
N/P 3–17	.... *Tetrastrum staurogeniaeforme*		Biomass 1–5%
Si/P 16–90	*Oocystis* sp.	*Merismopedia* sp.	
	Pteromonas angulosa	*Rhodomonas minuta*	
	Cyclotella atomus	*Tetrastrum* sp.	

Table 2e

Dystrophic reservoir: Bu			
	1 Jul–15 Aug	20 Aug–27 Sep	
Pmod	*Ceratium hirundinella*	*Pediastrum duplex*	Biomass > 20%
NO_3 high	*Scenedesmus disciformis*	*Lepocinclis salina*	
	Merismopedia sp.		
	.... *Oocystis* sp.		Biomass 5–20%
	Cryptomonas rostratiformis	*Ceratium hirundinella*	
N/P > 1000(ES)	*Lepocinclis salina*	*Phacus longicauda*	
N/P = 50(LS)	*Pediastrum duplex*	*Cryptomonas rostratiformis*	Biomass 1–5%
Si/P > 1000(ES)	*Phacus longicauda*	*Scenedesmus disciformis*	
Si/P = 300(LS)	*Closterium spp.*	*Pediastrum simplex*	

(ES) Early Summer; (LS) Late Summer.

be differentiated on the basis of nitrate concentration. In most of these reservoirs, the dominant algae, throughout the study period, were *Peridinium cinctum* and *Planctonema lauterbornii.* At the end of summer, *Microcystis* and euglenophytes became important in some of them.

The eutrophic group (Table 2c) embraces two reservoirs -Alarcón and Regajo- which receive diluted inflows of wastewater. That generates higher soluble phosphorus concentrations than those found in the meso-eutrophic group of reservoirs. These reservoirs are quite different in phytoplankton composition, however, both show an elevated number of typically eutrophic chlorococcal species. The dominants were species of *Pediastrum, Cryptomonas* and cyanobacteria, usually different species in each reservoir.

At the upper end of the trophic series, there were three very different reservoirs (Table 2d). Forata watershed intercepts irregular, short but intense rain events, which caused big fluctuations in nutrient availability. It has been included in the hypertrophic reservoirs: although very close to the eutrophic group, the reservoir supports larger biomass concentrations. Sometimes cyanobacteria were abundant, but after summer storms *Pediastrum simplex* was predominant altogether with flagellate species. On some occasions notable densities of centric diatoms, together with marked Si-depletion, were observed.

María Cristina proved to be a typically hypertrophic system, with absolute dominance of cyanobacteria, low concentrations of dissolved nutrients (N, P, alkalinity) and high pH. During the study, inflows were lacking except for a short and meagre rain event in its watershed at the end of summer; thus internal loads of nutrients from the sediment played a major role. Silica also became limiting in this reservoir in early summer, coinciding with centric diatom growth, simultaneously with some cyanobacteria and chlorococcales. At the end of summer, 95% of the phytoplankton biomass was made up by three cyanobacterial species.

Finally, the phytoplankton of Beniarrés is much different from the rest of reservoirs. *Scenedesmus acuminatus* is the algae which showed the highest percentage of total biomass, however in this reservoir unicellular volvocales of the genera *Pteromonas* and *Chlamydomonas* reached very high densities, which due to its small size are not so well represented when biomass is used. Dominance in number of individuals of these genera, as well as the abundance of the same main species in Table 2d have been observed in all seasons, since spring of 1994 to autumn 1995 (from our own data on bimonthly samples). This persuade us to propose a new 'holotrophic' category for such systems, where nutrients – phosphorus and ammonia (Table 1) – are always at high concentrations due to high inputs and easy mixing of the bottom anoxic waters.

The dystrophic Buseo reservoir, may be included in the eutrophic group, even though the composition of its phytoplankton was quite different, with an abundance of euglenophyta, likely attributable to the high levels of dissolved organic carbon present (Table 2e).

The graphic representation of diversity as a function of average biomass (Figure 3) shows a humpbacked relationship. Diversity is low at low biomass, gradually increases with increasing biomass in the oligo-mesotrophic reservoirs and it decreases again at the higher biomasses towards the category of hypertrophic reservoirs. Diversity reached highest values in the eutrophic reservoirs. However, it was unexpectedly low in the meso-eutrophic reservoirs.

The average percentage biomass represented by each of the main algal groups, together with standard deviation, is represented by histograms in Figure 4. In the oligo-mesotrophic reservoirs, dinophytes and bacillariophytes are the most important taxa and the proportion of chrysophytes is distinctly higher than in the other reservoirs, while that of cyanobacteria is low. In meso-eutrophic reservoirs, chlorophyta and dinophyta are clearly the most important phytoplankton groups. In eutrophic reservoirs chlorophyta is the dominant group, but other groups, such as cyanophyta are also well represented. Both chorophyta and cyanophyta had the higher biomass percentage in hypertrophic reservoirs. However the 'holotrophic' reservoir was characterised by an absolute predominance of chlorophyta. Euglenophyta had a high contribution to biomass exclusively in the dystrophic Buseo reservoir.

Figure 5 shows the distribution of phytoplankton species against the axes of the first two principal components (of PCA-1). The two components together accounted for 30.3% of the total variance. The first component (20% of total variance) indicates trophic level. At the negative extreme of this axis, the dinoflagellate *Ceratium hirundinella* and the filamentous ullotrichal *Planctonema lauterbornii* are found, while at the positive end there are eutrophy indicators, as several species of *Scenedesmus* (*S. acuminatus, S. acutum, S. intermedius*), unicellular volvocals (*Chlamydomonas, Pteromonas*) and some centric bacillariophyte species (*Cyclotella atomus* and *Stephanodiscus minutulus*). On the other hand, principal component 2, segregates life forms: so, we found large desmids (*Staurastrum, Cosmarium*), colonial greens (*Coelastrum* spp, *Pediastrum simplex*) and large dinoflagellates in the positive part, while negatively correlated are diatoms (*Fragillaria ulna var. acus* and small centrics) and small unicellular flagellates (chrysophytes and volvocales). Very important is the presence of large cells and colonial algae (*Coelastrum, Dictyosphaerium, Pediastrum* and *Staurastrum* species) in eutrophic and hypertrophic reservoirs, during late summer. These life forms may be the result of selective grazing pressure from herbivorous zooplanktonic species. Some of them are also favoured by destratification after late summer storms. On the

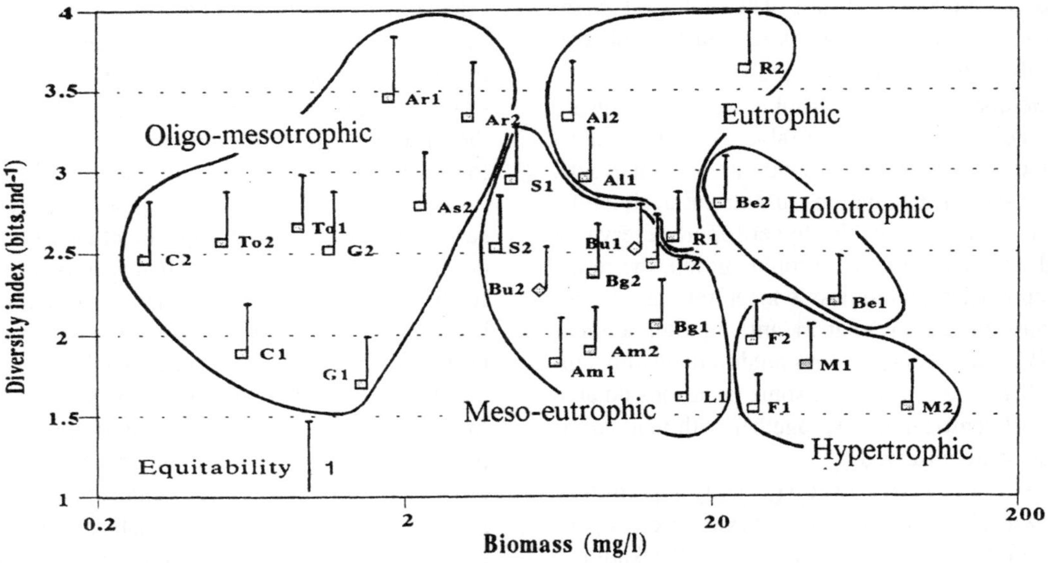

Figure 3. Relationship between biomass (biovolume in logarithmic scale) and Shannon-Wiener diversity index of phytoplankton species biomass. Both variables are the corresponding averages for samples of the photic zone in each of the reservoirs, during early and late summer, indicated respectively by 1 and 2 just after reservoir initials. Diversity is calculated with biovolume data, withdrawing spp $< 0.5\%$ of the biovolume. The equitability is indicated in the vertical line over each point.

other hand in oligo-mesotrophic reservoirs, sequential replacement of species due to nutrient deficiencies and grazing activities as summer advances, leads to an important growth of *Ceratium* and *Peridinium*, because these genera can survive with lower nutrient concentrations (Reynolds, 1984a) and are inedible by filter feeding grazers. It is also quite clear that congeneric or phylogenetically close species fall at extreme positions in the graph, e.g. species of the genus *Cyclotella, Scenedesmus, Pediastrum* and *Cryptomonas*.

To confirm the interpretation of these two first principal components, a second analysis, PCA-2, was performed with these two components from PCA-1, as variables together with the measured physico-chemical variables. Figure 6 shows the distribution of these variables in the space of the first two components of PCA-2. It confirms the relationship between the first component derived from phytoplankton abundance (Phy 1) and trophic indicators (Chlorophyll, TP, suspended solids, turbidity, BOD) and the relation with other abiotic variables more or less associated with primary production (pH. oxygen, temperature). At the other end of this axis, Eh and Margalef pigment index is found, as well as nitrates (which are consumed by algae and nitrifying bacteria)and silica (which is consumed in diatom growth). The second component derived from phytoplankton abundance (Phy 2) is related to the second line of variation of physical and chemical parameters; that is, the contrast between (1) photic waters in which production processes greatly exceed decomposition ones, with high oxygen concentrations and nutrient depletion (mainly phosphate, and silica) and (2) the photic waters of other reservoirs in which decomposition is important or the waters can be easily mixed vertically; thus becoming undersaturated in oxygen and rich in nutrients (phosphate, ammonia, nitrite and silica). Fe has a high correlation with this second source of variation, opposite to that of phosphate. In high trophic level reservoirs, rich in sulphate, like most those studied here, a sulfide-rich anoxic hypolimnion develops sulfide can precipitate as ferrous sulfide but phosphate is left in solution.

The means and standard deviations of the PCA-1 and PCA-2 factor scores for each reservoir are presented in Figures 7 and 8 respectively. The calculated factor score variances within reservoirs were consistently higher in respect of abiotic factors (PCA-2) than those with species abundance (PCA-1) except for Beniarrés and Alarcón reservoirs.

Reservoirs became ordered, along the first principal component in both analyses (PCA-1 and PCA-2) according to their trophic level, with a much higher differentiation among the eutrophic and hypertrophic reservoirs (which were widely distributed in the posi-

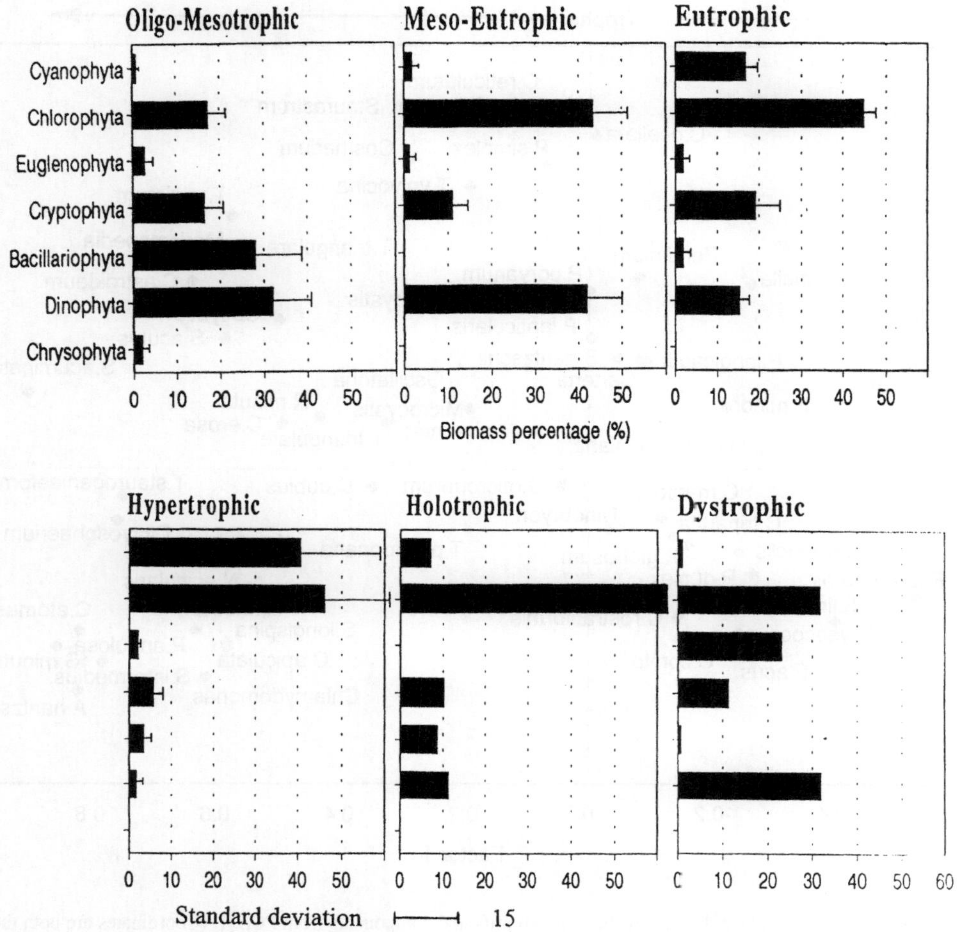

Figure 4. Histograms show the mean percent composition of main phytoplanktonic groups with respect to total biomass for each group of reservoirs. Standard deviations are also indicated.

tive part) than among the oligo-mesotrophic and meso-eutrophic examples (closely clustered in the negative part). The second component in both analyses differentiates subgroups of trophic categories, mainly also in the eutrophic-hypertrophic end. The analysis with abiotic variables (PCA-2) clearly separates the extremely polluted Beniarrés Reservoir, where decomposition processes prevail through most of the water column, from María Cristina Reservoir with its high biomass but low nutrients. On the other hand, when the analysis was performed with only phytoplankton densities (PCA-1), Regajo reservoir, which had the highest diversity, was still more differentiated from Beniarrés than from María Cristina. In PCA-1 this second component can be also associated to the influence of mixing upon phytoplankton. It clearly opposes the reservoirs with a higher separation between photosynthesis and respiration in the water column (due to a higher degree of stratification in Regajo and Forata or to shallowness in María Cristina), to those richer in dissolved phosphorus and ammonia, where destratification is frequent (Beniarrés, Alarcón and also Arquillo). The rest of the reservoirs fall close together in a more central position.

Only Buseo, the reservoir affected by forest fire, was positioned quite distinctly by the abiotic analysis (PCA-2, Figure 7), falling closer to the eutrophic-hypertrophic side. A massive inflow of ash and woody debris from the burned watershed determined an abnormal chemistry of the waters, with high Fe concentrations (the presence of humic substances has been described to efficiently adsorb iron, Vegas-Villarrubia, 1994). However, the phytoplankton composition was closer to meso-eutrophic reservoirs (Figure 8).

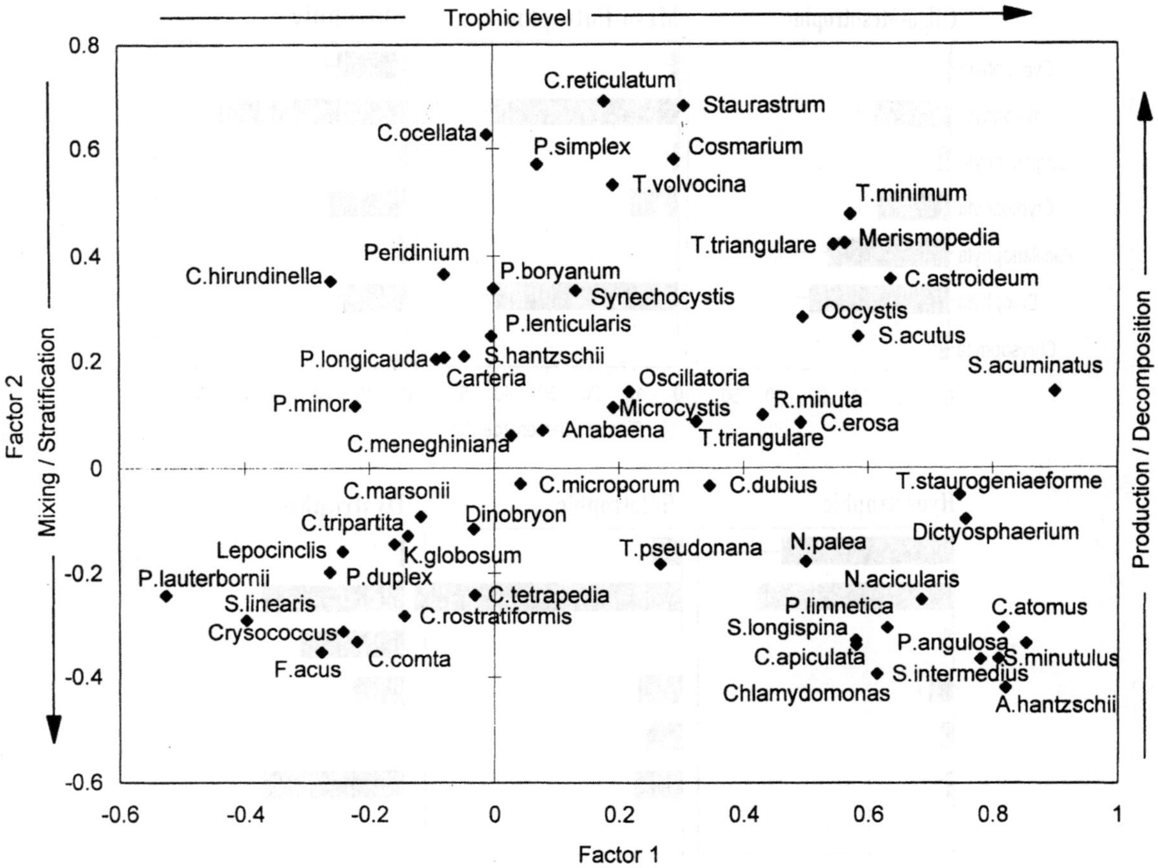

Figure 5. Position of the species in the plane defined by the first two principal components from PCA-1. (Coordinates are both the correlation coefficients and factor loading of each species with these axes). The first axis defines trophic level, the second segregates phytoplankton life-forms. The two left quadrants define the life-forms corresponding to lower trophic level and well established stratified epilimnetic conditions (top) or to the onset of stratification (bottom). The two right quadrants contain those life forms corresponding to a higher trophic level with a turbulent but well stratified epilimnion (top) or an epilimnion with frequent destratifications highly enriched in nutrients (bottom).

We also performed the same principal components analyses but using percentages of biomass to characterise the samples, instead of absolute densities, the results were much the same.

Discussion

We have found, as expected, significant relationships between phytoplankton biomass, Chlorophyll-*a* and total phosphorus. However, we have detected an overlap between the categories of mesotrophy and eutrophy when these variables were simply regressed (Figure 2). These categories became clearly distinct when multivariate analyses of phytoplankton were undertaken and not only when using phytoplankton composition alone (PCA-1, Figure 7), but also when main physical and chemical variables were considered (PCA-2, Figure 8). This means that the interplay of factors is complex, as is the template upon which to fit the trophic gradient. Such multivariate analyses are especially suited to distinguish among the environments of each of a series of lakes.

One of the more interesting results of our study is that the variability in species composition of oligo-mesotrophic and meso-eutrophic systems is small in comparison to that of eutrophic and hypertrophic systems (Figures 7 and 8). These results were not so evident in the study of 100 reservoirs with different geological substrates, which were selected among the biggest in size, and scarcely included hypertrophic reservoirs. The most significant factors accounting for

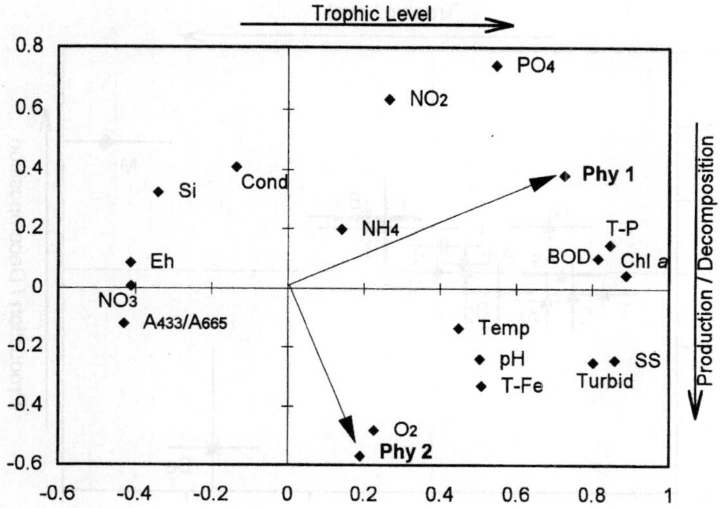

Figure 6. Position of physical and chemical variates and the first two principal components PHY1 and PHY2 from PCA-1 in the plane defined by the first two principal components of PCA-2. (Cond: conductivity), turbid: inverse of Secchi depth, A433/A665: absorbances ratio of Margalef pigment index, T–P: total P, T-Fe: total Fe, Temp: temperature, SS: suspended solids).

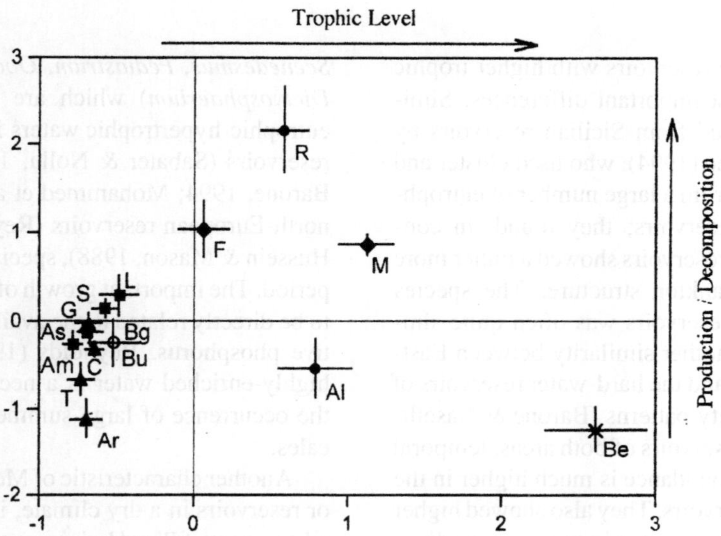

Figure 7. Position of the different reservoirs relative to the first two principal components from PCA-1. Each point is the mean of the factor scores corresponding to the samples from each reservoir. The standard variation is also plotted at each side of the corresponding mean. Abbreviations for each reservoir are indicated in Table 1. Reservoirs are ordered by trophic level in the first axis and the second axis is interpreted as the degree of separation between the zone where photosynthesis predominates and the zone were decomposition predominates.

the variability of the data, according to these authors, were firstly the alkalinity and mineralization of water, then eutrophy and with much smaller weight mixing degree (Margalef et al., 1982; Sabater & Nolla, 1991; Riera et al., 1992). The compact group of low and intermediate trophy in our results is quite similar to the group described previously by these authors for Eastern Spanish reservoirs, which have in common a high abundance of centric diatoms (*Cyclotella* spp.) and dinophyta (*Ceratium*, *Peridinium*), and also a high proportion of the Ulotrichal, *P. lauterbornii* (Table 2a and b, Figure 5). In our PCA results, we again found trophic level and the physical mixing of the water column to be the main factors influencing phytoplankton composition.

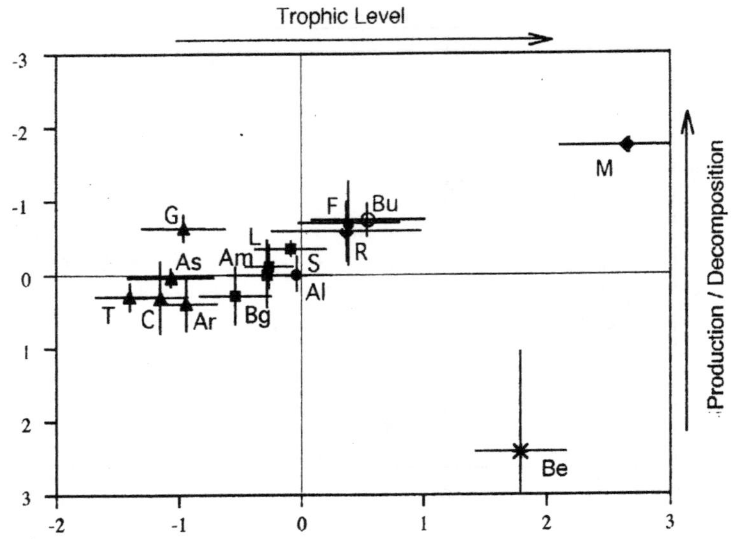

Figure 8. Position of the different reservoirs relative to the first two principal components from PCA-2. Each point is the mean of the factor scores corresponding to the samples from each reservoir. The standard variation is also plotted at each side of the corresponding mean. Abbreviatures for each reservoir are indicated in Table 1.

In our analyses, the reservoirs with higher trophic levels showed the most important differences. Similar results were reported from Sicilian reservoirs by Naselli-Flores & Barone (1994), who used cluster and nodal analysis of data from a large number of eutrophic and hypertrophic reservoirs; they found, in contrast, that mesotrophic reservoirs showed a rather more homogeneous phytoplankton structure. The species composition of these reservoirs was often quite similar to our findings. Another similarity between Eastern Spanish reservoirs and the hard-water reservoirs of Sicily was their diversity patterns (Barone & Naselli-Flores, 1990). In the reservoirs of both areas, temporal variability of species abundance is much higher in the eutrophic group of reservoirs. They also showed higher values of the diversity index than in most of our oligo-mesotrophic and meso-eutrophic reservoirs (Figure 3).

A factor that could account for the homogeneity and low diversity of the studied oligo-mesotrophic and meso-eutrophic reservoirs is their extremely calcareous hard-waters and so the low trophic level in most of them is maintained by phosphate precipitation. Calcite precipitation could be also intense during summer and essential micronutrients may also co-precipitate with it. All these could limit growth of a more diverse phytoplankton with more requirements.

On the other hand, the plankton of lakes of higher trophic status are characterised by an elevated number of chlorococcal species (*Coelastrum*, *Scenedesmus*, *Pediastrum*, *Oocystis*, *Tetraedron*, and *Dictyosphaerium*) which are considered typical of eutrophic-hypertrophic waters in other Mediterranean reservoirs (Sabater & Nolla, 1991; Naselli-Flores & Barone, 1994; Mohammed et al., 1989) as well as in north-European reservoirs (Reynolds, 1984a; Abdul-Hussein & Mason, 1988), specially during the summer period. The important growth of this algal group seems to be directly related to the availability of soluble reactive phosphorus. Reynolds (1984b) point out that a highly-enriched water is a necessary precondition to the occurrence of large summer-peaks of Chlorococcales.

Another characteristic of Mediterranean reservoirs, or reservoirs in a dry climate, is that they are subjected to severe filling/draining manipulations. This produces instabilities that may maintain a high number of species surviving in temporally heterogeneous and nutrient-rich environments, explaining the high diversity values found in our eutrophic reservoirs. All this might be linked to the intermediate disturbance hypothesis applied by other authors to phytoplankton ecology (Rojo & Alvarez-Cobelas 1993).

However, phytoplankton assemblages could be good descriptors also in perturbed systems, such as the studied reservoirs. The results of our analyses show that the dispersion of samples from the same reservoir is smaller when phytoplankton abundance are used than when physical and chemical data are involved (Figures 7 and 8). Phytoplankton composition is more conservative than physical and chemical parameters, then it reflects the underlying habitat template, temporary alterations notwithstanding. This is evidently enhanced by the founder effect of an inoculum of species reflecting dominant components in the recent past, which, if they have the capacity to survive the perturbations, will persist as a phytoplankton community indicative of a less variable type of water.

In our work, involving only summer phytoplankton data, the distribution of samples came with a high degree of coherence . However it is evident from our results (Figure 5) that life forms segregate notoriously across the second main axis of variation of the data, which is obviously associated to the temporal hydrological variations. But within life forms there is also segregation in the trophic spectrum, thus we have congeneric or closely related species far apart in our analysis. For instance *Stephanodiscus* spp. and *Cyclotella atomus* appear at the eutrophic end, while the rest of *Cyclotella* species are in the oligotrophic side. The same is true for species of *Scenedesmus*, most of the them lie in the eutrophic side, but *S. linearis* is positioned in the opposite extreme. Species of *Pediastrum* and *Coelastrum* showed also important segregation in their distributions. This indicates that some genera have evolved to have species occupying preferentially certain of the possible trophic situations.

It is clear from our study that the more marked differentiation of phytoplankton assemblages, as well as of physical and chemical factors, is found in the hypertrophic reservoirs (Figures 7 and 8). On one side, we have the case of the hypertrophic reservoir, María Cristina, where nutrients were rather low, pH was extremely high (> 10), whereas alkalinity showed the lowest measured values in the studied reservoirs (< 2 meq l^{-1}), partly as a consequence of apatite precipitation. Oxygen is oversaturated in the whole water column, thus photosynthesis greatly exceeds decomposition. Biomass is exported to sediments and decomposition becomes, to a great extent, separated from production. The phytoplanktonic community shifted to complete cyanophyta dominance in late summer with the presence of nitrogen-fixing Nostocaceae species which indicates depletion of nitrogen, probably because denitrification is important. Most authors have found N:P ratios to be negatively correlated with cyanophyta biomass in hypertrophic reservoirs (Abdul-Hussein & Mason, 1988).

At the other extreme, we find the Beniarrés reservoir, where inflowing sewage water continuously supplies inorganic nutrients and DOC. Nitrogen in the form of ammonia does not suffer denitrification and never becomes limiting. Phosphate and silica have very high concentrations. The reservoir has an important layer of anoxic water, which can be easily mixed with the overlying photic zone, in part because the outgoing flow is located near the bottom of the reservoir. Then, surface waters can be easily enriched by phosphate, ammonia and organic compounds, maintaining a dense biomass without nutrient depletion. If light becomes highly attenuated, mixotrophic algal groups could be also favoured. In this reservoir, flagellates – mainly unicellular volvocales – have special importance (*Chlamydomonas* sp. and *Pteromonas angulosa)*; although cryptophytes could be important as well. The group of chlorococcales (*Scenedesmus acuminatus, Dictyospherium* sp.) also have a high relative proportion as well as, in some moments, centric diatoms (*Stephanodiscus minutulus*).

In their review of the scattered literature on hypertrophic systems, Alvarez-Cobelas & Jacobsen (1992) described four clearly distinct types of hypertrophic phytoplankton seasonal succession. On the basis of the phytoplankton composition, Maria Cristina reservoir fits to type II of Alvarez-Cobelas & Jacobsen (1992), whereas Beniarrés reservoir is closer to the types III and IV of these authors. The latter types were characterised by these authors with examples of fish ponds and high rate oxidation ponds and they were associated to waters with higher concentrations of ammonia and dissolved organic matter.

Thus, today we categorise two very different aquatic environments within the same group of hypertrophic systems. We propose that the hypertrophic category should include the ecosystems with cyanobacterial predominance and rather low nutrients (types I and II of Alvarez-Cobelas & Jacobsen, 1992), and that we should separate with some other term, the ecosystems with abundance of flagellates, *Scenedesmus* and other C-strategists in the sense of Reynolds (1988, competitive species within r-strategists) and high concentration of nutrients (types III and IV of Alvarez-Cobelas & Jacobsen, 1992). These systems are beyond hypertrophy, because mechanisms to reduce forced feeding are less effective. We propose the name 'Holotrophic', to indicate the whole system in the trophic level. In these systems, the vertical separation between production and decomposition is reduced and the mixing of waters of different redox potentials enhances nutrient recycling by enriching the photic zone in phosphate and ammonia. The bottom layer of anoxic water prevents the total export of phosphorus to the sediment, moreover ammonia does not denitrify or nitrify, because oxygen is involved in processes with organic matter and the electrochemical system does not reach enough (redox) potential for nitrification. The divergence between hypertrophic and holotrophic has to be viewed in terms of processes, an eutrophic reservoir could develop to an hypertrophic or to an holotrophic one, depending on amount and quality of nutrients, itself a function of redox potential of the waters.

References

Abdul-Hussein, M. M. & C. F. Mason, 1988. The phytoplankton community of a eutrophic reservoir. Hydrobiologia 169: 265–277.

Alvarez-Cobelas, M. & B. A. Jacobsen, 1992. Hypertrophic phytoplankton: an overview. Freshwat. Forum 2: 184–199.

APHA - AWWA - WEF, 1992. Standard methods for the examination of water and wastewater. 18th edition. American Public Health Association. Washington D.C.

Armengol, J., J. L. Riera & J. A. Morguí, 1991. Major ionic composition in the Spanish reservoirs. Verh. int. Ver. Limnol. 24: 1363–1366.

Barone, R. & L. Naselli Flores, 1990. Osservazioni sulle comunità planctonique dei laghi artificiali siciliani: relazioni diversità – biomassa. Naturalista sicil. 14 (suppl.): 35–48.

Barone, R. & L. Naselli Flores, 1994. Phytoplankton dynamics in a shallow, hypertrophic reservoir Lake Arancio, Sicily. Hydrobiologia 289: 199–214.

Estrada, M., 1975. Statistical consideration of some limnological parameters in Spanish reservoirs. Verh. int. Ver. Limnol. 19: 1849–1859.

Golterman, H. L., R. S. Clymo & M. A. M. Ohnstad, 1978. Methods for physical and chemical analysis of fresh waters. IBP Handbook no. 8. Blackwell Sci. Pub., Oxford, U.K.

Margalef, R., M. D. Planas, J. Armengol, A. Vidal, N. Prat, A. Guisset, J. Toja & M. Estrada, 1976. Limnología de los embalses españoles. Dirección General de Obras Hidráulicas. MOP. Madrid. Pub. 123, 422 pp.

Margalef, R., M. Mir, & M. Estrada, 1982. Phytoplankton composition and distribution as an expression of properties of reservoirs. Can. W. Res. J. 7: 26–50.

Mohammed, A. A., A. M. Ahmed & A. M. El-Otify, 1989. Field and laboratory studies on Nile phytoplankton in Egypt. IV Phytoplankton of Aswan High Dam Lake (Lake Nasser). Int. Revue ges. Hydrobiol. 74: 549–578.

Moss, B., 1973. The influence of environmental factors on the distribution of freshwater algae: experimental study. IV. Growth of test species in natural lake waters and conclusion. J. Ecol. 61: 193–211.

Naselli Flores, L. & R. Barone, 1994. Relationship between trophic state and plankton community structure in 21 Sicilian dam reservoirs. Hydrobiologia 275/276: 197–205.

Norusis, M. J., 1990. SPSS Base System User's Guide. SPSS Inc. Chicago, 520 pp.

OECD, 1982. Eutrophication of Waters. Monitoring, Assessment, and Control. OECD, Paris, 154 pp.

Ortiz Casas, J. L. & R. Peña, 1984. Applicability of the OECD eutrophication models to Spanish reservoirs. Verh. int. Ver. Limnol. 22: 1521–1535.

Reynolds, C. S., 1980. Phytoplankton assemblages and the periodicity in stratifying lake ecosystems. Holarct. Ecol. 3: 141–159.

Reynolds, C. S., 1982. Phytoplankton periodicity: its motivation, mechanisms and manipulation. Report of the Freshwater Biological Association 50: 60–75.

Reynolds, C. S., 1984a. Phytoplankton periodicity: the interactions of form, function and environmental variability. Freshwat. Biol. 14: 111–142.

Reynolds, C. S., 1984b. The ecology of freshwater phytoplankton. Cambridge University Press, 384 pp.

Reynolds, C. S., 1986. Experimental manipulations of the phytoplankton periodicity in large, limnetic enclosures in Blelham Tarn, English Lake District. Hydrobiologia 138: 43–64.

Reynolds, C. S., 1988. Functional morphology and the adaptative strategies of freshwater phytoplankton. In Sandgren, C. D. (ed), Growth and survival strategies of freshwater phytoplankton. Cambridge University Press: 388–433.

Riera, J. L., D. Jaume, J. De Manuel, J. A. Morgui & J. Armengol, 1992. Patterns of variation in the limnology of Spanish reservoirs: a regional study. Limnetica 8: 111–124.

Rojo, C. & M. Alvarez-Cobelas, 1993. Hypertrophic phytoplankton and the intermediate disturbance hypothesis. Hydrobiologia 249: 43–57.

Rott, E., 1981. Some results from phytoplankton counting intercalibrations. Schweiz. Z. Hydrol. 43: 34–62.

Sabater, S. & J. Nolla, 1991. Distribution patterns of phytoplankton in Spanish reservoirs: first results and comparison after fifteen years. Verh. int. Ver. Limnol. 24: 1371–1375.

Shapiro, J., 1989. Current beliefs regarding dominance by blue-greens: The case of the importance of CO_2 and pH. Verh. int. Ver. Limnol. 24: 38–54.

Sommer, U., 1989. Nutrient status and nutrient competition of phytoplankton in a shallow, hypertrophic lake. Limnol. Oceanogr. 34: 1162–1173.

Utermöhl, H., 1958. Zur Vervollkommnung der quantitativen Phytoplankton-Methodik. Mitt. int. Ver. Theor. Angew. Limnol. 9: 38 pp.

Vegas Vilarrubia, T., 1994. Water chemistry of the Guri reservoir (rainy season 1989) – relationships between humic colour and aqueous iron and their limnological importance. Arch. Hydrobiol. 132: 69–94.

Appendix . List of planktonic algae, with authorities, used in the PCA, with their corresponding mean densities (ind/ml) in the photic zome of each reservoir during summer.

	Tous	Guadalest	Contreras	Arenós	Arquillo	Amadorio	Sichar	Benagéber	Loriguilla	Buseo	Alarcón	Regajo	Forata	M.Cristina	Beniarrés
Bacillariophyta															
Cyclostephanos dubius (Fricke) Round	0	0	0	0	0	0	0	0	0	0	69	0	0	644	0
Cyclotella atomus Hustedt	0	0	0	0	0	0	0	0	0	0	855	0	0	4511	3414
Cyclotella comta (Ehrenberg) Kützing	0	0	0	0	172	0	0	0	0	0	0	0	0	0	0
Cyclotella meneghiniana Kützing	0	0	1276	0	0	0	0	63	0	0	0	0	0	5156	0
Cyclotella ocellata Pantocsek	0	1271	0	338	171	4	0	7	0	0	0	823	428	0	0
Cyclotella tripartita Hakansson	1534	0	0	0	0	0	0	0	0	0	0	0	0	0	0
Fragilaria ulna var. acus (Kützing) Lange-Bertalot	17	8	0	0	29	0	0	13	6	0	6	0	0	0	0
Nitzschia acicularis (Kützing) W. Smith	0	0	1	8	29	0	0	3	0	0	42	19	0	397	3721
Nitzschia palea (Kützing) W. Smith	3	8	5	8	36	26	12	6	22	9	44	44	7	159	491
Stephanodiscus cf.hantzschii Grunow	0	0	0	0	0	0	0	0	0	0	0	0	214	0	0
Stephanodiscus minutulus (Kützing) Cleve & Möller	0	0	0	0	0	0	0	0	0	0	784	0	0	2578	17272
Thalassiosira pseudonana Hasle & Heimdal	0	0	0	0	0	0	0	0	0	0	17	0	0	0	0
Chrysophyta															
Chrysococcus spp.	29	0	0	0	336	0	0	0	87	0	14	0	0	0	0
Dinobryon sp.	4	2	0	75	0	9	0	0	0	0	38	0	0	0	0
Kephyrion cf. globosum (Hilliard) Starmach	0	0	0	0	240	0	0	0	0	0	0	19	0	0	0
Cryptophyta															
Cryptomonas erosa Ehrenberg	34	32	8	231	127	257	110	208	331	127	711	1249	2238	555	2236
Cryptomonas marsonii Skuja	52	23	5	64	253	44	0	0	75	26	288	275	27	0	0
Cryptomonas rostratiformis Skuja	4	0	7	0	39	73	0	153	140	114	148	0	0	29	0
Rhodomonas minuta Skuja	171	53	210	515	1066	79	220	787	445	128	721	1324	4534	1396	5048
Dinophyta															
Ceratium hirundinella (O.F. Müller) Dujardin	4	8	6	14	0	15	0	19	17	75	3	40	17	0	0
Peridinium spp.	2	13	0	40	24	34	87	90	219	0	20	194	7	49	46
Chlorophyta															
Actinastrum hantzschii Lagerh.	0	0	0	0	0	0	0	0	0	0	355	0	0	0	6207
Carteria sp.	0	0	0	16	0	0	208	766	188	0	61	600	3964	0	0
Chlamydomonas sp.2	0	0	0	0	37	0	4	0	3	22	0	0	6	0	7280
Coelastrum astroideum De-Not.	0	0	0	0	133	0	0	0	8	664	698	2021	729	1746	2242
Coelastrum microporum Näg. in A. Br.	0	0	0	0	0	0	0	204	162	0	0	0	173	0	767
Coelastrum reticulatum (Dang.) Senn	0	31	0	0	0	0	0	0	65	0	129	1846	459	0	0
Cosmarium spp.	0	0	0	6	0	13	81	37	280	0	15	2026	32	0	702
Crucigenia tetrapedia (Kirchn) W. & G. S. West	0	0	0	0	116	0	0	0	0	9	115	0	262	0	0
Crucigeniella cf. apiculata (Lemm.) Kom.	0	0	0	0	233	0	0	0	0	47	91	43	320	0	6732
Dictyosphaerium sp.	0	0	0	0	0	0	0	69	0	0	94	104	548	0	4519
Oocystis sp.	54	5	95	268	54	39	743	1372	1369	883	738	3266	9886	639	3029
Pediastrum boryanum (Turp.) Meneg.	0	0	0	0	0	7	47	16	46	1	30	34	425	10	0
Pediastrum duplex Meyen	0	0	0	0	40	11	0	27	25	87	63	0	42	0	0
Pediastrum simplex (simplex+echinulatum) Meyen	0	0	0	0	0	146	87	0	0	25	40	667	1107	452	0
Pedinomonas minor Korschikof	273	7	156	402	1312	212	614	605	816	0	2815	1802	21	0	0
Phacotus lenticularis (Ehrenberg) Stein	0	0	0	0	0	0	0	0	0	0	0	0	625	10	0
Planctonema lauterbornii Schmidle	268	98	32	456	391	8424	1354	4631	943	0	66	0	0	0	0
Pteromonas angulosa (Carter) Lemmermann	0	0	0	0	0	0	0	0	0	0	0	0	0	0	2586
Scenedesmus acuminatus (Lagerh.) Chod.	0	0	0	0	0	0	47	0	0	12	194	605	37	724	14759
Scenedesmus acutus Meyen	0	0	0	0	30	0	0	0	42	0	121	305	0	10	406
Scenedesmus cf. intermedius Chod.	0	0	0	0	0	0	0	0	0	0	363	0	0	118	2397
Scenedesmus cf. longispina Chod.	15	16	0	0	0	27	1	10	36	52	127	0	67	0	1350
Scenedesmus linearis Kom.	46	0	0	246	154	210	135	457	356	61	161	6	0	0	0
Staurastrum sp.	0	0	0	0	0	0	0	3	15	0	22	91	33	29	0
Tetraedron minimum (A. Br.) Hansg	0	0	0	0	3	0	0	0	41	0	7	352	7	238	409
Tetraedron triangulare Kors.	0	0	0	0	0	0	0	12	21	0	0	84	0	100	171
Tetrastrum staurogeniaeforme (Schröd.) Lemm.	4	0	0	0	0	0	0	0	0	10	29	37	0	142	856
Tetrastrum triangulare (Chod.) Kom.	0	0	0	0	0	0	0	0	0	61	108	164	4	0	81
Euglenophyta															
Lepocinclis sp.	0	0	0	0	38	0	8	15	2	471	1	5	3	0	0
Phacus cf. longicauda (Ehrenberg) Dujardin	0	4	0	0	4	0	16	0	0	69	2	31	0	0	0
Trachelomonas cf. volvocina Ehrenberg	0	0	0	0	14	0	16	3	0	18	7	113	56	356	0
Cyanobacteria															
Anabaenopsis circularis (G. S. West) Wol. et Miller	0	0	0	0	0	0	0	0	0	0	0	0	0	6513	0
Merismopedia sp.	0	0	0	0	0	0	0	0	0	175	0	3352	40	0	2546
Microcystis sp.	0	0	0	0	0	0	0	0	7	0	13	0	4	138	0
Oscillatoria sp.	0	0	0	0	0	0	0	0	0	0	2	0	0	3504	0
Pseudanabaena cf. limnetica (Lemm.) Kom.	10	4	1	0	0	0	0	0	0	0	0	0	0	71	451
Synechocystis sp.	23	0	0	0	0	0	0	0	0	0	0	126	0	177685	0

Hydrobiologia **369/370:** 45–57, 1998.
M. Alvarez-Cobelas, C. S. Reynolds, P. Sanchez-Castillo & J. Kristiansen (eds), Phytoplankton and Trophic Gradients.

Phytoplankton of two Araucanian lakes of differing trophic status (Argentina)

Mónica M. Diaz, Fernando L. Pedrozo & Pedro F. Temporetti
Centro Regional Universitario Bariloche, Unidad Postal Universidad del Comahue, (8400) Bariloche, Argentina

Key words: phytoplankton, trophic state, lakes, Patagonia

Abstract

The Araucanian lake district in southern South America encompasses many great lakes of glacial origin, as well as a large number of smaller lakes. In this study, we present data on two waterbodies, one large (Nahuel Huapi), and one shallow (Verde). The phytoplankton community structure and dynamics or either lake were monitored for a year, in relation to the physical and chemical fluctuation, as well as the morphometry of the respective basins. In the large lake, the phytoplankton biomass was strongly dominated by diatoms (*Aulacoseira granulata*, *Rhizosolenia eriensis* and *Cyclotella stelligera*) and by dinoflagellates (*Gymnodinium* and *Peridinium* spp.). In the small lake, Verde, the dominant algae were *Trachelomonas* spp. and *Cosmarium punctulatum,* during the summer biomass maximum, and *Rhodomonas lacustris*, *Chrysochromulina parva* and *Navicula* spp. during autumn. The maximum biomass value was 634 mg m^{-3} in Lake Nahuel Huapi in spring and 7800 mg m^{-3} in Lake Verde in summer. According to their phytoplankton and physical and chemical features, Lakes Nahuel Huapi and Verde are readily classifiable, as ultra-oligotrophic and mesotrophic, respectively.

Introduction

The structure and dynamics of the phytoplankton in lakes from temperate areas have been studied intensively in the northern hemisphere. Within the southern hemisphere, especially in the Argentine Patagonian region where there is a profusion of waterbodies, available information is scarce. These lakes have been sampled only occasionally or seasonally (Thomasson, 1959, 1963; García de Emiliani & Schiaffino, 1974; Bonetto et al., 1971; Izaguirre et al., 1990). Recently we published two papers, concerning phytoplankton community (Diaz & Pedrozo, 1993) and nutrient limitation (Diaz & Pedrozo, 1996), based on more frequent samples. Comparative studies on South American lakes and reports classifying lakes in this region are rare (Quirós, 1988). The few published accounts include works by Menu-Marque & Marinone (1985), Quirós & Drago (1985), Quirós (1988, 1990, 1991), Izaquirre et al. (1990) and Izaguirre (1993). Of these, only the last two (Izaguirre et al., 1990 and Izaguirre 1993) used phytoplankton structure to characterize lakes regionally.

The Andes Region of Patagonia has many deep water bodies of glacial origin, with a low salt content (conductivity around 30 $\mu S\ cm^{-1}$) and ranging between ultra-oligotrophic and oligotrophic in character. In this study, we set out to identify the main differences in the composition of phytoplankton between two Patagonian waterbodies of differing trophic state. In particular, we sought to:

- assess the composition of phytoplankton, both quantitatively and qualitatively.
- compare the classification of trophic state based on phytoplankton to that based on physical and chemical variables.
- identify species that could be used as indicators of the trophic spectrum of the phytoplankton.

Table 1. Environmental variables, average and range of chemical data for the studied lakes.

Site	Nahuel Huapi Lake	Lake Verde
Location (Lat, Long)	40° 50′ S; 71° 30′ W	40° 50′ S; 71° 50′ W
Area (A) (Ha)	52900.0	0.01
Ice cover period (day)	–	–
Volume (V) (Hm^3)	83053.0	0.22
Mean depth (V/A) (m)	157.0	2.6
Max depth (h) (m)	438.0	9.5
Residence time (T) (Year)	12.4	–
Mixing frequency	Monomictic temperate	Monomictic temperate
Secchi (m)	14.4	4.0
pH	6.8 (6.6–7.4)	6.9 (6.4–7.3)
Salinity ($mg\ l^{-1}$)	21.0	57.4
TP (range) ($\mu gP\ l^{-1}$)	5.0 (2.5–9.4)	11.0 (1.1–73.8)
SRP (range) ($\mu gP\ l^{-1}$)	3.5 (3.0–6.9)	5.3 (nd–21.7)
TN (range) ($\mu gN\ l^{-1}$)	72.2 (34.1–245.0)	–
DIN (range) ($\mu gP\ l^{-1}$)	< 5 (nd–5.0)	11.1 (0.4–28.6)
Alkalinity ($mg\ l^{-1}$)	18.7	22.0 (16.6–27.8)
CO_2 ($mg\ l^{-1}$)	5.9	5.7 (4.3–7.1)
Chl a (range) mg/m^3	0.94 (0.10–1.20)	0.80 (0.05–2.70)

The sites of study

Lake Nahuel Huapi is part of Nahuel Huapi National Park. The lake has a glacial origin, and was formed by damming behind a large frontal moraine (Cordini, 1950). Its maximum depth is 438 m, and mean depth 157 m. The total area is 529 km^2 (Cordini, 1939), embracing seven arms of various sizes. The long perimeter of the lake (357 km) results from its rather jagged shore line. The residence time is 12.4 years (Table 1). The lake is warm monomictic with a winter circulation (8 °C) and direct stratification in summer, when there is a well-developed thermocline.

Lake Verde (40° 50′S; 71° 50′W, 790 masl) is in Villa La Angostura county, about 90 km from Bariloche city. The lake is a natural sanctuary administered by the Dirección de Parques and Paseos of Neuquén Province. In spite of its proximity to an urban area, it is protected from direct anthropogenic influence. The lake is surrounded by woodland, dominated by *Nothofagus dombeyi* (Mirb.) Blume and *Austrocedrus chilensis* (Don) Flor.et Bout. The total area of the lake is 114 m^2, with maximum and mean depths of 9.5 m and 2.6 m, respectively. The shores are vegetated by the locally common species, *Schoenoplectus californicus* (Meyer) Sojak, *Myriophyllum elatinoides* Gand., *Potamogeton linguatus* Hangst. and *Nitella* sp.

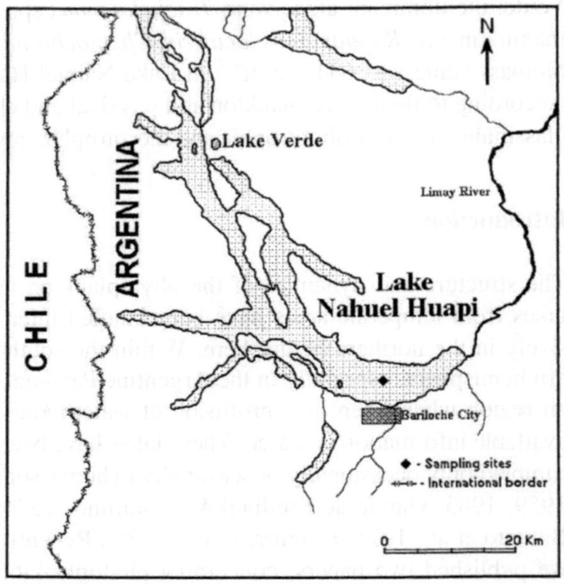

Figure 1. Location of the two sampled Araucanian Lakes in Patagonia, Argentina.

Lake Verde is also a warm monomictic system, which fully mixes during fall and winter. During most of spring and summer the lake remains stratified.

Materials and methods

Lake Verde was sampled every second week from February 1990 to February 1991. Lake Nahuel Huapi was sampled weekly during January and February 1995 and monthly from March 1995 to March 1996 (Figure 1).

Phytoplankton samples were collected using a Van Dorn bottle. Samples were taken at 1-meter intervals in lake Verde and at 5-meter intervals (0–25) in lake Nahuel Huapi and pooled into a single final sample. The final samples were fixed with acetic Lugol's solution and counted following the standard Utermöhl technique. The phytoplankton biomass was estimated as biovolume and expressed as fresh wet weight. Specific diversity and evenness were calculated on a biomass basis using the Shannon index. The picoplankton was not considered in these calculations.

The following variables were measured in the field: dissolved oxygen (Winkler method, APHA, 1985), pH and conductivity (specific electrodes with automatic temperature compensation); transparency (Secchi disk); temperature and thermal profiles (thermistor). Water samples were collected with a Van Dorn bottle. Aliquots of water were filtered in situ through GF/C, GF/F fiberglass and membrane filters for determination of chlorophyll-*a*, suspended solids and dissolved nutrients, respectively. All samples were preserved, transported and stored according to standard procedures (APHA, 1985).

The following variables were analyzed in the laboratory: on filters, chlorophyll-*a* by extraction with 90% acetone. Calculations followed the equation recommended in APHA (1985). On unfiltered water samples: total phosphorus (TP), by acid oxidation with potassium persulfate in 1 atm pressure for 2 h; the resulting solution was analysed for soluble reactive phosphorus (SRP: Golterman et al., 1978); and for total nitrogen (TN), by the alkaline oxidation method and reduction to nitrite in a cadmium column (Mackereth et al., 1978; Grasshoff et al., 1983). On filtered water samples: SRP by the ascorbic acid method (Golterman et al., 1978); N-ammonia ($N\text{-}NH_4$) by the indophenol blue method (APHA, 1985); N-nitrates + N-nitrites ($N\text{-}NO_3 + N\text{-}NO_2$) by reduction in a cadmium column (APHA, 1985); N-nitrites by diazotisation (APHA, 1985), and $N\text{-}NO_3$ by difference. $N\text{-}NO_3 + N\text{-}NO_2 + N\text{-}NH_4$ was considered as total disolved nitrogen (DIN). Silica was measured by the heteropoli-blue method (APHA, 1985).

Results

Lake Nahuel Huapi

Physical and chemical features

Table 1 summarizes the information on several properties pertaining to water quality in Lake Nahuel Huapi (Pedrozo et al., 1995). The salt content was low, the mean specific conductivity $\sim 30\ \mu S\ cm^{-1}$ and maximum $40\ \mu S\ cm^{-1}$. The pH varied from slightly acid to slightly alkaline (pH = 6.6–7.4), typical of the waterbodies in the region. Transparency was high, with a Secchi depth of 14 m, and with higher values recorded (22.5 m) under windless and clear-sky conditions. Dissolved oxygen values were typically high ($10.95\ mgO_2\ l^{-1}$), as a consequence of the trophic and thermal characteristics of the lake, with winter readings that exceed $11\ mgO_2\ l^{-1}$.

The average concentrations of nutrients (Figure 2a, 2b) in their various forms were: TP $5\ \mu g\ l^{-1}$, TN $60\ \mu g\ l^{-1}$; SRP $3\ \mu g\ l^{-1}$; $N\text{-}NO_3 + N\text{-}NO_2$ $29\ \mu g\ l^{-1}$ and $N\text{-}NH_4$ $20\ \mu g\ l^{-1}$. The relationship DIN:SRP was about 10.

The content of major ions was in the following sequence:

$$Ca^{2+} > Na^{+} > Mg^{2+} > K^{+} \text{ and } HCO_3^{-} > SO_4^{2-} > Cl^{-}$$

The average chlorophyll-*a* value was about $0.5\ mg\ m^{-3}$ and ranged between 0.1 and $1.2\ mg\ m^{-3}$ (Figure 2c).

Phytoplankton

Forty phytoplankton species were identified: 11 Chlorophyceae, 11 Bacillariophyceae, 6 Cyanophyceae, 6 Chrysophyceae, 3 Dinophyceae, 2 Cryptophyceae and 1 Prymnesiophyceae.

The sequence of species abundance in Lake Nahuel Huapi was initiated in spring by the dominance of *Chrysochromulina parva* Lackey and *Cyclotella stelligera* Cleve & Grunow. The latter was replaced by *Rhizosolenia eriensis* Smith and then by *Gymnodinium* sp., towards the end of the spring but still persisted during summer. At the beginning of the fall, a peak of a spherical species of Chlorococcalean was recorded. During winter there was a sequence of *Dactylococcopsis raphidioides* Hansgirg, *C. stelligera*, *Rhizosolenia eriensis* and *Synedra ulna* (Nitzsch) Ehrenb., all codominating with *C. parva*, which remained present all year.

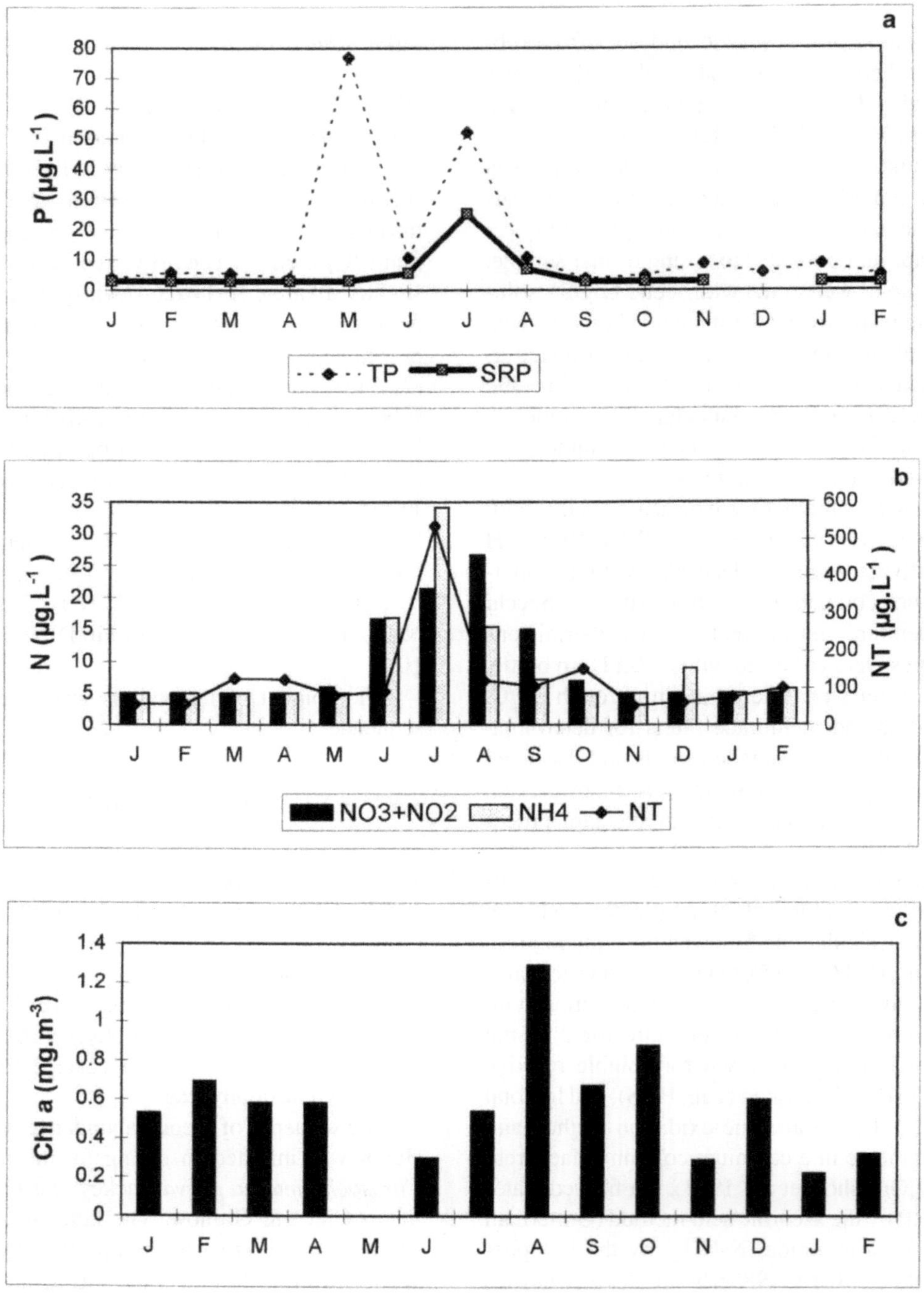

Figure 2. Variation of some physical and chemical variables in the Nahuel Huapi Lake, during 1995.

Populations varied between 129 and 831 cell ml^{-1}, with a minimum in March 1996 and maxima in March 1995 and September 1996. The mean annual density over the period January 1995-March 1996 was 362 cell ml^{-1}.

The groups of phytoplankton that contributed more than 50% to the total biomass were Prymnesiophyceans

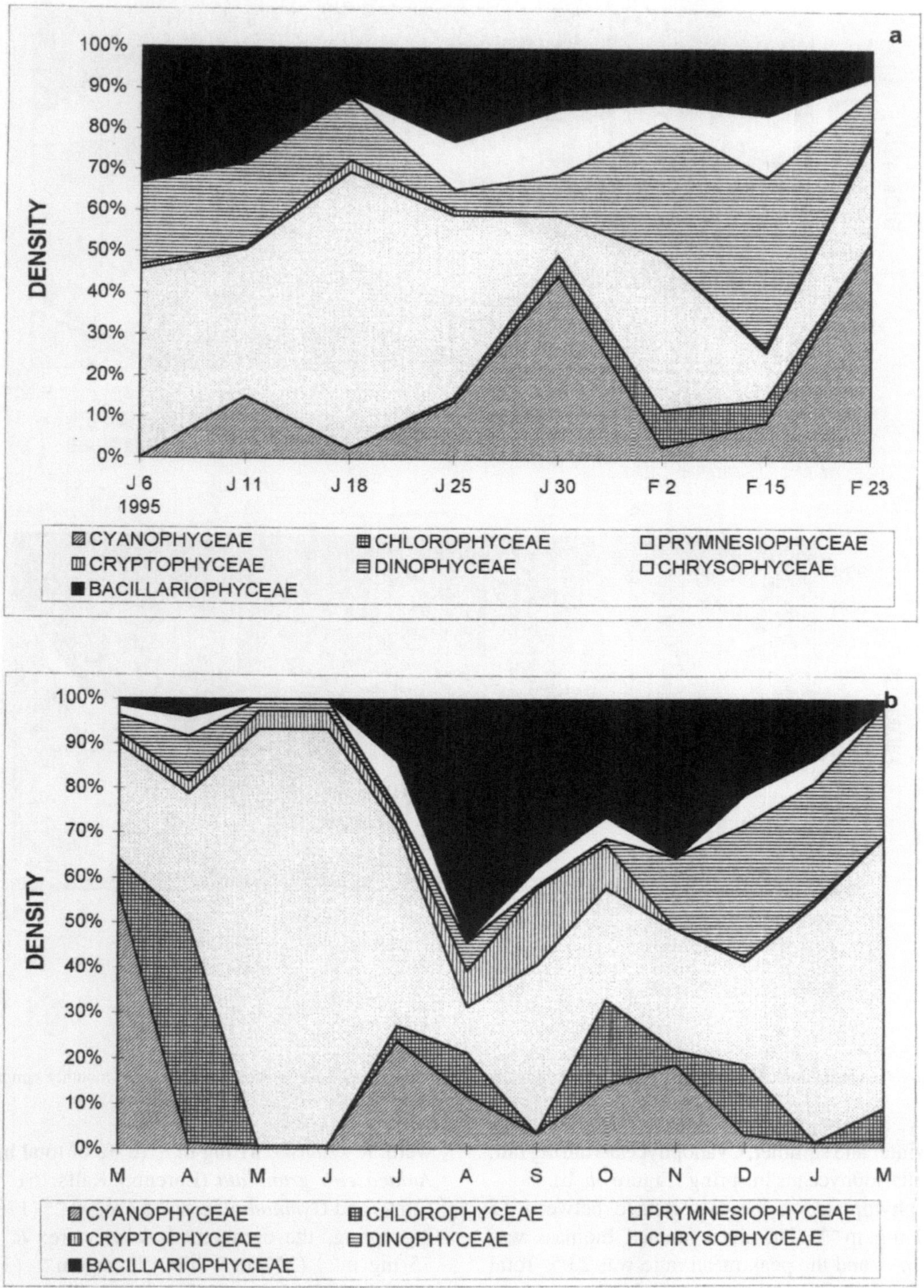

Figure 3. Composition of algal groups in the Nahuel Huapi Lake during 1995: a) weekly sampling; b) monthly sampling.

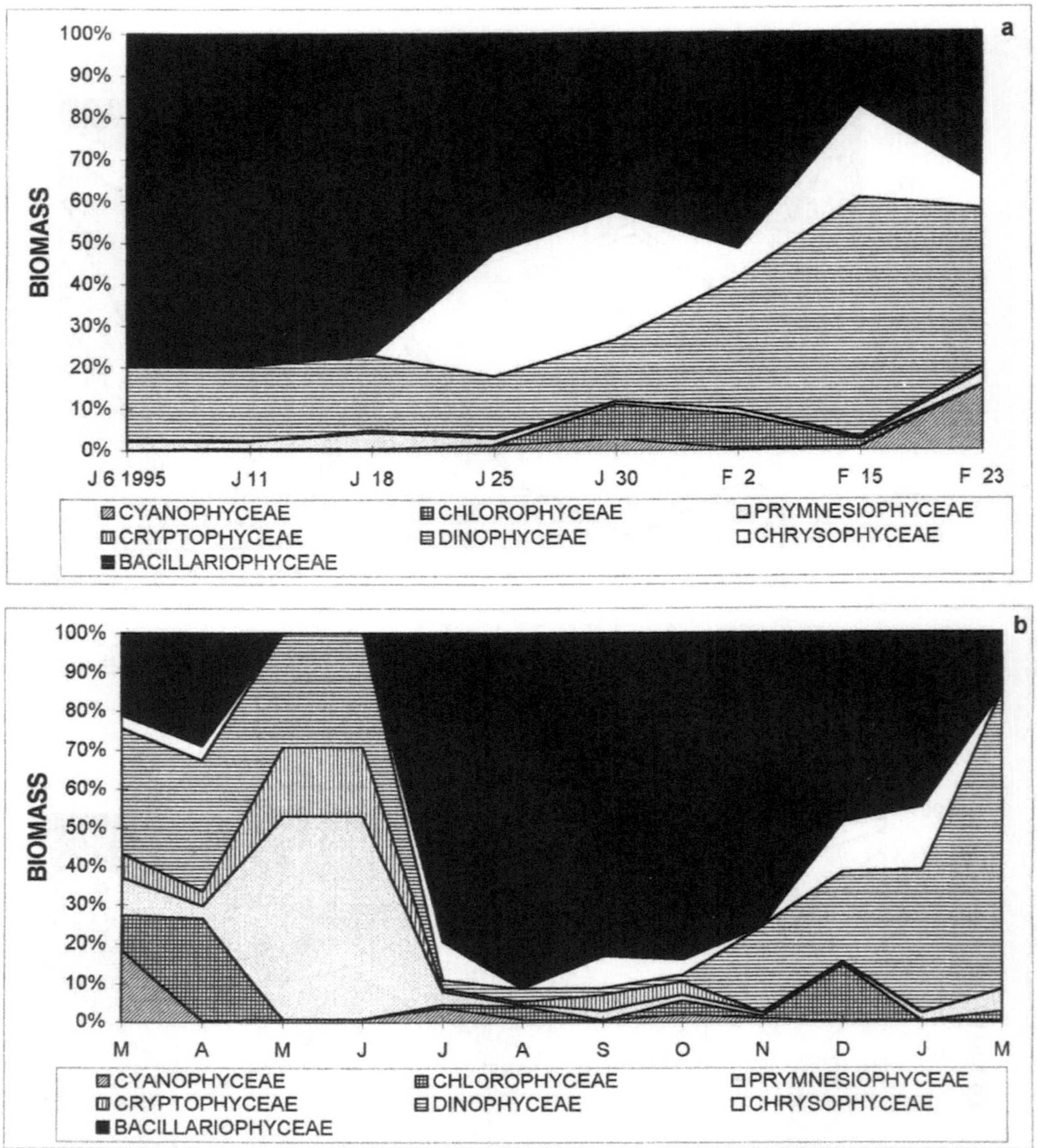

Figure 4. Contribution of different algal groups to total biomass in Nahuel Huapi Lake: a) weekly sampling; b) monthly sampling.

during winter and summer, Cyanophyceans during fall, and Bacillariophyceans in spring (Figure 3a, b).

The phytoplankton biomass ranged between 15 and 300 mg m^{-3}. The mean annual biomass was 147 mg m^{-3} and the peak:mean ratio was 2.05. Total biomass showed two peaks: one in summer and another in spring. The dominant species in summer (with the fraction of the total biomass they contributed) were: *R. eriensis*, 81 mg m^{-3} (27% of total biomass), *Aulacoseira granulata* (Ehrenb.) Ralfs, 61 mg m^{-3} (20%) and *Gymnodinium* sp., 57 mg m^{-3} (17%). During spring, the dominant species were: *R. eriensis* 75 mg m^{-3} (25%), *S. ulna* 76 mg m^{-3} (25%), *A. granulata* 70 mg m^{-3} (23%) and *Chrysosphaerella* sp. 23 mg m^{-3} (8%) (Figure 4a, b).

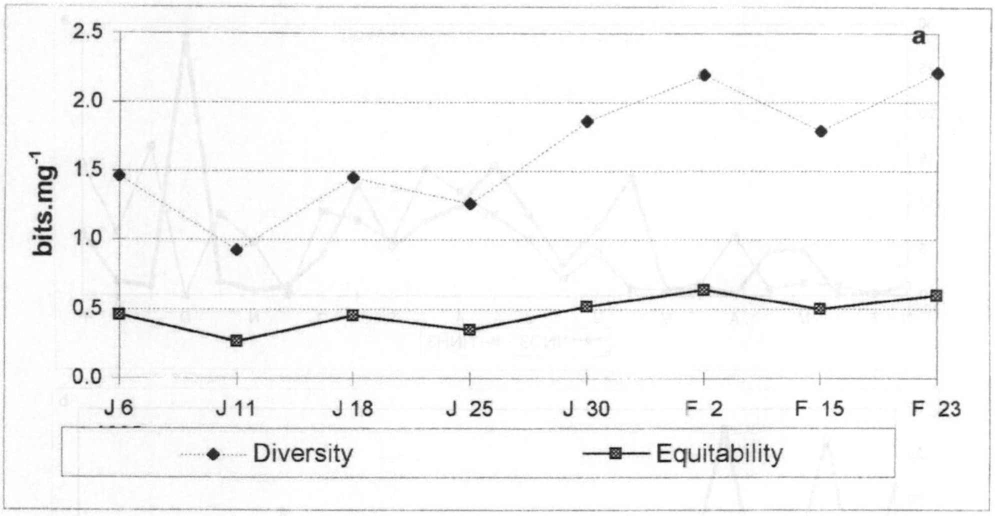

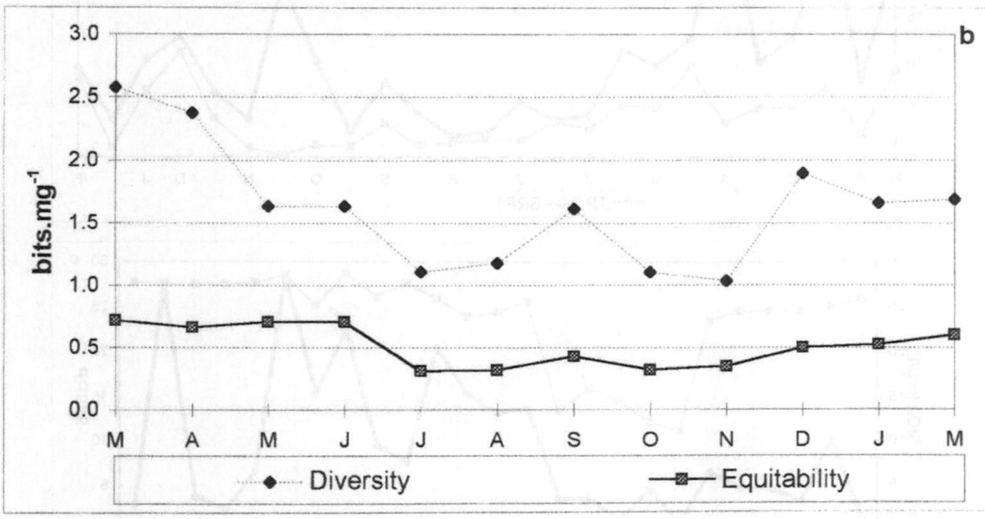

Figure 5. Variation of diversity and eveness during 1995 in Nahuel Huapi Lake: a) weekly sampling; b) monthly sampling.

Thirty one species out of a total of 40 recorded, contributed more than 0.5% to the total biomass on at least one occasion during the year, 19 species contributed more than 5%, and 6 more than 25%.

During early spring the lake was mixed, the levels of TP and SRP were low and the diatoms (*R. eriensis*, *A. granulata* and *S. ulna*) were the dominant species. Later (October to December), *R. eriensis* was the most important species in terms of biomass, a situation that continued during summer until the development of *Dinobryon divergens* Imhof and *Gymnodinium* sp. During summer, the specific diversity ranged between 1 and 2 and the evenness between 0.3 and 0.6 (Figure 5a, b). By fall, there was a species replacement and *C. parva* and *Rhodomonas lacustris* (Pascher et Ruttner) Javornicky became dominant in biomass and density. Simultaneously, there was a modest increase in the levels of TN and a more pronounced increase in TP and a decrease in the DIN:SRP ratio, which were coincident with the species replacement and the beginning of the mixing period. The maximum values of diversity and evenness (2.53 and 0.7, respectively) were recorded during this period. In winter, when values of phosphorus and nitrogen were higher, *R. eriensis* was dominant in terms of biomass and *C. parva* in terms of

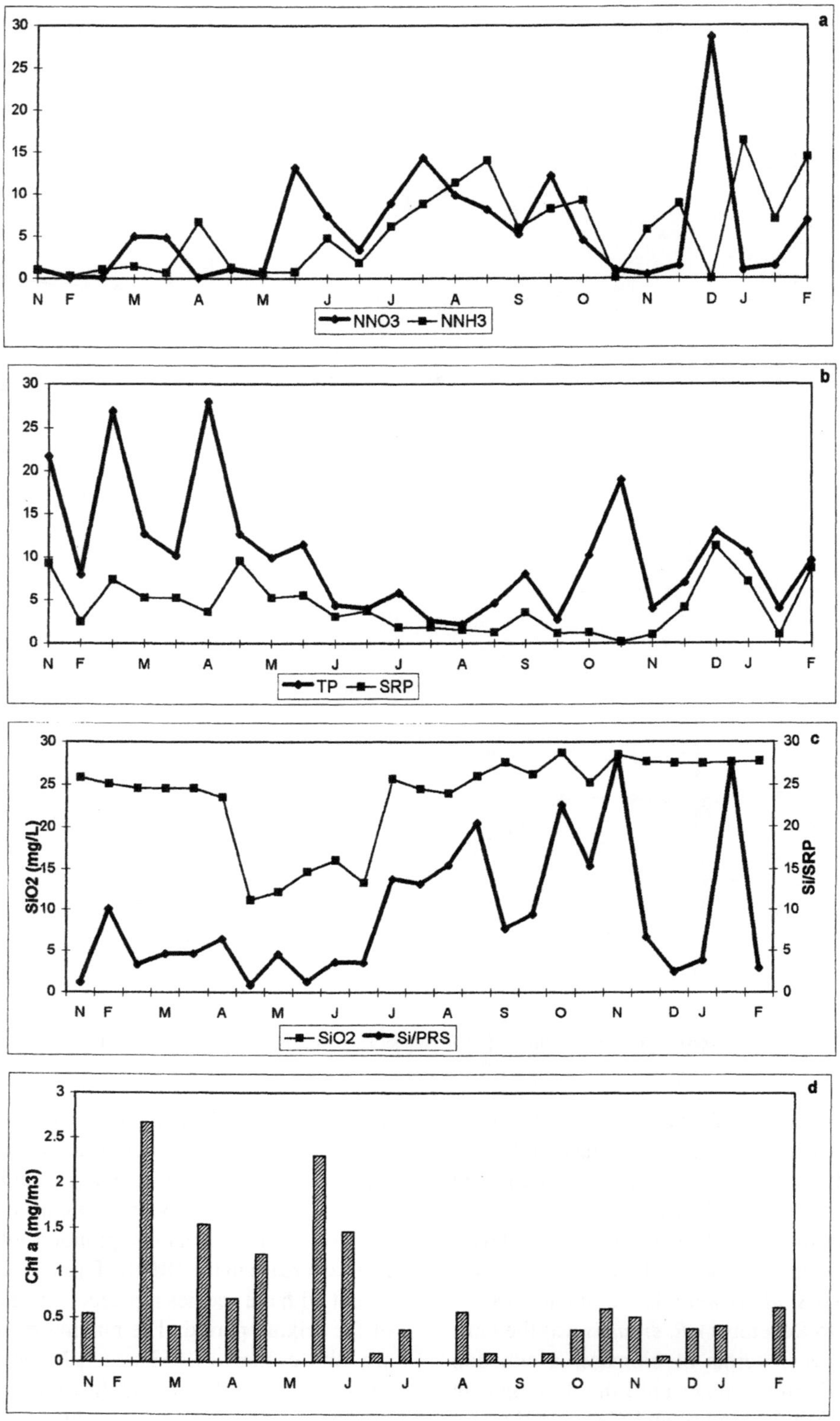

Figure 6. Variation of some physical and chemical variables in the Lake Verde, during 1990–1991.

density. Diversity and evenness showed the minimum values (1–1.5 and 0.3–0.6, respectively).

Lake Verde

Physical and chemical features

The information on physical and chemical variables is summarized in Table 1. Maximum depth during the study period was 9 m, while the transparency ranged between 3 and 5 m. The conductivity (mean: 82.1 μS cm^{-1}; range: 73–95 μS cm^{-1}) is higher than the average for Andean Patagonian lakes.

The pH ranged between 6.4 and 7.3 (mean = 6.9). The ionic composition had the following form:

$$Ca^{2+} > Mg^{2+} > Na^{+} > K^{+} \text{ and } HCO_3^{-} > SO_4^{2-} > Cl^{-}$$

The oxygen concentration ranged between 7.6 and 9.6 mg l^{-1}(avg = 8.9 mg l^{-1}). The nitrogen content was low. Ammonium contributed about 50% of the DIN, ranging between 0.4 and 28.6 μg N l^{-1} (Figure 6a). The soluble reactive phosphorus (SRP) ranged between undetectable values and 21.7 μgP l^{-1} (mean = 5.3 μgP l^{-1}), contributing in average 48% of the total phosphorus (Figure 6b). Silica concentrations varied seasonally, with a minimum of 11.2 mg l^{-1} (April 1990) and a maximum of 28.7 mg l^{-1} (October 1990) (mean = 24.1 mg l^{-1}) (Figure 6c).

The concentration of chlorophyll-*a* was low. In winter it ranged between 0.05 and 0.57 mg m^{-3}. Maximum values were observed during summer, when it peaked up to 2.7 mg m^{-3} (Figure 6d). The low spectrophotometric absorbance at 365 nm (between 0.1 and 0.27 units; average 0.05) indicate low levels of dissolved organic matter.

Phytoplankton

A total of 74 taxa were identified. Five groups were represented by ten or more species: Cyanophyceae (12 species), Chlorophyceae (13 species), Zygophyceae (13 species), Euglenophyceae (10 species) and Bacillariophyceae (15 species). On the other hand, Chrysophyceae, Dinophyceae and Prymnesiophyceae were represented by 5, 3 and 1 species, respectively.

A handful of species were responsable for the sequence of abundance that occurred in Lake Verde. In spring, the highest densities coincided with abundance of *Dinobryon divergens*, which was replaced by *Rhodomonas lacustris* and afterwards by *D. divergens*. The most abundant species in summer were *Chrysochromulina parva*, *R. lacustris* and *Kephyrion spirale* (Lack.) Cour. *Oscillatoria curviceps* C. Agardh peaked in the late summer, when it was replaced by *R. lacustris*, which was dominant during fall and part of winter. In winter, the maximum densities corresponded to *O. curviceps* and *O. agardhii,* which were followed by *D. divergens*.

The density values ranged between 121 and 12,183 cell ml^{-1}, with minimum on May 1990 and a maximum on July 1990. Mean annual density for the period February 1990-February 1991 was 2,158 cell ml^{-1}, with maxima on every season of the year.

The main pulses occurred during fall and winter (Figure 7). The highest densities corresponded to the Cryptophyceae and Prymnesiophyceae in fall and the Cyanophyceae and Chrysophyceae in winter. In spring the highest abundance corresponded to Chrysophyceae and in summer to Chlorophyta and Cyanophyceae.

The phytoplankton biomass ranged between 96 and 7800 mg m^{-3}. The mean annual biomass was 1517 mg m^{-3} and the peak:mean ratio was 5.05. The pattern of variation shows a maximum in summer and minor increases during fall. The maximal biomass of dominant species at those peaks were: *Cosmarium punctulatum* Bréb., 2,988 mg m^{-3} (39% of total biomass) and *Trachelomonas* spp. 3,320 and 3,440 mg m^{-3} (43 and 84%) during the summer peaks. Three species were dominant during the fall maximum: *R. lacustris* with a biomass of 1,395 mg m^{-3} (48%), *Cryptomonas erosa* Ehrenb. 758 mg m^{-3} (29% of total biomass) and *Navicula cocconeiformis* Greg. ex Grev. 640 mg m^{-3} (21%). Forty three species out of 74 contributed less 0.5%; 31 contributed more than 0.5% to the total biomass in at least one occasion during the year, 20 species contributed more than 5% and 13 more than 25% (Figure 8).

The species diversity varied in a seemingly random fashion across the year. There were two maxima, one in summer (2.95) and another in spring (2.28) and a very apparent minimum in autumn (0.40). The evenness ranged between 0.14 and 0.93, with a maximum value during the annual biomass peak (Figure 9).

By the end of winter (September 1990), the availability of dissolved inorganic nitrogen and silica, associated with the increase in light intensity and photoperiod favoured a change in the phytoplankton community. Chrysophyceae, which were the most abundant during the winter, were replaced by nanoplanktonic species (*Ch. parva* and *R. lacustris*, Figure 8). This change in community structure was not matched by variations in

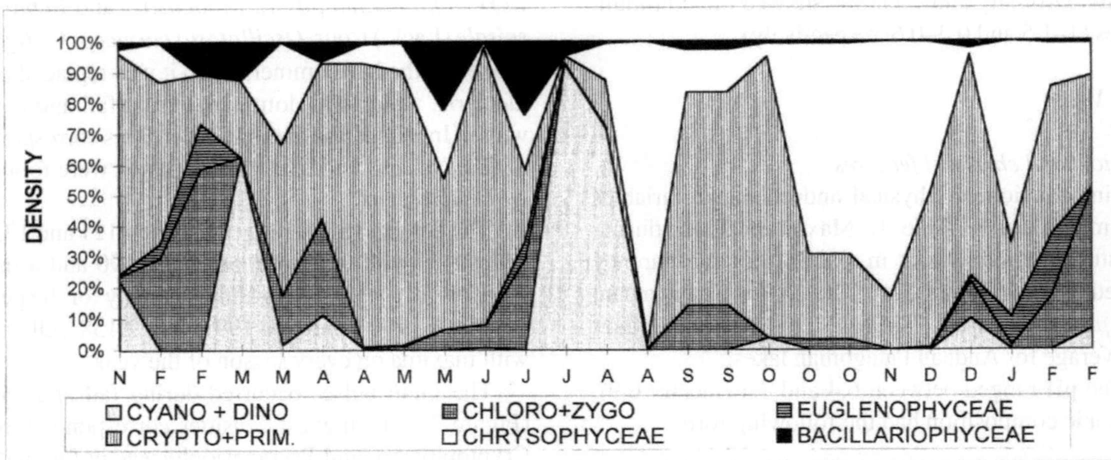

Figure 7. Composition of algal groups in the Lake Verde during 1990–1991.

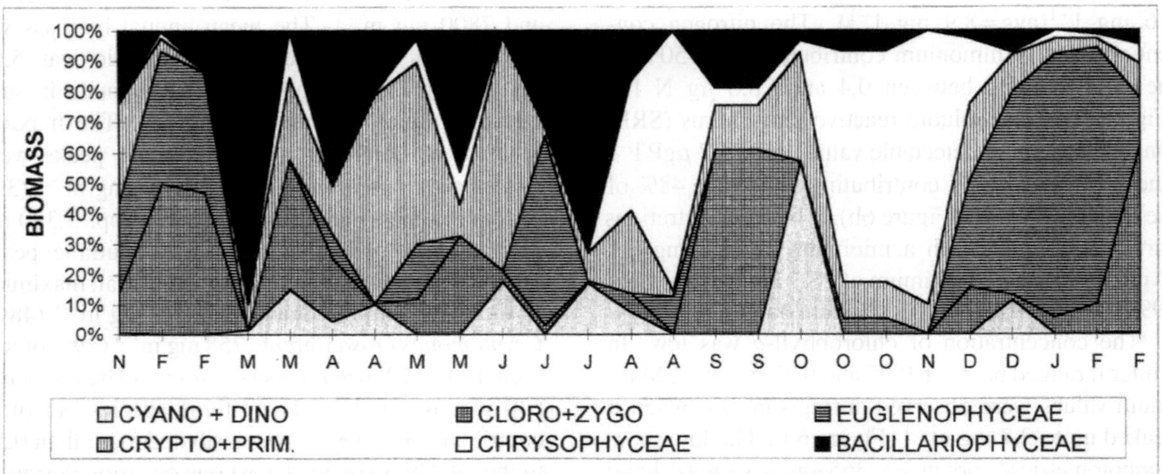

Figure 8. Contribution of different algal groups to total biomass in Lake Verde.

biomass or in total density, which continued unchanged until October.

Low biomass and density levels were observed during spring. Dinoflagellates represented 50% of total biomass by mid-September 1990 and were replaced again by Chrysophyceae (75–90% of total biomass) in October and November 1990. This replacement occurred simultaneously with the decrease in SRP, $N\text{-}NO_3$ and $N\text{-}NH_4$.

The nanoplankton was abundant and oscillated in relative abundance. It contributed between 20 and 90% to total density during spring, although its contribution to total biomass was lower (between 10 to 35%).

In summer, the warmer temperatures and higher nutrient concentrations favored the maximum in annual biomass. The diversity not only showed a species richness increase, but also a higher complexity and functional diversification of the phytoplankton structure, with a coexistence of species of small and large size. Small species (*Rhodomonas*, *Cryptomonas*, *Chrysochromulina* and *Monoraphidium*) occurred together with large species, such as *Elakatothrix americana* Wille, *C. punctulatum* and *Trachelomonas* spp.

There were differences in biomass, nutrient levels and phytoplankton composition between the summer

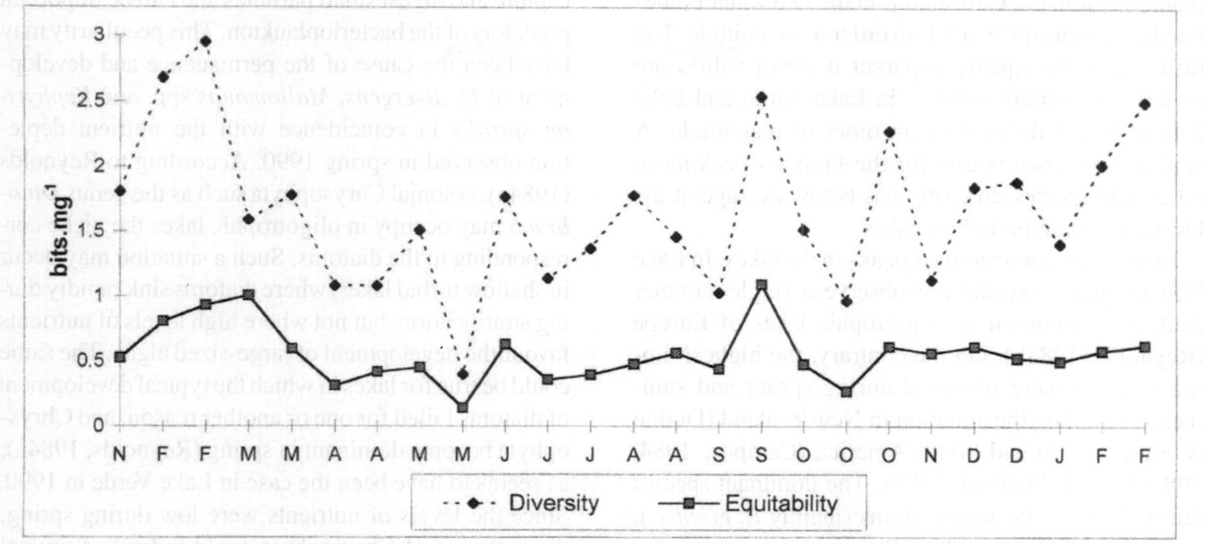

Figure 9. Variation of diversity and equitability during 1990–1991 in the Lake Verde.

1990 and 1991. The maximum biomass recorded during the study (almost 8 g m^{-3}) occurred during February 1990, when chlorophytes and euglenoids were dominant. The high biomass caused decreases in phosphorus and depletion of nitrogen.

In January 1991, total biomass was only half that of the previous year, though the levels of phosphorus and nitrogen were higher than in 1990, and the phytoplankton community was completely dominated by Euglenophyceae. The main difference between the two consecutive summers lies in the preceding periods, in other words, in the characteristics of the lake during spring. In November 1989, the SRP was higher than 20 μgP l^{-1} (DIN:SRP = 0.12) while in November 1990, the SRP was only 5 μgP l^{-1} (DIN:SRP = 4.8).

In March 1990, there was a sharp decrease in biomass and in the development of diatoms. Although there was an important decrease in silica during the fall, it was probably not due to the development of diatoms, which were present in low numbers, but, more likely to the effect of heavy rains.

As fall progressed, a shortening of the photoperiod and continuous circulation of the water column caused an increase in concentration of nutrients, and a new increase in biomass, with dominance of euglenophytes, green algae, Chrysophyceae and nanoplanktonic algae.

The winter biomass minimum was characterized by the presence of *R. lacustris*, *C. stelligera*, *A. granulata*, *Fragilaria crotonensis* Kitton and *Peridinium* sp. As winter progressed, the diatoms were replaced by *D. divergens* and *Mallomonas* spp. During this period an increase of the different forms of nitrogen was observed.

Discussion

Based on the total phosphorus content, total nitrogen, chlorophyll-*a*, maximum value of chlorophyll-*a* and Secchi disk depth, the OECD (1982) trophic categories allow us to classify Lake Nahuel Huapi as ultra-oligotrophic. On the same classification, Lake Verde should be considered mesotrophic. However, the values of mean chlorophyll-*a* and maximum chlorophyll-*a* for Lake Verde are within the ranges associated with oligotrophic waterbodies. According to Baigún & Marinone (1995), prediction models applied in the Northern Hemisphere overestimate chlorophyll concentration for Argentine Patagonian lakes, especially at higher trophic conditions.

If maximum phytoplankton biomass is used as the only indicator of trophic state, Lake Nahuel Huapi would lie in the same category given by Vollenweider (1968) but Lake Verde (7.8 g m^{-3}) would lie somewhere beyond mesotrophy (3 a 5 g m^{-3}).

The phytoplankton density in mesotrophic Lake Verde was about two orders of magnitude higher than in Lake Nahuel Huapi. The minimum values of phytoplankton biomass were similar in the two water bodies but the maxima differed by an order of magnitude. The differences are equally apparent if mean values are compared: average biomass in Lake Verde and Lake Nahuel Huapi differed by an order of magnitude. A similar difference is true for the biomass peak:mean ratio. The species diversity was below average at the biomass maximum in both lakes.

There were two biomass peaks in the lakes. In Lake Nahuel Huapi, we did not observe a single summer peak as is common in oligotrophic lakes of Europe (Reynolds, 1984b). On the contrary, the highest biomass values were observed during spring and summer, resembling the situation in New Zealand (Duthie & Stout, 1986) and South America (Campos, 1984, 1987; Diaz & Pedrozo, 1993). The dominant species during such peaks were diatoms (mainly *R. eriensis*), which benefit from high Si:P typical of oligotrophic lakes (Sommer et al., 1986). Analysing data from many authors, Reynolds (1984b) and Margalef (1983) concluded that this species is characteristic of oligotrophic environments with poorly mineralized waters. *Rhizosolenia* occurred together with *S. ulna* and *A. granulata* in spring. The latter two species are cited from mesotrophic environments (Rawson, 1956; Reynolds, 1984a).

Although *C. stelligera* was not important by biomass, it was nevertheless very abundant during spring. On the other hand, the spring maximum dominated by diatoms was followed by a summer maximum dominated by dinoflagellates (*Gymnodinium* sp.). These two events agree with the scenario suggested by Reynolds (1984b) for oligotrophic lakes.

The nanoplanktonic algae *C. parva* and *R. lacustris* were dominant during the autumn biomass minimum but *Rhizosolenia* was again dominant in winter.

The characteristics that distinguish Lake Verde as mesotrophic are the two biomass peaks: one in summer dominated by *Cosmarium punctulatum* and *Trachelomonas* spp.; and another in autumn, dominated by nanoplanktonic species (*R. lacustris* and *Cryptomonas erosa*) and *Navicula cocconeiformis*. Diatom dominance, which is typical of mesotrophic lakes during spring and fall (Reynolds, 1984b, Sommer et al., 1986), was not observed in Lake Verde, although there were subdominant populations of *N. cocconeiformis*. Nevertheless, *D. divergens* was dominant in spring and winter.

Bird & Kalff (1987) pointed out the importance of phagotrophy to the Chrysophyceae, particularly within the genus *Dinobryon*. These flagellates are able to capture and digest small particles and can be important predators of the bacterioplankton. This peculiarity may have been the cause of the permanence and development of *D. divergens*, *Mallomonas* spp. and *Kephyrion spiralis* in coincidence with the nutrient depletion observed in spring 1990. According to Reynolds (1984a), colonial Chrysophyta such as the genus *Dinobryon* may occupy in oligotrophic lakes the niche corresponding to the diatoms. Such a situation may occur in shallow turbid lakes, where diatoms sink rapidly during stratification, but not where high levels of nutrients favour the development of large-sized algae. The same could be true for lakes in which the typical development of diatoms failed for one or another reason, and Chrysophyta become dominant in spring (Reynolds, 1984a), as seems to have been the case in Lake Verde in 1990. Since the levels of nutrients were low during spring, the presence of Chrysophyta could indicate a crucial role for phagotrophy, and the advantage to survival it offers during periods of severe nutrient depletion.

Acknowledgements

To Bioq. Lucía Roselli and Lic. Gustavo Baffico who supported most of the chemistry analysis and Mrs. Leandro Castiñeira and Pablo Gonzalez Lucas for their help in the field work. Financial support for Nahuel Huapi study was provided by the Dirección Provincial de Aguas de la Provincia de Río Negro, Argentina. Wiener lab provided kits for ammonia analysis.

References

Apha, 1985. Standard Methods for the Examination of Water and Wastewater. 16th Ed. American Public Health Association, Washington, 1134 pp.

Baigún, C. & M.C. Marinone, 1995. Cold-temperate lakes of South America: do they fit northern hemisphere models? Arch. Hydrobiol. 135: 23–51.

Bird, D. F. & J. Kalff, 1987. Algal phagotrophy: Regulating factors and importance relative to photosyntesis in *Dinobryon* (Chrysophyceae). Limnol. Oceanogr. 32: 277–284.

Bonetto, A.A., W. Dioni & P. Depetris, 1971. Informe preliminar sobre las investigaciones limnológicas en la cuenca del río Manso y lago Mascardi (Rio Negro, Argentina). Publicación No. 4. Depto. de Recursos Naturales y Energía. Fundación Bariloche, 62 pp.

Campos, H., 1984. Limnological study of Araucanian lakes (Chile). Verh. int. Ver. Limnol. 22: 1319–1327.

Campos, H., W. Steffen, G. Agüero, O. Parra & L. Zuñiga, 1987. Limnology of lake Riñihue. Limnologica 18: 339–357.

Cordini, J. R., 1939. El lago Nahuel Huapi. Bol. Direcc. Min. Geol., B. Aires, 47 pp.

Cordini, J. R., 1950. Algunas características limnológicas del Lago Nahuel Huapi. Anales del Museo Nahuel Haupi Perito Dr. Francisco P. Moreno 2: 113–127.

Diaz, M. & F. Pedrozo, 1993. Seasonal succession of phytoplankton in a small Andean Patagonian lake (Rep. Argentina) and some considerations about the PEG model. Arch. Hydrobiol. 127: 167–184.

Diaz, M. & F. Pedrozo, 1996. Nutrient limitation in Andean-Patagonian lakes at latitude 40–41° S. Arch. Hydrobiol. 138: 123–143.

Duthie, H. C. & V. M. Stout, 1986. Phytoplankton periodicity of the Waitaki Lakes, New Zealand. Hydrobiologia 138: 221–236.

Garcia de Emiliani, M.O. & M. Schiaffino, 1974. Fitoplancton del lago Mascardi (Río Negro, Argentina). Bol. Soc. Arg. Bot. 15: 12–22.

Golterman, H. L., R. Clymo & M. Ohnstad, 1978. Methods for Physical and Chemical Analysis of Freshwater. IBP Handbook, 8, Blackwell, London, 140 pp.

Grasshoff, K., M. Ehrhardt & K. Kremling, 1983. Methods of seawater analysis. Verlag. Chemie, Second, revised and extended edition, 419 pp.

Izaguirre, I., P. del Giorgio, I. O'Farrell & G. Tell, 1990. Clasificación de 20 cuerpos de agua andino-patagónicos (Argentina) en base a la estructura del fitoplancton estival. Cryptogamie Algol. 11: 31–46.

Izaguirre, I., 1993. Comparative analysis of the phytoplankton of six lentic environments from the Province of Chubut (Argentina). Physis (Buenos Aires), Sec. B, 48: 7–23.

Mackereth, F. J. H., J. Heron & J.F. Talling, 1978. Water analysis. Freshwater Biological Association. Scientific Publication N° 36, 120 pp.

Margalef, R., 1983. Limnología. Omega, 1010 pp.

Menu-Marque, S. & C. Marinone, 1985. El zooplancton de seis lagos del Chubut (Argentina) y sus probables relaciones con la ictiofauna y algunos factores ambientales. Taller internacional sobre Ecología y manejo de peces en lagos y embalses. FAO, Roma: 158–178.

OECD, 1982. Eutrophication of waters, monitoring, assessment and control. OECD, Paris, 154 pp.

Pedrozo, F., M. Manuel, R. Alcalde & W. Lopez, 1995. Estado trófico del Lago Nahuel Huapi y estimación preliminar de su posible evolución trófica. Informe técnico del Centro Regional Universitario Bariloche, Universidad Nacional del Comahue y el Departamento Provincial de Aguas de la Provincia de Río Negro, 23 pp.

Quirós, R., 1988. Relationships between air temperature, depth, nutrients and chlorophyll in 103 Argentinian lakes. Verh. int. Ver. Limnol. 23: 647–658.

Quirós, R., 1990. Predictors of relative fish biomass in lakes and reservoirs of Argentina. Can. J. Fish. Aquat. Sci. 47: 928–939.

Quirós, R., 1991. Empirical relationships between nutrients, phyto- and zooplankton and relative fish biomass in lakes and reservoirs of Argentina. Verh. int. Ver. Limnol. 24: 1198–1206.

Quiros, R. & E. Drago, 1985. Relaciones entre variables físicas, morfométricas y climáticas en lagos patagónicos. Rev. Asoc. Cs. Nat. Litoral 16: 181–199.

Rawson, D. S., 1956. Algal indicators of trophic lake types. Limnol. Oceanogr. 1: 18–25.

Reynolds, C. S., 1984a. The ecology of freshwater phytoplankton. Cambridge University Press, Cambridge, 384 pp.

Reynolds, C. S., 1984b. Phytoplankton periodicity: The interactions of form, function and environmental variability. Freshwat. Biol. 14: 111–142.

Sommer, U., Z. M. Gliwicz, W. Lampert & A. Duncan, 1986. The PEG model of seasonal succession of planktonic events in fresh waters. Arch. Hydrobiol. 106: 433–471.

Temporetti, P., M. Alonso, G. Baffico, M. Díaz, W. Lopez & F. Pedrozo, ms. Estado trófico, comunidad íctica y producción intensiva de salmónidos en el embalse de Alicurá (Patagonia, Argentina). Serie COPESCAL, FAO, 26 pp, in press.

Thomasson, K., 1959. Nahuel Huapi. Plankton of same lakes in an Argentine National Park, with notes on terrestrial vegetation. Acta Phytogeogr. Suecica 42: 1–83.

Thomasson, K., 1963. Araucanian lakes. Plankton Studies in North Patagonia, with notes on Terrestrial vegetation. Acta Phytogeogr. Suecica 47: 1–139.

Vollenweider, R. A., 1968. Scientific fundamentals of the eutrophication of lakes and flowing waters, with particular references to nitrogen and phosphorus as factors in eutrophication. OECD Report, Paris.

Vollenweider, R.A., 1969. A manual on methods for measuring primary production in aquatic environments. IBP Handb. N° 12. Blackwell Scientific Publications, Oxford, 213 pp.

Hydrobiologia **369/370:** 59–71, 1998.
M. Alvarez-Cobelas, C. S. Reynolds, P. Sanchez-Castillo & J. Kristiansen (eds), Phytoplankton and Trophic Gradients.

Phytoplankton species composition is more sensitive than OECD criteria to the trophic status of three Brazilian tropical lakes

V. L. M. Huszar, L. H. S. Silva, P. Domingos, M. Marinho & S. Melo
Laboratory of Phycology, Dept. Botany, Museu Nacional, Universidade Federal do Rio de Janeiro, Quinta da Boa Vista, São Cristóvão, Rio de Janeiro, 20 940-040, Brazil

Key words: trophic gradient, phytoplankton, tropical freshwater environments, species assemblages, shallow lakes

Abstract

We compared the structure and function of the phytoplankton community in three tropical systems in Brazil: an oligo-mesotrophic floodplain lake; an eutrophic reservoir; and a hypereutrophic coastal lagoon. Phytoplankton biomass increased along this trophic gradient averaging 2.6, 6.2 and 31.6 $mm^3\ l^{-1}$ in the lake, reservoir and lagoon, respectively. Similarly, production increased from 427 to 918 gC $m^{-2}\ y^{-1}$. Along the trophic gradient, diversity and species richness of phytoplankton decreased. There was, however, no relationship between trophy and rates of change of phytoplankton communities or the ratio of maximum/mean biomass. Phytoplankton size varied between systems with small algae (2–20 μm) dominating in the hypereutrophic lagoon and larger algae (20–200 μm) dominating in the other systems. Many of the trophic indices developed for temperate areas could not be applied to these tropical systems since there was no clear relationship between Secchi depth, chlorophyll, or TP and trophic status. Phytoplankton species may be more useful than these trophic indices to the classification of the condition of tropical lakes.

Introduction

The structure of the phytoplankton associations and the abundance of species vary along trophic and seasonal gradients (Reynolds, 1984). Interactions between these gradients distinguish among the functional characteristics of species to influence the composition of phytoplankton assemblages (Seip & Reynolds, 1995).

Despite the considerable data available about phytoplankton ecology, new considerations have not emerged concerning the trophic gradient since the preliminary revision of Kalff & Knoechel (1978). Phytoplankton responses have been treated primarily in terms of concentrations of chorophyll *a*. The management of lakes and reservoirs has fostered an increasing interest on the use of assemblages of species that indicate trophic levels, but the attributes of these communities have been rarely used in these evaluations. Other aspects of the structure and function of the phytoplankton community, such as seasonality of biomass, growth rates, cell volume, etc., may also be related to the trophic gradient (Smith, 1978; Alvarez-Cobelas & Rojo, 1994; Seip & Reynolds, 1995).

The literature about distribution of species among trophic categories is wide, contradictory, and refers mainly to temperate-zone environments. The difficulty of transposing findings from temperate lakes to tropical ones was pointed out long ago by Thienemann (1955). The inadequacy of the criteria applied to lakes in Brazil has recently been given new emphasis by Esteves (1988), who pointed out that the indicators that are most used in typology of temperate lakes do not hold in tropical waters. Determinands such as the concentration of organic matter in the sediment, the concentration of oxygen in the hypolimnion, phytoplankton production (profiles and its magnitude) and diversity of species, often bias towards an inappropriate trophic classification when applied to tropical lakes. Higher temperature and differences in thermal stability and nutrient recycling are among the factors likely to influence the responses of phytoplankton and,

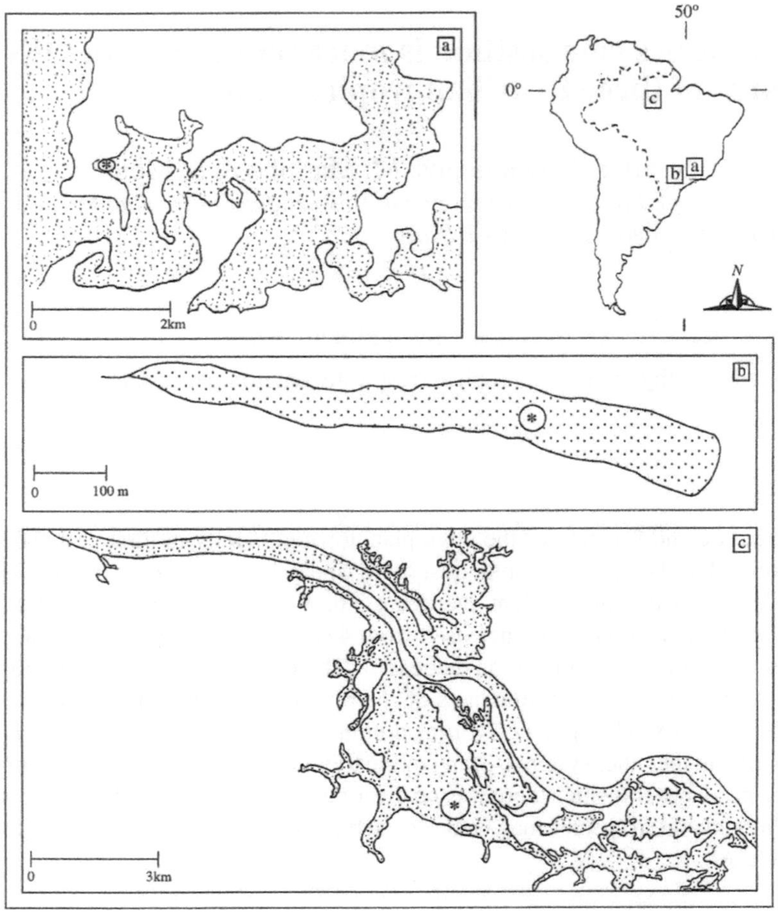

Figure 1. Maps of Barra Lagoon (a), Monte Alegre Reservoir (b) and Batata Lake (c), showing the geographic location and sampling stations.

hence, to distort the interpretation of the relation of the community composition to the trophic status.

The present study discusses some ecological aspects of the phytoplankton composition of three tropical environments, which are taken as representative of the Brazilian aquatic ecosystems, with respect to trophic state and seasonality.

Description of study sites

The main characteristics of the lakes are summarized in Table 1. Batata Lake (Figure 1c) is situated in the Amazon region, of northern Brazil; it is a floodplain lake with clear waters, is moderately acidic and is deficient in most nutrients. It is permanently connected to Trombetas River, a tributary of the Amazon. It is characterized by the distinctive annual fluctuation of its surface level (Huszar, 1994). There is no input of anthropogenic organic matter to the lake (Esteves et al., 1990). However, the lake received bauxite tailings from ore processing, between 1979 and 1989, which now cover 30% of the total lake area at high water (Roland & Esteves, 1993). The ratio N/P and the low concentrations of soluble reactive phosphorus indicate that capacity limitation by phosphorus is more likely. Mixing of the entire water column varies in frequency from daily to several days, depending on the lake depth (Melo, 1996).

Monte Alegre Reservoir (Figure 1b) is situated in the southeast of Brazil. Its waters drain areas of rich soils. It has a relatively long hydraulic residence time and is not managed for water supply. Its waters are slightly acid and the N/P ratio fluctuates, but neither N nor P was ever found to have fallen to growth-limiting

Table 1. Location, morphometrical data and water variables of the lakes (LW = low water; HW = high water).

	Batata Lake		Monte Alegre Reservoir		Barra Lagoon	
Latitude	56°14′W, 1°28′S		47°43′W, 21°11′S		42°43′W, 22°55′S	
Area (km^2)	LW–18; HW–31		0.07		6.2	
Volume (10^{6}m^3)	40		0.2		–	
Mean depth (m)	2.2		2.9		–	
Maximum depth (m)	12		5		1.6	
Residence time (days)	not estimated		45		30	
	Range	Mean	Range	Mean	Range	Mean
Secchi depth (m)	0.9–2.1	1.2	0.9–2.1	1.4	0.2–0.6	0.3
pH	5.1–6.4	5.7	5.3–7.2	6.2	8.1–9.4	9.0
Carbon dioxide (mM)	0.06–0.6	0.16	–	–	–	–
Alkalinity (mEq l^{-1})	0.006–0.111	0.045	–	–	–	–
Water temperature (°C)	26–33	30	20–28	24	22–30	28
Conductivity (μS cm^{-1})	7–16	9.9	66–142	81.2	6905–11392	8956
Soluble reactive phosphorus (μg l^{-1})	3.1–7.5	5.0	2.2 - 41.9	15.0	3.1–232.5	65.4
Dissolved inorganic nitrogen (μg l^{-1})	40–210	124	91–804	295	0–1235	110
Total phosphorus (μg l^{-1})	5–60	20	12–67	33	113–300	190
Total nitrogen (μg l^{-1})	380–1000	559	260 - 840	528	1163–288	1796

concentrations. Mixing of the full water column can occur continuously, daily or irregularly (Arcifa et al., 1990; Silva, 1995). Day-time stratification may occur more frequently in the hot-rainy period.

Barra Lagoon (Figure 1a) is an oligohaline, mildly alkaline coastal lagoon, situated in Southeastern Brazil. It is connected to the sea through two other lagoons. It receives a heavy load of untreated domestic sewage. Mass mortalities of fishes have occurred during summer months. It is a system with continuous or daily mixing, without persistent stratification. No tendency to limitation by phosphorus was registered during the seasonal cycle. However, this system was deficient in dissolved inorganic nitrogen concentrations during most of the year, the exception being in the first three weeks after a mass-mortality of fishes (Domingos et al., 1994).

Materials and methods

Phytoplankton samples were collected at weekly, at stations in the main basin of Batata Lake (Figure 1c), from September 1988 to October 1989, and in Lagoa da Barra (Figure 1a), from October 1990 to June 1991. Samples were taken biweekly at Monte Alegre Reservoir (Figure 1b), April 1988–March 1989). Samples were fixed with Lugol's solution and the phytoplankton were identified and enumerated following the sedimentation and inverted-microscope method of Utermöhl (1958). At least one hundred specimens of the dominant species were enumerated (Lund et al., 1958) in random fields (Uhelinger, 1964). Specific biomass was estimated from the product of the population and the mean unit volume (Edler, 1979). Species diversity (by biomass) was estimated according to the Shannon-Wiener index (Shannon & Weaver, 1963) and the rate of community compositional change (by biomass) was calculated according to Lewis' (1978) summed-difference method. More details about methodology can be found in Huszar (1994), Domingos et al. (1994) and Silva (1995).

Results

Attributes of the phytoplankton communities are presented in Figures 2–7, considering the following phases: Batata Lake-falling (decreasing of water level), low water, rising (increasing of level water), high water; Monte Alegre Reservoir – I and IV hot-rainy, II cold-dry, III hot-dry to hot-rainy (transition); Barra Lagoon – pre-mortality (A and B), mortality (C) and post-mortality (D). Further details are given in Huszar (1994), Domingos et al. (1994) and Silva (1995).

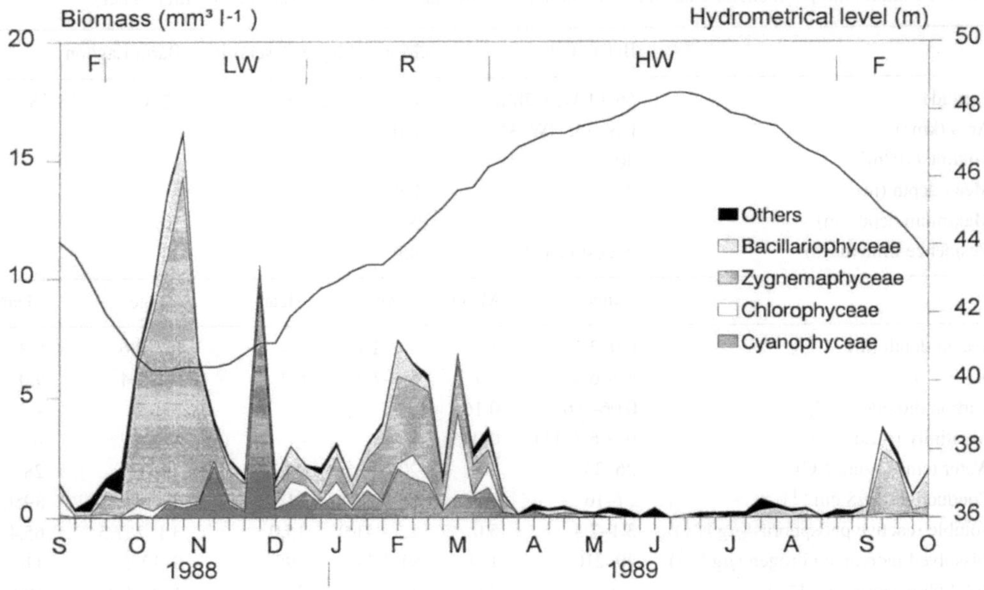

Figure 2. Fluctuations in the phytoplankton biomass (as fresh biovolume) in Batata Lake, subdivided according to phylogenetic affinities, in relation to the level of the Trombetas river and the hydrological phases of the lake. (F = falling; LW = low-water; R= rising; HW = high-water).

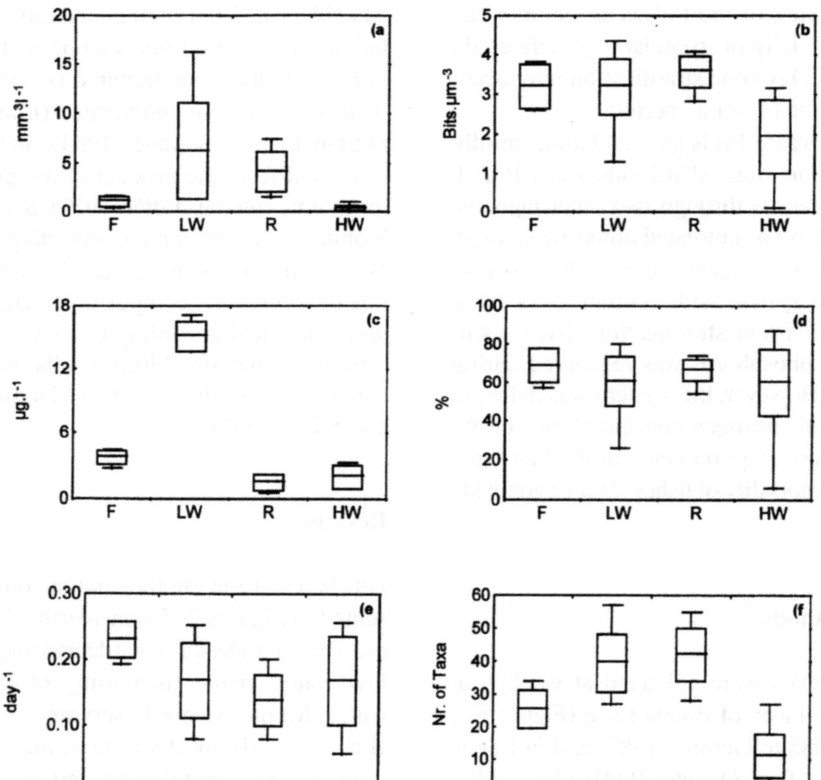

Figure 3. Phytoplankton attributes (a-biomass; b-diversity; c-chlorophyll; d-eveness; e- rate of community change; f-species richness;) of the Batata Lake (range, mean and standard deviation) by phases (F = falling; LW = low water; R = rising; HW = high water).

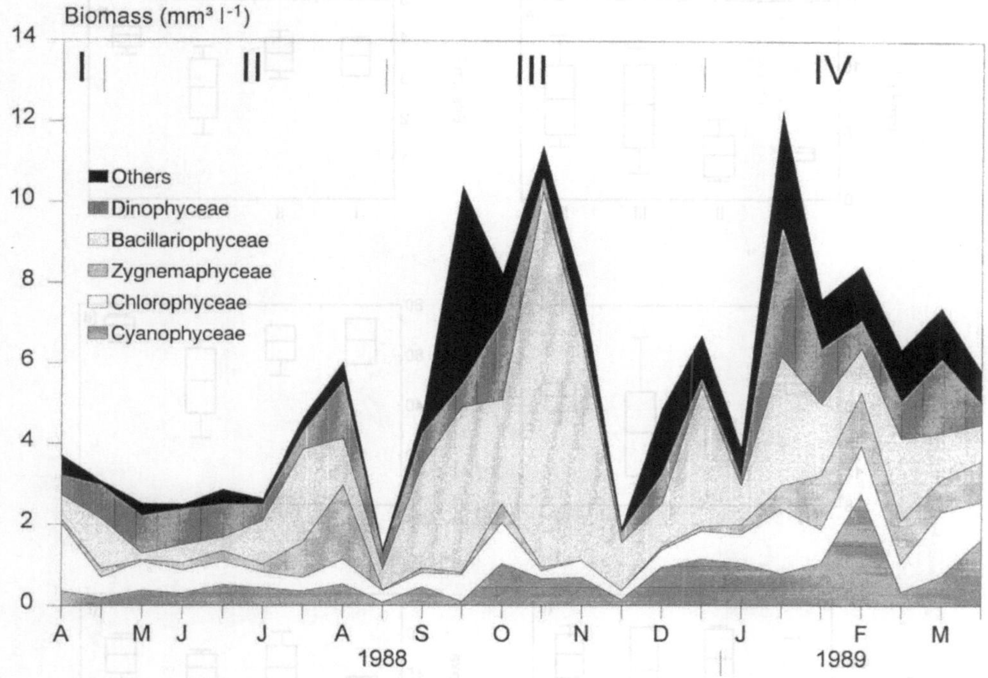

Figure 4. Fluctuations in the phytoplankton biomass (as fresh biovolume) in Monte Alegre Reservoir, subdivided according to phylogenetic affinities, in relation to the phases (I = hot-rainy; II = cold-dry; III = hot-dry to hot-rainy; IV = hot-rainy).

Batata Lake

Under the strong hydrological influence of its flood-plain cycle (Figure 2), Batata Lake was characterised by a season of net growth lasting about seven months (low- and rising- water). In the other months, when the lake was deeper (high- and falling water), biomass was low, mainly as a result of fluvial dilution. Moderate production 427 g C m^{-2} y^{-1} was detected by Roland (1995).

The maximum biomass was around 4.5 times greater than the annual average, and 2.5 times greater than the average for the season of net growth. Three biomass peaks were observed (Figure 2): two in the low water period, one of them with a dominance of desmids – *Staurodesmus triangularis* and *Pleurotaenium tenuissimum* – and the other with a dominance of blue-green algae – *Jaaginema geminatum* and *Oscillatoria* cf. *transvaalensis*. The third peak, with the same desmids plus *Cosmarium arctoum* and *Jaaginema geminatum*, occurred in the rising period. Only the first period could be considered equilibrial, in the sense of Sommer et al. (1993), when two species composed over 80% of the total biomass for about three weeks.

There were 174 species in 58 samples taken over the annual cycle. The phytoplankton assemblages, related to the main dimensions of the trophic spectrum, are expressed in Table 2. Both the species diversity and the species richness were high and these attributes in general fluctuated in a similar manner. Lower values occurred in the high water period. Eveness was almost constant throughout the annual cycle. The rates of community change were high and reached maximal values during the period of falling water (Figure 3).

Organisms < 20μm dominated the community for about seven months (falling and high water); in the rest of the period (low water and rising) species between 20 and 200 μm contributed more than in the other phases. On average, there were about 27 species per sample; 14 contributed over 0.5% to the total biomass, four over 5%, and one over 25%.

The adaptative strategies were separated according to Reynolds (1988). Falling and high water phases are characterized mainly by the dominance of robust and small organisms (< 20 μm; C competitor species); at low water, larger organisms with attenuated or elaborated shapes were more important (20–200μm; R ruderal species), whereas in period of the rising water, both groups occurred together.

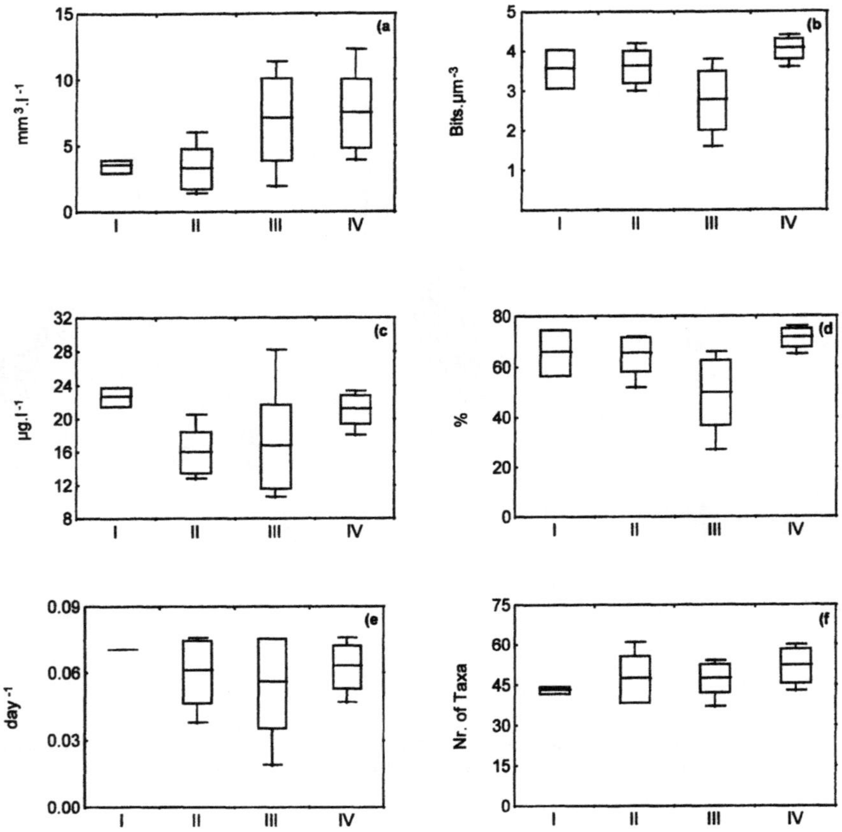

Figure 5. Phytoplankton attributes (a-biomass; b-diversity; c-chlorophyll; d-eveness; e- rate of community change; f-species richness;); of the Monte Alegre Reservoir (range, mean and standard deviation) by phases (I = hot-rainy; II = cool-dry; III = hot-dry to hot-rainy; IV = hot-rainy).

Monte Alegre Reservoir

High phytoplankton biomass (Figure 4) and high primary production (780 gC m^{-2} y^{-1}, Meschiatti et al., 1986) were supported by Monte Alegre Reservoir. The maximum biomass was about 2.2 times higher than the annual mean and there were three seasonal biomass peaks: 1) one in the cold-dry period, when many species contributed similar percentages to the total biomass; 2) one in the hot-dry to hot-rainy transition, composed of diatoms, euglenophyceae and cryptomonads; and 3) in the hot-rainy period, when many species contributed with similar biomass. It was not possible to identify an equilibrium period according to the criteria of Sommer et al. (1993).

A hundred and eighteen species were recorded in Monte Alegre Reservoir. The phytoplankton assemblages and the main dimensions of the trophic spectrum are shown in Table 3. Diversity and species richness were relatively high, in contrast to the rates of community change (Figure 5). The phytoplankton biomass was dominated by species of microplankton (20–200 μm). At an average of about 48 species per sample, 22 contributed more than 0.5% to the total biomass, four with more than 5% and only 1 species with more than 25% (*Aulacoseira granulata).* C and R species dominated, together, most of the year. When stratification was interrupted by mixing periods, the R species, *Aulacoseira granulata*, was dominant.

Barra Lagoon

The Barra Lagoon was found to be a very productive system (918 gC m^{-2} y^{-1}, Carmouze et al., 1994), experiencing a mass fish mortality in the summer, when the phytoplankton biomass reached its maximum (Figure 6). The maximum biomass was about 2.1 times bigger than the average biomass. Its seasonality could be defined on account of this characteristic (Domingos et al., 1994). Four peaks were identified: two in the pre-mortality period, both dominated by small blue-greens (*Synechocystis aquatilis* f. *salina*

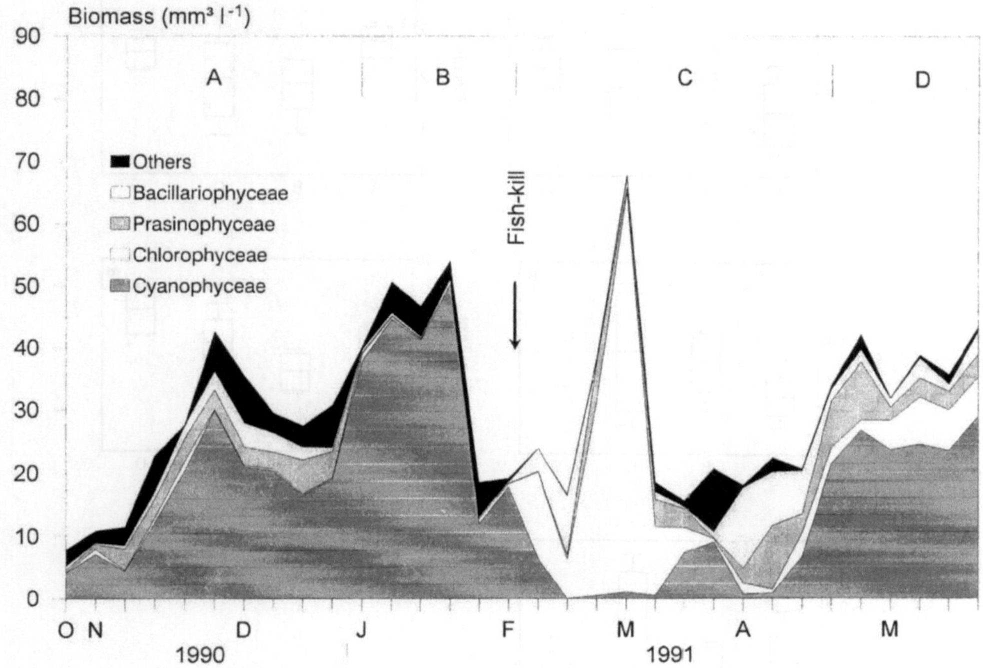

Figure 6. Fluctuations in the phytoplankton biomass (as fresh biovolume) in Barra Lagoon, subdivided according to phylogenetic affinities, in relation to the phases (A and B = pre-mortality; C = fish mortality; D = post-mortality).

Table 2. Biomass averages and phytoplankton assemblages related to the averages of the trophic spectrum in different water phases of Batata Lake.

	Falling	Low-water	Rising	High-water
Dissolved inorganic nitrogen ($\mu g\ l^{-1}$)	243.0	134.0	128.0	94.0
Soluble reactive phosphorus ($\mu g\ l^{-1}$)	3.1	4.4	5.0	7.5
pH	6.0	5.6	5.9	5.4
Alkalinity ($mEq\ l^{-1}$)	0.032	0.016	0.057	0.077
N/P	60	38	24	60
Biomass ($mm^3\ l^{-1}$)	0.49	7.63	2.23	0.12
Assemblages	1. – *Cyclotella stelligera* – *Urosolenia eriensis* – *Merismopedia tenuissima*	2. – *Staurodesmus triangularis* – *Pleurotaenium tenuissimum* – *Staurastrum pseudotetracerum* – *Aulacoseira granulata* 3. – *Jaaginema geminatum* – *Oscillatoria* cf. *transvaalensis*	4. – *S. triangularis* – *P. tenuissimum* – *Cosmarium arctoum* – *J. geminatum* – *M. tenuissima* 5. –*Botryococcus fernandoi* – *C. arctoum* – *J. geminatum* – *M. tenuissima* – *Peridinium umbonatum*	6. – *Cryptomonas* cf. *marsonii* – *C. stelligera* – *Chromulina gyrans* – *P. umbonatum*

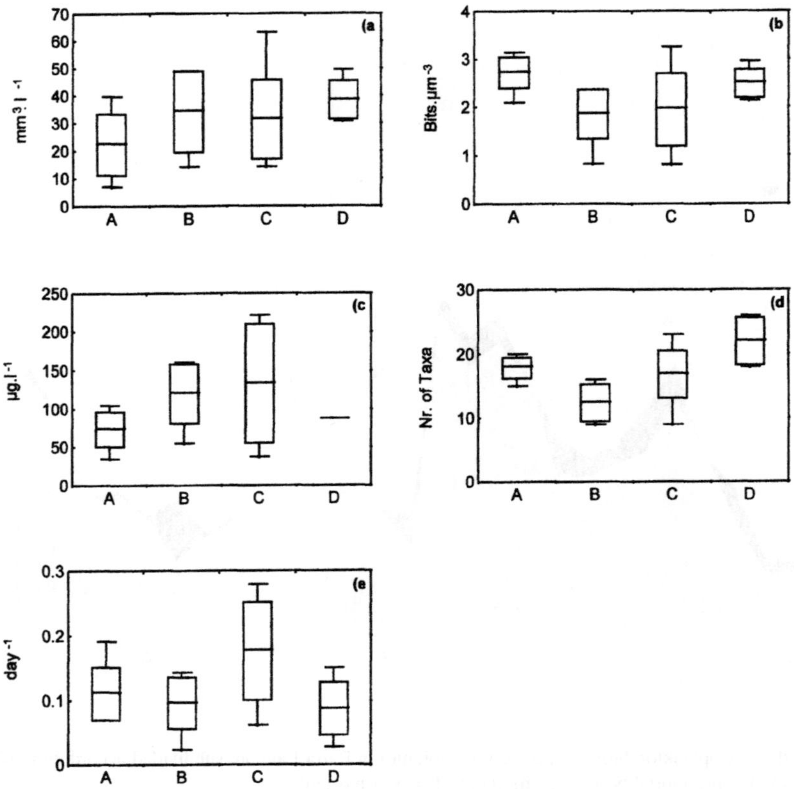

Figure 7. Phytoplankton attributes (a-biomass; b-diversity; c-chlorophyll; d-species richness; e-rate of community change) of the Barra Lagoon (range, mean and standard deviation) by phases (A and B = pre-mortality; C = fish mortality; D = post-mortality).

Table 3. Biomass averages and phytoplankton assemblages related to the averages of the trophic spectrum in different water phases of Monte Alegre reservoir.

	Cold-dry period	Hot-dry to hot-rainy period	Hot-rainy period
Dissolved inorganic nitrogen ($\mu g\ l^{-1}$)	445.2	128.0	219.3
Soluble reactive phosphorus ($\mu g\ l^{-1}$)	14.5	19.1	18.0
pH	6.1	6.3	6.1
N/P	–	13	10
Biomass ($mm^3\ l^{-1}$)	4.2	7.0	7.4
Assemblages	1. – *Aulacoseira granulata*	2. – *Lepocinclis ovum*	4. – *A. granulata*
	– *Peridinium umbonatum*	– *A. granulata*	– *P. umbonatum*
	– *P. volzii*	– *P. volzii*	– *Actinotaenium perminutum*
	– *Staurastrum tetracerum*		– *Cryptomonas marssonii*
	– *Staurodesmus cuspidatus*	3. – *A. granulata*	– *Phacus longicauda*
	– *Staurodesmus phimus*		– *Microcystis aeruginosa* f. *flos aquae*

and *Synechococcus elongatus*); only the second peak was considered to be an equilibrium period, according to the criteria of Sommer et al. (1993). One of the other two peaks occurred about one month after the mortality, when *Chlorella vulgaris* var. *autotrophica*, *C. minutissima* and *Scenedesmus intermedius* were dominant. The final peak was recorded three months after the fish mortality, and lasted until the end of the

Table 4. Biomass averages and phytoplankton assemblages related to the averages of the trophic spectrum in different water phases of Barra Lagoon.

	Pre-mortality		Fish-mortality	Post-mortality
	A	B	C	D
Dissolved inorganic nitrogen (μg l^{-1})	40.9	23.1	247.1	–
Soluble reactive phosphorus (μg l^{-1})	21.7	30.5	159.0	–
N/P	13.2	8.2	8.3	–
pH	9.1	9.2	8.8	–
Biomass (mm^3 l^{-1})	22.4	34.3	31.3	38.3
Assemblages				
	– *Synechocystis aquatilis* f. *salina* – *Synechococcus elongatus* – *Pavlova lutherii* – *Pyramimonas grossii*	– *S. aquatilis* f. *salina* – *S. elongatus* – *P. lutherii*	– *Chlorella minutissima* – *C. vulgaris var. autotrophica* – *Scenedesmus intermedius* – *Actinocyclus norweggii var. subsalsa*	– *S. aquatilis* f. *salina* – *C. vulgaris var. autotrophica* – *Oocystis submarina* – *P. grossii*

study, with dominance of *Synechocystis aquatilis* f. *salina*. The toxicity of *S. aquatilis* f. *salina* was tested in mice bioassays by Azevedo & Carmouze (1994). They suggested that this small blue-green, in high biomass during its senescent phase, could have produced haemolitic toxins leading to the mass mortality of the filter-feeding fish (*Brevoortia tyrannus*, Clupeideae).

The planktonic flora of the Barra Lagoon comprised 72 species during the year of study (Menezes & Domingos, 1994). Phytoplankton assemblages and their relevance to the trophic spectrum are set out in Table 4. The diversity, species richness and community change rates were relatively low through the year (Figure 7). The post-mortality change rate, however, was high.

Between 90 and 100% of the phytoplankton biomass comprised small and stout organisms < 20 μm, with 70 to 90% being < 10 μm. *Synechocystis aquatilis* f. *salina* – a characteristic species throughout the year – can be considered a C species because of its short generation time of (ca. 1.7 days, Nascimento, 1996). The small green-alga recorded after mortality can be also considered a C species. The mean of the net rate of increase, estimated according Sommer (1988) in natural population (29 °C, 12 hr d^{-1}) by daily sampling during six days, was 0.8 d^{-1} for *Chlorella minutissima*, and 1.4 d^{-1}, for *C. vulgaris* var. *autotrophica*.

Out of an average of 17 species per sample, 12 contributed more than 0.5% to the total biomass, four more than 5% and only one species more than 25% (*Synechocystis aquatilis* f. *salina*).

Discussion

The majority of the tropical lakes tends to be included among the eutrophic category because of the lack of obvious light limitation, of the small variations of the photoperiod and temperature through the year, and of the regularity of these factors (Thienemann, 1955; Esteves, 1988). Such characteristics impose metabolic patterns that differentiate systems in tropical and temperate regions. However, the concept of trophic status was evolved for temperate lakes and even this is almost wholly evaluated by phosphorus concentration (Moss, 1973; OECD, 1982). Many attributes of the phytoplankton community, however, can be related to the trophic status. With increasing trophy, phytoplankton biomass usually shows larger seasonal fluctuations (Marshall & Peters, 1989) and the fluctuations themselves may often be related to physical variability (Seip & Reynolds, 1995). In each of the present cases, there were three or four seasonal peaks per calendar year and the observed cyclicity is most clearly correlated to physical variables (Figures 2, 4 and 6).

The ratios of maximum biomass/mean annual biomass were similar in all the three environments. The maximum and mean biomass, and annual production per unit area were higher in the coastal lagoon than in the other systems; however, richness and diversity were smaller. Community change rates were more striking in the floodplain lake than in the other systems, which can be related mainly to the protracted flood pulse to which it is submitted. Relatively low rates occurred in the reservoir and in the coastal lagoon, except immediately after the fish mortality, when they were high (Figure 7).

The morphological/functional strategies of the phytoplankton organisms, established by Margalef (1978) and Reynolds (1988), seem to co-vary loosely with the trophic gradient. The S stress-tolerant species dominate when nutrients become depleted or when they are chronically in low concentrations; C species do the opposite; and R species are tolerant of low light and many live in eutrophic environments (Reynolds, 1988). We observed a dominance of C species throughout the year in the coastal lagoon; coexistence of C and R or dominance of only R species in the reservoir; and alternation of C species, during high water, with C and R species, in the other periods in the floodplain lake, seem to indicate this tendency. Our observations are consistent with the tendencies proposed.

The occurrence of smaller organisms in oligotrophic waters, and progressively larger organisms with increasing nutrient status has been noted by Harris (1994). A trend to dominance by smaller organisms in the oligotrophication process was observed by Willén (1985). No such trend was evident in the lakes. The coastal lagoon, with high levels of nutrients, presented dominance of small algae (pico/nanoplankton). On the other hand, in the reservoir and during the season of net growth in the floodplain lake, both with a moderate concentration of nutrients, microplankton algae contributed a bigger share to the biomass.

The criteria given in literature for the classification of the trophic states of lakes, including OECD (1982), introduce problems of interpretation, when interegional comparisons are made. Taking into account the OECD (1982) criteria regarding the total-P levels, Batata Lake and Monte Alegre Reservoir could be classified as mesotrophic and Barra Lagoon as hypereutrophic. Considering the concentration of chlorophyll *a*, the first two systems could be included in the eutrophic category and the coastal lagoon in the hypertrophic category. Considering the mean and minimum transparency of the water, all environments would be hypertrophic. The existence of a trophic gradient is a fact, but the delimitation of each level is not always clear. Furthermore, extreme situations are more easily defined than intermediate levels, because the latter are found sometimes in one category, sometimes in another, depending on the criteria.

The assemblages of species can be important to the resolution of the trophic status, since the species indicate the complexity of the processes in the system. In general, assemblages with oligo-mesotrophic tendencies were registered in Batata Lake (Huszar & Reynolds, 1997); for example, *Cyclotella/ Urosolenia/ Merismopedia*, are more frequently associated with oligotrophic and moderately eutrophic systems (Reynolds, 1984; Reynolds, 1998). The *Staurodesmus/ Pleurotaenium/ Staurastrum/ Cosmarium* grouping, largely known as dominants in phosphorus-poor environments. Eutrophic species however, were also represented as *Aulacoseira granulata*, which is typical of eutrophic environments of low latitudes (Kilham & Kilham, 1975). Despite its relatively broad distribution in the growth period, this species has always occurred in relatively low biomass. *Jaaginema* and *Oscillatoria*, recorded in the lake at the low water phase, belong to an assemblage considered to be eutrophic and characteristic of shallow, turbid and well-mixed environments (Alvarez-Cobelas & Jacobsen, 1992; Reynolds, 1994). The dominance of oligotrophic desmids was a clear feature of the growth period in the floodplain lake, coinciding with the high availability of free CO_2 (0.16 mM). These assemblages correspond to the drainage characteristics of Batata Lake and the nutrient-poor low alkalinity waters of the Trombetas river (Roland, 1995).

The species that contributed most to the phytoplankton biomass of Monte Alegre Reservoir featured assemblages of the type *Aulacoseira/ Staurastrum/ Staurodesmus/ Peridinium* in cold-dry period; *Lepocinclis/Aulacoseira* in the transition period, or almost exclusive dominance of the latter, after mixings of water column. *Aulacoseira/ Cryptomonas/ Phacus/ Actinotaenium/Peridinium/Microcystis* characterized the hot-rainy period with more prolonged stratification. Except for some desmids species, the other taxa registered in these assemblages are said to be characteristic of mesotrophic and eutrophic environments (Reynolds, 1997). In fact, the watershed of Monte Alegre Reservoir is located in naturally-rich agricultural soils. Periodical and irregular mixing happened throughout the annual cycle, which may explain the absence of dominance by species such as *Microcystis aeruginosa*, characteristic of the late summer in eutrophic environments of temperate areas (Reynolds, 1980), and in stable phases of hot-rainy periods in Brazilian tropical reservoirs (Tundisi, 1990). The mixing contributed to the dominance of *A. granulata*, typical species of well-mixed eutrophic environments (Lima et al., 1979).

In Barra Lagoon, non nitrogen-fixing Chroococcales (*Synechocystis/ Synechococcus*) extensively dominated the phytoplankton. High pH values and low free CO_2 may account for dominance by such species

(Shapiro, 1990). Phytoplankton assemblages reflected the richness of the drainage basin (high availability of nutrients). In addition, the acknowledged euryhalinity of *Synechocystis aquatilis* f. *salina*, *Synechococcus elongatus*, *Pyramimonas grossii* and *Pavlova lutherii* (Menezes & Domingos, 1994) suggest the selectivity of a marine influence. Assemblages found in the Barra Lagoon were formed by species of blue-green algae, flagellates and Chroococcales, dominant groups in hypertrophic environments (Alvarez-Cobelas & Jacobsen, 1992). The pre-mortality was characterized by the *Synechocystis / Synechococcus / Pavlova / Pyramimonas* assemblage and the mortality, by *Chlorella / Scenedesmus / Actinocyclus* or *Chaetoceros / Pyramimonas*. After the return of the similar initial conditions, the assemblages become similar too, with high biomass of *Synechocystis / Synechococcus* supported by some *Oocystis / Actinocyclus*. The exceptional fish mortality resulted in an important elevation in the concentration of NH_4^+ ions, from the decomposition of fish mass, and a fall in pH (Carmouze et al., 1994), promoting the immediate substitution of the blue-green dominance by *Chlorella / Scenedesmus / Actinocyclus* assemblage.

Blue-green algae at high biomass have often been considered characteristic of eutrophic environments. These algae, however, are not exclusive to productive environments. They can be an important part of the plankton of non-eutrophic environments as, for example, some large African lakes that are not very productive (Hecky & Kling, 1987). This is also shown in the trophic spectrum synthesized by Reynolds (1997). In the floodplain lake, blue-greens contributed with about 5 to 30% of the biomass in the season of net growth, dominating for a week.

In shallow tropical environments, nutrient availability does not provide a good indicator of the trophic level. The concentrations of dissolved inorganic nutrients fractions do not represent the true availability of resources to the organism metabolism, owing to high cycling and uptake rates (Esteves, 1988). In Brazilian coastal lagoons, for example, around 70% of the organic matter is recycled in the water column (Machado, 1989; Carmouze et al., 1994). Inadequacy of the transparency as an indicator of the trophic nature of these systems is directly influenced by resuspension of fine material from shallow sediments. Thus, the effect of the nutrient availability on the phytoplankton, usually defining the trophic nature of these systems, is greatly modified by extensive shallows. Besides, the effect of the nutrient availability over the phytoplankton is usually difficult to evaluate owing to the difficulty of its detection, especially of SRP, and to the low levels actually required to saturate the growth of algae (Reynolds, 1992).

It is important to point out that biomass, productivity, and chlorophyll *a*, all attributes which are also used to evaluate the trophic states of environments, are usually inadequate to characterise shallow tropical lakes. According to current terminology, based substantially on the system evolved for deep, temperate lakes, many shallow tropical lakes would be determined to be eutrophic.

Moss's (1973) diagram, which is based on the distribution of phytoplankton species and on environmental factors (especially the chemical composition of water), permits systems which are not easily classified by their nutrient concentration to be assimilated more satisfactorily. From his diagram, it is possible to include lakes with high productivity and high biomass made up of assemblages usually considered to be oligotrophic. The floodplain lake studied, considered by the OECD (1982) system as mesotrophic (by total-P) or eutrophic (by chlorophyll a), supports dominant assemblages of oligo-mesotrophic species. A similar situation occurred in the reservoir, whose classification by the OECD criteria, permits its inclusion it as mesotrophic (total-P) or eutrophic (chlorophyll *a*), but considering Moss's diagram, the reservoir classification was better defined as eutrophic, due to the larger contribution of species mentioned as typical of these environments (Reynolds, 1998).

The pH/CO_2 system contributes to the definition of the assemblages, clearing situations that were, at first sight, those other trophic states. The CO_2 system (Talling, 1976) and the competition for dissolved inorganic carbon (Williams & Turpin, 1987) play an important role determining the qualitative composition of phytoplankton and the capacity to use bicarbonate as a carbon source is a fundamental ability of eutrophic species (Moss, 1973). It is important to point out, however, the need for a clear definition of the concept of oligotrophic and eutrophic species. The approach of active inorganic carbon transport and its concentrating mechanisms (Williams & Turpin, 1987; Raven, 1991) perhaps can be invoked.

In this way, Moss's diagram seems to be the conceptual model that included more adequate attributes to the trophic classification of the Brazilian tropical lakes analyzed, since it takes into account both strategies of phytoplankton species, from the availability of free CO_2 and bicarbonates (Thienemann-Naumann's original sense), and the nutrient availability and the productivity of these systems (recent usage). For the shallow tropical environments, it is also important to consider the physical factors, such as the river influence and periodical mixing of the water column, that affect the variables used in the trophic determination.

Acknowledgements

To Dr. Francisco Esteves, Dr. Marlene Arcifa and Dr. Jean-Pierre Carmouze for the opportunity to study these interesting environments. To Dr. C.S.Reynolds for his comments on the manuscript. To CNPq, Brasília, Brasil; CAPES, Brasília, Brasil; FUJB-UFRJ, Rio de Janeiro, Brasil; MRN, Rio de Janeiro, Brasil; FAPESP, São Paulo, Brasil; UNESCO, Uruguay; and ORSTOM, France for financial support.

References

Alvarez-Cobelas, M. & B. A. Jacobsen, 1992. Hypertrophic phytoplankton: an overview. Freshwat. Forum 2: 184–199.

Alvarez-Cobelas, M. & C. Rojo, 1994. Spatial, seasonal and long term variability of phytoplankton photosynthesis in lakes. J. Plankton Res. 16: 1691–1716.

Arcifa, M. S., A. J. Meschiatti & E. A. T. Gomes, 1990. Thermal regime and stability of a shallow tropical reservoir: Lake Monte Alegre, Brazil. Rev. Hydrobiol. trop. 23: 271–281.

Azevedo, S. M. F. O. & J. P. Carmouze, 1994. Une mortalité de poissons dans une lagune tropicale (Brésil) durant une période de dominance de Cyanophycées. Coïncidence ou conséquence? Rev. Hydrobiol. trop. 27: 265–272.

Carmouze, J. P., B. Farias & P. Domingos, 1994. Evolution du metabolisme d'une lagune tropicale (Brésil) au cours d'une période marquée por une mortalité de poissons. Rev. Hydrobiol. trop. 27: 199–215.

Domingos, P., V. L. M. Huszar & J. P. Carmouze, 1994. Composition et biomasse du phytoplankton d'une lagune tropicale (Brésil) au cours d'une période marquée par une mortalité de poissons. Rev. Hydrobiol. trop. 27: 235–250.

Edler, L. (ed.), 1979. Recommendations for marine biological studies in the Baltic Sea. Phytoplankton and chlorophyll. (Unesco, Working Group 11, Baltic Marine Biologists), 38 pp.

Esteves, F. A., 1988. Considerações sobre a aplicação da tipologia de lagos temperados a lagos tropicais. Acta Limnol. bras. 2: 3–28.

Esteves, F. A., R. Bozelli & F. Roland, 1990. Lago Batata: um laboratório de limnologia tropical. Ciência Hoje 64: 26–31.

Harris, G. P., 1994. Pattern, process and prediction in aquatic ecology: a limnological view of some general ecological problems. Freshwat. Biol. 32: 143–160.

Hecky, R. & H. J. Kling, 1987. Phytoplankton ecology of the great lakes in the rift valleys of central Africa. Arch. Hydrobiol. 25: 197–228.

Huszar, V. L. M., 1994. Fitoplâncton de um lago amazônico impactado por rejeito de bauxita: estrutura da comunidade, flutuações espaciais e temporais. Doctoral thesis, Universidade Federal de São Carlos, São Carlos, 219 pp.

Huszar, V. L. M. & C. S. Reynolds, 1997. Phytoplankton periodicity and sequences of dominance in an Amazonian flood-plain lake (Lago Batata, Pará, Brazil): responses to gradual environmental change. Hydrobiologia 346: 169–181.

Kalff, J. & R. Knoechel, 1978. Phytoplankton and their dynamics in oligotrophic and eutrophic lakes. Annu. Rev. Ecol. Syst. 9: 475–495.

Kilham, S. S. & P. Kilham, 1975. *Melosira granulata* (Ehr.) Ralfs: morphology and ecology of a cosmopolitan freshwater diatom. Verh. int. Ver. Limnol. 19: 2716–2721.

Lewis, W.M. Jr., 1978. Analysis of succession in a tropical phytoplankton community and a new measure of succession rate. Am. Nat. 112: 401–414.

Lima, W. C., J. G. Tundisi & M. Marins, 1979. A systemic approach to the sensitivity of *Melosira italica* (Ehr.) Kutz. Revta. bras. Biol. 39: 559–563.

Lund, J. W. G., C. Kipling & E. D. LeCren, 1958. The inverted microscope method of estimating algal number and the statistical basis of estimating by couting. Hydrobiologia 11: 143–170.

Machado, E. C., 1989. Desoxigenação e regeneração de nutrientes na lagoa de Guarapina, Rio de Janeiro. Dissertation, Universidade Federal Fluminense, Niterói, 130 pp.

Margalef, R., 1978. Life-forms of phytoplankton as survival alternatives in an unstable environment. Oceanol. Acta. 1: 493–509.

Marshall, C. T. & R. H. Peters, 1989. General patterns in the seasonal development of chlorophyll *a* for temperate lakes. Limnol. Oceanogr. 34: 856–867.

Melo, S., 1996. Influência do ciclo hidrológico sobre as variações nictemerais do fitoplâncton de um lago amazônico (Lago Batata, Pará, Brasil). Dissertation, Universidade Federal do Rio de Janeiro, Rio de Janeiro, 61 pp.

Menezes, M. & P. Domingos, 1994. Flore planctonique d'une lagune tropicale (Brésil). Rev. Hydrobiol. trop. 27: 273-297.

Meschiatti, A. J., E. A. T. Gomes, L. V. Lang, R. A. Rodrigues & R. M. Hamasaki, 1986. Estudos limnológicos no lago da fazenda Monte Alegre, Ribeirão Preto, SP. Dissertation, Universidade de São Paulo, Ribeirão Preto, 133 pp.

Moss, B., 1973. The influence of environmental factors on the distribution of freshwater algae: an experimental study. IV. Growth of test species in natural lake waters and conclusion. J. Ecol. 61: 193–211.

Nascimento, S. M., 1996. Um estudo ecofisiológico de *Synechocystis aquatilis* f. *salina*: a razão N/P e sua relação com o crescimento e a atividade fotossintética. Dissertation, Universidade Federal Fluminense, Niterói, 74 pp.

OECD, 1982. Eutrophication of waters. Monitoring, assessment and control. OECD, Paris, 154 pp.

Raven, J. A., 1991. Implications of inorganic carbon utilization: ecology, evolution and geochemistry. Can. J. Bot. 69: 908–924.

Reynolds, C. S., 1980. Phytoplankton assemblages and their periodicity in stratifying systems. Holarct. Ecol. 3: 141–159.

Reynolds, C. S., 1984. Phytoplankton periodicity: the interactions of form, function and environmental variability. Freshwat. Biol. 14: 111–142.

Reynolds, C. S., 1988. Functional morphology and adaptive strategies of freshwater phytoplankton. In Sandgren C. D. (ed), Growth and survival strategies of freshwater phytoplankton. Cambridge Univ. Press, Cambridge: 388–433.

Reynolds, C. S., 1992. Eutrophication and the management of planktonic algae: What Vollenweider couldn't tell us. In Sutcliffe, D. W. & J. G. Jones (eds), Eutrophication: research and application to water supply. Freshwater Biological Association, Ambleside: 4–29.

Reynolds, C. S., 1994. The long, the short and the stalled: on the attributes of phytoplankton selected by physical mixing in lakes and rivers. Hydrobiologia 289: 9–21.

Reynolds, C. S., 1998. What factors influence the composition of phytoplankton in lakes of different trophic status. Hydrobiologia 369/370: 11–26.

Roland, F., 1995. Produção primária fitoplanctônica em um lago amazônico impactado por rejeito de bauxita, Lago Batata, Pará, Brasil. Doctoral thesis. São Carlos, Universidade Federal de São Carlos, 118 pp.

Roland, F. & F. A. Esteves. 1993. Dynamics of phosphorus, carbon and nitrogen in Amazonian lake impacted by bauxite tailing (Batata Lake, Pará, Brasil). Verh. int. Ver. theor. angew. Limnol. 25: 925–930.

Seip, K. L. & C. S. Reynolds, 1995. Phytoplankton functional attributes along trophic gradients and season. Limnol. Oceanogr. 40: 589–597.

Shannon, C. E. & W. Weaver, 1963. The mathematical theory of communication. Univ. Illinois Press, Urbana, 173 pp.

Shapiro, J., 1990. Current beliefs regarding dominance by bluegreens: the case for the importance of CO_2 and pH. Verh. int. Ver. Limnol. 24: 38–54.

Silva, L. H. S., 1995. Variabilidade temporal na estrutura da comunidade fitoplanctônica de um reservatório eutrófico (Lago Monte Alegre) Ribeirão Preto, São Paulo, Brasil. Dissertation, Universidade Federal do Rio de Janeiro, Rio de Janeiro, 141 pp.

Smith, V. H., 1978. Nutrient dependence of primary production in lakes. Limnol.Oceanogr. 4: 1051–1064.

Sommer, U., 1988. Growth and survival strategies of planktonic diatoms. In Sandgren C. D. (ed), Growth and survival strategies of freshwater phytoplankton. Cambridge Univ. Press, Cambridge: 227–260.

Sommer, U., J. Padisák, C. S. Reynolds & P. Juhász-Nagy, 1993. Hutchinson's heritage: the diversity-disturbance relationship in phytoplankton. Hydrobiologia 249: 1–7.

Talling, J. F., 1976. The depletion of carbon dioxide from lake water by phytoplankton. J. Ecol. 64: 79–121.

Thienemann, A., 1955. Die Binnengewässer in Natur und Kultur. Springer-Verlag, Berlin, 155 pp.

Tundisi, J. G., 1990. Distribuição espacial, seqüência temporal e ciclo sazonal do fitoplâncton em represas: fatores limitantes e controladores. Revta bras. Biol. 50: 937–955.

Uhelinger, V., 1964. Étude statistique des méthodes de dénobrement planctonique. Arch. Sci. 17: 121–123.

Utermöhl, H., 1958. Zur Vervollkommnung der quantitativen Phytoplankton-Methodik. Mitt. int. Ver. Limnol. 9: 1–38.

Willén, E., 1985. Phytoplankton and reversed eutrophication in lake Mälaren, central Sweeden, 1965–1983. Br. Phycol. J. 22: 193–208.

Williams, T. G. & D. H. Turpin. 1987. Photosynthetic kinetics determine the outcome of competition for dissolved inorganic carbon by freshwater microalgae: implications for acidified lakes. Oecologia 73: 307–311.

Hydrobiologia **369/370**: 73–87, 1998.
M. Alvarez-Cobelas, C. S. Reynolds, P. Sanchez-Castillo & J. Kristiansen (eds), Phytoplankton and Trophic Gradients.

Phytoplankton communities in relation to trophic status in lakes from Hope Bay (Antarctic Peninsula)

Irina Izaguirre*, Alicia Vinocur, Gabriela Mataloni & Mónica Pose
Depto. de Ciencias Biológicas, Facultad de Ciencias Exactas y Naturales, Universidad de Buenos Aires, 1428 Buenos Aires, Argentina

Key words: Antarctica, lakes, phytoplankton, trophic status, typology

Abstract

Nine lakes and ponds of Hope Bay (Antarctic Peninsula), with different trophic status were surveyed during four summer periods between 1991 and 1996. Some limnological features and the structure of their algal assemblages were strongly influenced by the eutrophication caused by sea-birds activity in their areas. Such differences among lakes were evidenced by the results of Principal Component Analyses, based on both physico-chemical and phytoplankton data. Oligotrophic lakes showed a scarce phytoplankton, commonly dominated by small flagellated Chrysophyceae, and a well developed phytobenthos. Water bodies enriched by ornithogenic nutrients revealed an abundant phytoplankton usually dominated by Chlorophyceae (Volvocales) and planktonic Cyanobacteria. A study on one particular lake showed that summer changes in the phytoplankton community were mainly affected by physical phenomena such as variation in light intensity, and thawing, mixing and freezing processes.

Introduction

Hope Bay area, located on the tip of the Antarctic Peninsula, encompasses Base Antártica Esperanza, the northernmost continental Antarctic station (63°23′ S, 56°59′ W). This area includes several lakes, ponds and streams. Algal communities of these environments have been studied from different points of view since 1991. Early research was focused on the phytoplankton of the main water body of Hope Bay, Boeckella Lake (Izaguirre et al., 1993). More recent papers included an ample floristic inventory of the phytoplanktonic and epilithic algae from lakes and streams (Vinocur & Izaguirre, 1994; Tell et al., 1995; Vinocur & Pizarro, 1995).

Information about most lakes is, nevertheless, scarce. Corte (1962) described some limnological features of Boeckella, Esperanza and Flora lakes and provided a general list of their algal species. Later on, Drago & Paira (1987) surveyed these lakes along with Chico Lake, in order to describe their morphometry and the geomorphological constitution of the area.

Similarly to what occurs in other Antarctic areas, the different trophic status of the lakes and ponds of Hope Bay depends mainly on their size and proximity to bird colonies. These water bodies show overall similarities with others subjected to comparable conditions. As described by Hawes (1990), the most oligotrophic lakes have reduced phytoplankton populations, and the vegetation is dominated by slow growing, long-lived benthos. Conversely, phytoplankton development is substantial in nutrient enriched lakes.

This paper provides a preliminary characterization of all main water bodies around Hope Bay, based on abiotic factors and the structure of their phytoplanktonic communities. The influence of different environmental conditions and the morphometric features of the lakes on the composition and abundance of phytoplankton is also discussed.

Study Area

Hope Bay is included in the Northern Province of the Maritime Antarctic Region, having a cold moist mar-

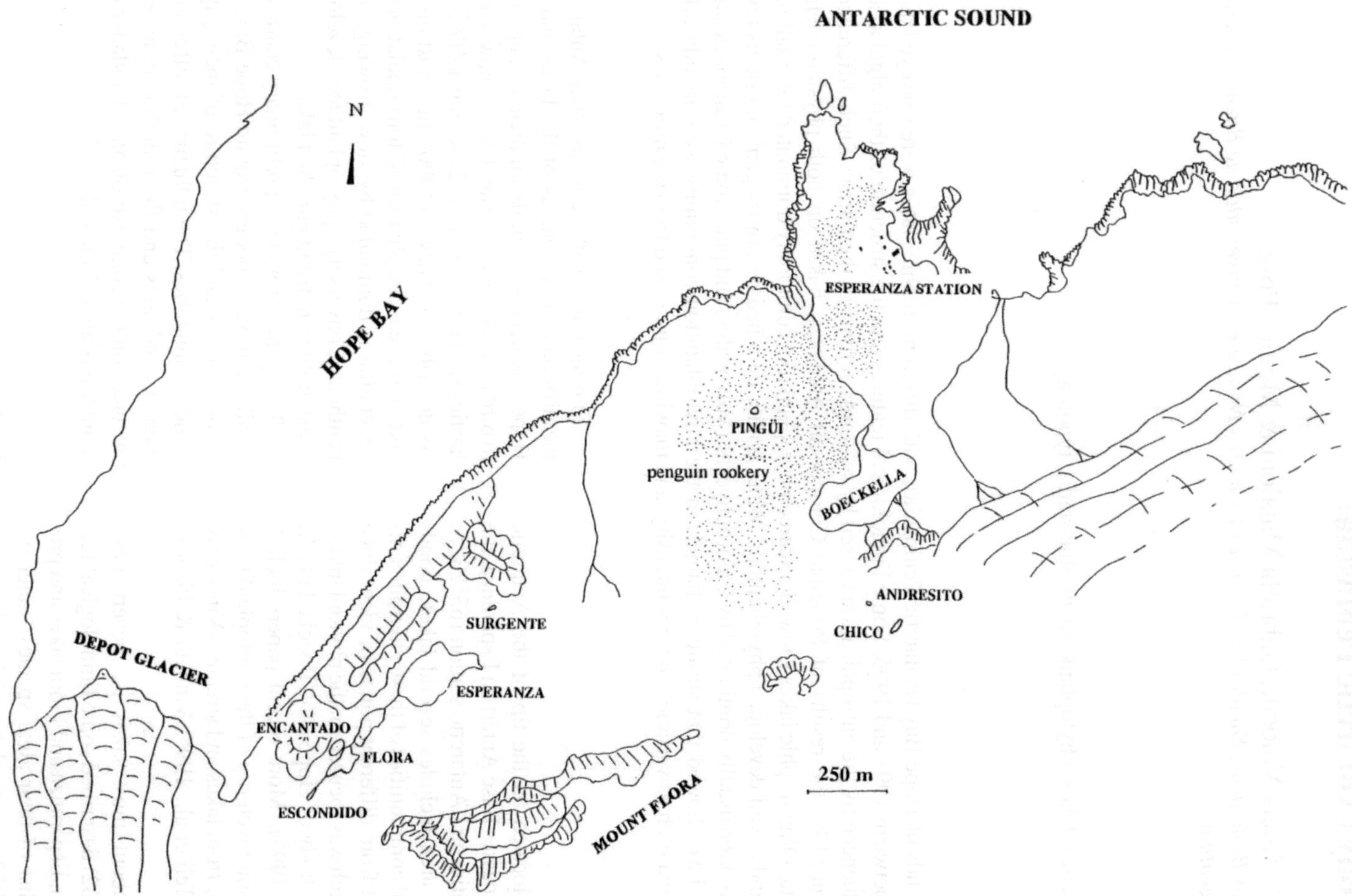

Figure 1. Geographic location of the lakes surveyed.

itime climate (Lewis Smith, 1984). The southwestern margin of the bay is characterized by the presence of many water bodies, mostly of glacial origin (Figure 1). Some of them can be considered typical lakes, since they do not freeze to the bottom and are generally free of ice in summer. On the other hand, some very shallow water bodies, which show a permanent ice-cover all year round, can be regarded as ponds or pools.

As in other areas of the west coast of Antarctic Peninsula, freshwater environments of Hope Bay support a relatively rich cryptogamic flora. In spite of the physical barrier of the Southern Ocean, the position of Hope Bay at the tip of the peninsula favours algal colonization (Tell et al., 1995).

Hope Bay hosts one of the largest Antarctic penguin rookeries (123,850 breeding pairs of *Pygoscelis adeliae* in 1986) (Myrcha et al., 1987). This fact stands for the soil properties of this area, described by Myrcha et al. (1987) and Tatur (1989). Soil profiles show considerable quantities of phosphate in different forms as a result of actual and past bird activities (relic soils). Nutrient contents of some aquatic environments of this area are also strongly influenced by these activities.

Most lakes studied herein are located in the Five Lakes Valley, about 3 km away from the Base, between Depot Glacier and Mount Flora. It includes four lakes (one of them composed of two lobes) and a pool. These environments have clear waters, and are variable in size and depth. The whole system is separated from the Adelie penguin rookeries by a large front morrain. Pingüi pond, in turn, is located within the rookeries. It is a small, shallow water body, strongly influenced by bird activities. There are two small lakes on Mount Flora shelf, at 100 m above sea level, whose surfaces remain ice-covered most of the summer. Boeckella Lake, the largest water body in Hope Bay, is located about 600 m away from Base Esperanza, to which it provides drinking water. It is influenced by the penguin rookery to a lesser extent than Pingüi pond. All of these lakes are monomictic, with a variable ice-free summer period.

The main features of the lakes surveyed are summarized in Table 1. Some morphometrical data were obtained from Corte (1962) and Drago & Paira (1987).

Materials and methods

Nine lakes and ponds of Hope Bay were monthly sampled during four Antarctic summer seasons between 1991 and 1996. Water temperature, pH and conductivity were measured 'in situ' with Luftman P300 and C400 combined electronic meters. Winkler method (APHA-AWWA-WPCF, 1975) was used to assess dissolved oxygen concentrations.

Water samples for nutrient and chlorophyll-*a* analyses were taken from the subsurface layer in acid-washed PVC flasks. At Esperanza Station, samples were sequentially filtered through Whatman GF/F filters (for chlorophyll-*a* analyses) and 0.2 μm Nucleopore membrane filters (for soluble reactive phosphorus -SRP-, nitrate and ammonia assays). GF/F filters were extracted into 8 ml of ethanol at 60–70 °C. Chlorophyll-a concentrations were spectrophotometrically estimated using the equations given by Marker et al. (1980).

Samples for nutrients analyses were immediately frozen and transported to Buenos Aires. SRP and nitrate concentrations were measured according to Mackereth et al. (1978), while ammonia concentrations were determined following APHA-AWWA-WPCF (1975).

Phytoplankton samples for qualitative and quantitative analyses were taken from the subsurface layer. Living specimens were observed and drawn at Esperanza Station for taxonomical identification. Samples for quantitative analyses were preserved with 1% Lugol's solution. Replicate chambers were left to sediment for 24 hours and counts were performed according to Utermöhl (1958). A maximum counting error of 25% was accepted in estimating the abundance of each major algal group (Venrick, 1978). In all cases individuals were counted and for colonial or filamentous algae the size and/or number of cells corresponding to a standard individual was established.

Both its size and human use make the analysis of the trophic status of Boeckella Lake particularly relevant. For this reason this water body has been intensively studied during two consecutive summers (1992, 1993). Two sites were established in opposite areas of the lake representing different degrees of eutrophication. These were weekly sampled at three depths (surface, 1 m and 2 m, close to the bottom) as described above. Cell volume was calculated for each species on the basis of the average cell dimensions using appropiate volume formulae. Species diversity was calculated according to Shannon-Wiener index (Magurran, 1988).

In order to analyze the differences among lakes and samples, a 2-way ANOVA was performed for each physical, chemical and biological variable using lakes and samples as factors.

Table 1. Main morphometric features of the nine water bodies. * Data from Drago & Paira (1987). ** Data from Corte (1962)

Lake or pond	Altitute (m)	Distance from the sea (m)	Z max (m)	Area (m^2)	Volume (m^3)	Ice-free period	Importance of sea birds
BOECKELLA* (BOECK)	49	650	4	67,454	124,097	summer	High
ESPERANZA (ESPER)	52	300	approx. 4.75**	27,500	–	summer	Very low
FLORA	52	270	approx. 6.95**	approx. 6250	–	summer	Very low
ENCANTADO (ENCA)	approx. 55	250	> 1	approx. 2500	–	summer	Low
ESCONDIDO (ESCON)	approx. 50	250	approx. 1	–	–	Short periods during summer	Very low
SURGENTE (SURG)	approx. 50	500	< 1	–	–	summer	Low
CHICO*	100	1100	5.48	2972	9082	Short periods during summer	Very low
ANDRESITO (ANDR)	100	approx. 1000	< 0.50	–	–	Short periods during summer	Medium
PINGUI (PING)	20	approx. 450	approx. 1	–	–	summer	Extremely high

Two ordinations of the lakes were obtained by Principal Component Analyses based on correlation matrices using physico-chemical and phytoplankton abundances respectively. The coincidence between such ordinations was verified by Pearson's correlation analysis between biotic and abiotic factors.

Results

a) Physical and chemical features

Present results show the limnological features of the studied lakes to be affected by three main factors: morphometry, ornithogenic nutrient inputs, and freezing-melting and mixing cycles.

The most oligotrophic lake was the closest to Depot Glacier (Escondido Lake), with undetectable values of SRP and very low quantities of nitrate and ammonia (Table 2). This water body had very constant low temperatures (Table 2).

Five Lakes Valley also hosts most of the oligotrophic lakes (Esperanza, Flora, Encantado and Surgente). Average SRP ranged from undetectable to 68 $\mu g\ l^{-1}$, while nitrate concentrations varied between 72 and 135 $\mu g\ l^{-1}$, and ammonia between 38 and 89 $\mu g\ l^{-1}$. Maximum temperatures of these lakes increased according to their distance from the glacier (1.1–11.1 °C). The low maximum temperature recorded in Surgente pond is related to its spring nature, being fed by an underground runnel. For all these lakes, conductivities are generally very low: 4.4–38.1 $\mu S\ cm^{-1}$ (Table 2), with the exception of Encantado Lake, which always showed higher values (66.1–130 $\mu S\ cm^{-1}$), probably due to its location in a small depresion surrounded by morraine debris. According to pH, these water bodies varied from slightly acid to circumneutral (5.2–7.2).

Lakes located on Mount Flora shelf (Chico and Andresito) strongly differ in their morphometry (Table 1). The former is a typical oligotrophic lake with nutrients and conductivity figures comparable to those previously described. Conversely, Andresito pond showed much higher nutrient levels (Table 2) due to its position in a small endorrheic depression, and their high inputs of ornithogenic matter, mainly during snow thawing.

Boeckella Lake characteristics, in turn, ranged from oligotrophic to mesotrophic, depending on the margin of the lake and the period of the summer. Thus, the margin adjacent to the penguin rookeries showed stronger signs of eutrophication. Mean concentrations of nutrients in this lake were 253 $\mu g\ l^{-1}$ SRP, 494 $\mu g\ l^{-1}NO_3$-N, and 370 $\mu g\ l^{-1}$ NH_4-N (Table 2). Since this water body was more frequently sampled, it was possible to detect fluctuations in some limnological

Table 2. Ranges and mean values of physical, chemical and biological variables of each water body recorded from 1991 to 1996. ND: non detectable, MV: mean values

	Boeckella	Esperanza	Flora	Encantado	Surgente	Escondido	Chico	Andresito	Pingüi
pH	5.7–6.3	5.2–6.9	5.2–6.2	5.2–6.5	5.7–6.6	5.2–7.2	5.5–8.6	5.3–8.5	7–7.8
	MV: 6.1	MV: 6.1	MV: 5.9	MV: 6.07	MV: 6.06	MV: 6.4	MV: 6.55	MV: 6.22	MV: 7.27
Temperature (°C)	0.5–4.7	0.3–11.1	0–6.4	0–10.9	0–3.2	0–1.1	0–2.5	(–2)–10	0–17
	MV: 2.46	MV: 3.95	MV: 2.2	MV: 3.59	MV: 2.03	MV: 0.6	MV: 0.66	MV: 1.14	MV: 5.93
Conductivity (μS cm^{-1})	11.6–88.1	13–38.1	12–26.5	66.1–130	18.3–32.3	4.4–36.5	16.2–40.8	45.4–189	182–3830
	MV: 33.1	MV: 24.7	MV: 19.3	MV: 90.9	MV: 26.8	MV: 20.8	MV: 32.5	MV: 113.6	MV: 1867.8
Dissolved oxygen (mg l^{-1})	9.2–15.3	8–10.1	8.5–13.9	7–13.1	7–16	13.7	10.6–15.3	6.7–16	6–16
	MV: 12.65	MV: 9.37	MV: 10.73	MV: 9.69	MV: 11.06		MV: 13.76	MV: 10.2	MV: 10.47
SRP (μg l^{-1})	61–644	ND–362.1	14.8–29.5	ND–16.1	ND–84.4	ND–7	ND–46.4	312.2–1348	847–7934
	MV: 253.1	MV: 67.9	MV: 10.8	MV: 5.9	MV: 18.4	MV: 5.3	MV: 16.9	MV: 882.3	MV: 4336.3
N-NO_3 (μg l^{-1})	201.4–809	23.1–201.4	46.8–124.2	68–200.6	48–151.1	53.2–83.8	125.4–261.2	959–2807	1767–4379
	MV: 494.2	MV: 112.6	MV: 72	MV: 135.7	MV: 93.4	MV: 68.5	MV: 195.2	MV: 1744.4	MV: 2917.2
N-NH_4 (μg l^{-1})	328.1–412	ND–76.1	59.2–118.1	ND–101.3	ND–76.1	ND–101.3	ND–81.3	3818–4614	$> 10,000$
	MV: 370.2	MV: 38	MV: 88.7	MV: 51.5	MV: 38.02	MV: 51.5	MV: 40.6	MV: 4216.1	
Chlorophyll-*a* (μg l^{-1})	ND–6.7	ND–0.82	ND–1.61	0.13–1.59	ND–1.59	ND–0.7	ND–2.2	0.26–7.83	3.53–549.4
	MV: 2.04	MV: 0.37	MV: 0.68	MV: 0.77	MV: 0.62	MV: 0.38	MV: 0.78	MV: 2.77	MV: 136.5

features in relation to ice melting and mixing processes. Melting periods were always accompanied by the lowest values for both conductivity and nutrient concentrations.

The shallow pond Pingüi shows a typical hypertrophic status, with extremely high amounts of all nutrients associated to its location amidst a large penguin rookery. This pond exhibited the widest ranges of temperature, dissolved oxygen and conductivity values, due to its small size and the temporal fluctuations of the bird activities.

b) Algal abundance and composition

A total number of 103 algal taxa was identified from the taxonomical analyses of the samples collected during the four seasons (Table 3). Strong differences in the trophic status of the lakes of Hope Bay were also evident from their algal abundance and composition, as well as from the predominance of either the planktonic or benthic communities.

At one extreme, Escondido Lake showed the lowest algal abundances (mean = 73 organisms ml^{-1}), and consequent low chlorophyll-*a* concentrations (Figure 2 and Table 2, respectively). This water body does not possess a true perennial phytobenthos, but algal assemblages mainly conformed by cryobiontic Chlorophyta such as *Chlamydomonas nivalis*, *Cylindrocystis crassa, Cylindrocystis brebissonii* and *Raphidonema nivale*.

A perennial phytobenthos dominates in other oligotrophic lakes. These water bodies showed a scarce phytoplankton with mean abundances between 181 and 520 organisms ml^{-1} (Figure 2), and chlorophyll-*a* concentrations ranging from undetectable to 1.61μg l^{-1} (Table 2). Phytoplankton is dominated by Chrysophyceae, frequently accompanied by some tychoplanktonic taxa removed from the benthos. Among these group of lakes, Encantado differs from the rest by having a well developed phytoplankton, mainly composed of small flagellated chrysophytes (Figure 3). Total phytoplankton densities in this lake were consequently higher, with an average value of 1707 organisms ml^{-1}.

The largest water body in Hope Bay, Boeckella Lake, has an intermediate trophic status, as mentioned earlier. The true phytoplankton observed in this lake was alternatively dominated by Chrysophyceae and Chlorophyceae (Volvocales). Algal density ranged from 56 to 2054 organisms ml^{-1} for the studied period (mean = 774 organisms ml^{-1}), and chlorophyll-*a* from undetectable to 6.7 μg l^{-1} (mean = 2.04 μg l^{-1}). Algal assemblages included some species belonging to the genus *Chlamydomonas, Ochromonas*, and other tychoplanktonic taxa such as *Cylindrocystis brebissonii, Pinnularia microstauron* var. *microstauron* and *Phormidium fragile*.

Table 3. Phytoplankton species collected in the studied lakes. Numbers refer to occurrences in each lake considering the four summer periods of study. See Table 1 for lake abbreviations.

	Boeck	Esper	Flora	Enca	Surg	Escon	Chico	Andr	Ping
CHLOROPHYCEAE									
Chamydomonas nivalis (Bauer) Wille	4		1	1	2	1	2	3	3
Chamydomonas subcaudata Wille	1				1				2
Chlamydomonas spp.	4	1	1	3	2	1	4	3	2
Coelastrum microporum Näg.									1
Coenochloris planoconvexa Hind.				2					
Koliella tatrae (Kol) Hindak	1		1					1	
cf. *Pleurococcus antarcticus* var. *antarcticus* W. *et* G. S. West	3	1	1	1	1		1	1	3
cf. *Pleurococcus antarcticus* var. *minor* Fritsch	1	1	1				1	1	
Prasiola crispa (Lightf.) Menegh.	4	2	1	1	2		2	3	1
Raphidonema nivale Lagerh.	1					2			
Scotiella antarctica Fritsch	2		1	1			1	3	1
cf. *Trochiscia antarctica* Fritsch	2								2
Ulothrix australis Gain		1	2						
Ulothrix moniliformis Kütz.	1	1			1		1	3	1
ZYGOPHYCEAE									
Cylindrocystis brebissonii (Menegh. *ex* Ralfs) de Bary	3		3			1	2		1
Cylindrocystis crassa De Bary	1		1			1			
Mougeotia sp.	2								
CYANOPHYCEAE									
Anabaena sp.	1								1
Aphanocapsa delicatissima W. *et* G. S. West	4	1						1	1
Aphanocapsa elachista var. *elachista* W *et* G. S. West	3	1	2			1	1	1	2
Chlorogloea purpurea Geitler	2		1				1		
Chroococcus minutus (Kütz.) Näg.	2								
Eucapsis alpina Clements *et* Schantz				1			1		
Eucapsis minuta Fritsch	3								1
Gloeocapsa dermochroa Näg.	1	1							
Gloeocapsa cf. *ralfsiana* (Harv.) Kütz.	2	2	1	1	2		1		1
Isocystis cf. *pallida* Woronichin	1								1
Leptolyngbya antarctica (W. *et* G. S. West) Anag. *et* Kom.	2	2						1	
Leptolyngbya fragilis (Gomont) Anag. *et* Kom.	4	2	2	1	3		3	2	3
Leptolyngbya frigida (Fritsch) Anag. *et* Kom.	3	3	1	1	2		2	3	2
Leptolyngbya tenue (Gomont) Anag. *et* Kom.	2		2	2					1
Lyngbya digueti Gomont	1	1	1	1	1				1
Merismopedia tenuissima Lemm.	2								
Myxosarcina concinna Printz	2		1	1	1	1	1	1	3
Oscillatoria chlorina Kütz.	4		1		1		3	1	3
Oscillatoria lacustris (Kleb.) Geitler		2							
Oscillatoria limnetica Lemm.	2						1	2	
Oscillatoria sancta (Kütz.) Gomont	1		1						
Phormidium acutum (Brühl *et* Biswas) Anag. et Kom.	1						1		
Phormidium aerugineo-coeruleum (Gomont) Anag. *et* Kom.	1	2	1			1			
Phormidium autumnale (Ag.) Gomont	1	1	2	1	2		1		
Phormidium deflexum (W. *et* G. S. West) Anag. *et* Kom.	3	2		1	1		1		1
Phormidium pseudopriestleyi (W. *et* G. S. West) Anag. *et* Kom.	2			2					1
Planktothrix agardhii (Gomont) Anag. *et* Kom.		2	2				2		
Pseudanabaena catenata Lauterb.	1	1						1	1
Synechococcus elongatus Näg.	2						3		1
Synechococcus cf. *linearis* (Schmidle *et* Lauterb.) Kom.	1								1

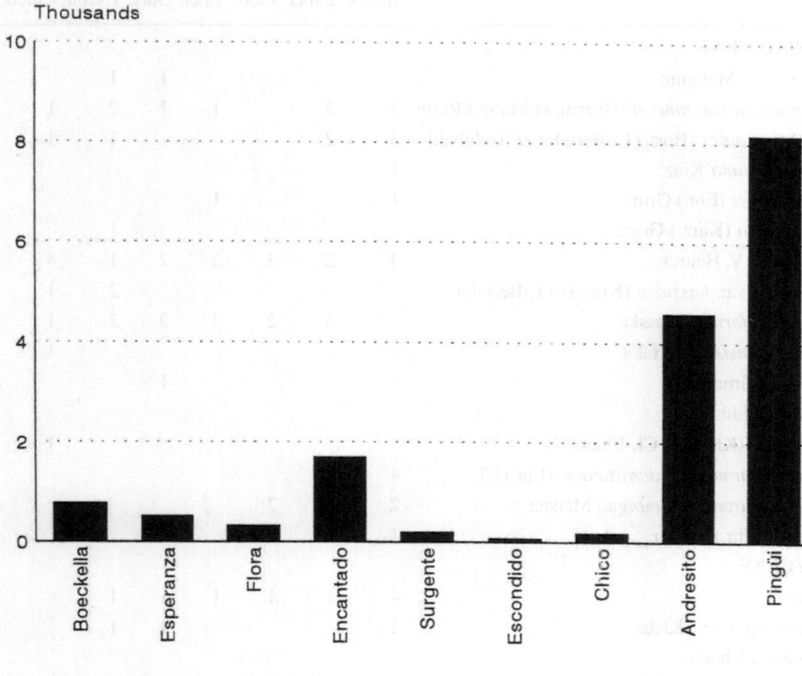

Figure 2. Average values of total phytoplankton density for each lake (in organisms ml^{-1}).

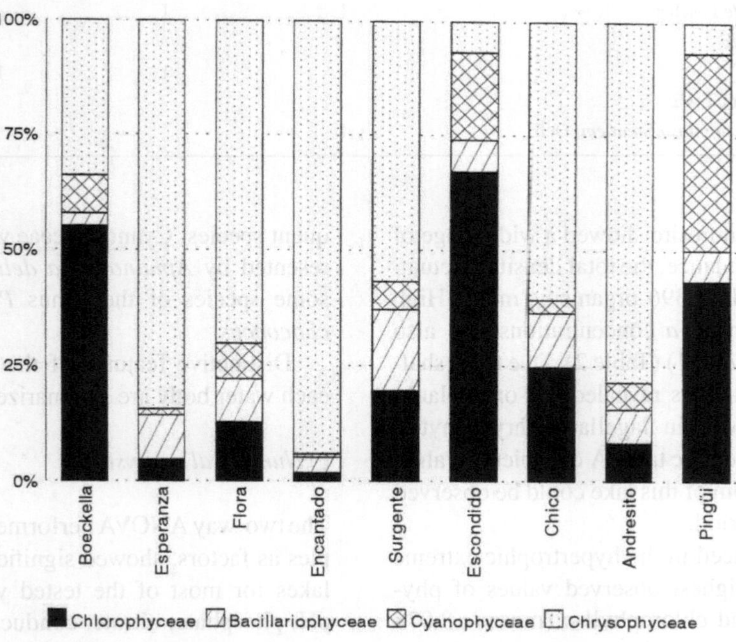

Figure 3. Relative proportion of the different algal groups in each water body.

Table 3. (continued)

	Boeck	Esper	Flora	Enca	Surg	Escon	Chico	Andr	Ping
BACILLARIOPHYCEAE									
Achnanthes germanii Manguin					1	1			
Achnanthes lapponica var. *ninckei* (Guerm. *et* Mang.) Reim.	3	2		1	1	2	1	1	
Achnanthes subatomoides (Hust.) L.-Bertalot *et* Archibald	3	2				1	1	1	
Cyclotella meneghiniana Kütz.	1							1	
Hantzschia amphioxys (Ehr.) Grun.	1			1					
Navicula confervacea (Kütz.) Grun.						1			1
Navicula muticopsis V. Heurck	4	2	3	2	2	1	4	2	4
Navicula soehrensis var. *hassiaca* (Krasske) L-Bertalot						2	1	1	
Navicula tabellariaeformis Krasske		3	2	3	3	2	1		
Nitzschia homburgiensis L-Bertalot	1						1		
Nitzschia paleacea Grun.	1				1				
Pinnularia borealis Ehr.								2	
Pinnularia ignobilis (Krasske) Cl.-Euler							1	1	
Pinnularia microstauron var. *microstauron* (Ehr.) Cl.	4	2							
Pinnularia microstauron var. *ambigua* Meister	2		2						
Rhopalodia gibba (Ehr.) Müller	1	1							
CHRYSOPHYCEAE									
Ochromonas sp.	4	1	1	1	1	1			
cf. *Chrysococcus rufescens* Klebs	1				1	1		1	
Phaeogloea mucosa Chodat								1	
cysts	4	1	1	1	1				
TRIBOPHYCEAE									
Characiopsis anabaenae var. *major* Ettl	1						1		
Characiopsis sphagnicola var. *sphagnicola* Pascher	1			1			1		
Tribonema utriculosum (Kütz.) Hazen.	4						2		
Tribonema viride Pascher	1			1					
DINOPHYCEAE									
Peridinium sp.				1			1		
EUGLENOPHYCEAE									
Trachelomonas rugulosa Stein *em.* Defl.	1							1	1

The small pond Andresito showed a wide range of variation in algal abundance, the total density fluctuating between 296 and 17,396 organisms ml^{-1}. High variability of chlorophyll-*a* concentrations was also recorded (0.26–7.83 $\mu g\, l^{-1}$) (Table 2). Due to its shallowness, the algal samples revealed not only planktonic species such as certain flagellated chrysophytes, but also many phytobenthic taxa. A conspicuous algal felt covering the bottom of this lake could be observed during the ice-free period.

Pingüi pond is placed in the hypertrophic extreme of the range, with highest observed values of phytoplankton density and chlorophyll-*a* (means = 8,078 organisms ml^{-1} and 136.5 $\mu g\, l^{-1}$ respectively). Phytoplankton was dominated by Volvocales, among which *Chlamydomonas subcaudata* resulted the more frequent species. Cyanophyceae were subdominant, represented by *Aphanocapsa delicatissima*, along with some species of the genus *Phormidium* and *Synechococcus.*

Distinctive features of the algal assemblages for each water body are summarized in Table 4.

c) Numerical analyses

The two-way ANOVA performed, using lakes and samples as factors, showed significant differences among lakes for most of the tested variables: temperature, pH, phosphate, nitrate, conductivity ($p < 0.005$) and chlorophyll-*a* ($p < 0.01$). However, differences among samples of each particular lake were not significant for any of the variables related with the trophic status.

Table 4. Tentative classification of algal composition in lakes of Hope Bay

Lakes	Escondido	Flora Esperanza Chico Surgente	1) Encantado 2) Boeckella	Andresito	Pingüi
Characteristics	Young lake proglacial Close to Depot Glacier Many species proceed from the snow	Typical oligotrophic lakes With perennial phytobenthos Scant phytoplankton	Oligotrophic to mesotrophic lakes With well developed phytoplankton	Very small pond Moderate bird activities Almost permanent ice-cover	Very enriched pond by ornithogenic matter Hypertrophic
Dominant species	*Chlamydomonas nivalis* *Raphidonema nivale* *Cylindrocystis crassa* *C. brebissonii*	Small flagellated Chrysophyceae (e.g. *Ochromonas* sp.) Tychoplanktonic species from benthic algal mats (*Leptolyngbya fragilis* *L. frigida, L. tenue* *Gloeocapsa* sp.)	1) *Ochromonas* sp. 2) *Ochromonas* sp. *Chlamydomonas* spp.	Small flagellates (Chrysophyceae when the pond is ice-covered) *Phormidium* spp. *Synechococcus* spp.	Volvocales (*Chlamydomonas subcaudata*, *Chlamydomonas* spp.) *Aphanocapsa delicatissima* *Phormidium* spp.

Ultra-oligotrophic ——————————————————————> Hypertrophic

Based on these results, PCA analyses were performed using the average values over both time and samples for each variable and lake.

The first two factors of the PCA based on physical and chemical data account for 93.5% of the total variance. The first principal component resulted highly correlated with nutrients, conductivity and pH, whereas the second factor with dissolved oxygen and temperature. Figure 4 shows the ordination of the lakes in the space defined by the first three factors. The first axis, which expresses the overall trophic status of the lakes, evidences the complete separation of Pingüi pond from the rest. In relation to the second axis, lakes are ordinated mainly by their mean temperatures and dissolved oxygen; thus the coldest and more oxygenated ones (Chico and Escondido) are placed at the extreme of this axis.

The results of the PCA based on the abundance of the different algal groups show a similar ordination pattern. The first two factors of this PCA account for 99.5% of the total variance, with a strong load of chlorophytes, diatoms and cyanobacteria on the first component and chrysophytes on the second one. Figure 5 represents the ordination of the lakes in relation to the first three axis.

Similarity between the two ordinations obtained is corroborated by the results of a correlation analysis between the first axis of both PCA ($r = -0.984$, $p < 0.001$).

d) Summer fluctuations of the phytoplankton populations

Studies on phytoplankton carried out in Boeckella Lake during two consecutive summer periods allowed a first characterization of the summer fluctuations of this community. Changes in phytoplankton populations were mainly regulated by thawing of the ice cover of the lake and affluents, mixing and nutrients input from the neighbouring penguin rookery. Variations in density for the dominant taxa are represented in Figure 6.

In January 1992, while the lake surface was still covered with ice, phytoplankton succession proceeded with a peak in the density of a small flagellated chrysophyte (*Ochromonas* sp.), which declined soon after the ice cover broke and mixing occurred. Abundance of *Chlamydomonas* spp. also decreased as a consequence of dilution during the maximum of snow thawing and stream discharge. Presence of these algal taxa was, though in low numbers, rather constant during the

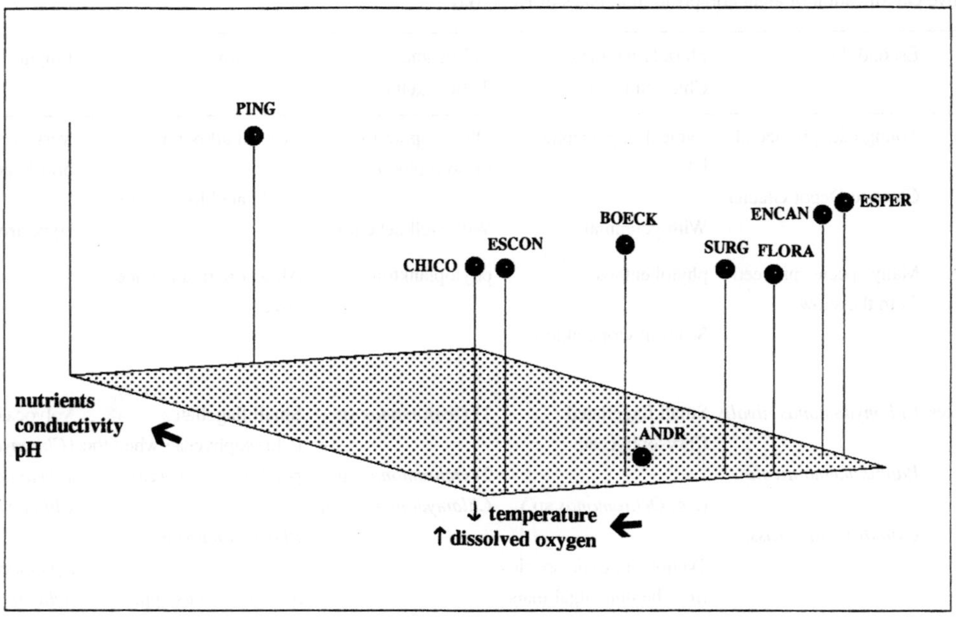

Figure 4. Ordination of the lakes corresponding to the PCA based on physical and chemical variables. See Table 1 for lake abbreviations.

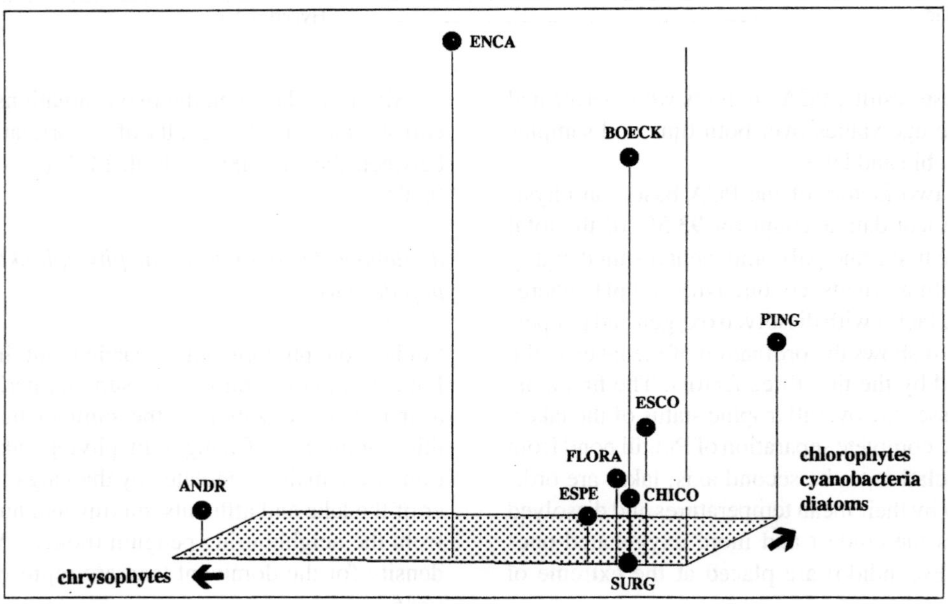

Figure 5. Ordination of the lakes corresponding to the PCA based on phytoplankton data. See Table 1 for lake abbreviations.

whole summer. Its highest density was recorded in late February 1993, when it constituted the dominant taxa in the lake.

Typical cryobiontic species such as *Chlamydomonas nivalis* and *Cylindrocystis brebissonii* peaked (27/1/92 and 26/12/92) by snow thawing. Conversely, the benthic diatom *Pinnularia microstauron* var. *microstauron* reached a maximum on 3/2/93, in coincidence with low water level in the lake.

Volvocales represented by *Chlamydomonas* spp. were dominant in density during both summer peri-

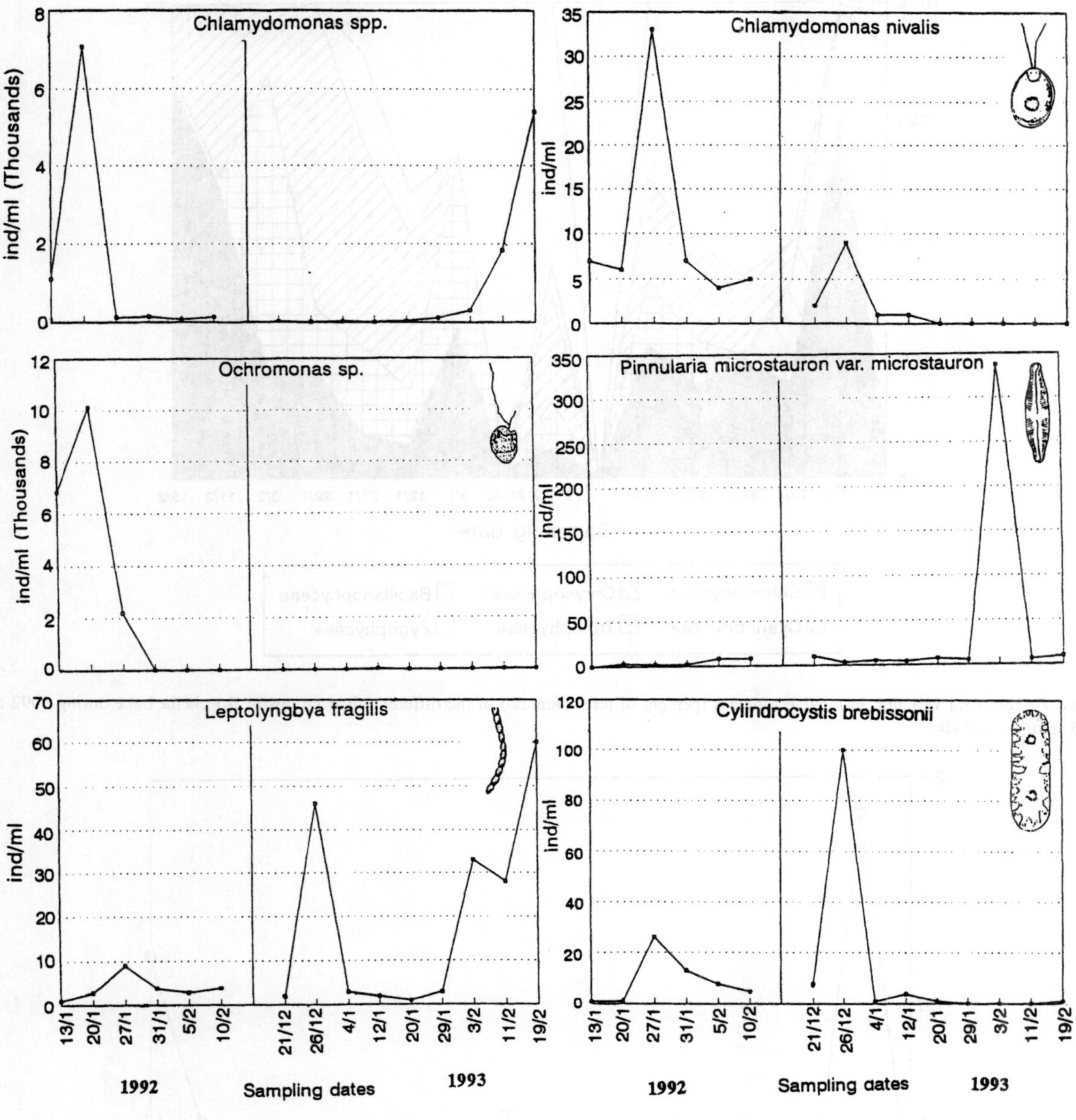

Figure 6. Temporal fluctuations of the dominant algal taxa in Boeckella Lake during 1992 and 1993 summer periods.

ods, except from 13/1 to 27/1/92 (lake with ice-cover), when chrysophytes took over this role.

In terms of the total biomass, the dominant group varied along the summer period. The most constant taxa were *Chlamydomonas* spp. Nevertheless, after the thawing periods, the Zygophyceae contributed with a high percentage to the bulk of the algal biomass; this group was mainly represented by some typical cryobiontic species such as *Cylindrocystis brebissonii* and *C. crassa.* Occasionally, after windy days, some Cyanophyceae, which are very abundant in the benthic algal mats of the lake, are removed from the bottom, thus contributing to the total algal biomass (Figure 7).

Total phytoplankton biomass varied between 0.031 and 2.346 $mm^3\ l^{-1}$ considering the two years of this study. Mean biomass was 0.685 and 0.536 $mm^3\ l^{-1}$ in summers 1992 and 1993 respectively (Figure 8). In the summer succession of phytoplankton, two or three biomass peaks can be distinguished. One of them usually occurred at the begining of the summer period, when

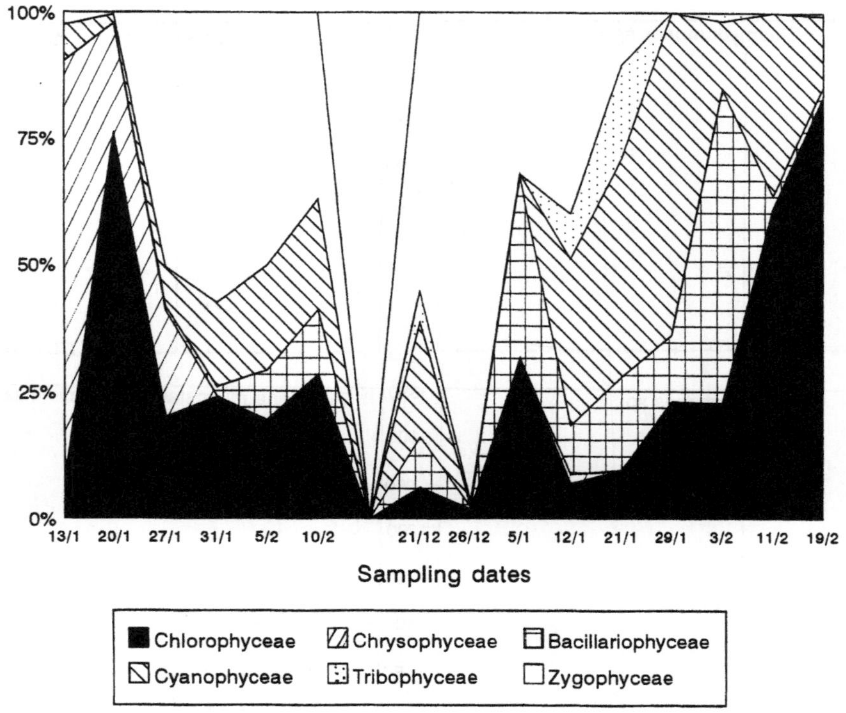

Figure 7. Temporal variation in relative biomass (percent of total biomass) of the different algal groups in Boeckella Lake during 1992 and 1993 summer periods.

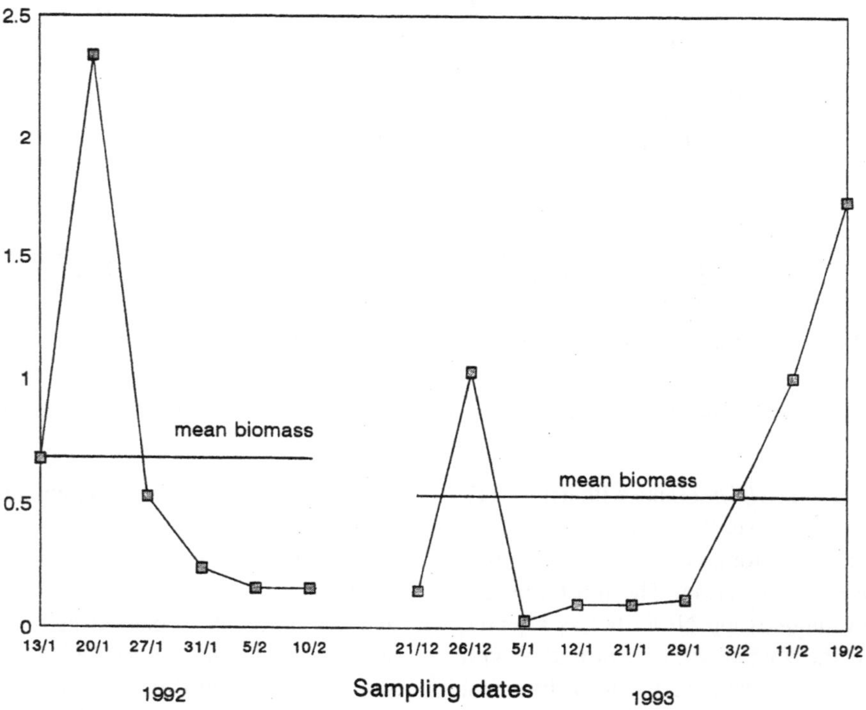

Figure 8. Temporal fluctuation in total phytoplankton biomass ($mm^3\ l^{-1}$) in Boeckella Lake (summers 1992, 1993). Mean biomass for each summer period is also indicated.

the lake was still covered by ice, being the Chrysophyceae the dominant algal group. Another peak normally occurred at the end of February, when retention time increases and the nutrient inputs were maximal; in this case *Chlamydomonas* spp. were the dominant taxa. Moreover, another minor biomass peak could be observed in relation to the increasing of cryobiontic species during thawings, such as *Cylindrocystis brebissonii* and *C. crassa*.

Shannon-Wiener species diversity index calculated from the phytoplankton samples of Boeckella Lake was generally low (0.21–1.6), accordingly to most of the Antarctic freshwater bodies (Figure 9).

Discussion

The wide range of trophic status of the lakes of Hope Bay was confirmed by these results. According to them, all of these water bodies can be placed in different categories along a gradient ranging from ultra-oligotrophic to hypertrophic stages. The typology obtained is similar to that described for other maritime Antarctic lakes (Hawes, 1990; Oppenheim, 1990).

On one hand, ultra-oligotrophic Escondido Lake can be regarded as proglacial following Oppenheim (1990). A distinct feature of this lake is the direct influence of the glacier leading to low and constant temperature and conductivity values. It has extremely low levels of all nutrients and phytoplankton chlorophyll-*a* concentrations. The influence of the glacier also accounts for the high frequency of cryobiontic species.

Older and nearby lakes, which are not directly influenced by ornithogenic nutrient inputs, display conductivity and nutrient levels typical of oligotrophic water bodies. Vegetation comprises mainly a well developed, perennial phytobenthos. The scant phytoplankton is dominated by small flagellated chrysophytes. These characteristics are similar to those reported for other oligotrophic Antarctic lakes. As discussed by Priddle (1980) and Hawes (1990), in extremely oligotrophic water bodies the phytoplankton development is minimal and an extensive cover of algae grows on the lake floor. Priddle et al. (1986) pointed out that the addition of nutrients in oligotrophic lakes from Signy Island promoted algal growth.

As lakes become more affected by sea birds, the increasing degree of trophism is reflected by higher nutrient concentrations, and the dominant community shifts from phytobenthos to phytoplankton.

Boeckella Lake stands as a clear example of a water body subjected to enrichment by ornithogenic matter. As reported in previous studies (Izaguirre et al., 1993; 1996), phytoplankton in this lake is alternatively dominated by Chlorophyceae (Volvocales) and Chrysophyceae classes. During ice cover periods, the community is usually dominated by a small chrysophyte (*Ochromonas* sp.), which is well adapted to low light intensities. Under-ice prevalence of small flagellates had been previously reported by Priddle et al. (1986). According to Sandgren (1988), flagellated chrysophytes are nutritionally opportunistic, and their demographic pattern has usually brief episodes of rapid growth followed by subsequent rapid decline. In terms of the adaptative strategies proposed by Reynolds (1988) they can be included in the C-category.

The small pond Pingüi is an extreme case of enrichment by sea birds. Chlorophyll-*a* values recorded in this pond are comparable with those reported for the most eutrophic Antarctic lakes (Hawes, 1990). Under such conditions, phytobenthos is seriously light-limited. Phytoplankton includes a high proportion of flagellated chlorophytes, particularly *Chlamydomonas subcaudata*, and some cyanobacterial species.

Although our studies did not include an analysis of the different size fractions of phytoplankton, picoplankton abundance was found to be very high at the most eutrophic ponds. Studies carried out in other Antarctic lakes have stressed the importance of this size fraction in the primary production of freshwater environments (Andreoli et al., 1992; Hawes, 1990) under oligotrophic as well as hypertrophic conditions.

In relation to the phytoplankton dynamics, the lakes of Hope Bay show the same pattern than other maritime Antarctic freshwater ecosystems. At this high latitude, phytoplankton has usually an active growth cycle extending from spring to late summer. Algal growth is initiated by flagellated, shade-adapted species (Burch, 1988). Present studies on phytoplankton dynamics in Boeckella Lake comprise only the summer period. Nevertheless, well-defined blooms of small chrysophytes were found both at the beginning and at the end of the season, when the lake surface was frozen (Izaguirre et al., 1993; Pose & Izaguirre, 1998).

Dominance of flagellates in ice-covered lakes is not only allowed by their shade adaption, but also by their capability to migrate and hence mantain their position in a water column not subjected to wind induced turbulence (Priddle et al., 1986).

Decline in phytoplankton density after the ice melting is associated to a dillution effect and a decrease in

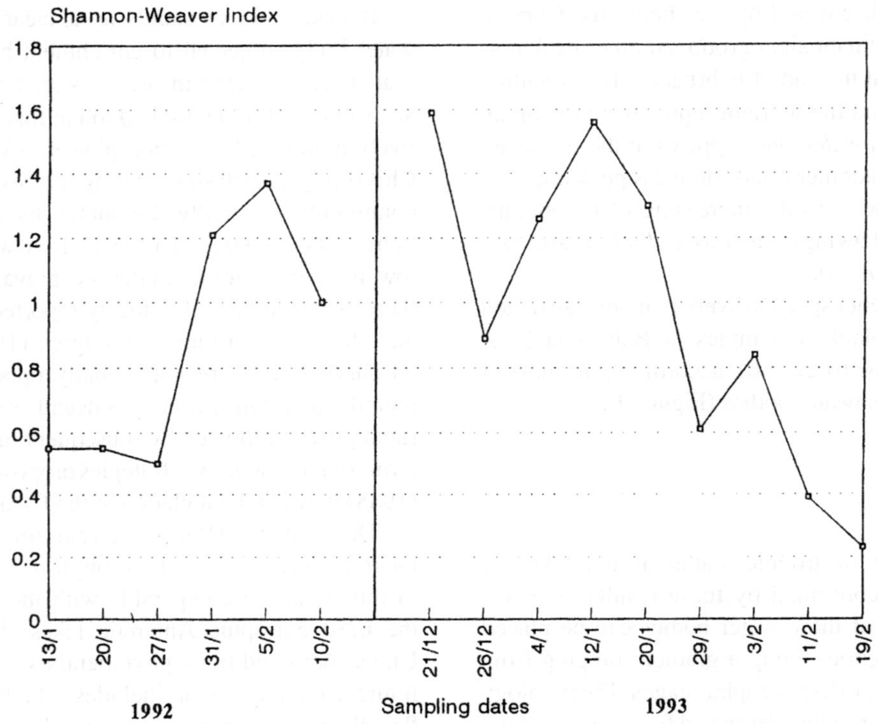

Figure 9. Summer changes of density-based, species diversity in Boeckella Lake.

the retention time of the water body. In particular, this can be as low as a few days (2–3) in Boeckella Lake during the streams discharge peak (Izaguirre et al., 1996). It has also been demonstrated that phytoplankton abundance increases in the lake towards the end of the summer, in accordance with increasing retention time and nutrient levels, being the latter affected by fluctuations in the activity of the penguin colony.

Acknowledgements

The authors wish to thank Instituto Antártico Argentino (DNA), for supporting the Antarctic Expeditions. We are indebted to the members of the Esperanza Station for the logistic support. We also thank F. Unrein and J. Masello for their assistance with field work, R. Lombardo for his critical comments on numerical analyses, and J. Giani for his technical cooperation in the laboratory work.

References

Andreoli, C., L. R. Scarabe, C. Grassi, F. Dalla Vecchia, S. Spini, C. Tolomio, N. Rascio, Y. Moro & L. Masiero, 1992. The photoautotrophic picoplankton of Terra Nova Bay and of some lakes of Northern Victoria Land (Antarctica). Proceedings of the second meeting on Antarctic Biology, Padova. Dipartamento di Biologia dell'Universita: 47–53.

APHA-AWWA-WPCF, 1975. Standard methods for the examination of water and waste-water. American Public Health Association, Washington D.C., 1193 pp.

Burch, M. D., 1988. Annual cycle of phytoplankton in Ace Lake, an ice covered, saline meromictic lake. Hydrobiologia 165: 59–75.

Corte, A., 1962. Algas de agua dulce en lagos semicongelados de Bahía Esperanza, Peninsula Antártica. Contribución del Instituto Antártico Argentino 69: 1–38.

Drago, E. & A. Paira, 1987. Informe de la campaña antártica de verano 1986/87. Instituto Antártico Argentino, Buenos Aires.

Hawes, I., 1990. Eutrophication and vegetation development in Maritime Antarctic Lakes. In K. R. Kerry & G. Hempel (eds), Antarctic Ecosystems. Ecological change and conservation. Springer Verlag, Berlin: 83–90.

Izaguirre, I., G. Mataloni, A. Vinocur & G. Tell, 1993. Temporal and spatial variations of phytoplankton from Boeckella Lake (Hope Bay, Antarctic Peninsula). Antarctic Science 5: 137–141.

Izaguirre, I., A. Vinocur & G. Mataloni, 1996. Summer changes in the vertical distribution of chlorophyll-*a* in Boeckella Lake (Hope Bay, Antarctic Peninsula). Nankyoku Shiryo (Antarctic Record) 1: 43–52.

Lewis Smith, R. Y., 1984. Terrestrial plant biology of the sub-Antarctic and Antarctic. In R. Laws (ed.), Antarctic Ecology, 1. Academic Press, London: 61–162.

Mackereth, J. F. H., J. Hiron & J. F. Talling, 1978. Water analysis: Some revised methods for limnologists. Freshwater Biology Association, Sc. Publ. 36, 117 pp.

Magurran, A. E., 1988. Ecological diversity and its measurements. London: Chapman and Hall, Princeton Univ. Press, 179 pp.

Marker, A. F. H., E. A. Nusch, H. Rai & B. Riemann, 1980. The measurement of photosynthetic pigments in freshwater and standardization of methods: conclusions and recommendations. Arch. Hydrobiol. Beih. Ergebn. Limnol. 14: 91–106.

Myrcha, A., A. Tatur & R. del Valle, 1987. Numbers of Adélie penguins breeding at Hope Bay and Seymour Island rookeries (West Antarctica) in 1985. Polar Res. 8: 411–422.

Oppenheim, D. R., 1990. A preliminary study of benthic diatoms in contrasting lake environments. In K. R. Kerry & G. Hempel (eds), Antarctic Ecosystems. Ecological change and conservation. Springer Verlag, Berlin: 91–99.

Pose, M. & I. Izaguirre, 1998. Sucesión estival de las poblaciones fitoplanctónicas en el lago Boeckella (Bahía Esperanza, Península Antártica). Contrib. Inst. Antártico Argentino, 449, 27 pp.

Priddle, J., 1980. The production ecology of benthic plants in some Antarctic lakes. I. In situ studies. J. Ecol. 68: 141–153.

Priddle, J., I. Hawes & J. C. Ellis-Evans, 1986. Antarctic ecosystems as habitats for phytoplankton. Biol. Rev. 61: 199–238.

Reynolds, C. S., 1988. Functional morphology and the adaptive strategies of freshwater phytoplankton. In C. D. Sandgren (ed.), Growth and Reproductive Strategies of Freshwater Phytoplankton: 388–433.

Sandgren, C. D., 1988. The ecology of chrysophyte flagellates: their growth and perennation strategies as freshwater phytoplankton. In C. D. Sandgren (ed.), Growth and Reproductive Strategies of Freshwater Phytoplankton: 9–104.

Tatur, A., 1989. Ornithogenic soils of the maritime Antarctic. Polish Polar Res. 10: 481–532.

Tell, G., A. Vinocur & I. Izaguirre, 1995. Cyanophyta of lakes and ponds of Hope Bay, Antarctic Peninsula. Polar Biol. 15: 503–509.

Utermöhl, H., 1958. Zur Vervollkommnung der quantitativen Phytoplankton-Methodik. Mitt. int. Ver. Limnol. 9: 1–38.

Venrick, E. L., 1978. How many cells to count ?. In A. Sournia (ed.), Phytoplankton Manual, Paris, Unesco: 167–180.

Vinocur, A. & I. Izaguirre, 1994. Freshwater algae (excluding Cyanophyta) from nine lakes and pools of Hope Bay, Antarctic Peninsula. Antarctic Science 4: 483–489.

Vinocur, A. & H. Pizarro, 1995. Periphyton flora of some lotic and lentic environments of Hope Bay (Antarctic Peninsula). Polar Biol. 15: 401–414.

Hydrobiologia **369/370:** 89–97, 1998.
M. Alvarez-Cobelas, C. S. Reynolds, P. Sanchez-Castillo & J. Kristiansen (eds), Phytoplankton and Trophic Gradients.

The most typical phytoplankton taxa in four types of boreal lakes

Liisa Lepistö & Ulla Rosenström
Finnish Environment Institute, P.O. Box 140, SF - 00251 Helsinki, Finland

Key words: phytoplankton, boreal lakes, acidic, oligotrophic, dystrophic, eutrophic

Abstract

The monitoring of phytoplankton quantities and species composition in Finnish lakes started in the early 1960s. In addition, a substantial number of physical and chemical variables have been measured and recorded from the monitored lakes. The purpose of the present evaluation is to differentiate the phytoplankton taxa of the lakes being classified as acidic, oligotrophic, dystrophic and eutrophic. At present, these four lake types are differentiated on the bases of concentration of total phosphorus ($\mu g\ l^{-1}$), water colour ($mg\ l^{-1}$ Pt), and Secchi-disc transparency. A total of 38 lakes with 163 samples from the years 1984 to 1994 were selected for this study.

Cyanoprokaryota *Merismopedia warmingiana* (*M. tenuissima*) was abundant in clear and dystrophic acidic lakes whereas *Dinobryon divergens* (Chrysophyceae) was dominant only in the clear lake types. *Oocystis rhomboidea* (Chlorophyceae) was also an important species, especially in the dark acidic lakes. In oligotrophic lakes, the phytoplankton biomass was low because of the combined effect of low organism numbers and small cell size of the occurring species. Additional chrysophyceans like *Uroglena americana*, unidentified Ochromonadales and *Dinobryon divergens* were especially abundant, as well as *Merismopedia warmingiana*. In the dystrophic lakes flagellated taxa, such as *Cryptomonas* spp., *Uroglena americana*, Ochromonadales and *Pseudopedinella* spp. dominated. In eutrophic lakes, the cyanoprokaryotes *Microcystis aeruginosa*, *M. viridis* and *M. wesenbergii*, as well as *Anabaena flos-aquae* and *Aphanizomenon flos-aquae* dominated both cell numbers and biomass. Some identification problems are also discussed.

Introduction

'No two lakes are alike, and so no two floras are alike, although there are certain broad categories of lakes in which the basic conditions are similar' (Round, 1981). There is, however, a latitudinal grouping of lakes into polar/alpine, temperate and tropical lakes. Nevertheless, in each geographical region there are many different lake types due to variations in catchment characteristics, causing variations in physical and chemical conditions, leading to acidic, oligotrophic, dystrophic, mesotrophic and eutrophic states each with characteristic phytoplankton assemblages (e.g. Hutchinson, 1967; Round, 1981). In acidic lakes, the phytoplankton biomass is generally sparse and the diversity of species is low. Sometimes, however, a temporary phytoplankton biomass increase can be observed as a result of zooplankton decrease due to higher mortality (Arvola et al., 1992). Phytoplankton species diversity is also rather poor in dystrophic lakes, as light is available only in the top few metres and a large portion of the nutrients remain in the monimolimnion, and therefore only selected species with special adaptations survive (Arvola, 1985). Eutrophic lakes are known for their large phytoplankton biomass and high species diversity, although the diversity may decrease in hypereutrophic lakes. Very typical for eutrophic lakes are also blooms of blue-green algae (e.g. Reynolds, 1984; Trifonova, 1988) that may lead to poisoning of animals as well as humans (Carmichael et al., 1985, Sivonen, 1990).

Consequently, phytoplankton quantity, as well as its species composition, will reflect the trophic state of lakes (Järnefelt 1952, 1956; Heinonen, 1980; Trifonova, 1988; Brettum, 1989; Willén, 1992). In the Nordic countries, at least, data on phytoplankton abundance

Table 1. Physical and chemical characteristics (mean and standard deviation) of the different lake types in 1984 to 1994.

Variables		The lake types Acidic n=11	Oligotrophic n=47	Dystrophic n=52	Eutrophic n=53
pH	$\bar{x}$	5.5	7.1	6.9	7.8
	s	0.46	0.21	0.23	0.64
Colour (mg l^{-1} Pt)		29	18	59	73
		47.77*	9.29	6.87	24.45
Conductivity (mSm^{-1})		2.3	4.7	4.3	11.8
		0.78	1.63	1.88	3.63
Total N (μg l^{-1})		280	330	560	1090
		95.30	111.78	236.59	253.56
Total P (μg l^{-1})		8	5	13	87
		4.51	1.88	2.68	33.16
Biomass (mg l^{-1})		0.29	0.33	0.75	11.96
		0.14	0.13	0.30	11.41
Secchi depth (m)		**	4.9	2.2	1.0

n = number of samples per lake type, $\bar{x}$ = mean and s = standard deviation.
* Some of the acidic lakes were also dystrophic.
** No results from the acidic lakes.

have been applied in this way: a lake where the maximum phytoplankton biomass remains below 0.5 mg l^{-1} is considered to be oligotrophic, and eutrophic when the maximum biomass exceeds 2.5–3 mg l^{-1} (Heinonen, 1980; Brettum, 1989; Willén, 1992).

The purpose of this study was to classify the various monitored Finnish lakes as either acidic, oligotrophic, dystrophic or eutrophic based on the available data material, and to determine their phytoplankton composition.

Description of the lake types

Acidic lakes
Many lakes in Finland are slightly acidic and often also dystrophic due to the dissolved humic substances that are derived from the surrounding soils. In the past few decades the pH of more and more lakes have become successively lower due to the acid precipitation (Kenttämies, 1994). Acidification of lakes changes the food webs as only the most acid tolerant species remain.

Oligotrophic lakes
Oligotrophic lakes of types other than the above are characterized by high transparency and low nutrient concentrations. The food web is more versatile than in the acidic lakes, and the circulation of nutrients is efficient (Järnefelt, 1956).

Dystrophic lakes
Dystrophic lakes are often very demanding habitats for phytoplankton. Due to the dark water colour (usually > 50 mg l^{-1} Pt), the epilimnion and photosynthetic layers are usually very thin. Many of the dystrophic lakes are also meromictic which leads to nutrient depletion in the productive layer as nutrients are quickly used up by primary producers.

Eutrophic lakes
Eutrophic lakes are very rich in nutrients, mainly as a result of anthropogenic loading (eg. Thompson & Rhee, 1994). After several years of enhanced nutrient loading, lakes begin to be self supporting due to internal nutrient loading (Ryding & Forsberg, 1977). Eutrophication leads to low transparency as the flora increases.

Materials and methods

The monitoring of phytoplankton quantities and taxa richness in Finnish lakes was started in the early 1960s, also several physical and chemical variables were included in the monitoring programme. The present study was conducted using environmental data from the water quality database of the Finnish Environment Institute (FEI). A total of 38 lakes including 163 samples taken during 1984 to 1994 was selected using the amount of total phosphorus (μg l^{-1}), the colour of the water (mg l^{-1} Pt), and transparency measured as Secchi-disc depth. The acidic lakes were chosen from those investigated during the HAPRO project – Acidification in Finland (Kauppi et al., 1990). All the lakes studied were situated between the latitudes 60 ° and 69 ° N. Some background data on the sites are given in Table 1.

Only samples between 1984 and 1994 were evaluated so as to avoid effects attributable to of long-term changes in lake ecosystems and phytoplankton communities. The samples were all taken in July because of the empirically stated stability of phytoplankton communities at that time (Heinonen, 1982).

The corresponding phytoplankton compositions were obtained from the phytoplankton database, which is also maintained by FEI. The data represents the phytoplankton of the upper 2 metres of the water column. Phytoplankton species were determined and counted under an inverted microscope by phase-contrast illumination from samples preserved with Lugol's solution supplemented with acetic acid. The counting units were cells, colonies and trichomes with a length of 100 μm (Lepistö, 1995).

The biomass of different phytoplankton groups was logarithmically transformed. This enabled the different groups of lakes to be compared even though e.g. the biomass of the eutrophic lakes was of a totally different magnitude than the others. Table 2 shows both the average biomass of phytoplankton groups in the lakes, as well as the standard deviation within the lake group. The number of identified taxa is also given.

The phytoplankton species of each lake were studied separately. The frequency of a species within a group, i.e. the number of samples where the particular species was found in relation to the total number of samples from a given lake type, as well as the biomass, and the number of cells in a given volume of water were taken into account (Tables 3–6). The biomass and cell numbers are shown as sum values from all lakes, because average values were very small. The number of cells was important because the effect of certain small sized species are never considered when only comparing biomass values.

Results and discussion

The phytoplankton groups

The distribution of phytoplankton among the different lake types showed some clear and expected results. The total biomass was greatest in the eutrophic lakes (Table 2), as was the standard deviation. This category included both eutrophic and hyper-eutrophic lakes. The other three lake groups showed similar but low biomasses, although that of dystrophic lakes was slightly larger. This is probably due to the larger cell size of dystrophic lake phytoplankton compared to the other less productive lake groups. The phytoplankton biomass level corresponded to the nutrient level of the lake types.

Cyanoprokaryota (Cyanophyceae) was the most important phytoplankton group in the eutrophic lakes. The high standard deviation implies that there was a large variation between the individual lakes. This result was expected as most blue-green algae prefer nutrient rich and neutral lakes (Round, 1981; Trifonova, 1988).

Cryptophyceae were abundant in dystrophic and eutrophic waters which corresponds to the observations of Ilmavirta (1983). According to Brettum (1989) and Willén (1992), Cryptophyceae are found in different types of water bodies, with a tendency for small sized cells to occur in less productive waters. The identification of cryptomonads is difficult and it is also a very heterogenous group with a wide ecological spectrum. In dystrophic waters the flagellated Cryptophyceae are able to migrate between the illuminated epilimnion and the nutrient rich hypolimnion (Salonen et al., 1984). The pigmentation of Cryptophyceae, like that of Cyanophyceae (van den Hoek, 1984), also benefit survival in low light conditions. Dystrophic lakes have low light conditions because of the very dark water, and eutrophic lakes because of the massive biomass production which decreases light penetration. Consequently, it is essential for phytoplankton in these types of waters to be able to maximize their light harvesting by moving nearer to the surface.

Dinophyceae produced by far the largest biomass in eutrophic lakes, partly due to the quite large cell size of the species favouring eutrophy. Some Dinophyceae species have been reported as indicator species of eutrophy (Järnefelt, 1952), but according to Brettum (1989), this group in general prefers oligotrophy. In the other three lake groups dinoflagellates were most important in acidic lakes implying that there are also some acid-tolerant species.

Table 2. The distribution of phytoplankton taxa, logarithmically transformed total biomass and biomass of algal groups (mg l^{-1} fresh weight) in different lake types.

Phytoplankton taxa	The lake types			
	Acidic n = 11	Oligotrophic n = 47	Dystrophic n = 52	Eutrophic n=53
Total number of taxa	70	292	363	520
The total biomass $\bar{x}$	0.270	0.280	0.674	4.335
s	0.068	0.057	0.164	0.784
Cyanophyceae	0.018	0.044	0.067	1.780
	0.031	0.057	0.084	1.030
Cryptophyceae	0.034	0.060	0.175	0.594
	0.045	0.050	0.133	0.510
Dinophyceae	0.064	0.025	0.027	0.320
	0.104	0.029	0.040	0.395
Chrysophyceae	0.095	0.071	0.121	0.152
	0.097	0.061	0.077	0.220
Diatomophyceae	0.006	0.054	0.254	0.910
	0.015	0.040	0.245	0.581
Tribophyceae	0.001	0.001	0.001	0.025
	0.002	0.002	0.002	0.038
Raphidophyceae	–	< 0.001	0.003	0.024
	–	< 0.001	0.006	0.095
Euglenophyceae	–	<0.001	0.001	0.182
	–	0.001	0.003	0.196
Chlorophyceae	0.051	0.023	0.033	0.483
	0.047	0.018	0.030	0.394
Conjugatophyceae	0.001	0.003	0.012	0.053
	0.002	0.010	0.060	0.079

n = number of samples per lake type, $\bar{x}$ = mean and s = standard deviation.

Chrysophyceae were common in all types of lakes but they seem to prefer more oligotrophic lakes as observed also by other authors such as Round (1981), Ilmavirta (1983), Eloranta (1995), Trifonova (1988) and Willén (1992). This algal group is complicated and heterogenous (Hutchinson, 1967), making generalisations quite impossible.

Diatomophyceae and Euglenophyceae were both clearly sensitive to acidity by showing a very small biomass in acidic lakes or by being totally absent. Diatoms are common in all kinds of waters (Hutchinson, 1967; Heinonen, 1980), but in acidic waters especially planktonic diatoms are rare (Kippo-Edlund & Heitto, 1990). Euglenophyceae are almost entirely restricted to eutrophic lakes, mainly to waters with a high organic content (Hutchinson, 1967; Tikkanen & Willén, 1992).

The species of Chlorophyceae seem to be quite common in eutrophic lakes, although there are clearly some species preferring acidic, as well as oligotrophic lakes. These species occur in quite the same amounts in all the three lake groups with low nutrient levels.

In addition, there were also fairly rare phytoplankton classes, such as Tribophyceae, Raphidophyceae, and Conjugatophyceae. These algae were found mainly in eutrophic lakes.

Phytoplankton species

Acidic lakes

The amount of samples from the acidic lake category was very small. The total number of taxa was only 70, which is quite a common phenomenon for many acidic lakes (Arvola et al., 1990). A typical characteristic of the acidic lakes investigated was their small size, which allowed very little turbulence. This normally excludes all the non-flagellated species such as Diatomophyceae. The eleven lakes were further divided into clear acidic (7 lakes) and humic lakes (4 lakes) and studied separately.

Table 3. The most important phytoplankton taxa in acidic lakes (total of samples: = 11).

Phytoplankton taxa	(Cell ml^{-1} 10^3) (sum)	Biomass (mg l^{-1}) (sum)	(n)
Clear acidic lakes (7 samples)			
Merismopedia warmingiana Lagerh.	9.46	0.09	4
Glenodinium spp.	0.15	0.15	5
Peridinium umbonatum Stein	0.21	0.51	5
Dinobryon divergens Imhof	3.09	0.47	2
Dinobryon pediforme (Lemm.) Steincke	0.04	0.01	2
D. sociale v. *americanum* (Brun.) Bach.	0.15	0.02	1
Botryococcus sp.	0.03	0.14	3
Oocystis rhomboidea Fott	0.26	0.01	4
Humic lakes (4 samples)			
Merismopedia warmingiana Lagerh.	11.83	0.11	2
Cryptomonas spp.	0.30	0.32	4
Dinobryon acuminatum Ruttner	0.09	0.01	2
Mallomonas akrokomos Ruttner	0.14	0.14	2
Stichogloea doederleinii (Schmidle) Wille	0.19	0.03	1
Uroglena americana Calkins	0.61	0.06	2
'Sphaerocystis schroeteri' Chod.	0.06	0.12	3
Oocystis rhomboidea Fott.	0.87	0.04	3

n = the number of samples where the respective species have been observed.

There were eight common species in the clear acidic lakes (Table 3). The Cyanoprokaryote *Merismopedia warmingiana* (*M. tenuissima* Lemm.) was clearly acidophilic, as observed also by Morling & Willén (1990). It was rare in lakes with phosphorus concentrations greater than 30 μg l^{-1}, Brettum (1989) has even recorded a limiting concentration of 10 μg l^{-1} for this species. It has also been found in humic oligotrophic lakes (Ilmavirta 1980). In more eutrophic waters *M. minima* G. Beck may be misinterpreted as *M. warmingiana* (*M. tenuissima*) according to Hindák (1992).

Glenodinium spp. and *Peridinium umbonatum* (*P. inconspicuum* Lemm.) represent the dinoflagellates. *Glenodinium* species were found in slightly acidic lakes and with very low phosphorus concentrations. The genus was scarce in other lake types. *P. umbonatum*, on the other hand, was most abundant in the lake with the lowest pH. *Dinobryon divergens* was quite common in all lake types except for the eutrophic ones. Interestingly it was found in lakes that were very acidic and supported few other taxa (less than ten). According to Eloranta (1989), *D. pediforme* is the best acidic lake indicator of the genus *Dinobryon*.

The common phytoplankton species in humic lakes were a mixture of those found in dystrophic and in clear acidic lakes. *Merismopedia warmingiana* and *Oocystis rhomboidea* represented the acidic lakes and were indifferent to water colour. Dystrophy-tolerant species such as *Cryptomonas* spp., *Mallomonas akrokomos*, and *Uroglena americana* showed indifference to acidity. Other investigations have also shown that cryptomonads are indifferent in relation to acidity (Ilmavirta, 1983; Brettum, 1989). The identification of cryptomonads is still incomplete.

Oligotrophic lakes

A total of 292 taxa were recognised from the oligotrophic lake samples. Most of the common species were chrysophyceans, such as *Dinobryon borgei*, *D. divergens* and *Uroglena americana*, or unidentified Ochromonadales (Table 4). Both *Dinobryon* species seemed to have a competitive advantage in oligotrophic lakes as well as in slightly acidic surroundings as noticed by Rawson (1956), and Tikkanen & Willén (1992). They are obviously favoured by their alternative mixotrophy (Straškrabová & Šimek, 1993). According to Eloranta (1989), *D. borgei* prefers oligotrophic waters whereas *D. divergens* can also be found in more eutrophic waters. *Uroglena americana* was more numerous in dystrophic lakes than in oligotrophic ones, but was also identified in several (33) samples from purely oligotrophic lakes.

A diatom, *Rhizosolenia longiseta*, was abundant in the oligotrophic lakes. Brettum (1989) agrees with

Table 4. The most important phytoplankton taxa in oligotrophic lakes (total of samples: = 47).

Phytoplankton taxa	(Cell ml^{-1} 10^3) (sum)	Biomass (mg l^{-1}) (sum)	(n)
Aphanothece clathrata (W. & G.S. West) Kom.	0.52	0.27	21
Merismopedia warmingiana Lagerh.	2.61	0.02	25
Ochromonadales	4.53	0.24	40
Bitrichia chodatii (Rev.) Chod.	0.08	0.02	25
Dinobryon borgei Lemm.	0.86	0.01	37
D. divergens Imhof	2.63	0.40	34
Kephyrion ovale (Lack.) Hub.-Pest.	0.27	0.01	31
Mallomonas akrokomos Ruttner	0.14	0.14	31
Spiniferomonas spp.	0.53	0.04	29
Uroglena americana Kalkins	5.35	0.56	33
Cyclotella kuetzingiana Thwaites	0.04	0.18	16
Rhizosolenia longiseta Zach.	0.26	0.47	32
Tabellaria flocculosa (Roth) Kütz.	0.17	0.26	31
Botryococcus sp.	0.03	0.23	38

n = the number of samples where the respective species have been observed.

Table 5. The most important phytoplankton taxa in dystrophic lakes (total of samples: = 52).

Phytoplankton taxa	(Cell ml^{-1} 10^3) (sum)	Biomass (mg l^{-1}) (sum)	(n)
Cryptomonas marssonii Skuja	0.25	0.27	6
Cryptomonas spp.	5.48	6.63	51
Cryptomonadales	0.46	0.04	17
Mallomonas akrokomos Ruttner	1.46	1.48	38
Ochromonadales	5.63	0.31	44
Pseudopedinella spp.	2.06	0.12	42
Uroglena americana Calkins	13.00	1.39	37
Aulacoseira italica (Kütz.) Sim.	0.63	1.98	26
Centrales	0.91	0.12	24
Rhizosolenia eriensis H.L. Smith	0.74	1.25	17
R. longiseta Zach.	1.54	2.70	42
Tabellaria flocculosa (Roth.) Kütz.	3.08	4.71	38
Gonyostomum semen (Ehr.) Dies.	0.008	0.14	17
Ankyra lanceolata (Korsh.) Fott	0.22	0.08	18
Botryococcus sp.	0.03	0.27	38

n = the number of samples where the respective species have been observed.

this observation, although it is supposed to be fairly common in all lake types according to Tikkanen & Willén (1992). A green alga, *Botryococcus* sp. (earlier identified as *B. braunii* Kütz.), was also frequent in this lake type. *B. braunii* is known to be a dominant taxa only in oligotrophic lakes (Järnefelt, 1952), yet the identification of this taxa may be incomplete (Komarék and Marvan, 1992).

Dystrophic lakes

In this investigation the water colour of the dystrophic lakes ranged from 50 to 70 mg l^{-1} Pt. These lakes were fairly deep and their productive layer may also have been deep. The total phosphorus concentration was higher than in oligotrophic or acidic lakes. A total of 363 phytoplankton taxa were identified which is approximately 25% more than in oligotrophic lakes. The results are shown in Table 5.

Table 6. The most important phytoplankton taxa in eutrophic lakes (total of samples: = 53).

Phytoplankton taxa	(Cell ml^{-1} 10^3) (sum)	Biomass (mg l^{-1}) (sum)	(*n*)
Chroococcus limneticus Lemm.	2.71	4.90	27
Microcystis aeruginosa Kütz.	3.62	29.13	47
M. viridis (A.Br.) Lemm.	5.83	47.95	32
M. wesenbergii (Kom.) Starm.	2.10	28.99	32
Snowella lacustris (Chod.) Kom. & Hind.	9.26	12.72	40
Anabaena flos-aquae Bréb.	33.74	42.74	42
A. solitaria v. *planctonica* (Brun.) Kom.	4.74	35.72	25
A. spiroides Kleb.	6.43	22.40	36
Aphanizomenon flos-aquae (Linné) Ralfs	50.76	120.50	32
Acanthoceras zachariasii (Brun.) Sim.	0.53	7.30	23
Aulacoseira ambigua (O. Müll.) Sim.	17.66	18.63	47
A. granulata (Ehr.) Sim.	2.83	20.49	30
A. granulata v. *angustissima* (O. Müll.) Sim.	10.1	83.84	29
Fragilaria crotonensis Kitton	37.11	16.70	34
Actinastrum hantzschii Lagerh.	1.63	0.27	16
Dichotomococcus curvatus Korshikov	16.45	0.66	19
Pediastrum duplex Meyen	1.16	16.31	49
Scenedesmus subspicatus Chod.	1.16	0.34	19
Selenastrum gracile Reinsh	0.56	0.11	20
Tetraëdron caudatum (Corda) Hansg.	1.50	0.24	32
Trachelomonas volvocinopsis Swirenko	0.40	1.66	30

n = the number of samples where the respective species have been observed.

The most common taxon was *Cryptomonas* spp. which was found in 51 samples (out of 52). This taxon is known to prefer dystrophic and nutrient rich waters (Tikkanen & Willén, 1992), but also oligotrophic humic waters (Ilmavirta, 1980). Larger *Cryptomonas* species favour dystrophic lakes (Ilmavirta, 1990), which seems to agree with our data as their biomass is large.

Several chrysophyceans were common in dystrophic lakes, such as *Mallomonas akrokomos*, *Uroglena americana* and *Pseudopedinella* spp. All taxa are flagellated and survive in very low nutrient concentrations; traits that are useful in dystrophic waters (Arvola, 1983; Ilmavirta, 1990). According to Tikkanen & Willén (1992), the genus *Pseudopedinella* is common in nutrient rich environments, which strongly disagrees with our data.

Some diatoms were also abundant in dystrophic lakes. These species were *Aulacoseira italica*, *Rhizosolenia longiseta*, and *Tabellaria flocculosa*. A plausible reason for their occurrence is that these particular lakes were not very dystrophic, but fairly large in size, which enabled turbulence to occur. On the other hand, *A. italica* has been reported to correlate positively with dystrophy ($p = 0.01$) (Pinel-Alloul et al., 1990), due to its low light demands (Talling, 1957).

Eutrophic lakes

The diversity of phytoplankton species was greatest in the eutrophic lakes. When the lakes became hyper-eutrophic the phytoplankton diversity diminished. A total of 520 taxa were identified from the samples. Table 6 shows 21 of the most common species.

As mentioned earlier, the blue-green algae were a prominent phytoplankton group in eutrophic lakes. *Microcystis aeruginosa* was identified in all the eutrophic lakes and from every single sample in the hyper-eutrophic lakes. Due to taxonomical changes *M. flos-aquae* (Wittr.) Kirchner has been included in *M. aeruginosa*. *M. viridis* was also common, and was not found in any other lake type. *Anabaena flos-aquae* Brébisson (earlier also identified as *A. circinalis* Kützing, not Rabenhorst), was common in the most eutrophic lakes, but not in the hyper eutrophic lakes. *Aphanizomenon flos-aque* is a typical eutrophy indicating species (Watanabe, 1991) and, in this study, it could definitely be classified as a hyper-eutrophic species. In fact it was the most important species in the hyper-eutrophic lakes. In less eutrophic waters, the abundant *Aphanizomenon* species is possibly *A. yezoense* described by Watanabe (1991).

Three diatoms, *Aulacoseira ambigua*, *A. granulata* and *Fragilaria crotonensis* were also numerous in eutrophic lakes. These species preferred eutrophic lakes, not hyper-eutrophic ones, and are known as indicators of eutrophy (Järnefelt, 1952; Lepistö, 1988; Trifonova, 1988). Diatoms normally form spring peaks, but may also dominate during the whole growing season in shallow and moderately nutrient-rich conditions (Willén, 1992), as they benefit from mixing of the water column during summer (Lund, 1971).

Pediastrum duplex is a green alga showing many modifications, as exemplified by *P. duplex* v. *gracillimum* W. & G.S. West and *P. limneticum* Thunmark, due to changes in its environment. *P. duplex* was fairly common in all lake types. It was also identified from almost all eutrophic lake samples (49/53) and it is known as a species indicative of eutrophy (Järnefelt 1952, 1956; Brettum, 1989), in both natural and cultural eutrophied lakes (Mantere & Heinonen, 1983).

Conclusions

The results show a distinctive pattern of distribution of key phytoplankton taxa among the four types of lake. Furthermore, the number of taxa identified varied among lake groups; most taxa were identified from the eutrophic lakes, whereas the numbers in acidic lakes were small.

Euglenophyceae were very rare in oligotrophic lakes and non-existent in acidic lakes. Members of classes Tribophyceae, Raphidophyceae and Conjugatophyceae were observed mainly in eutrophic waters, but only in low densities. Some species of Chlorophyceae were found to occur in acidic and dystrophic lakes though, in general, they are most common in eutrophic waters.

The species found in eutrophic lakes were in general those which are known to favour nutrient-rich conditions. However, in more extreme conditions, as in acidic and humic lakes, some species were observed regularly such as: *Merismopedia warmingiana*, *Glenodinium* spp., *Peridinium umbonatum* and *Dinobryon divergens*. *Merismopedia warmingiana* was the only abundant representative of the Cyanoprokaryota in humic and acidic lakes. *Glenodinium* spp., *Peridinium umbonatum* and *Dinobryon divergens* seem to tolerate acidic conditions very well. Members of the genus *Pseudopedinella* were not as common in eutrophic waters as was expected, but were instead found in dystrophic waters, where also *Rhizosolenia longiseta* was abundant. This species was often found in oligotrophic waters as well. *Oocystis rhomboidea* represented the chlorophyceans in the acidic lakes and also in the humic lakes.

This investigation demonstrates once more the importance of correct species identification for the interpretation of the results especially when considering environmental impacts. In order to facilitate the interpretation of the results of the environmental phytoplankton studies, we suggest that the uncertain species names should be presented with a question mark or cf. if possible, rather than plain sp. or spp. The phytoplankton taxa presented in this study are not necessarily indicator species, but they were numerous in the particular types of lake studied and, so, characteristic of them.

References

Arvola, L., 1983. Primary production and phytoplankton in two small, polyhumic forest lakes in Southern Finland. Hydrobiologia 101: 105–110.

Arvola, L., 1985. On the factors affecting phytoplankton species composition in highly coloured small lakes. Lammi Notes 12: 1–4.

Arvola, L., T. R. Metsälä, A. Similä & M. Rask, 1990. Phyto- and zooplankton in relation to water pH and humic content in small lakes in Southern Finland. Verh. int. Ver. Limnol. 24: 688–692.

Arvola, L., K. Salonen, P. Kankaala & A. Lehtovaara, 1992. Vertical distributions of bacteria and algae in deeply stratified humic lake under high grazing pressure from Daphnia longispina. Hydrobiologia 229: 253–269.

Brettum, P., 1989. Algae as indicators of water quality. Phytoplankton. Norsk institut for vannforskning. NIVA-Rapport 0-86116: 1–111. (In Norwegian).

Carmichael, W. W., C. L. A. Jones, N. A. Mahmood & W. C. Theiss, 1985. Algal toxins and water-based diseases. CRC Crit. Rev. envir. Control 15: 275–313.

Eloranta, P., 1989. On the ecology of the genus *Dinobryon* in Finnish lakes. Beiheft zur Nova Hedwigia 95: 99–109.

Eloranta, P., 1995. Phytoplankton of the national park lakes in Central and Southern Finland. Ann. bot. fenn. 32: 193–209.

Heinonen, P., 1980. Quantity and composition of phytoplankton in Finnish inland waters. Publ. Water Res. Inst. 37: 1–91.

Heinonen, P., 1982. On the annual variation of phytoplankton biomass in Finnish inland waters. Hydrobiologia 86: 29–31.

Hindák, F., 1992. On the taxonomy of the genus *Merismopedia* and related genera (Cyanophyta). Algol. Stud. 67: 3–19.

Hoek van den, H., 1984. Einführung in die Phycologie. Stuttgart, 479 pp.

Hutchinson, G.E., 1967. A Treatise in Limnology, 2. New York, 1115 pp.

Ilmavirta, V., 1980. Phytoplankton in 35 Finnish brown-water lakes of different trophic status. Hydrobiologia/Dev. Hydrobiol. Dokulil, M., H. Metz & D. Jewson (eds) 3: 121–130.

Ilmavirta, V., 1983. The role of flagellated phytoplankton in chains of small brown-water lakes in Southern Finland. Ann. bot. fenn. 20: 187–195.

Ilmavirta, V., 1990. Occurrence of phytoplankton species along nutrient, pH and colour gradients in Eastern Finnish lakes. Verh. int. Ver. Limnol. 24: 702–706.

Järnefelt, H., 1952. Plankton als Indikator der Trophiegruppen der Seen. Ann. Acad. Sci. Fenn. A IV: 1–29.

Järnefelt, H., 1956. Zur Limnologie einiger Gewässer Finnlands. XVI. Ann. Zool. Soc. 'Vanamo' 17: 1–201.

Kauppi, P., P. Anttila & K. Kenttämies (eds), 1990. Acidification in Finland. Springer-Verlag, Berlin, Heidelberg, 1237 pp.

Kenttämies, K., 1994. The effect of acidic deposition on waters. In Acidification of inland waters. The third Soviet-Karelian symposium on water problems, Joensuu, Finland, 3-7 June 1991. Publications of the National Board of Water and the Environment, ser. A 188. Helsinki: 9–41.

Kippo-Edlund, P. & A. Heitto, 1990. Phytoplankton and acidification in small forest lakes in Finland. In Kauppi, P., P. Anttila & K. Kenttämies (eds), Acidification in Finland. Springer-Verlag, Berlin, Heidelberg: 973–984.

Komarék, J. & P. Marvan, 1992. Morphological differences in natural populations of the genus Botryococcus (Chlorophyceae). Arch. Protistenkd. 141: 65–100.

Lepistö, L., 1988. Some centric diatoms as indicators of water quality in Finnish lakes. In Simola, H. (ed), Proceedings of the tenth international diatom symposium. Joensuu, Finland, August 28–September 2, 1988: 499–504.

Lepistö, L., 1995. Phytoplankton succession from 1968 to 1990 in the subarctic Lokka reservoir. Publ. Water Envir. Res. Inst. 19: 1–42.

Lund, J. W. G., 1971. An artificial alteration of the seasonal cycle of the plankton diatom *Melosira italica* subs. *subarctica* in an English lake. J. Ecol. 59: 521–533.

Mantere, R. & P. Heinonen, 1983. The quantity and composition of phytoplankton, particularly chlorophyta, in lakes of different trophy levels. Publ. Water Res. Inst. 49: 58–63.

Morling, G. & T. Willén, 1990. Acidification and phytoplankton development in some West-Swedish lakes 1966–1983. Limnologica 20: 291–306.

Pinel-Alloul, B., G. Méthod, G.Verrault & Y. Vigneault, 1990. Phytoplankton in Quebec lakes: variation with lake morphometry, and with natural and anthropogenic acidification. Can. J. Fish. aquat. Sci. 47: 1047–1057.

Rawson, D. S., 1956. Algal indicators of trophic lake types. Limnol. Oceanogr. 1: 18–25.

Reynolds, C. S., 1984. The ecology of freshwater phytoplankton. Cambridge University Press, 384 pp.

Round, F. E., 1981. The ecology of algae. Cambridge University Press, 653 pp.

Ryding, S. O. & C. Forsberg, 1977. Sediments as a nutrient source in shallow polluted lakes. In Golterman, H. (ed), Interactions between sediments and freshwater. W. Junk Publishers, The Hague: 229–234.

Salonen, K., R.I. Jones & L. Arvola, 1984. Hypolimnetic phosphorus retrieval by diel vertical migrations of lake phytoplankton. Freshwat. Biol. 14: 431–438.

Sivonen, K., 1990. Toxic cyanobacteria in Finnish fresh waters and the Baltic sea. Publications of department of microbiology, University of Helsinki 39: 1–70.

Straškrabová, V. & K. Šimek, 1993. Microbial loop in lakes and reservoirs related to trophy and metazooplankton development. Verh. int. Ver. Limnol. 25: 1183–1186.

Talling, J. F., 1957. Photosynthetic characteristics of some freshwater plankton diatoms in relation to underwater radiation. New Phytol. 56: 29–50.

Thompson, P.-A. & G.-Y. Rhee, 1994. Phytoplankton Responses to Eutrophication. Ergebn. Limnol. 42: 126–166.

Tikkanen T. & T. Willén, 1992. Växtplanktonflora. Naturvårdsverket. Eskilstuna, 280 pp. (In Swedish).

Trifonova, I. S., 1988. Oligotrophic-eutrophic succession of lake phytoplankton. In Round, F. E. (ed), Algae in the aquatic environment. Biopress Ltd., Bristol: 107–124.

Watanabe, M., 1991. Studies on the Planktonic Blue-green Algae 3. Some *Aphanizomenon* Species in Hokkaido, Northern Japan. Bull. Natn. Sci. Mus., Tokyo, Ser. B. 17: 141–150.

Willén, E., 1992. Long-term changes in the phytoplankton of large lakes in response to changes in nutrient loading. Nord. J. Bot. 12: 577–587.

Hydrobiologia **369/370**: 99–108, 1998.
M. Alvarez-Cobelas, C. S. Reynolds, P. Sanchez-Castillo & J. Kristiansen (eds), Phytoplankton and Trophic Gradients.

Phytoplankton composition and biomass structure in relation to trophic gradient in some temperate and subarctic lakes of north-western Russia and the Prebaltic

I.S. Trifonova
Institute of Lake Research RAS, Sevastianova 9, 196105 St.Petersburg, Russia

Key words: lake phytoplankton, algae, succession, trophic gradient,biomass, structure, diversity, environmental factors

Abstract

A comparative analysis of the phytoplankton of 27 lakes of three Lake Districts (Karelian Isthmus, Latgalian Elevation and Bolshezemelskaja Tundra), all situated at high temperate (60 and 56 ° N) and subarctic (67–68 °N) latitudes with a wide range of the trophy, is presented. A direct correlation between TP content and phytoplankton biomass was observed in all three regions. As the mean annual TP concentrations range from 10 to 137 mg m^{-3}, the variation of average biomass for a growing season was 0.4–20 g m^{-3}. The total number of phytoplankton species found in individual lakes varied from 25 to 160, increasing along the trophic gradient. The largest number of phytoplankton species was found in lakes supporting mean biomass in the range 10–20 g m^{-3}. Taxonomic diversity decreased in some hypertrophic polluted lakes and in acidic lakes. In lakes of all regions, the number of species of chlorophyte, blue-green and euglenophyte algae increased with lake trophy. Analyses of distribution and ecology of mass species shows that in spite of the influence of latitudinal factors, such as the mineral content of lake waters the succession of dominant phytoplankton species show consistent trends: more cosmopolitan species with wide ecological amplitudes accompany the trophic gradient. Step-wise regression analyses showed that trophic factor had the greatest influence on the total phytoplankton biomass and biomass of blue-greens, chlorophytes and euglenophytes. Biomass of dinoflagellates and chrysophytes was influenced considerably more by total ionic concentration than by the trophic factor. The most important factor for diatoms was lake morphometry. Biomass of nanoplankton algae was influenced by trophic factors but also by lake depth.

Introduction

The oligotrophic-eutrophic spectrum identifies the main type of lake evolution throughout the temperate zone (Odum, 1971). The trophic type of any given lake can be considered as a stage of its evolution, and each stable change leading to an increase of autotrophic production as a step towards eutrophication (Rodhe, 1969).The main factor in lake eutrophication is an increase in nutrient loading, especially of phosphorus, that stimulates the growth of phytoplankton (Edmondson, 1969; Lund, 1969; Vollenweider & Kerekes, 1980).

Phytoplankton composition changes as a result of nutrient enrichment were first diagnosed by Pearsall (1932). He described a sequence from extremely poor plankton with predominance of desmids and chrysophytes, through a transitional stage with predominance of diatoms, to highly productive plankton in which bluegreen algae completely dominate. Approximately similar patterns of phytoplankton composition parallel with an increase in productivity were traced for lakes in other regions of the Europe and America (Järnefelt, 1953; Nygaard, 1956; Brook, 1965; Hutchinson, 1967; Lund, 1969; Edmondson, 1969; Willen, 1979; Kristiansen, 1977; Spodniewska, 1978; Skulberg, 1980; Reynolds, 1984; Rott, 1984) including lakes of North-

Western Russia and the Prebaltic (Trifonova, 1979, 1988).

Based on numerous observations, an evaluation of phytoplankton biomass for lakes of different trophic type has been made (Mikheeva, 1975; Spodniewska, 1978; Trifonova, 1979) and several scales of lake trophy according to phytoplankton biomass have been suggested (e.g. Kalff & Knoechel 1978; Trifonova, 1979, 1989; Heinonen, 1980). Differentiation of the phytoplankton seasonal succession against the trophic gradient has been traced for the Great American Lakes (Stoermer, 1978) for some lakes of Central Europe (Rott, 1984; Sommer, 1986), and for some Russian and Prebaltic lakes (Trifonova, 1979, 1986). I have shown also the direct correlation of the type of biomass seasonal dynamic to the mean annual Total Phosphorus (TP, hereafter) content in lake water (Trifonova, 1988, 1989).

The direct correlation between phytoplankton biomass and phosphorus loading allows the first one to be predicted in relation to trophic state. It is not possible, however, to predict exactly the succession of algae (Lund, 1969). All authors noted that it is difficult to distinguish species-indicators of different levels of trophy. The majority of algae occur in all types of lakes, but their population numbers vary greatly. Planktonic indices have been suggested in an attempt to relate the composition of phytoplankton to lake trophic status (Nygaard, 1956), but these are not indicative, except for extreme trophic types. Transitional stages in the limits of the same trophic type are unclear (Hutchinson, 1967; Reynolds, 1984; Trifonova, 1990). Hutchinson (1967) suggested a classification of phytoplankton assemblages in relation to trophic state, using only the dominant species. Developing this direction several schemes of mass species succession along the trophic gradient have been suggested for lakes of different regions (Reynolds, 1984; Rott, 1984; Trifonova, 1990). These schemes have much in common as to general trends of assemblage change but differ in individual species. So, in spite of the possibility of tracing and predicting the general pattern of successional change in phytoplankton assemblages, it is still unclear which factors (regional and interegional) influence the succession of individual species in lakes of different trophic status.

This paper attempts to compare changes in phytoplankton communities along the trophic gradient in lakes situated in temperate and Subarctic zones. The lakes of three Lake Districts, encompassing a wide range of trophy have been used both for comparative analysis and analysis of factors affecting the composition and structure of lake phytoplankton in the series as a whole.

Characteristics of lakes

Year-round investigations were carried out in 1972–1983, at the Karelian Isthmus lakes, in 1981–1985 at the Latgalian Elevation lakes and in 1986–1989 at the Bolshezemelskaja Tundra lakes. Detailed descriptions of three regions and their lakes have been given in several publications (Trifonova, 1988, 1990). Here it is necessary only to focus on the main differences connected with the latitudinal situation and anthropogenic influence.

The Karelian Isthmus lies between the Gulf of Finland and Lake Ladoga, at about 60 °N in the boreal zone. The Latgalian Elevation is situated on the border of Latvia and Russia, at about 56 °N, in the subboreal zone. The origin of lakes in both regions is connected to glacial activity. Their mutual differences relate to the character of the bedrocks. In the Karelian Isthmus they are mainly granites and, in the Latgalian Elevation,the bedrocks are predominantly carbonate. As a result, the mineral content of the Karelian Isthmus waters is very low, rarely exceeding 60–80 mg l^{-1} and the total ionic concentration in Latgalian lakes is as high as 200–400 mg l^{-1}.

Despite it proximity to St. Petersburg, the Karelian Isthmus can be still considered as a forest region. Forests occupy about 60% of its territory. The content of TP in the great majority of lakes is much lower than 50–60 μg l^{-1} and very rarely exceeds 100 μg l^{-1}. The Latgalian Elevation is a district of intensive agriculture. About 80% of its territory is ploughed and cultivated. There is a high phosphorus content in lakes water. In the majority of lakes it varies between 60–100 μg l^{-1}.

The lakes of the Bolshezemelskaja Tundra are situated in the North-East of the European part of Russia, north of the Polar Circle (67-68 °N) in the Subarctic. The determining factor here is perennial freezing, promoting the formation of the thermokarst landscapes. Shallow thermokarst lakes (Arctic ponds) are the most common type of tundra lakes, although deep glacial lakes are also found. The total ionic concentration of lake water in natural pure lakes is low -20–30 mg l^{-1}, and the content of TP varies from 20 to 50 μg l^{-1}. Some of the lakes lie in an area of intensive oil and gas prospecting. Some are polluted by boring solutions, oil products and also by domestic wastes. In

Table 1. Limnological characteristics of the lakes. S – water area; H – depth; Q – water exchange rate; SD – transparency.

	S km²	H (m) max	mean	Q	SD (m)	pH	Δ_{ions} (mg l^{-1})	TP (mg m^{-3})	TN (mg l^{-1})	N:P
The Latgalian Elevation										
Baltas	0.15	14.0	3.8	–	3.5	7.0–8.0	205	24	3.76	157
Lapetis	0.13	3.5	1.8	6.40	1.8	6.6–7.4	100	26	0.93	36
Udrinka	0.19	26.0	7.8	1.10	2.6	6.8–7.9	307	32	2.16	68
Ilzas	0.12	5.4	2.4	0.46	1.7	7.3–8.8	96	33	0.87	26
Grizhanku	0.15	7.0	4.5	0.56	1.3	7.5–8.7	232	62	1.23	20
Rudushku	0.06	8.2	4.2	0.86	1.7	7.0–8.4	273	72	1.15	16
Laborzu	0.09	4.0	2.5	5.20	0.9	7.1–8.5	308	80	1.58	20
Kazimirowo	0.08	3.8	1.7	–	0.7	7.6–9.1	314	137	0.91	7
The Karelian Isthmus										
Ochotnichje	0.07	10.0	4.7	0.06	4.0	4.4–4.6	12	11	0.40	36
Snetkovskoje	0.67	8.0	4.0	0.07	4.0	6.8–7.6	60	15	0.77	52
Mitchurinskoje	5.82	15.2	8.0	0.09	4.0	6.8–7.4	44	24	0.56	10
Pravdinskoje	6.95	14.3	9.5	1.00	1.8	6.2–8.0	49	52	0.89	17
Nachimovskoje	14.20	22.0	6.5	0.28	2.0	6.0–8.0	46	34	0.83	24
Lugovoe	0.19	4.0	2.0	5.80	0.5	5.4–7.2	36	51	0.63	12
Krasnoje	9.36	14.6	7.2	0.92	1.7	6.7–8.5	63	46	0.65	15
Razliv	10.30	4.6	2.2	10.00	0.6	6.5–8.5	64	75	1.18	16
Borisovskoje	1.38	8.0	3.6	0.55	0.8	6.8–8.5	57	37	0.75	21
Wishnevskoje	10.52	4.0	2.0	0.25	0.5	6.9–9.4	69	130	2.15	16
The Bolshezemelskaya Tundra										
Izjato	3.35	7.0	2.4	0.23	4.0	6.9–7.3	34	24	0.95	39
Taleyto	1.05	1.4	0.9	–	2.6	6.3–6.6	32	28	–	–
Nahyto	3.26	6.0	2.4	–	1.7	7.0–7.5	43	46	0.75	–
Mitrofan	0.31	20.0	6.1	0.34	3.2	6.4–7.0	46	30	0.56	19
Pjartavto	0.51	2.0	1.1	0.67	0.9	6.6–7.2	58	56	0.89	16
Tibeyto	0.06	4.5	2.3	–	2.5	7.1–8.2	82	74	1.16	16
Inzerey Lakes (2, 4, B)	0.44-0.88	0.9-1.8	1.0-1.5	–	0.2-0.8	6.2–8.5	27-59	72-110	1.60-2.50	17-20

polluted lakes, the mineral content increased to up to 200–300 mg l^{-1} and TP content up to 70–100 $\mu g\ l^{-1}$.

The period of ice-free, open water is about 180 days in the Karelian Isthmus, about 200 days on the Latgalian Elevation but only about 90 days in the Bolshezemelskaja Tundra.

The main limnological characteristics of the lakes investigated are given in Table 1. In this table, lakes are arranged in increasing order of TP characterizing their potential productivity. For all three regions, maximal TP content was in lakes undergoing intensive anthropogenic eutrophication. Parallel with increasing nutrient content, the N:P ratio and transparency decreased. For lakes of all three regions direct correlation between mean annual TP content and average for the growing season phytoplankton biomass has been observed. The regression equation expressing this relationship is as follows: $\log B_{phyto} = 1.34 \log TP-1.62$ (Trifonova, 1990). This direct correlation permits biomass to be used as a an index of the trophic gradient and a base for classifying phytoplankton communities.

Methods

All the lakes were studied over two years. Samples were taken in the deepest parts with a Ruttner sampler every 10 days or biweekly. They were fixed with 4% formalin, concentrated by sedimentation and counted in a Najotte chamber (Guseva, 1959). The biomass was determined as algal volume for each lake, and converted to fresh weight assuming a specific gravity of 1 g cm^{-3}. Algae with cell dimensions $\leq 30\ \mu m$ were considered as nannoplankton. For evaluation of

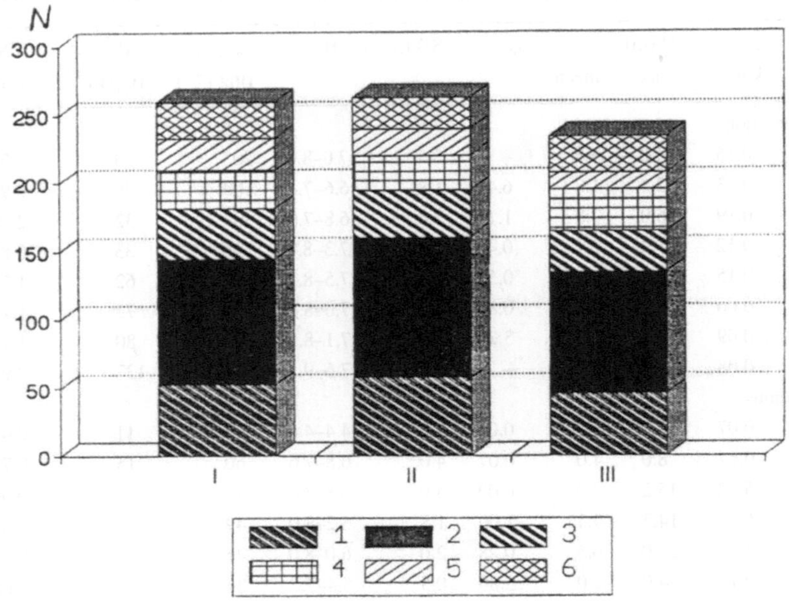

Figure 1. Species richness of the lake phytoplankton in lakes of 3 regions: I – Latgalian Elevation; II – Karelian Isthmus; III – Bolshezemelskaja Tundra. 1 – Bacillariophyta; 2 – Chlorophyta; 3 – Cyanophyta; 4 – Chrysophyta; 5 – Euglenophyta; 6 - Others.

phytoplankton diversity the Shannon index, based on biomass, was used (Odum, 1971).

The influence of the environmental factors on phytoplankton structure was evaluated by stepwise multivariate regression and principal component analysis.

Results and discussion

Species composition and succession of mass species

Comparison of phytoplankton species composition presents considerable difficulties and the total number of species is generally accepted to be of relative value. For this comparison I have taken only species of algae found regularly in quantitative samples of the pelagic phytoplankton and excluded accidental species and taxa of lower rank than species. Such methods are familiar to those working on the comparative floristics of higher plants (for example Safonova & Ermolaev, 1969).

It was found that total number of phytoplankton species for all three regions is almost equal (235–264). Most numerous were species of diatoms (20%), chlorophyceae (40%) and bluegreens (15%) (Figure 1). The phytoplankton of tundra lakes is not less diverse than that of lakes of the temperate zone.

The total number of phytoplankton species in individual lakes varied from 25 to 160 (Figure 2). This number is quite comparable with data from other regions. For example, the number of species phytoplankton of Finnish lakes varied from 3 to 160, the average being 52–62 (Heinonen, 1980). In eutrophic Danish lakes, it was 160–196 (Kristiansen, 1986). Even in tropical African lakes it ranged from 9 to 143 (Kalff & Watson, 1986). Probably the most usual number of species for lake phytoplankton is 50–150, although in large lakes it may be more than 300 (Willén, 1979).

Taxonomic diversity increased from oligotrophic to mesotrophic and eutrophic lakes but decreased in some hypertrophic, polluted and acidic lakes. Such pattern of changes in phytoplankton species diversity along the trophic gradient is well known (Pavoni, 1963; Eloranta, 1978). Different authors give different values for the level of trophy at which the number of species increases. For example, for Finnish lakes level was a phytoplankton biomass of 5 mg l^{-1} (Eloranta, 1978; Heinonen, 1980). Our results show that the biggest species diversity (158–160) is characteristic of lakes with average biomass 12–20 mg l^{-1} (Trifonova, 1990), particularly shallow highly-eutrophic lakes. Obviously, the decrease of phytoplankton species diversity observed in hypertrophic lakes is connected with toxic effect of pollution. This conclusion has been supported

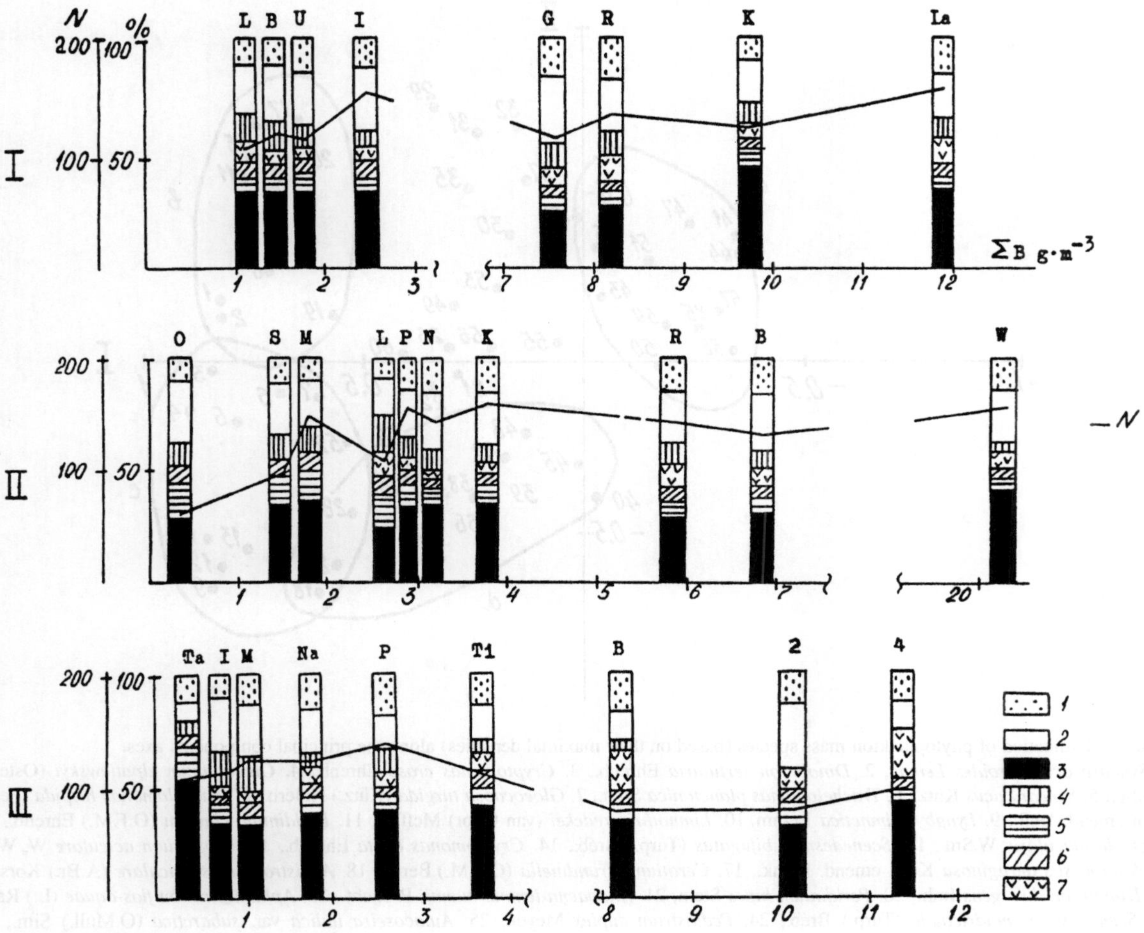

Figure 2. Phytoplankton species richness along the gradient of its mean biomass in lakes of 3 regions: I – Latgalian Elevation; II – Karelian Isthmus; III – Bolshezemelskaja Tundra. 1 – Cyanophyta; 2 – Bacillariophyta; 3 – Chlorophyta; 4 – Chrysophyta; 5 – Dinophyta; 6 – Cryptophyta; 7 – Euglenophyta. Bold letters are the first letters of lake names.

by the low species diversity (20–30 species) in phytoplankton of tundra lakes undergoing oil pollution. However, the strongest factor depressing phytoplankton species diversity is acidification (see also Nygaard, 1956).

In lakes of all regions the number of species of chlorophyceae, bluegreens and euglenophyte algae increased with lake trophy (Figure 2).

Analyses of similarity of lake phytoplankton communities revealed that the effect of trophic conditions on composition of phytoplankton mass species was much stronger than on its total species composition. About 80 species can be considered as mass species composing from 20 to 80% of total phytoplankton biomass. About 60 of them are common to lakes of all three regions.

The influence of the trophic conditions on the composition of the phytoplankton dominants was also substantiated by ordination of 60 mass species by principal component analysis. In the figure 3 axis 1 is responsible for 46% of the total variance and is connected with TP content and water mineralization; axis 2 explains 24% and is connected with lake area and depth . The distribution of algae along the axes of principal components permits complexes of species to be distinguished, characteristic for oligotrophic deepwater lakes (a), eutrophic stratified (b) and eutrophic shallow lakes (c) as well as a complex of algae characteristic for shallow lakes of pond type (d) (Figure 3).

Analyses of distribution and ecology of mass algae in the selected lakes showed that, in spite of the influence of latitudinal factors, particularly water miner-

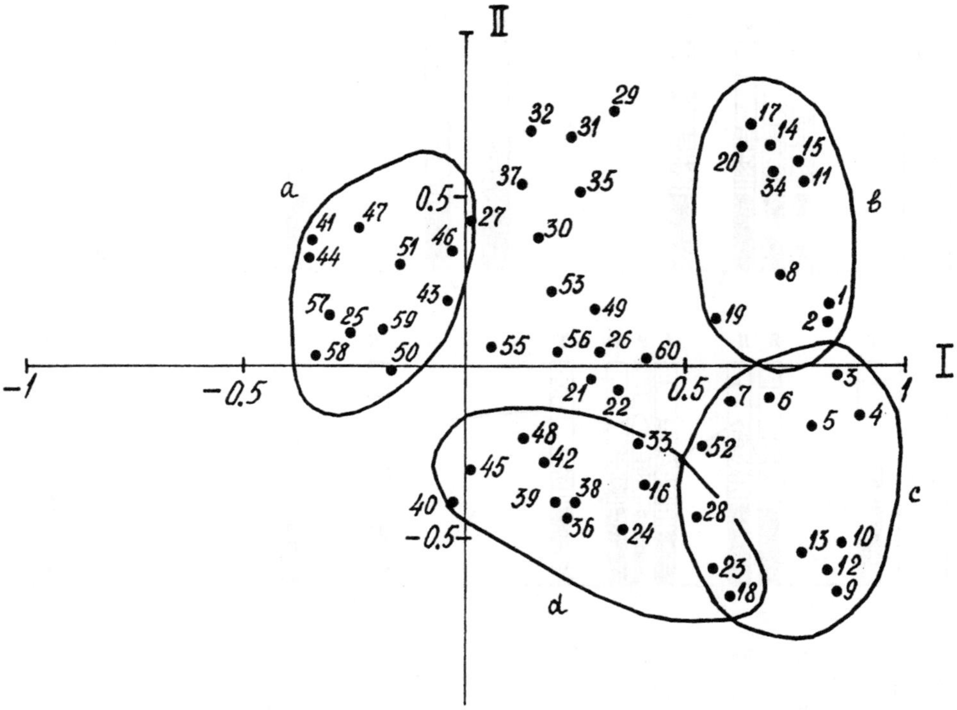

Figure 3. Ordination of phytoplankton mass species (based on their maximal densities) along the principal components axes.
1. *Synedra actinastroides Lemm.*, 2. *Dinobryon sertularia* Ehrenb., 3. *Cryptomonas erosa* Ehrenb., 4. *Glenodinium elpatiewskyi* (Ostenf.) Schiller, 5. *Synedra acus* Kütz., 6. *Trachelomonas planctonica* Swir., 7. *Gloeocapsa turgida* (Kütz.) Holerb., 8. *Trachelomonas hispida* (Perty) Stein emend. Defl., 9. *Lyngbya limnetica* Lemm.,10. *Limnothrix redekei* (van Goor) Meffert, 11. *Peridinium cinctum* (O.F.M.) Ehrenb., 12. *Nitzschia acicularis* W.Sm., 13. *Scenedesmus bijugatus* (Turp.) Bréb., 14. *Cryptomonas ovata* Ehrenb., 15. *Closterium aciculare* W. West, 16. *Microcystis aeruginosa* Kütz. emend. Elenk., 17. *Ceratium hirundinella* (O.F.M.) Bergh, 18. *Ankistrodesmus aciculare* (A.Br.) Korsch., 19. *Dinobryon divergens* Imh., 20. *Peridinium bipes* Stein, 21. *Anabaena lemmermannii* P. Richt., 22. *Aphanizomenon flos-aquae* (L.) Ralfs, 23. *Scenedesmus quadricauda* (Turp.) Bréb., 24. *Pediastrum duplex* Meyen, 25. *Aulacoseira italica* var. *subarctica* (O.Mull.) Sim., 26. *Diatoma elongatum* (Lyngb.) J. Ag., 27. *Crucigenia irregularis* Wille, 28. *Euglena polymorpha* Dang., 29. *Fragilaria crotonensis* Kitton, 30. *Mallomonas tonsurata* Teil., 31. *Closterium acutum* (Lyngb.) Bréb., 32. *Stephanodiscus rotula* (Kütz.) Hendey, 33. *S. hantzschii* Grun., 34. *Phacotus lenticularis* Ehrenb., 35. *Asterionella formosa* Hass., 36. *Chlorogonium maximum* Skuja, 37. *Cyclotella radiosa* (Ehrenb.) Kütz., 38. *Dictyosphaerium pulchellum* Wood, 39. *Chlamydomonas monadina* Stein, 40. *Peridinium aciculiferum* Lemm., 41. *Aulacoseira islandica* (O.Mull.) Sim., 42. *Anabaena hassalii* (Kütz.) Wittr., 43. *A. solitaria* Kleb., 44. *Tabellaria flocculosa* (Roth) Kütz., 45. *Pandorina morum* (O.Mull.) Bory, 46. *Gomphosphaeria lacustris* Chod., 47. *Sphaerocystis schroeteri* Korsch., 48. *Dinobryon bavaricum* Imh., 49. *Aulacoseira granulata* (Ehrenb.) Sim., 50. *Microcystis pulverea* (Wood) Forti emend. Elenk., 51. *Dinobryon sociale* Ehrenb., 52. *Synedra ulna* var. danica (Kütz.) Grun., 53. *Oocystis submarina* Lagerh., 54. *O. solitaria* Wittr., 55. *Aulacoseira ambigua* (Grun.) Sim., 56. *Euglena acus* Ehrenb., 57. *Tabellaria fenestrata* (Lyngb.) Kütz., 58. *Dinobryon cylindricum* Imh., 59. *Rhizosolenia longiseta* Zachar., 60. *Planktothrix agardhii* (Gom.) Anagnostidis et Komárek.

al content, the succession of phytoplankton dominant species for all 3 regions conforms to a common pattern. In the course of eutrophication,the role of cosmopolitan species with wide ecological amplitude increases (Table 2). Among the algae of all taxonomic groups, the species which become most abundant are species capable both of planktonic and benthic way of life and mixotrophic species capable of growing in water rich in nutrients and organic matter. The convergence of phytoplankton structure is a result of lake eutrophication (Trifonova, 1990).

Phytoplankton biomass and structure

The level of phytoplankton biomass was connected strictly with concentration of TP and, generally, it increased along the trophic gradient. The mean biomass for the growing season varied between 0.2–0.7 g m^{-3} in oligotrophic lakes, and 12–20 g m^{-3} in highly eutrophic lakes (Figure 4). In some lakes (for example, Lake Kazimirovo in Latgalia), phytoplankton biomass was much lower than could have been anticipated from TP content. Probably in this case phy-

Table 2. Seasonal and general succession of lake phytoplankton mass species.

Trophic type	Spring	Summer		Autumn
Oligo-	*Aulacos. islandica*	← *Cyclotella* spp. →		*Aulacos. islandica*
trophic	*Dinobryon* spp. (*cylindricum,*		*Peridinium aciculif.*	
	sociale)	*Asterion. formosa*	*Ceratium hirund.*	
		Rhizosolenia spp.		
	Rhodomonas spp.	*Sphaerocystis* spp.	*Oocystis* spp.	
			Crucigenia spp.	
		Anabaena spp., *Gomphosph. lacustris*		
Meso-	*Aulacos. islandica*	← *Aulacos. granulata* →		*Aulacos. islandica*
trophic	*Peridin. aciculif.*	*Fragilar. crotonensis*		*Peridin. aciculif.*
	Aulacos. italica var.		*Aulacos. Italica*	
	subarctica →	*Cerat. hirundin.*		*Asterion. formosa*
	Asterion. formosa →			*Stephanod. rotula*
	Stephanod. rotula →	*Pandorina morum*		*Cyclot. radiosa*
	Cyclot. radiosa			*Dinobryon spp.*
	Dinobryon spp. (divergens), *Gomphosph. lacustris, Gloeotrichia* spp.			
	← *Anabaena* spp., *Aphanizomenon* spp., *Microcystis* spp. →			
Eutro-	*Cryptomonas* spp.	← *Aulacosira granulata* →		*Chlamydom.* spp.
phic	*Chlamydomonas* spp.	*Fragilar. crotonensis*		*Synedra* spp.
	Synedra spp.	*Cerat. hirundin.*		*Asterion. formosa*
	Aulacos. ambigua →	*Peridinum* spp.		*Stephanodiscus* spp.
	Stephanodiscus spp.	*Trachelomonas* spp.		*Aulacos. ambigua*
	Asterion. formosa	*Pandorina morum*		
	Dinobryon spp.	*Scenedesmus* spp.		
	(*sertularia*)	*Pediastrum* spp.		
	Anabaena spp., *Aphanizomenon* spp., *Gloeotrichia* spp., *Microcystis* spp. (*aeruginosa*)			
	← *Planktothrix agardhii,*		*Limnothrix redekei* →	
Hyper	*Chlamydomonas* spp.	← *Nitzschia* spp. →		*Chlamydom.* spp.
trophic	*Chlorogonium* spp.	← *Fragilaria* spp. →		*Chlorogon.* spp.
	Cryptomonas spp.	*Pediastrum* spp.	*Glenodinium* spp.	*Cryptom.* spp.
		Scenedesmus spp.	*Tetraedron* spp.	*Peridinium* spp.
	Trachelom. spp.	*Ankistrodesmus* spp.		*Trachelom.* spp.
	Synedra spp.	*Euglena* spp.	*Synedra* spp.	
	Stephanod. hantzschii	*Phacus.* spp.	*Stephanod. hantzschii*	
		← *Planktothrix* spp., *Limnothrix* spp. →		

toplankton growth was limited by nitrogen deficiency. It is known that nitrogen may become a limiting factor at a ratio N:P < 10 (Sakamoto, 1966; Smith, 1982). In some other lakes, especially those with predominance of *Planktothrix agardhii*, high productivity is realized by means of all-year-round vegetation. The mean annual biomass in such lakes may be much higher than in other lakes (Trifonova, 1988, 1989).

Figure 4 shows the change of phytoplankton biomass structure along the trophic gradient. In lakes of all three regions, the role of bluegreens, green algae and euglenophytes in the total biomass increased with increasing lake trophy. Absolute predominance of blue-greens was observed only in lakes Grizanku, Wishnevskoe and eutrophic shallow tundra lakes, where as much as 75% of the total biomass was of bluegreen algae. According to the literature, the most intensive development of bluegreens usually occurs in water bodies, with a N:P ratio (by weight) about 20 (Guseva, 1965) or 29 (Smith, 1982). In general our data supported these observations. Yet it is obvious that other factors such as water exchange rate, concentration

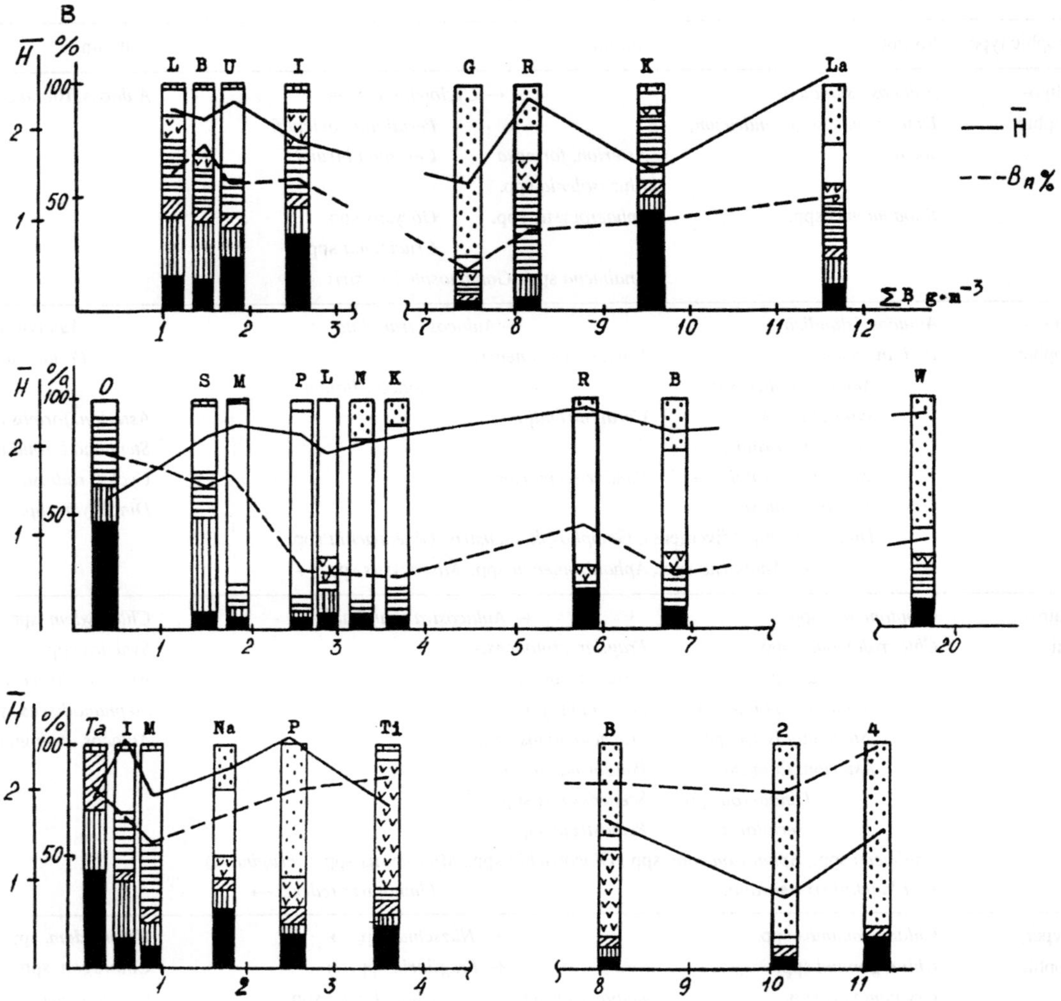

Figure 4. Change of phytoplankton structure and diversity in lakes of 3 regions along the gradient of its mean biomass. H – Shannon diversity; B_H – nannoplankton biomass. Others legends as in Figure 2.

of organic nitrogen are also important (Guseva, 1952; Edmondson, 1969; Lund, 1969; Reynolds, 1984).

Euglenophytes played a leading role not only in small eutrophic lakes but also in polluted tundra lakes and in macrophyte-dominated lakes, especially small Latgalian lakes.

Although the biomass of green algae increased along the trophic gradient, the highest percentage contributions of this group to phytoplankton were in the least productive lakes and in shallow hypertrophic and polluted lakes.

Dinoflagellates were most important in the phytoplankton of comparatively deep lakes, especially the stratified Latgalian lakes. They were practically absent in shallow tundra lakes.

The role of diatoms and chrysophytes in total phytoplankton biomass increased from oligotrophic to mesotrophic lakes but decreased in eutrophic and hypertrophic although their number and biomass in these lakes could be large. In general diatoms were most important in phytoplankton of large mesotrophic and eutrophic lakes of the Karelian Isthmus and chrysophytes in mesotrophic lakes of the Latgalian Elevation with higher total ionic concentration.

Step-wise regression analysis showed that trophic factors had the greatest influence on total phytoplankton biomass and biomass of chlorophytes, bluegreens and euglenophytes, being responsible of 86% of general dispersion (Table 3). Biomass of dinoflagellates and chrysophytes was influenced considerably more,

Table 3. Contribution of different environmental factors (R^2%) to the overall variability of phytoplankton biomass as ascertained by step-wise regression analysis. * – negative relationship. Cyan – Cyanophyta; Eug – Euglenophyta; Chl – Chlorophyta; Bac – Bacillariophta; Din – Dinophyta; Chr – Chrysophyta; B_n – nannoplankton biomass.

Factors R^2	B_{tot}	B_{Cyan}	B_{Eugl}	B_{Chl}	B_{Bac}	B_{Din}	B_{Chr}	B_n	B_n (% B_{tot})
Water area	–	–	–	–	37.6	–	–	–	–
Mean depth	–	–	17.9*	–	–	–	8	15.9*	8.7
Water exchange rate	–	1.6*	–	–	–	–	–	–	–
Transparency	–	–	–	–	23.3*	–	–	13.4	44.3
Water colour	–	–	–	–	-	–	–	–	–
Total ions	–	–	–	–	–	32.4	44	–	–
TP	86.5	76.1	57.5	62.0	–	28.6	–	62.7	–
TN	–	–	–	6.0	–	–	–	–	–
N:P	–	–	–	–	–	–	–	1.7	–
Total R^2	86.5	77.7	75.3	68.0	61.0	61.0	52	93.7	53.0

by total ionic concentration (32% and 44%) than by the trophic factor (28.6%). Most important for diatoms was lake area (38%) and trophic factor (23.3%).

Biomass of nannoplankton also increased along the trophic gradient. Besides the trophic factor (62% of dispersion), nannoplankton biomass was influenced also by lake depth (13%). The biggest values of nannoplankton biomass were in shallow eutrophic lakes. Its contribution to phytoplankton structure was greatest in oligotrophic, slightly-mesotrophic and hypertrophic lakes (Figure 4).

No regularities were revealed in the change of the Shannon diversity index along the trophic gradient (Figure 4). A sharp decrease was observed only in some hypertrophic lakes and lakes with extreme environmental conditions – polluted and acid.

In the course of lake eutrophication not only phytoplankton productivity and community structure change but also its seasonal succession (Trifonova, 1986, 1988, 1989). Lake phytoplankton passes a number of stages from oligotrophy with predominance of cold-water flagellates (chrysophyte and dinoflagellates) and one spring maximum of biomass, through several levels of mesotrophy with diatoms and chrysophytes predominance and 2–3 maxima in seasonal dynamic, to eutrophy with predominance of bluegreens, dinoflagellates, euglenophytes or green algae (depending on lake morphometry and water mineralization) and absence of biomass depression owing to nutrient limitation. The similarity of phytoplankton communities of the same trophic type (convergence of features) at the level of biomass structure was even more pronounced than at the level of individual species.

The critical role of nutrient loading in the succession of lake phytoplankton mass species and changes of its biomass structure is also supported by the results of lake restoration (Edmondson, 1969; Ahlgren, 1978; Gibson,1986). In all cases the decrease in nutrient loading leads to the decrease of blue-green algae abundance and total summer biomass.

Acknowledgements

I am grateful to Dr. Miguel Alvarez for the possibility to participate in the Workshop and to present this paper. I thank Dr. Marina Potapova for computer treatment of the data and Mrs. Olga Verzbitckaja for help in translating. Special thanks are due to Dr. Colin Reynolds for critical review and improvement of English.

References

Ahlgren, G., 1978. Response of phytoplankton and primary production to reduced nutrient loading in Lake Norrviken. Verh. int. Ver. Limnol. 20: 840–845.

Brook, A. J., 1965. Planktonic algae as indicators of lake types with special reference to the Desmidiaceae. Limnol. Oceanogr. 10: 403–411.

Eloranta, P., 1978. Effects of size of sample counted in phytoplankton analyses. Ann. bot. fenn. 15: 169–176.

Edmondson, W. T., 1969. Eutrophication in North America. In Eutrophication, causes, consequences, correctives. Washington: 124–149.

Gibson, C. E., 1986. Preliminary results on phosphorus reduction in Lough Neagh – Assessing the effect against a background of change. Hydrobiol. Bull. 20: 173–182.

Guseva, K. A., 1952. 'Water-bloom', its causes, prediction and measures of prevention. Tr. Vses. Gidrobiol. Obsch. Moscow: 3–94 (In Russian).

Guseva, K. A., 1959. Methods of phytoplankton counting. Tr. Inst. Biol. Vodochr. 2: 44–51 (In Russian).

Guseva, K. A., 1965. Role of bluegreen algae in water-bodies and factors of their mass development. In Ecology and physiology of bluegreen algae. Nauka Publ. House, Moskow: 36–42 (in Russian).

Heinonen, P., 1980. Quantity and composition of phytoplankton in Finnish Inland waters. Vesihallitus – National Board of waters, Finland, Helsinki. 91 pp.

Hutchinson, G. E., 1967. A Treatise on Limnology, II. Introduction to lake biology and the limnoplankton. Wiley Intersci. Publ., New York, 1115 pp.

Järnefelt, H., 1953. Plankton als Indikator der Trophiegruppen der Seen. Ann. Acad. Sci. Fenn, 18: 1–29.

Kalff, J. & R. Knoechel, 1978. Phytoplankton and their dynamics in oligotrophic and eutrophic lakes. Ann. Rev. Ecol. Syst. 9: 475–495.

Kalf, J. & S. Watson, 1986. Phytoplankton and its dynamics in two tropical lakes: a tropical and temperate zone comparison. Hydrobiologia 138: 161–176.

Kristiansen, J., 1977. A survey of investigation of freshwater phytoplankton in Denmark. Folia Limnol. Scand. 17: 37–44.

Kristiansen, J., 1986. Phytoplankton in eutrophic lakes – community structure and succession. In Riemann B. & M. Sondergaard (eds), Carbon Dynamics in Eutrophic Temperate Lakes. Elsevier: 28–40.

Lund, J. W. G., 1969. Phytoplankton. In Eutrophication, causes, consequences, correctives. Washington: 306–330.

Mikheeva, T. M., 1975. Evaluation of phytoplankton biomass values in lakes of the whole world. Hydrobiol. J. Kiev 11: 90–104 (In Russian).

Nygaard, G., 1956. Phytoplankton. Folia Limnol. scand. 8: 32–144.

Odum, E. P., 1971. Fundamentals of ecology. Saunders, Philadelphia.

Pavoni, M., 1963. Die Bedeutung des Nannoplanktons im Vergleich zum Netzplankton. Schweiz. Z. Hydrol. 25: 219–341.

Pearsall, W. H., 1932. Phytoplankton in the English lakes. II. The composition of the phytoplankton in relation to dissolved substances. J. Ecol. 20: 241–262.

Reynolds, C. S., 1984. The ecology of freshwater phytoplankton. Cambridge University press, Cambridge, 384 pp.

Rodhe, W., 1969. Crystallisation of eutrophication concept in northern Europe. In Eutrophication, causes, consequenses, correctives. Washington: 50–64.

Rott, E., 1984. Phytoplankton as biological parameter for the trophic characterisation of lakes. Verh. int. Ver. Limnol. 22: 1078–1085.

Sakamoto, M., 1966. Primary production by phytoplankton community in some Japanese lakes and its dependence on lake depth. Arch. Hydrobiol. 62: 1–28.

Safonova T. A. & V. I. Ermolaev, 1983. Algae of water-bodies of the lake Tchani system. Nauka Publ. House, Novosibirsk (in Russian).

Smith, V. N., 1982. The nitrogen and phosphorus dependence of algal biomass in lakes: an empirical and theoretical analysis. Limnol. Oceanogr. 27: 1101–1112.

Sommer, U., 1986. The periodicity of phytoplankton in Lake Constance (Bodensee) in comparison to other deep lakes of central Europe. Hydrobiologia 138: 1–7.

Skulberg, O. M., 1980. Blue-green algae in lake Mjφsa and other Norwegian lakes. Progr. Water Technol. 12: 121–140.

Spodniewska, I., 1978. Phytoplankton as the indicator of lake eutrophication. I. Summer situation in 34 Masurian lakes in 1973. Ekol. Polska 26: 53–70.

Stoermer, E. F., 1978. Phytoplankton assemblages as indicators of water quality in Laurentian Great Lakes. Trans. am. Microsc. Soc. 97: 2–16.

Trifonova, I. S., 1979. Species composition and productivity of phytoplankton in Karelian Isthmus lakes of different type. Nauka Publ. House, Leningrad, 168 pp. (In Russian).

Trifonova, I. S., 1986. Seasonal and general succession of lake phytoplankton. Hydrobiol. J. Kiev, 23: 21–28 (In Russian).

Trifonova, I. S., 1988. Oligotrophic-eutrophic succession of lake phytoplankton. In: Round F. (ed), Algae and aquatic environment. Biopress, Bristol: 107–124.

Trifonova, I. S., 1989. Changes in community structure and productivity of phytoplankton as indicators of lake and reservoirs eutrophication. Arch. Hydrobiol. Beih. Ergebn. Limnol. 33: 363–371.

Trifonova, I. S., 1990. Ecology and succession of lake phytoplankton. Nauka Publ. House, Leningrad, 174 pp. (In Russian).

Vollenweider, R. A. & J. Kerekes, 1980. The loading concept as basis for controlling eutrophication phylosophy and preliminary results of the OECD programme on eutrophication. Progr. Wat. Technol. 12: 5–38.

Willén, E., 1979. Water quality and phytoplankton in two large Swedish lakes. Acta Bot. Fenn. 110: 81–85.

Willén, T., 1975. Biological long-term investigation of Swedish lakes. Verh. int. Ver. Limnol. 19: 1117–1124.

Hydrobiologia **369/370:** 109–116, 1998.
M. Alvarez-Cobelas, C. S. Reynolds, P. Sanchez-Castillo & J. Kristiansen (eds), Phytoplankton and Trophic Gradients.

Ecophysiological characteristics of two planktonic desmid species originating from trophically different lakes

Elly Spijkerman & Peter F. M. Coesel
Department of Aquatic Ecology, University of Amsterdam, Kruislaan 320, NL-1098 SM Amsterdam, The Netherlands

Key words: alkaline phosphatase activity, desmids, light limitation, phosphorus limitation, phytoplankton, P uptake rates

Abstract

Two planktonic desmid species were compared in some of their ecophysiological characteristics. *Staurastrum chaetoceras*, well-known for its abundant occurrence in eutrophic lakes, showed a higher photosynthetic capacity and a higher maximum (intrinsic) growth rate than *Cosmarium abbreviatum* var. *planctonicum*, a taxon only encountered in oligo-mesotrophic habitats. The two taxa are comparable in cell size. When grown under a stringent continuous inorganic phosphorus (P_i) limitation *C. abbreviatum* realized a higher growth rate, due to a higher affinity for the uptake of P_i, than *S. chaetoceras*. On the other hand, under those conditions, *S. chaetoceras* displayed a two times higher maximum P_i uptake rate (V_{max}). Regarding cellular alkaline phosphatase activity (hydrolysis of the organic P substrate MFP) *C. abbreviatum* showed both a higher affinity and maximum rate than *S. chaetoceras*.

In a way, these characteristics reflect the distribution pattern of the two species in the field. For in eutrophic lakes, during the summer algal bloom, species often have to compete for light as the growth limiting factor, whereas species occurring in oligo-mesotrophic lakes usually face permanently growth-limiting P concentrations. Since in eutrophic lakes during summer algal bloom dissolved inorganic P concentrations can also be low, the ability of phytoplankton to acquire P_i from short-lived pulses (e.g. excretion of P by zooplankton or fish) has to be considered an important additional characteristic in view of competition. Concerning the two desmid species under discussion, *S. chaetoceras* will have a competitive advantage when P_i is supplied in distinct pulses, due to its higher V_{max} values. On the other hand, *C. abbreviatum* possibly will be superior in competition for organic P substrates.

In the species studied, different strategies were found to benefit optimally from the resource conditions inherent in the trophic state of their habitat.

Introduction

The ability of a given phytoplankton species to sustain a population in an oligotrophic environment requires ecophysiological characteristics different from those to be expected in a eutrophic habitat. When the summer species composition of phytoplankton of trophically distinct hard-water lakes from the temperate zone are compared, specific differences in the growth and nutrient-uptake characteristics of the main species are anticipated (Hecky & Kilham, 1974; Tilman, 1977; Kilham & Tilman, 1979). Conversely, nutrient uptake and growth kinetics may help to explain the distribution of algal species in the field. Cell size is found to be an important parameter determining ecophysiological parameters (Smith & Kalff, 1982), but opposite opinions exist about the relation between predominant algal cell size and nutrient load of a lake. Smaller cells, with their high surface area-to-volume ratio, generally have high nutrient uptake affinities and thus would benefit more from a low nutrient load in a lake (e.g. Herbland et al., 1985; Watson & Kalff, 1981; Wehr, 1993). On the other hand it is said that nutrient-depleted waters favour the occurrence of large cell-sized phytoplankton species because of their lower maximum growth rate (μ_{max}) (Maestrini & Bonin, 1981), which would be of survival value (Lund, 1965). This is inconsistent with the conclusion of Reynolds (1987) that species with similar cell size and cell shape demonstrate similar growth and survival strategies. Guillard et al. (1973) compared two clones of the diatom *Thalassiosira pseudonana* Hasle et Heimdal in their silicate-limited growth characteristics and found that the affinity of the clone from the oligotrophic site was higher and its μ_{max} was significantly lower than the clone from the eutrophic site. The two planktonic desmid taxa used in the present study, *Cosmarium abbreviatum* Rac. var. *planctonicum* W. & G. S. West (characteristic of an oligo-mesotrophic environment) and *Staurastrum chaetoceras* (Schr.) G.M. Smith (peaking in eutrophic water bodies: Coesel 1994) are comparable in cell size. Possible differences in kinetic characteristics, caused by cellular surface-to-volume ratios are thus reduced to a minimum. Both species amass their maximum population densities in the summer period when algal biomass is relatively high and dissolved nutrient concentrations are low. The desmid taxa were compared in their maximum growth rate, growth under light or continuous phosphorus- (P_{i-}) limited conditions, P_{i-} uptake rates and alkaline-phosphatase activity as possible indicators for their pronounced distribution pattern in the field.

High maximum P-uptake rates and maximum growth rates enable a species to invade an environment with (temporary) high ambient nutrient concentrations (r-strategist). Low maximum growth rates and high affinity for P uptake reflect the survival strategy of a species growing under low nutrient conditions (K-strategist, Sommer, 1981). During a summer bloom, low light intensities will benefit the growth of algal species with a high photosynthetic capacity. As for phosphorus, usually acting as the limiting nutrient, Hantke et al. (1996b) found that more than half of the total P content of a eutrophic lake was found in particles and the dissolved P fraction contained twice as much DOP (dissolved organic P) as SRP (soluble reactive P). Since it can be assumed that 60% of DOP consists of suitable substrates for phosphatases (Hantke et al., 1996b), alkaline phosphatase activity (APA) reflects the ability of algae to use organic P sources for growth.

The present paper is a concise review of previously presented findings about the growth and nutrient uptake kinetics of *C. abbreviatum* and *S. chaetoceras*, complemented with recently obtained data concerning alkaline phosphatase activities.

Materials and methods

Algal isolates

The experiments were performed with *Staurastrum chaetoceras*, clone AO 36, isolated from the alkaline eutrophic Lake IJmeer (Berger & Sweers, 1988) and *Cosmarium abbreviatum* var. *planctonicum*, clone AO 116, isolated from the alkaline, oligo-mesotrophic Lake Maarsseveen (I) (Swain et al., 1987). Characteristics of the lakes are listed in Table 1. Both clones were taken from the desmid collection in the Department of Aquatic Ecology, University of Amsterdam.

Culturing conditions in a temperature-light gradient

The species were batch-cultured in glass dishes with 50 ml Woods Hole medium. The growth rates at different combinations of temperature and incident irradiance were assessed in an apparatus that combines a temperature gradient with a light gradient at right angles. Algal cells were counted with a Coulter Multisizer. All experiments were performed under a 16 h : 8 h light-dark cycle. For more detailed information, see Coesel & Wardenaar (1990).

Culturing conditions in turbidostat and chemostat

The species were cultured at 20 ± 1 °C in 1 l continuous-flow culture vessels. The inflow medium for the light-limited cultures (turbidostat) contained 50 μM P, while for continuously P-limited cultures (chemostat) it contained only 5 μM P. For details of the continuous flow device and turbidostat experiments, see Coesel & Wardenaar (1994); of the composition of the P-limited medium, see Spijkerman & Coesel (1996a). Average irradiance in the culture vessel was calculated according to Van Liere et al. (1978). Illumination of the uni-algal cultures was continuous. Cultures were not axenic but bacterial biomass (estimated by acridine staining and counting under a fluorescence microscope) was negligible ($< 1\%$ of algal biomass). Chemostat cultures were considered to be in a steady state (dilution rate D = growth rate μ) when algal biomass varied less than 5% over at least 2 renewals of the culture vessel volume. For details about calcula-

Table 1. Some characteristics of Lake Maarsseveen I (Lingeman et al., 1987; Swain et al., 1987) and Lake IJmeer (Berger & Sweers, 1988; unpublished data of the Ministry of Public Works, directorate of IJsselmeer area). The IJmeer is part of the IJsselmeer.

	Lake Maarsseveen I	Lake IJmeer
Location (Lat., Long.)	5° 04′ E–52° 09′ N	5° 04′ E–52° 21′ N
Surface area (ha)	70	4500
Max. length (m)	1,760	20,000
Max. depth (m)	30.8	25 (in local pits)
Mean depth (m)	12.1	4.5
Average period of stratification	May–October	No stratification period in 90% of the area
Average depth of thermocline (m)	8–13	–
Range of vertical extinction coeff. (m^{-1})	0.4–0.6	1.0–4.6
Range of Secchi disc transparencies (m)	3.5–8.0	0.2–0.9
P-total ($mg\ l^{-1}$)	0.009–0.012	0.07–0.32
N-total ($mg\ l^{-1}$)	0.1–0.6	1.1–2.0
Conductivity ($\mu S\ cm^{-1}$)	350–500	600–900
pH	7.5–8.3	7.8–9.2
Chl_a maximum ($\mu g\ l^{-1}$)	4	47
Phytoplankton characterisation	Chrysophyceae	Chlorococcales

tions of chemostat culture parameters, see Spijkerman & Coesel (1996a).

P-uptake experiments

Uptake experiments were performed with culture material from steady-state conditions, which was twofold diluted with P-free culture medium and pulsed with different inorganic phosphorus (P_i) concentrations containing ^{32}P as described in Spijkerman & Coesel (1996a). Initial P_i concentrations ranged from 0.5 to 30 μM. Cellular ^{32}P contents were determined at t = 0, 30, 60 and 300 s, and initial uptake rates (V) for every initial P_i concentration were calculated from linear regression, following Riegman & Mur (1984). By curve fitting to the Michaelis-Menten equation

$$V = V_{max}[P_i / (K_m + P_i)]$$

maximum uptake rate (V_{max}) and half saturation constant for uptake (K_m) were computed.

Alkaline phosphatase activity

The alkaline phosphatase activity (APA) was determined on cells from continuous-flow cultures with an inflow medium free from P, adapted to a pulse of 10 μM P_i every two weeks ($D = 0.007\ h^{-1}$). Starved cells were harvested right before the pulse was given to the culture, washed and resuspended in P-free Woods Hole medium without Fe-EDTA (see Jansson et al., 1988). The medium was brought to a pH 8.5 to optimize the enzyme activity. APA was measured by adding 3-0-MFP (methyl fluoresceine phosphate) to the cells and measuring the release of MF (methyl fluoresceine) in time on a recorder connected to a Perklin Elmer fluorimeter (Perry, 1972). The initial concentrations of MFP ranged from 0.05 to 4.5 μM. The determination (in duplicate) was repeated 10 weeks later in another sample from the continuous flow culture in a comparable starved condition.

Results and discussion

One of the main results observed was the higher intrinsic growth rate of *Staurastrum chaetoceras* as compared to *Cosmarium abbreviatum*, this in spite of the fact that the two species are comparable in cell size. As is shown in Figure 1, the difference in growth rate in favour of *S. chaetoceras* increases with increasing

Table 2. Maximum uptake rates (V_{max}) in μmol P mg protein^{-1} h^{-1}, half saturation constants (K_m) in μmol P l^{-1} and affinities (V_{max}/K_m). Mean values ± standard deviation of *C. abbreviatum* and *S. chaetoceras* at different steady-state growth rates (μ). Number of replicates in parentheses. (After Spijkerman & Coesel, 1996b).

	Cosmarium abbreviatum				*Staurastrum chaetoceras*			
μ (h^{-1})	V_{max}	K_m	V_{max}/K_m	n	V_{max}	K_m	V_{max}/K_m	*n*
0.003	0.45 ± 0.10	1.54 ± 0.76	0.33 ± 0.14	(4)				
0.005	0.47	1.60	0.30	(1)	0.81 ± 0.22	4.09 ± 1.27	0.20 ± 0.05	(6)
0.007	0.34 ± 0.01	1.51 ± 0.84	0.27 ± 0.14	(3)				
0.010	0.40 ± 0.11	0.29 ± 0.54	0.33 ± 0.07	(3)	0.91 ± 0.22	6.44 ± 1.56	0.14 ± 0.02	(3)
0.015	0.27 ± 0.10	1.97 ± 1.02	0.17 ± 0.12	(3)	0.95 ± 0.09	5.80 ± 1.08	0.17 ± 0.03	(4)
0.020	0.65 ± 0.10	12.44 ± 1.76	0.05 ± 0.10	(3)	0.98 ± 0.21	6.95 ± 2.10	0.15 ± 0.02	(3)
0.030					0.57 ± 0.12	2.95 ± 0.72	0.20 ± 0.02	(3)
0.035					0.53 ± 0.34	6.94 ± 2.61	0.07 ± 0.04	(3)

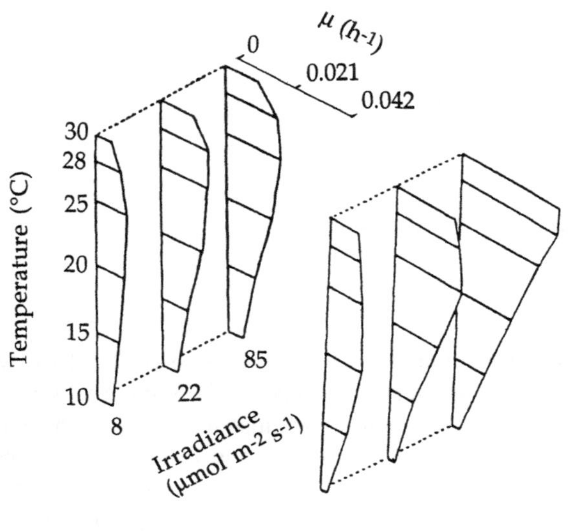

Figure 1. Growth response (μ, h^{-1}) of *Cosmarium abbreviatum* and *Staurastrum chaetoceras* tested in the light-temperature gradient apparatus. (After Coesel & Wardenaar, 1990)

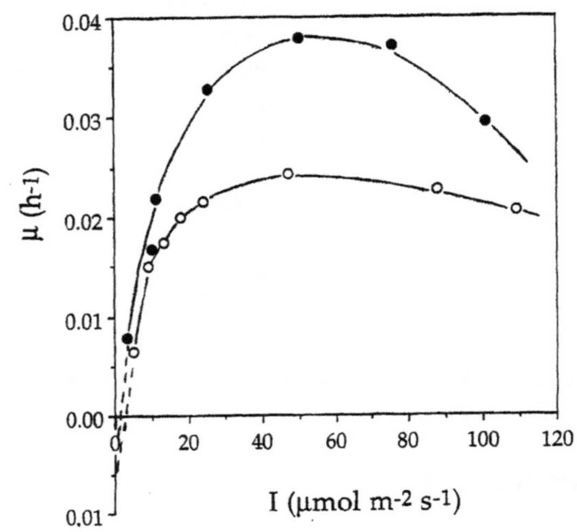

Figure 2. Growth rate (μ, h^{-1}) in relation to irradiance (I) in *Cosmarium abbreviatum* (○) and *Staurastrum chaetoceras* (•) grown in continuously illuminated turbidostat cultures. (After Coesel & Wardenaar, 1994).

light intensity and temperature. Actually, the maximum light intensity as offered by the light-temperature apparatus appeared to be still suboptimal for *S. chaetoceras*. Growth rate of the latter species could be raised to $\mu > 0.042$ h^{-1} by providing higher incident irradiance (not shown). The higher growth response to light intensity in *S. chaetoceras* as compared to *C. abbreviatum* (Figure 1) was confirmed in the turbidostat experiments at a fixed temperature of 20 °C (Figure 2) and could be related to a higher photosynthetic capacity in *S. chaetoceras* (Coesel & Wardenaar, 1994). From these data, it is likely that *S. chaetoceras* will have a relatively high competitive ability in shallow lakes where light climate and temperature in summer are favourable to realize near μ_{max} values, provided of course, that there are enough nutrients available. In this connection, it is important to may benefit from temporary high concentrations of essential nutrients. Indeed, the maximum uptake rate (V_{max}) for phosphate (usually a growth limiting nutrient in fresh waters) in *S. chaetoceras* appeared to be distinctly higher than that in *C. abbreviatum* (Table 2). These characteristics (high V_{max} and μ_{max}) can be considered adaptations to a dynamic environment (Zevenboom, 1986).

Growth responses under continuous P limitation (Figure 3) show that, under stringent P-limited conditions, *C. abbreviatum* can realize a higher growth

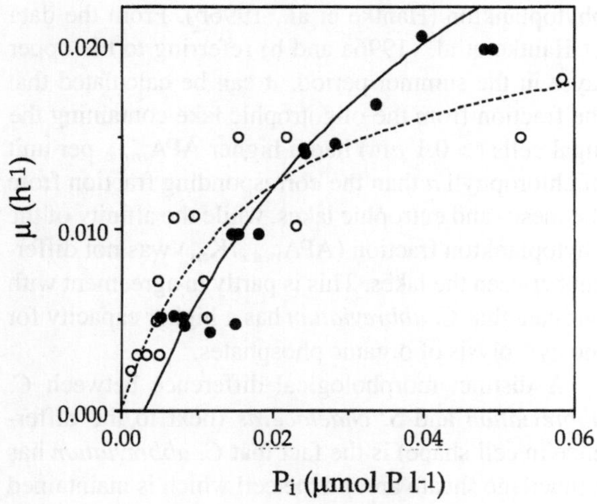

Figure 3. Growth rate (μ, h^{-1}) in relation to the inorganic phosphorus concentration in the culture vessel (P_i) of *Cosmarium abbreviatum* (○) and *Staurastrum chaetoceras* (●). Curves fitted to the Monod model. (From Spijkerman & Coesel, 1996b).

rate, due to a higher affinity for P than *S. chaetoceras* (Table 2). The results from Figure 2 and 3 concerning two particular growth rates, i.e. $\mu = 0.007\ h^{-1}$ and $\mu = 0.015\ h^{-1}$ are combined in Figure 4. This figure shows isoclines that indicate the minimum resource levels needed to sustain a rate of growth adequate to exceed the loss rate imposed (in steady-state continuous flow culture: loss rate = dilution rate = growth rate). This rendering is based on a general concept by Phillips (1973), which concept was extended by Taylor & Williams (1975) for growth of micro-organisms under substrate limitation. The theory in question has been worked out in more detail and applied for phytoplankton studies by Tilman (1980) and Tilman et al. (1982). Huisman & Weissing (1994, 1995) were the first to extend this theory to a combination of light and nutrient limitation. This is more complicated in view of the light gradient caused by shading and mutual interference of light- and nutrient-limited growth responses. Healey (1985) found that the internal concentration of the limiting nutrient of the blue-green alga *Synechococcus linearis* (Naeg.) Kom. increased with decreasing irradiance. This effect was the most pronounced with nitrate and also found for *Pseudoanabaena sp.* (see also Rhee & Gotham, 1981). Knoechel & deNoyelles (1980) found that *Dinobryon sertularia* from natural phytoplankton responded to changes in both light and P supply rates. Because of these interactions, the isoclines in Figure 4 most probably have round corners but, because no data concerning this situation have been obtained, only straight, right-angled isoclines are inserted here, for simplicity. Huisman & Weissing (1995) showed in their model that it is not just the ratio of nutrient supply to light supply, but also the absolute supply rate of the light that determines the outcome of competition. This is a fundamental difference with the nutrient ratio theory of Tilman (1982). Despite all simplifications applied to the isoclines in Figure 4, it is shown that at a low loss rate (in ecological terms, a weakly dynamic environment) *C. abbreviatum* is better adapted to utilize low P concentrations, whereas *S. chaetoceras* is in advantage at low light intensities (Figure 4a). In contrast to that, at higher loss rates (as met with in a more dynamic environment) *C. abbreviatum*

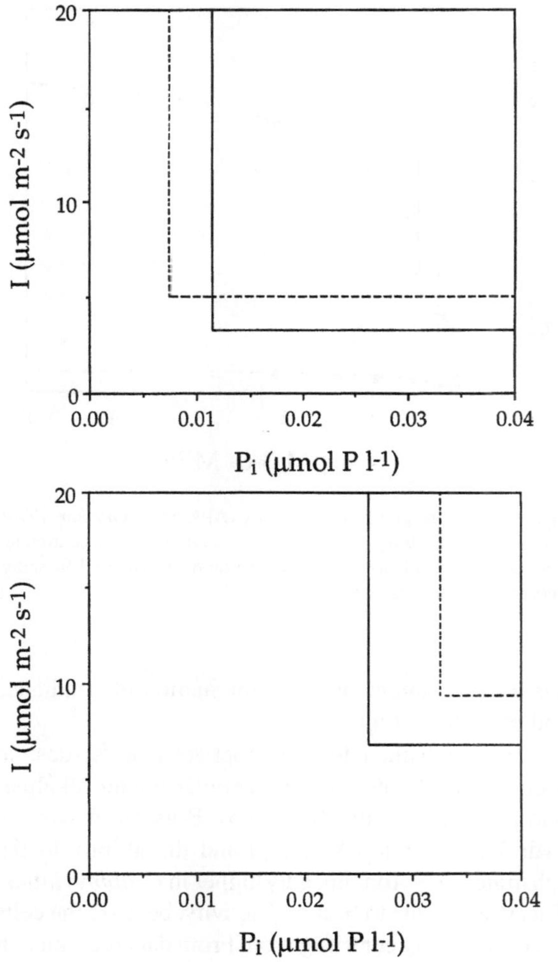

Figure 4. Growth isoclines of *Cosmarium abbreviatum* (– –) and *Staurastrum chaetoceras* (—) at a loss rate of (a) $D = 0.007\ h^{-1}$ and (b) $D = 0.015\ h^{-1}$ indicating their competitive ability for light (I) and phosphorus (P_i) limiting conditions.

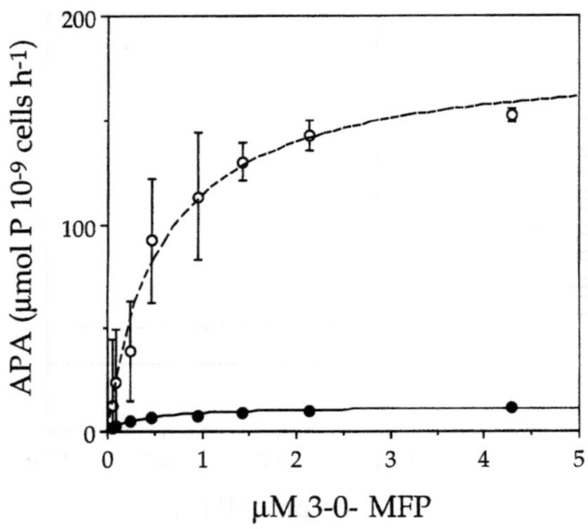

Figure 5. Alkaline phosphatase activity (APA) of *Cosmarium abbreviatum* (○) and *Staurastrum chaetoceras* (●) from starved continuous flow cultures in relation to the concentration of added 3-0-methyl fluoresceine phosphate (MFP).

has a disadvantage at any combination of irradiance and phosphate (Figure 4b).

Another distinct difference between the two desmid species was found in their cellular-bound alkaline-phosphatase activities (Figure 5). Both the maximum hydrolysation rate (APA_{max}) and the affinity to the substrate were pronouncedly higher in *C. abbreviatum*. This was not due to bacterial activity, because the cells were washed by centrifugation. From data presented in Figure 5 and Table 2 it can be calculated that *C. abbreviatum*, with its cell-bound phosphatases can free more than enough P_i from an organic P substrate (APA_{max}), to meet its maximum P_i uptake rate (V_{max}). This in contrast to *S. chaetoceras* where APA_{max} was much lower and V_{max} much higher than in *C. abbreviatum*. It is well-known that phosphatases are excreted by different planktonic organisms (e.g. Kuenzler & Perras, 1965; Berman, 1970) and our data suggest that *C. abbreviatum* may sequester enough P_i to satisfy its own requirements by hydrolysation of organic P compounds via cellular APA, whereas *S. chaetoceras* will benefit more from temporarily high concentrations of P_i released by dissolved AP (supplied by other planktonic organisms). Our findings can not be directly linked to field studies on APA. Hantke et al. (1996b) found increasing total phosphatase activities with higher eutrophication level, and the algal fraction in general contributed less than half of the total APA. In the oligotrophic lake they studied, 50% of the total hydrolysis was realized by the dissolved fraction and only 13% by phytoplankton (Hantke et al., 1996b). From the data of Hantke et al. (1996a and b) referring to the upper layer in the summer period, it can be calculated that the fraction from the oligotrophic lake containing the algal cells (> 0.1 μm) had a higher APA_{max} per unit of chlorophyll *a* than the corresponding fraction from the meso- and eutrophic lakes, while the affinity of the phytoplankton fraction (APA_{max}/K_m) was not different between the lakes. This is partly in agreement with our data that *C. abbreviatum* has a higher capacity for the hydrolysis of organic phosphates.

A distinct morphological difference between *C. abbreviatum* and *S. chaetoceras* (next to the difference in cell shape) is the fact that *C. abbreviatum* has a mucilage sheath around the cell which is maintained under all possible physiological conditions (Coesel, 1994). The higher APA of *C. abbreviatum* could be linked to this phenomenon, if the mucilage sheath surrounding the cells of *C. abbreviatum* acted as a storage for AP, but we were unable to demonstrate this by shaking or sonification, because these treatments released the cell-bound APA of *S. chaetoceras* as well. We failed to detect bacteria in the sheath as a possible source for APA, so we have to conclude that the algal cells themselves were responsible for the high enzyme production. That the algal cell sheath might act as a primary or secundary location for AP was also suggested by Doonan & Jensen (1977), but they could not prove it either.

The observed species-specific differences in the ecophysiological characteristics fit very well to the origin of the algal strains used. For *Staurastrum chaetoceras*, characterized by a relatively high photosynthetic capacity, high maximum P-uptake rate and high maximum growth rate, was isolated from a eutrophic lake, where, in summer time, light climate and dissolved nutrient concentrations are usually highly variable. In contrast, *Cosmarium abbreviatum*, marked by a relatively high affinity for inorganic P uptake and a high alkaline-phosphatase activity originated from an oligo-mesotrophic clearwater lake, where phosphorus is supposed to be the main algal growth limiting factor. So, although cell size may be a major factor in predicting competition between phytoplankton species for a limiting nutrient, the potential variation in uptake parameters among species of comparable size should not be underestimated.

Acknowledgements

We are grateful to Prof. Dr. J. Ringelberg for the fruitful discussions about the ecological relevance of the results and Dr. R. Riegman for scientific support during this research. We also thank J. Huisman for his comments on the 'Tilman' concept.

References

Berger, C. & H. E. Sweers, 1988. The IJsselmeer and its phytoplankton – with special attention to the suitability of the lake as a habitat for *Oscillatoria agardhii* Gom. J. Plankton Res. 10: 579–599.

Berman, T., 1970. Alkaline phosphatases and phosphorus availability in Lake Kinneret. Limnol. Oceanogr. 15: 663–674.

Coesel, P. F. M. 1994. On the ecological significance of a cellular mucilaginous envelope in planktic desmids. Algol. Stud. 73: 65–74.

Coesel, P. F. M. & K. Wardenaar, 1990. Growth responses of planktonic desmid species in a temperature-light gradient. Freshwat. Biol. 23: 551–560.

Coesel, P. F. M. & K. Wardenaar, 1994. Light-limited growth and photosynthetic characteristics of two planktonic desmid species. Freshwater Biol. 31: 221–226.

Doonan, B. B. & T. E. Jensen, 1977. Ultrastructural localization of alkaline phosphatase in the blue-green bacterium *Plectonema boryanum*. J. Bact. 132: 967–973.

Guillard, R. R. L., P. Kilham & T. A. Jackson, 1973. Kinetics of silicon-limited growth in the marine diatom *Thalassiosira pseudonana* hasle and heimdal (= *Cyclotella nana* hustedt). J. Phycol. 9: 233–237.

Hantke, B., I. Domany, P. Fleischer, M. Koch, P. Pleß, M. Wiendl & A. Melzer, 1996a. Depth profiles of the kinetics of phosphatase activity in hardwater lakes of different trophic level. Arch. Hydrobiol. 135: 451–471.

Hantke, B., P. Fleischer, I. Domany, M. Koch, P. Pleß, M. Wiendl & A. Melzer, 1996b. P-release from DOP by phosphatase activity in comparison to P excretion by zooplankton. Studies in hardwater lakes of different trophic level. Hydrobiologia 317: 151–162.

Healey, F. P., 1985. Interacting effects of light and nutrient limitation on the growth rate of *Synechococcus linearis* (Cyanophyceae). J. Phycol. 21: 134–146.

Hecky, R. E. & P. Kilham, 1974. Environmental control of phytoplankton cell size. Limnol. Oceanogr. 19: 361–366.

Herbland, A., A. Le Bouteiller & P. Raimboult, 1985. Size structure of phytoplankton biomass in the equatorial Atlantic Ocean. Deep-Sea Res. 32: 810–836.

Huisman, J. & F. J. Weissing, 1994. Light-limited growth and competition for light in well-mixed aquatic environments: An elementary model. Ecology 75: 507–520.

Huisman, J. & F. J. Weissing, 1995. Competition for nutrients and light in a mixed water column: A theoretical analysis. Am. Nat. 146: 536–564.

Jansson, M., H. Olsson & K. Pettersson, 1988. Phosphatases; origin, characteristics and function in lakes. Hydrobiologia 170: 157–175.

Kilham, P. & D. Tilman, 1979. The importance of resource competition and nutrient gradients for phytoplankton ecology. Ergebn. Limnol. 13: 110–119.

Knoechel, R. & F. deNoyelles, 1980. Analysis of the response of hypolimnetic phytoplankton in continuous culture to increased light or phosphorus using track autoradiography. Can. J. Fish. aquat. Sci. 37: 434–441.

Kuenzler, E. J. & J. P. Peras, 1965. Phosphatase of marine algae. Biol. Bull. 128: 271–284.

Lingeman, R., F. Heinis & A. Veen, 1987. Time series of physical, chemical and plankton parameters in Lake Maarsseveen I: 1980–1986. Hydrobiol. Bull. 21: 25–38.

Lund, J. W. G., 1965. The ecology of freshwater phytoplankton. Biol. Rev. 40: 231–293.

Maestrini, S. Y. & D. J. Bonin, 1981. Competition among phytoplankton based on inorganic macronutrients. In Platt T. (ed), Physiological basis of phytoplankton ecology. Can. Bull. Fish. aquat. Sci. Dept. of Fisheries and Oceans, Ottawa: 264–278.

Perry, M. J., 1972. Alkaline phosphatase activity in subtropical Central North Pacific waters using a sensitive fluorometric method. Mar. Biol. 15: 113–119.

Phillips, O. M., 1973. The equilibrium and stability of simple marine biological systems. I. Primary nutrients consumers. Am. Nat. 107: 73–93.

Reynolds, C. S., 1987. The response of phytoplankton communities to changing lake environments. Schweiz. Z. Hydrol. 49: 220–235.

Rhee, G-Y. & I. J. Gotham, 1981. The effect of environmental factors on phytoplankton growth: temperature and the interactions of temperature with nutrient limitation. Limnol. Oceanogr. 26: 635–648.

Riegman, R. & L. R. Mur, 1984. Regulation of phosphate uptake kinetics in *Oscillatoria agardhii*. Arch. Microbiol. 139: 28–32.

Smith, R. E. H. & J. Kalff, 1982. Size-dependent phosphorus uptake kinetics and cell quota in phytoplankton. J. Phycol. 18: 275–284.

Sommer, U., 1981. The role of r- and K-selection in the succession of phytoplankton in Lake Constance. Acta Oecologia 2: 327–342.

Spijkerman, E. & P. F. M. Coesel, 1996a. Phosphorus uptake and growth kinetics of two planktonic desmid species. Eur. J. Phycol. 31: 53–60.

Spijkerman, E. & P. F. M. Coesel, 1996b. Competition for phosphorus between planktonic desmid species in continuous flow culture. J. Phycol. 32: 939–948.

Swain, W. R., R. Lingeman & F. Heinis, 1987. A characterization and description of the Maarsseveen Lake system. Hydrobiol. Bull. 21: 5–16.

Taylor, P. A. & P. J. LeB. Williams, 1975. Theoretical studies on the coexistence of competing species under continuous-flow conditions. Can. J. Microbiol. 21: 90–98.

Tilman, D., 1977. Resource competition between planktonic algae: An experimental and theoretical approach. Ecology 58: 338–348.

Tilman, D., 1980. Resources: a graphical-mechanistic approach to competition and predation. Am. Nat. 116: 362–393.

Tilman, D., 1982. Resource competition and community structure. Princeton.

Tilman, D., S. S. Kilham & P. Kilham, 1982. Phytoplankton community ecology: The role of limiting nutrients. Ann. Rev. Ecol. Syst. 13: 349–372.

Van Liere, L., J. G. Loogman & L. R. Mur, 1978. Measuring light-irradiance in cultures of phototrophic micro-organisms. FEMS Microbiol. Letters 3: 161–164.

Watson, S. & J. Kalff, 1981. Relationships between nannoplankton and lake trophic status. Can. J. Fish. aquat. Sci. 38: 960–967.

Wehr, J. D., 1993. Effects of experimental manipulations of light and phosphorus supply on competition among picoplankton and nanoplankton in an oligotrophic lake. Can. J. Fish. aquat. Sci. 50: 936–945.

Wynne, D. & M. Gophen, 1981. Phosphatase activity in freshwater zooplankton. Oikos 37: 369–376.
Zevenboom, W., 1986. Ecophysiology of nutrient uptake, photosynthesis and growth. In Platt T. & W.K.W. Li (eds), Photosynthetic picoplankton. Can. Bull. Fish. aquat. Sci. 214: 391–422.

Hydrobiologia **369/370:** 117–125, 1998.
M. Alvarez-Cobelas, C. S. Reynolds, P. Sanchez-Castillo & J. Kristiansen (eds), Phytoplankton and Trophic Gradients.

Freshwater picocyanobacteria along a trophic gradient and light quality range

Lajos Vörös, Cristiana Callieri[1], Katalin V.-Balogh & Roberto Bertoni[1]
Balaton Limnological Research Institute, Tihany, POB 35, H-8237, Hungary
[1] *C.N.R. – Istituto Italiano di Idrobiologia, Largo Tonolli 50/52, I-28048 Pallanza, Italy*

Key words: freshwater picocyanobacteria, vertical attenuation coefficients, underwater light quality, trophic gradient, *Synechococcus* spp.

Abstract

Picocyanobacterial abundance and their contribution to the total phytoplankton biomass, estimated as chlorophyll *a*, was investigated in 32 deep and shallow lakes. The lake series covered a wide range of natural and artificial lakes, from high altitude clear lakes and deep, large subalpine lakes through large shallow lakes, small reservoirs and fish ponds. The chlorophyll *a* concentrations ranged between 0.2 and 390 μg l^{-1}. No simple relation existed between trophic state and the abundance of picocyanobacterial, mainly represented by *Synechococcus* spp. Below 10 μg Chl *a* l^{-1} the percentage contribution of picocyanobacteria biomass to total phytoplankton biomass exceeded 70% in some cases, whereas above 100 μg Chl *a* l^{-1}, the highest contribution was only 10%. At low chlorophyll *a* concentrations phycoerythrin-rich picocyanobacteria (PE) dominated almost exclusively but their contribution never exceeded 10% of the total picocyanobacterial abundance when chlorophyll *a* concentration was higher than 50 μg l^{-1}. Above this value there was high light attenuation and a shift of the maximum light penetration from the blue-green towards the red portion of the spectrum. In this underwater light climate phycocyanin-rich picocyanobacteria (PC) dominated. In the lakes chosen for their representation of ranges in trophy and light quality, PC cells prevailed over the PE when vertical attenuation coefficient of PAR was higher than 2.25 m^{-1} and red light penetrated farthest. PE cells were 100% when the vertical attenuation coefficient of PAR was lower than 0.55 m^{-1} and the green and blue were the most penetrating lights.

Introduction

Several studies have noted that photosynthetic picoplankton abundance and production increase with the increasing trophic state while their contribution to total phytoplankton biomass and production decreases both in marine and freshwater ecosystems (Stockner, 1988; Søndergaard, 1990; Stockner, 1991). However in the eutrophic part of the trophic spectra, the existence of this relationship seems uncertain (Stockner, 1991). Hawley & Whitton (1991) observed a significant negative relationship between $\log_{10}$ of maximum picoplankton cell density and maximum total chlorophyll concentration in ten lakes in Northern England. Similarly, Burns & Stockner (1991) reported a negative but weak correlation between the density of prokaryotic picoplankton and the level of chlorophyll *a* in six New Zealand lakes. Takamura & Nojiri (1994) found significantly higher picoplankton biomass in eutrophic than in oligotrophic lakes but they suggested that the picoplankton contribution was influenced by the total N:P ratio rather than by lake trophic state.

The effect of eutrophication on phytoplankton is not a simple response of algae to the increased nutrient supply. On the course of eutrophication one characteristic phenomenon is the change of the underwater light intensity and quality due to the self-shading effect of algal cells. In lakes the underwater light quality rapidly changes with depth because of the spectral selectivity of water itself and of the dissolved and particulated material (tripton/phytoplankton) suspended in it (Kirk, 1983; 1992). The presence of 'Gelbstoff' and of algal material results in a more rapid attenuation of blue and ultraviolet light due to absorption and scattering. Water molecules cause attenuation mainly in the red spectral range (Kishino et al., 1984). As for energy supply, phototrophs strongly depend on light and many studies have been done on the relation between freshwater phytoplankton and spectral extinction coefficient (Talling, 1971; Atlas & Bannister, 1980; Tilzer et al., 1995), or underwater light climate and production and/or pigment content (Krogman, 1973; Wyman & Fay, 1987). On the other hand few papers exist on freshwater picocyanobacteria and their presence and abundance in relation to the underwater light climate. From laboratory experiments on *Synechococcus* marine strains (Glover et al., 1986; Waterbury et al., 1986) and freshwater strains (Wyman & Fay, 1987; Hauschild et al., 1991; Callieri et al., 1996), it appears that phycoerythrin-rich cells grew best in green light. These results suggest that light quality, through its effects on growth rate, may be an important factor controlling the occurrence of picocyanobacteria with different accessory pigments in natural environment. Pick (1991) found an inverse relationship between light attenuation and freshwater picocyanobacterial abundance in lakes of varying trophy. She also demonstrated that in highly coloured lakes non-phycoerythrin cells dominated numerically whereas in the oligotrophic hard-water lakes phycoerythrin-rich cells were the most abundant; this shift in dominance was interpreted as due to changes in ambient light quality.

Our research was planned to study the occurrence, abundance and contribution to total phytoplankton biomass of phycoerythrin-rich (PE) and phycocyanin-rich (PC) picocyanobacteria in lakes, covering a wide range of trophic conditions and light climate. The novelty of our research was to study *in situ* the light quality as affecting the PE and PC relative abundance. The effect of spectral quality in regulating the pigment synthesis, already observed in laboratory (Hauschild et al., 1991), could determine the prevalence in natural systems of the picocyanobacteria types more adapted to those light conditions.

Table 1. List of study lakes.

Sampling sites	Area (km^2)	Av. depth (m)	Max. depth (m)
Hungary			
L. Balaton	596	3.2	
L. Balaton (east b.)*	228	3.7	
L. Balaton (west b.)*	38	2.3	
L. Balaton (Zala river)			
Kis-Balaton (upper res.)*	18	1	
Kis-Balaton (lower res.)*	16	0.8	
Zalavári víz	0.06		1.8
Vörsi viz	0.13		1.5
Marcali reservoir*	4	1.8	
Monostorapáti reservoir	0.3		2
L. Kornyi	0.25		0.8
L. Vadása	0.15	2	
L. Orfüi	0.1	2	
L. Pécsi*	0.75	3.3	
L. Herman Otto*	0.29	1	
L. Kovácsszénájai	0.16	2.2	
L. Darvas	0.02		2
Csíkos fenék	0.002		1
Deseda reservoir*	2.2	2.9	
L. Fertö	300	1.3	
Kis Herlakni pond	0.001		2
Italy			
L. Como*	146	154	410
L. Maggiore*	212	177	370
L. Garda*	368	133	350
L. Iseo*	62	123	251
L. Orta*	18	72	143
L. Mergozzo*	1.8	45	73
L. Varese*	15	11	26
L. Candia*	1.3	5.9	8
L. Paione Superiore*	0.014	5.1	11.7
L. Paione Inferiore*	0.014	7.3	13.5
L. Azzurro*	0.003		2
L. Devero*	1	20	40
Nepal			
L. Piramide Superiore*	0.6		8.2
L. Piramide Inferiore*	1.7		14.8

* Sites where underwater radiation profiles were measured.

Materials and methods

Underwater light intensity was measured in 24 sites, chosen according to a scale of increasing trophy. Samples were taken in 1994 and 1995 at different station and time of the year. For the deep lakes an integrating sampler (Bertoni, 1996, patent pending) was used in the first 20 m of the water column, whereas for shallow lakes, ponds and reservoirs samples were collected with a Van Dorn sampler every 1 m or less (depending upon the lake depth) and mixed together. Chlorophyll *a* concentrations and estimations of picocyanobacterial abundances, distinguishing between phycoerythrin and phycocyanin-rich cells, were performed in 32 lakes.

In addition, physical and chemical parameters were measured in those environments where no previous data were available to have a basic limnological background. In Table 1 a list of the sampling sites in lakes, ponds and reservoirs is reported together with their main morphometric characteristics.

In the oligotrophic and mesotrophic lakes the chlorophyll *a* content of phyto- and picoplankton was determined fluorometrically by Perkin-Elmer LS-2 Filter Fluorometer (Stainton et al., 1974) after extraction with methanol (Talling & Driver, 1961; Holm-Hansen & Riemann, 1978). The phytoplanktonic chlorophyll *a* concentration in the eutrophic and hyper-eutrophic Hungarian lakes was determined spectrophotometrically after hot methanol extraction (Iwamura et al., 1970). Total phytoplanktonic chlorophyll *a* was collected using GF/C filters whereas APP chlorophyll *a* was separated by differential filtration through 2 μm (gravity filtration) and 0.2 μm Nuclepore filters. In the oligotrophic and mesotrophic lakes studied the APP were mainly represented by picocyanobacteria so that the chlorophyll content refers to the prokaryotic component of APP. A different situation was encountered in some eutrophic Hungarian lakes where the particles of picoplanktonic size were abundant thus making the differential filtration for direct chlorophyll *a* measurement problematic. Therefore picoplanktonic chlorophyll *a* was estimated from picocyanobacterial biomass measurements following the relation reported by Vörös & Padisák (1991).

Picocyanobacterial abundance was estimated by direct counting with epifluorescence microscopy under green or blue excitation following the method described in Callieri & Pinolini (1995) for Italian and Himalayan lakes and Vörös et al. (1991) for Hungarian lakes.

The downwelling fluxes of photosynthetically active radiation (PAR) and of red, green and blue part of the visible spectra were measured by using LiCor (LI/185B) underwater radiometer equipped with a flat (2π) sensor and coloured glass filters (Callieri et al., 1996). The blue and the green were pass band filters with a maximum transmittance at 484 nm and 520 nm respectively; the red was a high pass filter with a maximum transmittance over 600 nm. From depths readings of PAR and of different wavebands the vertical extinction coefficients (K_{PAR}, K_{RED}, K_{GREEN}, K_{BLUE}) were calculated.

Results and discussion

Picocyanobacteria along the trophic gradient

The freshwater environments we considered in our research constitute a wide range of different trophic situations. To compare the large deep lakes with shallow ponds and reservoirs we considered the chlorophyll *a* as the best descriptive and general parameter to indicate a change in the trophic state. Chlorophyll *a* is strictly related to the algal biomass therefore a range going from 0.2 to 390 μg chlorophyll *a* l^{-1} can be easily recognised as a trophic gradient, wide enough to cover the overwhelming majority of standing waters. In Figure 1 the total picocyanobacterial abundance is plotted against the total chlorophyll *a* concentrations. The picocyanobacterial numbers were extremely high in the eutrophic and hypertrophic Hungarian lakes often exceeding 10^6cells ml^{-1}. At low chlorophyll *a* concentration (i.e. less then 10 μg l^{-1}) the picoplankton abundance tend to rise with increasing chlorophyll *a* concentration. At higher chlorophyll *a* concentration picocyanobacterial abundance cannot be predicted at all being high or low independently from the total phytoplankton chlorophyll. These results demonstrate that the pattern of picocyanobacterial abundance across trophic gradients cannot be inferred from a simple model such as that one proposed by Stockner (1991).

If we consider instead the picocyanobacterial contribution to total chlorophyll concentration along the trophic gradient, Stockner's model applies quite well but some comments are necessary (Figure 2). At chlorophyll *a* concentrations higher than 100 μg l^{-1} the picocyanobacteria contribution to total phytochlorophyll never exceeds 10%. Similar results have been obtained from seven Danish lakes in a chlorophyll concentration gradient from 7 to 200 μg l^{-1} (Søndergaard, 1991). In Figure 2 the different marks indicate lakes with a dominance of picocyanobacteria of PC type, of PE type and of PE and PC coexistence. In this last situation the maximum contribution of picocyanobacteria to total phytoplanktonic biomass, expressed as chlorophyll *a*, was 87%. In lakes with PC cell dominance the picocyanobacteria contribution to the total phyto-chlorophyll never exceeds 48% whereas PE can contribute till 79%. As it will be illustrated in the following section, PC cell dominance is common in the eutrophic and hypertrophic lakes whereas PE cell dominance is characteristic of oligotrophic lakes. It is worth noting that the data fill an area more than describe a line, showing the intrinsic weakness of sim-

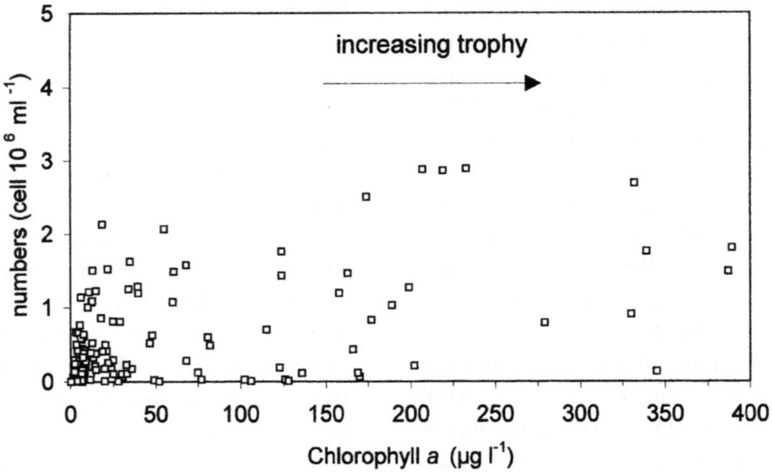

Figure 1. Pycocyanobacteria abundance versus euphotic chlorophyll *a* concentration in lakes of increasing trophic state.

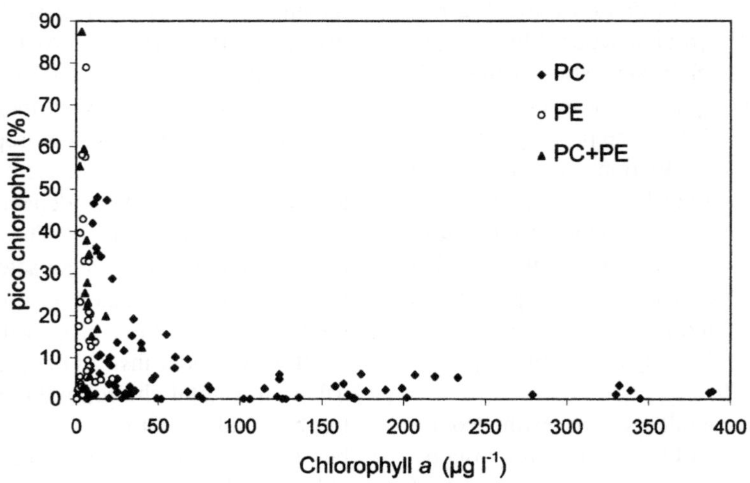

Figure 2. Relative contribution (%) of picocyanobacteria chlorophyll to total chlorophyll *a* concentration. Lakes in which PC cells (♦) or PE cells (○) dominate and lakes in which PC and PE cells are both present (▲).

ple univocal relations between trophic state and relative picoplankton contribution to total chlorophyll.

The change of the composition in PE and PC cells of picocyanobacterial assemblages across the trophic gradient is illustrated in Figure 3. In this graph the percentage of PE on total (PE + PC) picocyanobacterial numbers is plotted against chlorophyll concentrations. From Figure 3 is clear that the phycoerythrin-rich picocyanobacteria (PE) are absent in those lakes where the chlorophyll *a* content ranges between 200 and 400 μg chlorophyll *a* l^{-1}. PE cells begin to appear, but in very low percentage of the total picocyanobacteria, when chlorophyll *a* ranges between 50 and 200 μg chlorophyll *a* l^{-1}. Going further on towards oligotrophic conditions, PE cells became more than 50% of the total picocyanobacteria, reaching the 80%–100% in most cases. The increasing relative importance of PE is obviously mirrored by a decrease of PC cells percentage. But the relation between the two pico types is a nonlinear one, as it is evident from the plot of the abundance of PE and PC, in logarithmic scale, shown in Figure 4. In those lakes where the two pico types can coexist they have a stable point at high cell density and then a bifurcation with the decreasing cell abundance. This demonstrates that there are environments which are pico-type selective, i.e. in which the PE and PC cells cannot coexist. The chromatic adaptation of uni-

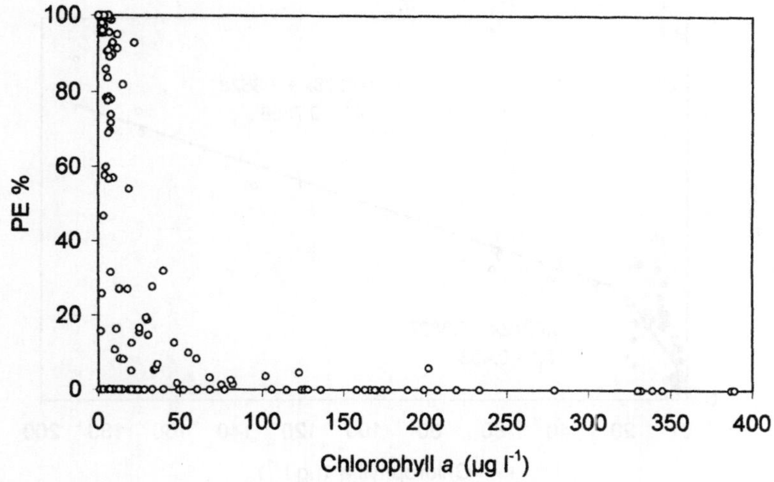

Figure 3. Relative abundance of PE cells in total picocyanobacteria assemblages in relation to increasing chlorophyll *a* concentration.

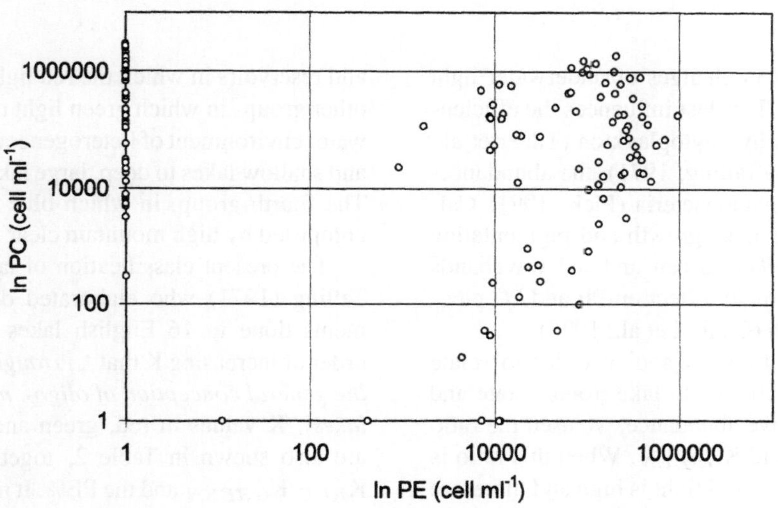

Figure 4. Relationship between PE and PC cell numbers transformed into natural logarithm.

cellular *Chroococcales* (Hauschild et al., 1991), could produce the above results.

Picocyanobacteria and the underwater light climate

Before discussing the picocyanobacteria-underwater light quality relation, it is important to evaluate the light quality dependence from the vertical attenuation coefficient of PAR (K_{PAR}) and to see how it changes along a trophic gradient. For this purpose a regression analysis of K_{PAR} versus chlorophyll concentration was performed on 73 pairs of data from the 24 lakes studied. To determine pigment-specific light attenuation and non-algal light extinction we calculated the intercept of the regression line with the Y axis (Tilzer, 1983). Our results show (Figure 5) that, among the different freshwater systems, two main groups of lakes can be singled out: one in which the light attenuation is attributable mainly to chlorophyll (y intercept = 0.3827) and the other in which the non-algal substances and the water itself are affecting the light attenuation (y intercept = 1.8628). At low chlorophyll concentration, in the oligotrophic clear lakes, the light attenuation strictly depend on phytoplankton. With the rising in algal biomass, in eutrophic shallow lakes, suspended particles and dissolved organic matter greatly contribute to the measured light attenuation. The difference in substances determining the light attenua-

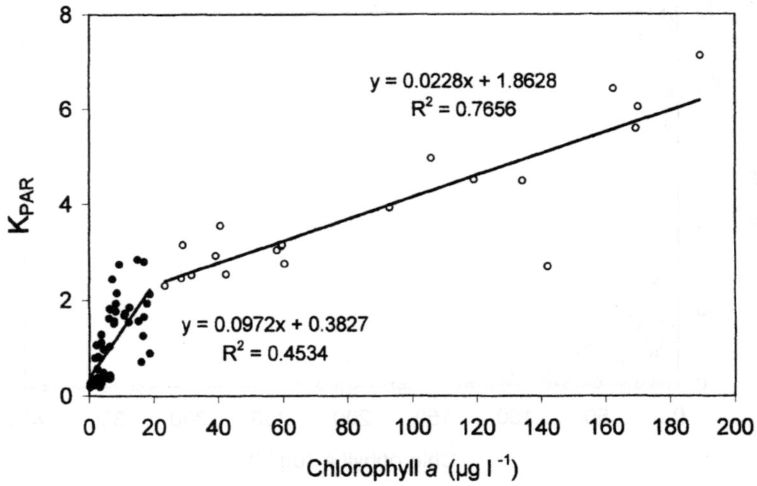

Figure 5. Relationship between K_{PAR} and chlorophyll *a* in lakes with light attenuation mainly (dark circles) or poorly (open circles) dependent from chlorophyll.

tion will likely result in changes of underwater light spectral composition. This last influences the efficiency of light utilization by phytoplankton (Tilzer et al., 1995), its production (Talling, 1971), the abundance-composition of picocyanobacteria (Pick, 1991; Callieri et al., 1996) and their growth and pigmentation (Hauschild et al., 1991). Green and red wavebands appeared to have the major effect on PE and PC picocyanobacterial growth (Callieri et al., 1996).

In light of these findings and in order to relate the underwater light climate to lake trophic state and with PE and PC relative abundance, we used the ratio between the K_{RED} and K_{GREEN}. When this ratio is > 1 the extinction of the red light is high and therefore the dominant underwater light is green (and blue). On the other hand very low values of this ratio indicate a red dominant underwater radiation. In lakes with chlorophyll concentrations lower than 10 μg l^{-1} the green light dominate in the euphotic zone whereas red light prevails in more eutrophic lakes (Figure 6). We found that the PE percentage in the total picocyanobacteria community increased with the increasing K_{RED}/K_{GREEN} ratio (Figure 7). Simultaneously the total chlorophyll *a* decreases and the waters become more transparent (low K_{PAR}) and less productive.

With all these information we summarised in Table 2 four groups of lakes with PAR extinction coefficients > 2.25 m^{-1}, between 2.25–0.55 m^{-1}, between 0.55–0.25 m^{-1} and < 0.25 m^{-1}. These four lake categories cover a wide range of underwater light climate with the first group composed by shallow lakes, ponds and reservoirs in which the red light is prevailing; two other groups in which green light dominate the underwater environment of heterogeneous lakes: from small and shallow lakes to deep, large lakes and alpine lakes. The fourth group, in which blue light dominates, is composed by high mountain clear lakes.

The present classification of lakes follows that of Talling (1971) who elaborated data from measurements done in 16 English lakes arranging them in order of increasing K that '*... roughly correspond with the general conception of oligo-, meso- and eutrophic lakes*'. K values in red, green and blue wavelengths are also shown in Table 2, together with the ratios K_{RED}/K_{GREEN} and the PE%. It must be stressed that along with the increasing percentage of PE type the cell abundance tend to decrease. This tendency is far from being predicted by a simple relationship nevertheless one possible speculation can be that PC cells which better exploit red light and have higher growth rate in respect to the PE cells (Callieri et al., 1996) can reach very high abundance in eutrophic and hypereutrophic lakes.

Conclusions

The relation between picocyanobacterial abundance and the trophic state of lakes, estimated from chlorophyll *a*, cannot be inferred from a simple model such as that one proposed by Stockner (1991) although the two variables are positively and significantly correlat-

Table 2. Four groups of lakes with different underwater light quality ordered by decreasing K_{PAR}. Values of total chlorophyll *a* concentration and relative contribution of PE cells to total picocyanobacteria abundances are also shown.

Location	K_{PAR} (m^{-1})	K_{RED} (m^{-1})	K_{GREEN} (m^{-1})	K_{BLUE} (m^{-1})	K_{RED}/K_{GREEN}	Chl*a* ($\mu g\ l^{-1}$)	PE (%)
$K_{BLUE} > K_{GREEN} > K_{RED}, K_{PAR} > 2.25$							
Monostorapáti reservoir	6.04	5.93	4.60	11.31	1.3	170	4
Deseda reservoir	4.96	4.60	4.88	7.11	0.9	106	2
Kis-Balaton (upper res.)	4.49	4.04	5.13	11.51	0.8	134	0
Marcali reservoir	3.04	2.64	3.16	5.95	0.8	58	0
L. Herman Ottó	2.93	2.30	3.41	4.46	0.7	39	32
Kis-Balaton (lower res.)	2.75	1.79	2.66	5.94	0.7	9.3	0
L. Balaton (Zala river)	2.54	2.32	2.58	5.67	0.9	42	6
$K_{BLUE} > K_{RED} > K_{GREEN}, 0.55 < K_{PAR} < 2.25$							
L. Balaton (east. b.)	1.85	2.03	1.88	2.83	1.1	12.5	57
L. Balaton (west. b.)	1.65	1.58	1.57	2.98	1.0	16.9	73
L. Pécsi	1.04	1.55	1.25	2.01	1.2	6.4	78
L. Varese	0.89	0.84	0.78	1.21	1.1	18.8	54
L. Candia	0.72	0.82	0.66	1.48	1.2	16.1	50
$K_{RED} > K_{BLUE} > K_{GREEN}, 0.25 < K_{PAR} < 0.55$							
L. Iseo	0.39	0.67	0.32	0.57	2.1	6.5	99
L. Maggiore	0.39	0.49	0.33	0.39	1.5	2.7	96
L. Orta	0.36	0.65	0.30	0.55	2.2	1.9	100
L. Como	0.31	0.57	0.26	0.41	2.2	3.3	98
L. Paione Superiore	0.30	0.59	0.24	0.24	2.5	1.1	100
L. Devero	0.29	0.66	0.33	0.29	2.0	0.5	100
L. Garda	0.28	0.58	0.21	0.26	2.8	3.8	98
L. Mergozzo	0.27	0.58	0.23	0.36	2.5	2.4	99
$K_{RED} > K_{GREEN} > K_{BLUE}, K_{PAR} < 0.25$							
L. Azzurro (Alpe Devero)	0.25	0.46	0.25	0.25	1.8	0.3	100
L. Piramide Inferiore	0.22	0.41	0.13	0.16	3.2	0.3	100
L. Paione inferiore	0.20	0.44	0.10	0.08	4.5	3.3	100
L. Piramide Superiore	0.19	0.53	0.30	0.12	1.8	0.4	100

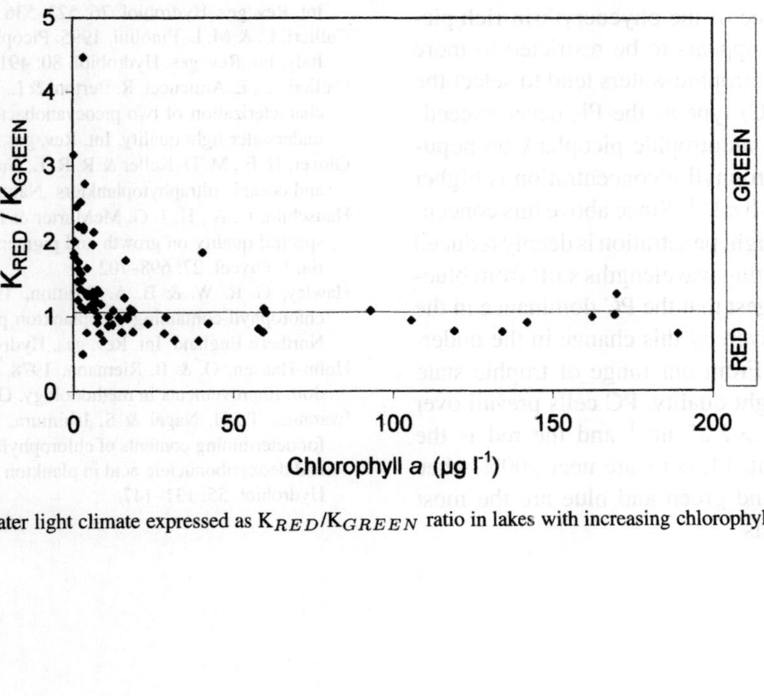

Figure 6. Underwater light climate expressed as K_{RED}/K_{GREEN} ratio in lakes with increasing chlorophyll *a* concentration.

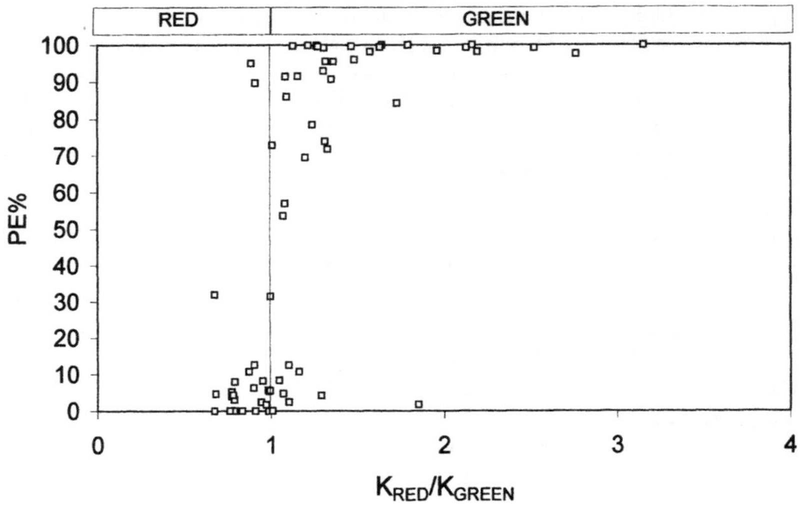

Figure 7. Relative contribution of PE cells to total picocyanobacteria abundance in lakes with underwater light climate expressed as K_{RED}/K_{GREEN}.

ed ($P < 0.001$) in our data set. Many lakes with low quantity of picocyanobacteria in spite of their high chlorophyll *a* concentration suggest that other variables important in determining the picoplankton abundance should be taken into account.

More evident is the effect of the increasing trophy on the relative contribution of picocyanobacteria to the whole phytoplankton community. Below 10 μg chlorophyll *a* l^{-1} picocyanobacteria can be more than 70% of the total phytoplankton biomass. Above the threshold of 100 μg chlorophyll *a* l^{-1} the contribution of the pico fraction does not exceed the 10%.

The trophic state of the lake is also crucial in setting the most abundant picocyanobacterial pigment composition. The presence of the phycoerythrin-rich picocyanobacteria (PE) appears to be restricted to more oligotrophic lakes. Eutrophic waters tend to select the phycocyanin-rich (PC) type as the PE never exceeds the 10% of the total autotrophic picoplankton population when the chlorophyll *a* concentration is higher than 50 μg chlorophyll *a* l^{-1}. Since above this concentration threshold the light penetration is deeply reduced and the more penetrating wavelengths shift from blue-green to red, we suggest that the PC dominance in the eutrophic lakes is driven by this change in the underwater light climate. From our range of trophic state and of underwater light quality, PC cells prevail over the PE when $K_{PAR} > 2.25$ m^{-1} and the red is the most penetrating light. PE cells are near 100% when $K_{PAR} < 0.55$ m^{-1} and green and blue are the most penetrating wavebands.

Acknowledgements

This work was partly supported by the Hungarian Research Fund (OTKA No T 016348) and Strategic Project EVK2-CNR. Assistance with fieldwork was given by Mr. M.Contesini.

References

Atlas, D. & T. T. Bannister, 1980. Dependence of mean spectral extinction coefficient of phytoplankton on depth, water-colour, and species. Limnol. Oceanogr. 25: 157–159.

Burns, C. W. & J. G. Stockner, 1991. Picoplankton in six New Zealand lakes: Abundance in relation to season and trophic state. Int. Rev. ges. Hydrobiol. 76: 523–536.

Callieri, C. & M. L. Pinolini, 1995. Picoplankton in Lake Maggiore, Italy. Int. Rev. ges. Hydrobiol. 80: 491–501.

Callieri, C., E. Amicucci, R. Bertoni & L. Vörös, 1996. Fluorometric characterization of two picocyanobacteria strains from different underwater light quality. Int. Rev. ges. Hydrobiol. 81: 13–23.

Glover, H. E., M. D. Keller & R. R. L. Guillard, 1986. Light quality and oceanic ultraphytoplankters. Nature 319: 142–143.

Hauschild, C. A., H. J. G. McMurter & F. R. Pick, 1991. Effect of spectral quality on growth and pigmentation of picocyanobacteria. J. Phycol. 27: 698–702.

Hawley, G. R. W. & B. A. Whitton, 1991. Seasonal changes in chlorophyll-containing picoplankton populations of ten lakes in Northern England. Int. Rev. ges. Hydrobiol. 76: 545–554.

Holm-Hansen, O. & B. Riemann, 1978. Chlorophyll *a* determination: improvements in methodology. Oikos 30: 438–447.

Iwamura, T., H. Nagai & S. Ishimura, 1970. Improved methods for determining contents of chlorophyll, protein, ribonucleic acid and deoxyribonucleic acid in plankton populations. Int. Rev. ges. Hydrobiol. 55: 131–147.

Kirk, J. T. O., 1983. Light and photosynthesis in aquatic ecosystems. Cambridge University Press, London, New York, 401 pp.

Kirk, J. T. O., 1992. A nature and measurement of the light environment in the ocean. In P.G. Falkowski & A. D. Woodhead (eds), Primary production and biochemical cycles in the sea. Plenum Press, New York: 9–29.

Kishino, M., C. R. Booth & N. Okami, 1984. Underwater radiant energy absorbed by phytoplankton, detritus, dissolved organic matter and pure water. Limnol. Oceanogr. 29: 340–349.

Krogman, D. K., 1973. Photosynthetic reactions and components of thylakoids. New York, 200 pp.

Pick, F. R., 1991. The abundance and composition of freshwater picocyanobacteria in relation to light penetration. Limnol. Oceanogr. 36: 1457–1462.

Søndergaard, M., 1990. Picophytoplankton in Danish lakes. Verh. int. Ver. Limnol. 24: 609–612.

Stainton, M. P., M. J. Capel & F. A. J. Armstrong, 1974. The chemical analysis of fresh waters. Envir. Can. Spec. Publ. No. 5, Ottawa, 119 pp.

Stockner, J. G., 1988. Phototrophic picoplankton: An overview from marine and freshwater ecosystems. Limnol. Oceanogr. 33: 765–775.

Stockner, J. G., 1991. Autotrophic picoplankton in freshwater ecosystems: the view from the summit. Int. Rev. ges. Hydrobiol. 76: 483–492.

Takamura, N. & Y. Nojiri, 1994. Picophytoplankton biomass in relation to lake trophic state and the TN:TP ratio of lake water in Japan. J. Phycol. 30: 439–444.

Talling, J. F. & D. Driver, 1961. Some problems in the estimation of chlorophyll *a* in phytoplankton. 10th Pacific Sci. Cong. Honolulu: 142–146.

Talling, J. F., 1971. The underwater light climate as a controlling factor in the production ecology of freshwater phytoplankton. Mitt. int. Ver. Limnol. 19: 214–243.

Tilzer, M. M., 1983. The importance of fractional light absorption by photosynthesis pigments for phytoplankton productivity in Lake Constance. Limnol. Oceanogr. 28: 833–846.

Tilzer, M. M., N. Stambler & C. Lovengreen, 1995. The role of phytoplankton in determining the underwater light climate in Lake Constance. Hydrobiologia 316: 161–172.

Vörös, L. & J. Padisák, 1991. Phytoplankton biomass and chlorophyll *a* in some shallow lakes in central Europe. Hydrobiologia 215: 111–119.

Vörös, L., P. Gulyás & J. Németh, 1991. Occurrence, dynamics and production of picoplankton in Hungarian shallow lakes. Int. Rev. ges. Hydrobiol. 76: 617–629.

Waterbury, J. B. S., S. W. Watson, F. W. Valois & D. G. Franks, 1986. Biological and ecological characterization of the marine unicellular cyanobacterium *Synechococcus*. Can. Bull. Fish. Aquat. Sci. 214: 71–120.

Wyman, M. & P. Fay, 1987. Acclimation to the natural light climate. In P. Fay & C. Van Baalen (eds), The cyanobacteria. Elsevier, Amsterdam: 347–376.

Hydrobiologia **369/370:** 127–131, 1998.
M. Alvarez-Cobelas, C. S. Reynolds, P. Sanchez-Castillo & J. Kristiansen (eds), Phytoplankton and Trophic Gradients.

Responses of the phytoplankton to a deliberate attempt to raise the trophic status of an acidic, oligotrophic mountain lake

C. S. Reynolds, G. H. M. Jaworski, J. V. Roscoe, D. P. Hewitt & D. G. George
NERC Institute of Freshwater Ecology, Windermere Laboratory, GB-LA 22 0LP Ambleside, U.K.

Key words: phytoplankton, trophic status, phosphorus, eutrophication, species composition

Abstract

The paper reports the impact of a sharp artificial enrichment of the available phosphorus in a small, acidic and oligotrophic corrie lake and its effects upon the phytoplankton supported. Annual average chlorophyll increased tenfold within two years, from $\sim$ 1.2 to 12.6 μg chl *a* l^{-1}, but the species represented by large populations are the same as previously. Chrysophyte species, however, make up a smaller fraction of the total crop.

Introduction

Seathwaite Tarn is a small upland, acidic and oligotrophic water body in the English Lake District. It is a classic corrie lake, although its level has been raised behind a concrete dam, in order to provide compensation water for the River Duddon. Its level is accordingly subject to artificially enhanced fluctuations. Other limnological properties of the lake are given in Table 1.

The catchment (4.80 km^2) is based on Ordovician intrusives (tuffs and andesites) and receives an annual rainfall of 2.0–2.2 m. It is classically barren: the rough *Festuca-Nardus* grassland alternates with loose screes and local pockets of *Sphagnum* bog. There is precious little neutralising capacity in the scant soil so the inflowing water to the Tarn is acidic (pH 5.1–5.4), base-poor (bicarbonate alkalinity, zero) and deficient in biologically available phosphorus (< 1 μg P l^{-1}). Accordingly, the lake is normally limpid and free of significant turbidity due to algae.

The site was selected for a large-scale field experiment to test the concept of reversal of acidification by use of phosphate fertilser, rather than the more usual lime treatment. The rationale for this was proposed by Davison (1987) and authenticated in an experimental pilot restoration of a water-filled former sand quarry (Davison et al., 1989). Essentially, the biological production promoted consumes acidity (carbon dioxide, nitrate) and generates base. From March 1992, a solution of a specially formulated phosphate fertiliser was dosed at fortnightly intervals to the Tarn in order to give a notional average concentration of 20 μg P l^{-1}. In 1993, the dosing rate was increased to a maximum of 28 μg P l^{-1}. In both years, almost all the phosphate was assimilated biologically with the consequent stimulus to pelagic production. The pH was lifted substantially (> 6) as hypothesised.

Apart from a short period at the end of 1992, the measured concentration of dissolved phosphorus remained < 0.6 μg P l^{-1} throughout 1993, confirming that the addition had been insufficient to saturate the potential phosphorus consumption of the phytoplankton. After July, 1993, however, free nitrate remained at levels below analytical detection. These results have already been reported in summary form (Davison et al., 1995). This paper is concerned only with the responses of the phytoplankton with special reference to the changes in the species-structure of the assemblage. In particular, it seeks to review the relevance of some suppositions about the influence of nutrients on the assembly of planktonic communities and upon the tenancy of the concept of a trophic spectrum.

Methods

Samples for chlorophyll determination and algal enumeration have been collected at approximately fort-

Table 1. Limnological data of the study site

Seathwaite Tarn, Cumbria, UK		
Latitude:	54°23′ N	Longitude: 3°9′ W
Altitude:	375 m	Ice cover: typically 10–20 d annum^{-1}
Area:	0.267 km^2	Volume: 3.28×10^6 m^3
Max depth:	26 m	Mean depth: 7.2 m
Residence time:	120–200 d	Mixing frequency: dimictic with winter ice
pH:	5.1–5.4*	Secchi disc: 8–15 m*
Max Total P:	~ 2 μg P l^{-1}*	Max and Min SRP always < 1 μg P l^{-1}*
Alkalinity	0*	Salinity $\sim$ 0.001%

* natural state; please see text for experimentally-imposed changes

Table 2. Comparison of trophic-state indicators before (1991) and during experimental enrichment of Seathwaite Tarn (1993)

1991	1993
1. Sequence of relative species abundance	
Chrysolykos/Chlorella/Dinobryon → *Chlorella/Monochrysis*	*Monochrysis/Chlorella* → *Chlorella* → *Chlorella/Monochrysis* → *Cryptomonas*
2. Species accounting for:	
> 0.5% 35	18
> 5% 7	4
> 25% 2	2
3. Phases of dominance per year.	
±1	2
4. Peaks of biomass per year.	
2, March-April and Oct-Nov	3–4
5. Annual mean biomass and the peak:mean ratio	
1.16 μg chl *a* l^{-1}, 1.61	12.4 μg chl *a* l^{-1}, 2.95
6. Frequency of mixing to	
> 0.5 m daily	
> 2.5 m ± daily	
> 10 m $\sim$ 200 days (Oct–May)	
7. Maximum Shannon-Weaver diversity index, H'' (ignoring spp. < 0.5% of the biomass)	
1.029	0.832
8. Equitability at maximum diversity	
0.257	0.263

nightly intervals throughout the three-year study period, from the open water at the deepest part of the lake, by means of a 5-m column sampling hose (Lund & Talling, 1957). Chlorophyll was extracted from GF/C-filtration residues by the hot methanol method (Talling, 1974). Phytoplankton were counted from appropriate volumes of lake water, fixed on collection with Lugol's Iodine, after concentration by sedimentation, either by the method of Lund et al. (1958) or according to the WRc modification to Lund's (1959) method (Youngman, 1971). All counts were expressed per unit volume of the original sample.

Diversity of the assemblage collected in individual samplings were tested by the specialised Shannon Weaver function (Shannon, 1948):

$$H'' = -\Sigma n_i/N \log_2(n_i/N)$$

where n_i is the measure of the ith species (as specific wet biomass calculated as the product of a cell-volume approximation and the number of cells counted; unfortunately, no direct measurements are available) and N is $\Sigma\, n_i$. The greater the number of species present in a sample then the greater is the value of H''. The theoretical maximum is $H''_{\max} = \log_2 s$ and s is the number of species in the sample with specific mass accounting for > 0.5% of the total. The equitability (or evenness) of the species representation is given from $H''/H''_{\max}$.

Results

Biomass

The plot tracing the routine measurements of chlorophyll concentrations through the three-year period, 1991–1993 (Figure 1), shows, as was broadly predicted, a significant increase in the size of the algal standing crop, from being always < 2 μg chl *a* l^{-1} throughout the first year, to maximum levels of 9 μg chl *a* l^{-1} in 1992 and to 36 μg chl *a* l^{-1} in 1993. The greater variability through time should also be noted, as should the emergence of a typical seasonality of biomass in enriched lakes (low winter minima and short periods of low biomass in summer).

Species dominance in 1991

Some 35 taxa were recorded from the routine algal counts in 1991. The largest number of these were referrable to the Chrysophyta; several species of green algae, the Cryptophytes *Rhodomonas, Cryptomonas* and dinoflagellates conforming to *Peridinium inconspicuum* Lemm. and to *Gymnodinium luteofaba* Javorničky made up most of the rest. It is striking that almost all are unicellular in habit or produce only simple coenobia (*Actinastrum, Quadrigula, Scenedesmus*); *Dinobryon sertularia* Ehrenberg, a filamentous green alga (*Mougeotia*), occasional encounters with the diatom, *Aulacoseira distans* Ehrenberg and with a coenobial cyanobacterium (*Merismopedia*) together represented the structurally more-complex algae. Several genera of benthic diatom were also noted in the plankton from time to time. Such a list is reasonably congruous within the understanding of mildly acidic, oligotrophic phytoplankton.

The species accounting for the largest proportion of the planktonic biomass were a *Chlorella* (cf. *C. ellipsoidea* Gerneck), which was present throughout but peaked at an estimated 1250 cells ml^{-1} in May and came back with about 2200 ml^{-1} in September, *Chrysolykos* sp. (1100 cells ml^{-1} in March), *Kephyrion* (maximum: 85 ml^{-1} in April), *Dinobryon sertularia* (380 cells ml^{-1} in May) and *Monochrysis* (maximum c. 70 cells ml^{-1} in June). Larger algae like *Peridinium* scarcely exceeded 500 l^{-1}, although a maximum of *Cryptomonas* cf. *ovata* Ehrenberg (300 l^{-1}) was noted in August.

Species dominance in 1992

The species composition of the Tarn phytoplankton was not substantially altered in 1992, even after the phosphorus additions were commenced in March. At that time, the dominant species comprised *Chlorella* (~ 300 cells ml^{-1}), *Kephyrion* and *Monochrysis* (each < 100 cells ml^{-1}). The first effect of fertilisation was that nearly everything increased in mass, with the strongest performance coming from *Monochrysis* (~ 700 cells ml^{-1} by late April). With further fertilisation, *Chlorella* became dominant during June (reaching 7000 cells ml^{-1}); later peaks were overwhelmingly dominated by *Chlorella* (September: 43 000 cells ml^{-1}; October: 29 000 cells ml^{-1}) when *Cryptomonas* species reached > 10 ml^{-1} for the first time. Chrysophytes persisted in similar numbers as previously, though the proportion of the biomass was greatly diminished.

Species dominance in 1993

The accelerated phosphorus addition in 1993 predictably led to the much larger algal production and standing crop. *Monochrysis* (reaching 7 500 cells ml^{-1} during March) again dominated initially but it was *Chlorella,* building to a remarkable 1.2 million cells ml^{-1} which contributed most to the mid-May biomass maximum. This population abruptly declined during June, leaving the water relatively very clear. A sec-

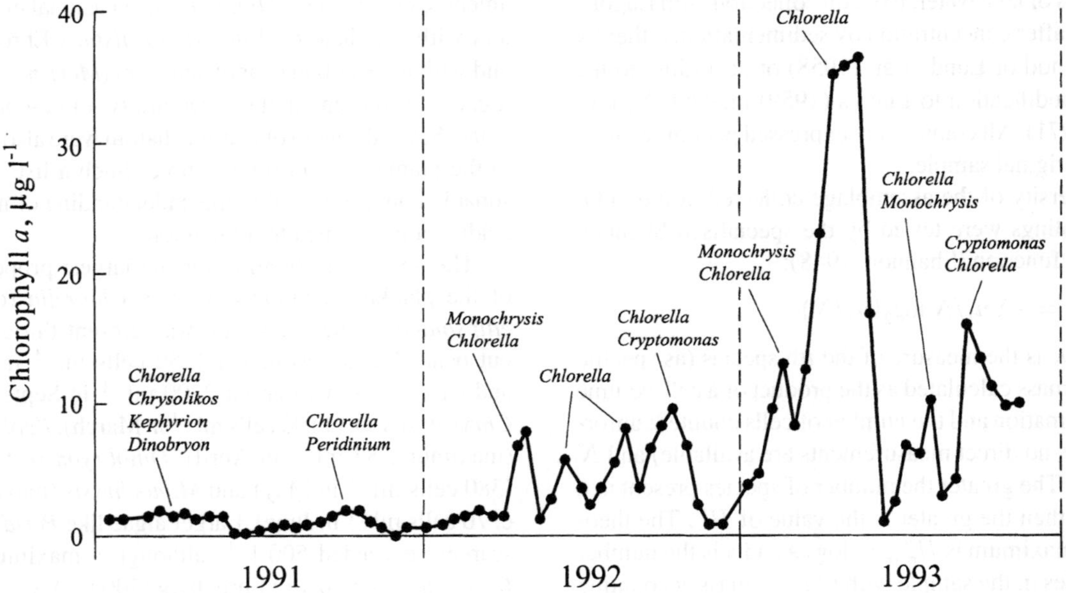

Figure 1. The seasonal variation in the concentration of chlorophyll *a* in Seathwaite Tarn, 1991–1993. The annotations refer to the dominant species and their maximum abundance.

ond, though lesser, maximum then began to develop, comprising some 20 000 *Chlorella* cells ml^{-1}, with subdominant Chrysophytes and *Cryptomonas*. In both cases, the rapid reduction in nanoplanktonic biomass coincided with a phase of intense feeding and recruitment of *Bosmina corregoni* Baird. Interestingly, the microplanktonic *Cryptomonas* (cf. *C. rostratiformis* Skuja) continued to increase, eventually to dominate the October chlorophyll maximum with 300 cells ml^{-1}, before it declined to overwintering levels.

Discussion

The expected promotion of production of phytoplankton in response to phophorus additions has occurred, with populations broadly scaled to the phosphorus loads. There is no question that the supportive capacity of the lake was restricted by phosphorus and there is no indication that the demand of the potential biomass was ever saturated in the experiments. In 1993, at least, the biomass assembled matched well the phosphorus made available to support it (see regression of Reynolds, 1992). The scale of the 1993 phosphorus-addition was also sufficient to promote the biological uptake of nitrate as intended, with the consequent elevation of pH.

The exercise was not billed as an experiment in eutrophication, though it is perfectly permissible to interpret it in that way. It is also useful to reflect upon the fact that the Tarn has not become beset with species of eutrophic lakes: any anticipation that the enriched conditions might have led to the dominance by the diatoms, colonial green algae or even the cyanobacteria common in the more enriched nearby waters of the English Lake District will, so far, have been quite erroneous. Instead, the flora has continued to comprise more or less the same algal species, albeit in altered relative proportions, but in much greater numbers.

That such little compositional change should be surprising at all perhaps says more about the shallowness of our understanding of pelagic systems than our knowledge of the organisms. Intuition anticipates species which are more or less efficient at obtaining nutrients, operate at higher or lower N:P ratios or use differentially the various sources of carbon. If these are dimensions upon which substantial change could have occurred, are we sure that two growing seasons, spanning little more than one year, are sufficient to have allowed other, potentially more successful species to have 'arrived' and established competing inocula, once the habitat was rendered supposedly more amenable? We also note that enrichment has not made the tarn unsuitable for any of its former planktonic plant species. Could it be that, since the lake is still acidic, it has not changed sufficiently for there to be any measurable change anyway? Both explanations have some probability but, if so, both emphasise that the common association among phytoplankton species structure and lake trophic status, as formalised by such as Rodhe (1948), Rawson (1956) and Moss (1973), is not a simple permutation of the nutrient content of the water.

Our deduction is that the development of associations of phytoplankton and the organisation of communities are influenced by many factors which act at differing scales and it takes time and a history of events to bring some degree of similarity (cosmopolitanism) among the species structures of chemically-similar but geographically-isolated lakes. Merely raising the phosphorus content may not by itself have overcome the conditions in Seathwaite Tarn which, hitherto, may have been perceived by such as *Asterionella*, *Eudorina* or *Anabaena* as being hostile (which could include its temperature range or the suitability of the benthic refuges). We suspect that the promotion of 'more of the same' (Reynolds, 1992) and the relative distance from other sites where eutrophic species are already well-established may well reflect the time that it takes for species to be transmitted from isolated lake to isolated lake. Many cases have been documented in the English Lake District (see Reynolds, 1998) where, during environmental alteration of lakes through enrichment, the first occasion that species, supposedly suited to the new conditions, have been noted in routine counts may occur 5–10 years later; this is very short against the scale of palaeolimnological investigations but nevertheless finite and of interest to the prediction of change in water bodies.

References

Davison, W., 1987. Internal elemental cycles affecting the long-term alkalinity status of lakes: implication for lake restoration. Schweiz. Z. Hydrol. 49: 186–200.

Davison, W., D. G. George & N. J. A. Edwards, 1995. Controlled reversal of lake acidification by treatment with phosphate fertilizer. Nature 377: 504–507.

Davison, W., C. S. Reynolds, E. Tipping & R. F. Needham, 1989. Reclamation of acid waters using sewage sludge. Envir. Poll. 57: 251–274.

Lund, J. W. G., 1959. A simple counting chamber for nannoplankton. Limnol. Oceanogr. 4: 57–65.

Lund, J. W. G., C. Kipling & E. D. Le Cren, 1958. The inverted microscope method of estimating algal numbers and the statistical basis of estimations by counting. Hydrobiologia 11: 143–170.

Lund, J. W. G. & J. F. Talling, 1957. Botanical limnological methods with special reference to the algae. Bot. Rev. 23: 489–583.

Moss, B., 1973. The influence of environmental factors on the distribution of freshwater algae: an experimental study. IV. Growth of test species in natural lake waters and conclusion. J. Ecol. 61: 193–211.

Rawson, D. S., 1956. Algal indicators of trophic lake types. Limnol. Oceanogr. 1: 18–25.

Reynolds, C. S., 1992. Eutrophication and the management of planktonic algae: what Vollenweider couldn't tell us. In D. W. Sutcliffe & J. G. Jones (eds), Eutrophication; research and application to water supply. Freshwater Biological Association, Ambleside: 4–29.

Reynolds, C. S., 1998. What factors influence the species composition of phytoplankton in lakes of different trophic status? Hydrobiologia 369/370: 11–26.

Rodhe, W., 1948. Environmental requirements of freshwater plankton algae: experimental studies in the ecology of phytoplankton. Symb. bot. Upsaliensis 10: 5–149.

Shannon, C. E., 1948. A mathematical theory of communication. Bell Syst.tech. J. 27: 623–656.

Talling, J. F., 1974. General outline of spectrophotometric methods. In R. A. Vollenweider (ed.), A manual on methods for measuring primary production in aquatic environments. Blackell Scientific Publications, Oxford: 22–25.

Talling, J. F. & D. Driver, 1963. Some problems in the estimation of chlorophyll-*a* in phytoplankton. In U.S. Atomic Energy Commission (ed.), Proceedings, Conference on primary productivity measurements, marine and freshwater, Hawaii, 1961. USEAC, Washington: 142–146.

Youngman, R. E., 1971. Algal monitoring of water supply reservoirs and rivers. Tech. Memo. Wat. Res. Ass. 63: 1–26.

Hydrobiologia **369/370:** 133–137, 1998.
M. Alvarez-Cobelas, C. S. Reynolds, P. Sanchez-Castillo & J. Kristiansen (eds), Phytoplankton and Trophic Gradients.

Comparative study of phytoplankton in an oligotrophic soft water lake under different pH-phosphate ranges

Susana Romo
Area de Ecología, Facultad de Biología, Universidad de Valencia, Burjasot-46100, Valencia, Spain

Key words: Scotland, shallow lake, phytoplankton, phosphate fertilization, acid rain

Abstract

This work reports phytoplankton changes in a shallow, acidic and oligotrophic lake (Loch Rusky, Scotland) from two different periods: one with artificial phosphate fertilization and neutral pH (SRP 3–18 μg l^{-1}, pH 6.6–7.8, 1972–73) and the other with low phosphate and acidic pH values (SRP < 1.5 μg l^{-1}, pH 5.6–6.9, 1985–86). Phosphate enhanced markedly total algal biomass and favoured the presence of cyanophytes, especially in the summer epilimnion, while cryptophytes dominated at this time during unfertilized periods. *Anabaena* sp. peaked during summer and a higher density of *Ceratium hirundinella* and *Peridinium* sp. in mid-summer was also observed during the fertilized period. Otherwise, algal composition and periodicity remained similar for the two comparative periods. In both, diatoms (*Asterionella formosa* and *Aulacoseira italica*) dominated during spring.

Introduction

Moss (1973) discerned a general pattern of distribution of phytoplankton species among lakes which separated the flora of oligotrophic, soft water lakes and those affected by acidification. Thus, reduction of free CO_2 levels, induced by increase in pH, should decrease the growth potential of oligotrophic species in soft waters. Implicitly, the converse is that the lowering of pH should help to maintain them. Equally, reduction of pH in hard waters should permit the development of oligotrophic species. Furthermore, in the presence of adequate algal nutrients, eutrophic species should be able to grow in both soft and hard waters. The present work attempts to test these hypotheses against data gathered in a study of an acidic oligotrophic lake.

The study site

The study was carried out in Loch Rusky, an oligotrophic, monomictic and shallow lake in the central part of Scotland (National Grid Reference NN615035), just to the south of the Highland Boundary Fault. The geology of the area is dominated by Devonian red sandstone, although the loch itself lies on superficial drift deposits. The catchment area is of mainly softwood forestry plantation (85%) and high moorland grazed by sheep (15%). It receives a main inflow from the north, Letter Burn, which water is acidic (pH 5.6–6.8), base-poor (alkalinity 0.1–0.9 meq l^{-1}) and deficient in SRP (< 2.5 μg l^{-1}), but not in nitrate (35–265 μg l^{-1}) or silicate (2–5 mg l^{-1}). Precipitation is the second main input into the lake.

Some morphometric characteristics of the lake are outlined in Table 1. Since 1964, the loch has been managed as a game fishery for gamefishing. Every 3–4 weeks, between April and September, the loch is restocked with juvenile rainbow trout (*Salmo gairdneri*) and brown trout (*Salmo trutta*). It also has natural populations of brown trout and minnows (*Phoxinus phoxinus*). Fish populations restrict the abundance of large zooplankters in the lake and rotifers dominate through most of the year, save in summer, when small cladocerans (*Daphnia hyalina* Leybig) and copepods become more abundant (Romo, 1990).

Materials and methods

During 1972–73, the lake was artificially fertilized with phosphate by the Forestry Commission, and physicochemical parameters, as well as phytoplankton com-

Table 1. Loch Rusky morphometric characteristics.

Area (Ha)	20
Length (m)	800
Breath (m)	500
Volume (Hm^3)	1.45
Maximum depth (m)	15
Catchment area (km^2)	3
Altitude (m)	136

Table 2. Comparative limnological and phytoplanktonic features of Loch Rusky between phosphate fertilized (1972–73; Maulood, 1974) and unfertilized (1985–86) periods.

Seasonal year	1972–73	1985–86
pH	6.6–7.8	5.6–6.9
Alkalinity (meq/l)	0.15–0.28	0.05–0.31
SRP (μg/l)	3.1–18.3	0.1–1.5 (3.75)
Nitrate (μg/l)	46–291	50–265
Ammonia (μg/l)	–	0–128
Silicate (mg/l)	0.1–3.32	0.08–6.72
Phytoplankton abundance (ind/ml)		
Range	36–17 000	4–310
Annual mean	3000	48
Shannon-Wiener diversity (bits/ind)	–	0.5–4.5

position and periodicity were reported in detail by Maulood (1974). This period is compared with a period, 1985–86, when the lake was in a more natural condition of phosphorus scarcity. This comparative study has special interest because phytoplankton responses to fertilization can be viewed on a larger scale that is sensitive to the time lag required in recolonisation (Hörnström et al., 1993). In this study, as in that of Maulood (1974) samples were collected from the deepest part of the lake at different depths and approximately fortnight intervals. The standard limnological and phytoplankton counting methods used are reported elsewhere (Romo, 1987).

Results

Some abiotic and biotic limnological variables from the two periods are compared in Table 2. As expected, pH increased during the 1972–73 fertilization to circumneutral values, although alkalinity remained within soft water ranges. Strong acid salts, such as chloride and especially sulphate (10–20 mg l^{-1}), were present in high concentrations in the lake and the inflow, and increased during rain periods in autumn. Dissolved organic carbon (D.O.C: 6–20 mg l^{-1}) was also related to seasonal wetness and mostly comprised humic compounds, which severely contributed to a truncation of the photic zone in the lake to < 4 m. During the fertilization period, SRP concentrations steeply increased (Table 2). Natural values for 1985–86 were below 0.5 $\mu g\ l^{-1}$ during the algal growth season (end-March to September) and remained below 1.5 $\mu g\ l^{-1}$through the rest of the year. Nitrate was the dominant nitrogen form having concentrations normally > 80 $\mu g\ l^{-1}$; silicate concentrations were depressed during summer (< 0.5 mg l^{-1}, Table 2).

The natural phytoplankton assemblage of the lake comprised 85 species. In 1985–86, three algal peaks were observed (Figure 1). The spring assemblage was dominated by *Aulacoseira italica* (Ehr.) Simonsen and *Asterionella formosa* Hassall, amounting to more than 50% of the total abundance (up to 135 ind ml^{-1}). Subdominant vernal species were *Tabellaria fenestrata* (Lygnb.) Kütz., *Synedra acus* Kütz and *Eunotia pectinalis* var. *minor*, with populations two orders of magnitude lower.

The summer epilimnion was dominated in July by *Rhodomonas minuta* var. *nannoplanctica* Skuja (up 300 ind ml^{-1}, 98% total abundance) and in early August by *Cryptomonas ovata* Ehr. (20 ind ml^{-1}, 75% total abundance). Subdominant epilimnetic species in July were *Coelastrum microporum* Naegeli and *Dinobryon divergens* Imhof, and in the thermocline *Chroococcus dispersus* (Keissl.) Lemm. and *Coelosphaerium naegelianum* Unger. An autumnal peak occurred in November, coincidental with an input of phosphate (3.75 $\mu g\ l^{-1}$). Phytoplankton was overwhelmed, specially at 3 and 10 m (180–90 ind ml^{-1}), by an unidentified colonial chroococcal cyanophyte.

Discussion

Comparing the newer data with the sequence of phytoplankton and total abundance obseved during the 1972–73 fertile period, some important differences can be observed (Figure 1). First, total algal abundance was greater by a factor of 63 with phosphate fertilization (from 48 to 3000 ind ml^{-1}). Second, algal composition differed during summer, with the dominance of *Anabaena* sp. from July to October (81–99% total abundance). The maximum population of *Anabaena* sp. in late August 1973 (16.5 × 10^3 ind ml^{-1}) coin-

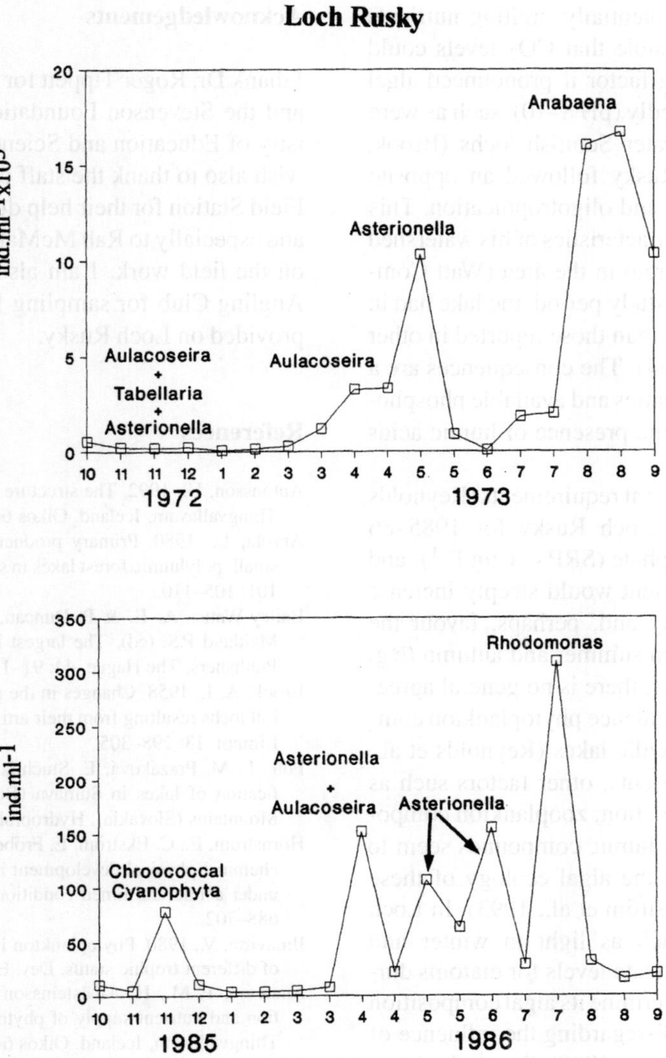

Figure 1. Total phytoplankton abundance and dominant species in Loch Rusky during fertilized (1972–73) and unfertilized (1985–86) periods. Abbreviations of the dominant species at their maximum abundances: *Aulacoseira = Aulacoseira italica, Asterionella = Asterionella formosa, Tabellaria = Tabellaria fenestrata, Anabaena = Anabaena sp., Rhodomonas = Rhodomonas minuta* var. *nannoplanctica*. Notice the change of scale on the *y*-axis. For more details see text.

cided with a considerable increase of SRP in the lake (18 $\mu g\ l^{-1}$ at surface). In mid-summer, a larger presence of *Ceratium hirundinella* (Müll.) Dujardin and *Peridinium* sp. was also observed. Third, diatoms dominated in late autumn, and during spring. The populations of the dominant species (*A. italica, A. formosa*) were two orders of magnitude higher than those of the unfertilized period. Otherwise, phytoplanton composition in the lake was qualitatively quite similar during the two periods.

According to OECD (1982) criteria, Loch Rusky could be classified as oligtrophic in the period 1985–86, on the basis of its chlorophyll-a content, and ultra-oligotrophic, on the basis of its phosphate availability. During 1972–73, the phosphate levels indicate a mesotrophic state. It is unlikely that the differences in algal composition between the fertilized and unfertilized periods, were caused only by changes in pH. The present data agree with Moss' (1973) model, insofar as his last statement relates the presence in soft waters of so-called eutrophic species to adequate concentrations of phosphate and other potentially limiting nutrients being available. It is possible that CO_2 levels could become an algal selecting factor if pronounced algal blooms increase pH markedly (pH 8–10), such as were observed in some soft water Scottish lochs (Brook, 1958). However, Loch Rusky followed an opposite trend toward acidification and oligotrophication. This is caused in part for the characteristics of his watershed and the influence of acid rain in the area (Watt Committee, 1983). During the study period, the lake had in fact larger sulphate values than those reported in other acid lakes (Fott et al., 1994). The consequences are a decrease in total algal densities and available phosphorus, which precipitate in the presence of humic acids (Vyhnálek et al., 1994).

According to algal nutrient requirements (Reynolds, 1992), phytoplankton in Loch Rusky for 1985–86 would be limited by phosphate (SRP $< 1\ \mu g\ l^{-1}$), and fertilization with this nutrient would steeply increase the algal carrying capacity and, perhaps, favour the presence of cyanophytes in summer and autumn (e.g. November 1985). However, there is no general agreement on how phosphate influence phytoplankton composition in oligotrophic acidic lakes (Reynolds et al., 1998). Together with nutrients, other factors such as lake morphometry and isolation, zooplankton composition and the presence of humic compounds seem to play an important role in the algal ecology of these freshwater systems (Hörnström et al., 1993). In Loch Rusky, other factors, such as light in winter and reduced turbulence and silicate levels for diatoms during summer, could also determine its algal composition and periodicity, without disregarding the influence of zooplankton predation (Romo, 1990). Some planktonic changes in Loch Rusky could be partially explained by some of the statements of the PEG-model (Sommer et al., 1986). Similar phytoplankton and zooplankton composition to that observed in Loch Rusky during the unfertilized period has been reported for some other oligotrophic and soft water Scottish lochs (Bailey-Watts & Duncan, 1981; Maitland et al., 1981) and northern deep lakes (Jónasson et al., 1992; Antonsson, 1992), as well as in some polyhumic shallow Norwegian and Finish waterbodies (Arvola, 1980, Ilmavirta, 1980). Some authors state that acification affects more to higher trophic levels, such as fish and large zooplankton (Leivestad, 1976; Fott et al., 1994), than phytoplankton composition, where acidic and oligotrophic conditions only seem to enhance the presence of unicelular flagellates (Pugnetti et al., 1993; Fott et al., 1994).

Acknowledgements

I thank Dr. Roger Tippett for his help during this work, and the Stevenson Foundation and the Spanish Ministry of Education and Science for financial support. I wish also to thank the staff of the Glasgow University Field Station for their help during my stay in Scotland, and especially to Rab McMath for his useful assistence on the field work. I am also grateful to Loch Rusky Angling Club for sampling facilities and information provided on Loch Rusky.

References

Antonsson, U., 1992. The structure and function of zooplankton in Thingvallavatn, Iceland. Oikos 64: 188–221.

Arvola, L., 1980. Primary production and phytoplankton in two small, polyhumic forest lakes in southern Finland. Hydrobiologia 101: 105–110.

Bailey-Watts, A. E. & P. Duncan, 1981. The phytoplankton. In Maitland P.S. (ed), The largest lochs in Scotland. Dr. W. Junk Publishers, The Hague, 44: 91–118.

Brook, A. J., 1958. Changes in the phytoplankton of some Scottish hill lochs resulting from their artificial enrichment. Verh. int. Ver. Limnol. 13: 298–305.

Fott, J., M. Prazáková, E. Stuchlík & Z. Stuchlíková, 1994. Acification of lakes in Sumava (Bohemia) and in the High Tatra Mountains (Slovakia). Hydrobiologia 274: 34–47.

Hörnström, E., C. Ekström, E. Fröberg & J. Ek, 1993. Plankton and chemical-physical development in six Swedish West Coast lakes under acidic and limed conditions. Can. J. Fish. aquat. Sci. 50: 688–702.

Ilmavirta, V., 1980. Phytoplankton in 35 Finnish brown-water lakes of different trophic status. Dev. Hydrobiol. 3: 121–130.

Jónasson, P. M., H. Adalsteinsson & G. Jónsson, 1992. Production and nutrient supply of phytoplankton in subarctic, dimictic Thingvallavatn, Iceland. Oikos 64: 162–187.

Leivestad, I., G. Hendrey, I. Muniz & K. E. Snekvi, 1976. Effects of acid precipitation on freshwater organisms. In Braekke F. (ed), Impact of acid precipition on forest and fresh-water ecosystems in Norway. S.N.S.F. project FR 6/76: 87–111, Oslo.

Maitland, P. S, B. D. Smith, G. M. Dennis, 1981. The crustazean zooplankton. In Maitland, P.S. (ed), The largest lochs in Scotland. Dr. W. Junk Publishers, The Hague: 135–154.

Maulood, B., 1974. Studies on the phytoplankton of Loch Lomond and of Neighbouring Lochs North and South of the Highland Boundary Fault. PhD Thesis, Glasgow University, 249 pp.

Moss, B., 1973. The influence of environmental factors on the distribution of freshwater algae: an experimental study. IV. Growth of test species in natural lake waters and conclusion. J. Ecol. 61: 193–211.

OECD, 1982. Eutrophication of waters. Monitoring, assessment and control, Paris, 154 pp.

Pugnetti, A., M. Manca, A. M. Nocentini, A. Boggero, R. Bettinetti, M. Bonardi & P. Cammarano, 1993. Ecological aspects of two alpine lakes (Lakes Paione, Italy). In Giussani, G. & C. Callieri (eds), Strategies for lake ecosystems beyond 2000. Stresa, Italy: 440–443.

Reynolds, C. S., 1992. Eutrophication and the management of planktonic algae: what Vollenweider could not tell us. In Sutcliffe, D. W. & J. G. Jones (eds), Eutrophication: research and application to water supply, Freshwat. Biol. Ass.: 4–29 pp.

Reynolds, C. S., G. H. M. Jaworski, J. V. Roscoe, D. P. Hewitt & D. G. George, 1998. Responses of the phytoplankton to a deliberate attempt to raise the trophic status of an acidic, oligotrophic mountain lake. Hydrobiologia 369-370: 127–131.

Romo, S., 1987. The annual cycle of nutrients and phytoplankton dynamics in a shallow monomictic lake in Scotland (Loch Rusky). MSc. Thesis, Glasgow University, 133 pp.

Romo, S., 1990. Seasonal zooplankton patterns in a shallow oligotrophic lake: Loch Rusky (Scotland). Annls Limnol. 26: 11–17.

Sommer, U., Z. M. Gliwicz, W. Lampert, & A. Duncan, 1986. The PEG-model of seasonal succession of planktonic events in fresh waters. Arch. Hydrobiol. 106: 433–471.

Vyhnálek, V., J. Fott & J. Kopácek, 1994. Chlorophyll-phosphorus relationship in acified lakes of the High Tatra Mountains (Slovakia). Hydrobiologia 274: 171–177.

Watt Committee, 1984. Acid rain. Watt Committe on Energy, report n. 14. London, 55 pp.

Hydrobiologia **369/370:** 139–152, 1998.
M. Alvarez-Cobelas, C. S. Reynolds, P. Sanchez-Castillo & J. Kristiansen (eds), Phytoplankton and Trophic Gradients.

Fish stock as a variable modifying trophic pattern of phytoplankton

J. Komárková
Hydrobiological Institute ASCR, Na sádkách 7, CZ - 370 05 České Budějovice, Czech Republic

Key words: phytoplankton, biomass, fish stock, trophic gradient, diversity, equitability

Abstract

Fish stock, zooplankton, chlorophyll *a* concentration, algal biomass (fresh mass), phytoplankton composition, species dominance and diversity (Shannon-Weaver Index and equitability) were evaluated to demonstrate their relevance in the trophic gradient at the levels of eutrophy and hypertrophy. Two patterns of food-web connectedness between phytoplankton and fish stock can be detected. One is direct (phytoplankton is mainly consumed directly by herbivorous or omnivorous fish) another one indirect (phytoplankton is consumed by zooplankton, which is regulated by the fish stock). Release of nutrients from fecal pellets and from the bottom due to the feeding activity of fish in shallow waters provide further feedbacks to the enhancement of phytoplankton growth. Changes in the predation pressure and bioturbation, as a consequence of shallowness, may be a source of intermediate disturbance which might promote phytoplankton diversity. Results from managed fish ponds (Komárková, 1983; Komárková et al., 1986), from reservoirs before and after biomanipulation (Komárková et al., 1995) and examples from a tropical lake Catemaco (Komárková & Tavera, 1996) confirm that the composition of phytoplankton is influenced, to varying degrees, by the nature and intensity of predation. Changes in fish stock can have both positive and negative effects on the transparency, concentration of nutrients and abundance of phytoplankton. Taxonomic composition and diversity of phytoplankton are affected differently, the effect being often the opposite from that predicted. Thus, in a water body with an unbalanced fish stock, using phytoplankton characteristics as the only criteria for identification of the trophic status can give incorrect information.

Introduction

The extent to which the composition of the phytoplankton of a water body conforms to a trophic classification based on other properties of that water body is complicated by a number of factors. The question, 'Which species of algae living in what kind of environment are typical for that particular trophic state?', still lacks reliable answers. Some generalisations are better than others, for instance, that the chrysophytes are typically associated with rather oligotrophic waters, while heavy blooms of cyanobacteria are largely confined to eutrophic or hypertrophic waters. Despite objections to the validity of species indicative of a given trophic level or of specific (see Washington, 1979), it is not less apparent that the assembly of species in given sites combining particular physical and chemical properties is far from random. The OECD (1982) Report on Eutrophication of freshwaters suggested criteria on which the trophy of a water body might be assessed: 1. Nutrient supply. 2. Supply of inorganic carbon. 3. Biomass, taxonomical composition and diversity of phytoplankton. The ranges of the variables and species of both algae and animals characterising oligotrophic, mesotrophic and eutrophic waters have been defined (Wetzel, 1975; Reynolds, 1984). Unreliabilities in this system are highlighted by Rojo (this volume).

An important regulatory role in the control of the biomass and species composition of the algal assemblage, at any trophic level, is played by their consumers, i.e. the zooplankton and, mostly in the tropics and sub-tropics, the fish. The impact of zooplankton on the phytoplankton composition, biomass and production has been studied by many authors (Porter, 1973, 1976; Reynolds et al., 1982; Lampert et al., 1986; Sommer, 1988). Sterner (1989) thoroughly summarized and evaluated all important papers concerning the problem. Even though the number of studies is already large, serious problems remain in understanding and managing interrelations among fishstock, zooplankton and phytoplankton interrelations to attain the optimal balance. The problems are especially important in eutrophic and hypertrophic systems, where a good knowledge of the mutual relations is necessary to a proper approach to improving water quality (Hrbáček, 1994). Sustainability of the system is complicated by the life cycles of the consumers, by seasonal periodicity of the phytoplankton, by feedback changes in phytoplankton composition and by other consequences of the presence of consumers (bioturbation, release of the nutrients from the bottom, excretion and fecal pellets decomposition: see Sterner, 1989; Persson & Hamrin, 1994). Structural changes in the assemblage may also arise as intermediate disturbance responses and through the competitive interactions among the phytoplankters (Padisák et al., 1993).

The seasonal succession of limnetic phytoplankton broadly follows the pattern of the PEG model (Sommer et al., 1986). Invasive, small-sized species grow fast in spring, until they are either grazed out by zooplankton or they are limited by nutrients (Vyhnálek et al., 1991). After a clear-water phase, they are replaced by slow-growing large or colonial 'canopy' species; these accumulate biomass but the net production/biomass rate decreases (Sommer, 1981; Sommer et al., 1986; Reynolds, 1994). In many cases, the reason for the switch from nanoplankton dominance to the dominance of net plankton is the increase of filter-feeding zooplankton. Even though this pattern holds for most meso- to eutrophic waters with a balanced fish stock, any shift in the balance of the components can alter the pattern substantially (Fott et al., 1980; Komárková et al., 1986; Jeppesen et al. 1990).

Ranges of the variables and the scheme of the trophic gradient worked out for the temperate zones do not always apply in tropical and subtropical waters (see Huszar et al., this volume). The ranges are often strongly modified by the type of the food chain in operation, the selection of which is tied into the nature of the phytoplankton and its suitability as food for zooplankton (Porter, 1977) and for planktivorous fish (Drenner et al., 1984) or for both. All the fish ponds in Central Europe are eutrophic to hypertrophic, however; according to the fish stock, they display a whole range of nutrient- and chlorophyll concentrations, transparency values and composition of biocoenoses.

According to the rules of community assembly (Odum, 1971), the diversity of the assemblage should also indicate its position in the trophic gradient. The diversity of the assemblage should decrease from oligotrophy toward eutrophy, being the lowest in hypertrophic systems. However, the list by Alvarez-Cobelas & Jacobsen (1992, Table 1),for instance, shows almost all possible values of the Shannon index for hypertrophic water bodies (0.0–4.5). This suggests that the biomass and composition of phytoplankton is controlled also by mechanisms other than nutrient supply. With the increasing trophic level, the predation impact becomes a most serious top-down control of the phytoplankton assemblage. When the control by zooplankton is missing, Hutchinson's 'paradox of phytoplankton' becomes still more evident.

The aim of this paper is to demonstrate several examples from previous work, directed to other questions, which nevertheless demonstrate relationships between the phytoplankton composition and diversity and between phytoplankton biomass and its consumers. Managed fishponds offer most extreme examples. Their shallowness and simple management, involving a re-initiation every first or second year, following a complete drainage, provide good conditions for experimentation. The mutual relationships can be best followed here, as precise information on the quality and amount of fish stock is readily available (such information is frequently less well known in lakes and reservoirs). The application of the food web theory led to the biomanipulation of the lakes (Björk, 1972; Cronberg, 1982; Benndorf, 1987) and reservoirs (Hrbáček et al.,1986; Hrbáček, 1994; Matěna & Vyhnálek, 1994). The biomanipulation of canyon-shaped reservoirs with a medium retention time was not fully successful (Komárková et al., 1995), however, the impact of the changed fishstock on the zooplankton and phytoplankton composition was evident. The characteristics of phytoplankton in a tropical lake Catemaco are cited as an example of direct predation of fish on the phytoplankton. All the investigated localities belong to the eutrophic and hypertrophic levels of the trophic gradient.

Localities and methods

The fishpond Ruda is a typical shallow, polymictic and hypertrophic pond for breeding the common carp (*Cyprinus carpio*) in South Bohemia (mean depth 1.3 m, bottom with a lot of soft sediment, see Table 1).

Table 1. Data on the localities. Biological and chemical characteristics in Ruda fishpond and Římov reservoir are averages for summer seasons (Apr.–Sept.). Data from expedition on 46 fish ponds are averages of samples taken roughly in the summer maximum of phytoplankton. Averages and ranges from Catemaco lake were calculated from samples taken both in dry and wet seasons. Residence time and flow are yearly means and ranges. X signs that no data are available.

	Ruda 1977	Ruda 1978	Třeboň 1977 46 fish ponds	Římov res. (1986–1995)	Catemaco (1993–1995)
Geographic location	South Bohemia, CZ	dtto	South Bohemia, CZ	South Bohemia, CZ 471 m a.s.l.	18°24′ N 95°04′ W 332 m a.s.l.
	temperate	dtto	temperate	temperate	tropic., volcanic
Area	0.71 km^2	dtto	1.5 to 350 ha	2.1 km^2	72.5 km^2
Volume	0.96 mill m^3	dtto	X	34.5 mill m^3	0.523 km^3
Watershed area	18.7 km^2	dtto	X	490 km^2	244 km^2
Residence time, days	602 (507–635)	895 (101–1916)	X, > 300	92	317
Flow, m^3 sec^{-1}	0.32 (0.02–0.62)	0.1 (0.006–0.11)	X	4.34	X
Mean depth, m	1.3	dtto	roughly 1 m	24	7.6
Max. depth, m	2.4	dtto	(2.5–3.5)	43	11
Trophic gradient	hypertrophy	eutrophy	eu-to	meso- to eutrophy	
	(polymictic)	(polymictic)	hypertrophy	(dimictic)	(polymictic,
(Mixing regime)			(polymictic)		temporal str.)
pH (range)	7.95 (7.1–10.8)	8.42 (6.8–9.55)*	8.7 (6.8–10.8)	8.5 (5.7–10.3)	(7.8–9.5)
Alkalinity, meq. l^{-1}	1.23	1.57	0.8	0.42	
(range)	(0.98–1.36)	(1.3–1.62)	(0.5–125)	(0.26–0.67)	(0.7–1.1)
Transp., m (range)	0.40 (0.2–0.5)	0.82 (0.4–> 1.3)	0.3 (0.1–> depth)	2.9 (1.2–6.5)	(0.3–1.1)
z_{eu}/z_{mix}	0.25 (0.15–0.4)	0.63 (0.2–0.8)	X	1.6 (0.3–5.5)	(0.5–0.1–0.03)
Tot. P, μg l^{-1} (range)	93 (52.3–165.9)	173 (69.2–208.2)	X	24.0 (5.0–70.8)	250 (59–484)
SRP, μg l^{-1} (range)	15.3 (5.0–26.9)	24.4 (12–32.5)	X	3.3 (1.7–4.9)**	12 (2.5–32)
Tot. N, mg l^{-1} (range)	1.77 (1.56–2.03)	1.78 (0.98–2.30)	X	3.03 (1.46–4.85)	454 (231–910)
DIN, mg l^{-1} (range)	0.55 (0.45–0.80)	0.43 (0.31–0.78)	X	2.57 (0.75–4.72)	X
Chlorophyll *a*, μg l^{-1}	94.7	48.4	100.1	18.5	49
(range)	(56.5–154.9)	(25.3–95.2)	(10.0–149.5)	(0.65–139.5)	(18–66)
Biomass, mg l^{-1}	24.6	11.4	24.8	4.31	48.4
(range)	(7.2–51.4)	(5.0–15.2)	(2.2–102.0)	(0.07–47.00)	(20–60)
Phytopl., main representatives	Chlorococcales Euglenophyceae Centr. diatoms	*Microcystis* *Aphanizom. f.-a.* *Pediastrum,*	Chlorococcales *Aphanizom.* *Microcystis*	*Asterionella* *Fragilaria* *Aphanizom. f.-a.*	*Cylindrosperm.* *Aulacoseira* *Pediastrum*
Zooplankton main representatives	*Bosmina,* Rotatoria *Daph. galeata*	*Daph. galeata* (large specimens) *Daph. pulicaria*	Different composition	Calanoids *Daphnia galeata*, diff. sizes	*Bosmina,* Rotatoria, *Macrocyclops,* *Diaptomus*
Fish species	Common carp 2–3 years age	Common carp 1 year age + little K_2.	Common carp diff. age	Bream, perch	Shads, cichlids, tilapia
Yield of fish stock or estim. fish stock	567 kg ha^{-1} m^{-1}	274 kg ha^{-1} m^{-1}	50–700 kg ha^{-1} m^{-1}	600–100 kg ha^{-1} m^{-1}	Yield estimation 140 kg ha^{-1}

* The bottom was overgrown by *Chara fragilis* and the pond was limed more than usually.
** Average only from 1994–95.

60% of the watershed is covered with coniferous forests. The rest of the area consists of meadows and fields. In the years studied, the pond was fertilized and the fish were fed by ray from the age of 1-year. The samples were taken with a Friedinger sampler at six sites and at 14-days' intervals during two successive seasons. Water was collected at several depths in order to obtain a representative sample. The samples for analysis of chemical variables, chlorophyll *a*, phytoplankton biomass and species composition were separated before filtering the water. In 1977, the pond was stocked with 2-year-old carp and the summer fishstock (a mean of the spring stock and the yield) was 365 kg ha^{-1} m^{-1}, i.e. 0.027 kg m^{-3}. In autumn 1977, the fishpond was drained and fished. Next spring (1978), the pond was again filled and stocked with 1-year old carp (length approx. 10 cm); later, it was supplied additionally with a small number of 2-year old carp (12% of total biomass), because of the poor health of the younger fish. In summer 1978, the fishstock reached 227 kg ha^{-1} m^{-1} (0.019 kg m^{-3}). It was calculated that natural production amounted 68 kg ha^{-1} m^{-1} in 1977, while it was only 10 kg ha^{-1} m^{-1}in 1978, (total production 413 and 116 kg ha^{-1}, respectively). For further information on the methods applied and results obtained, see Table 1 and Komárková et al. (1986).

In 1978, 46 fish ponds were sampled in the frame of a two-week-long expedition in the Třeboň region (Komárková, 1983). For a summary of the features of the ponds see Table 1. All ponds were similar in character, i.e. shallow man-made water bodies with a rich soft sediment and more or less distinct littoral region, used for raising carp. However, the sizes of the fish ponds and the stocks of carp were varied (Table 1). All the ponds were sampled in the same way, during the period of maximal and most stable development of phytoplankton, in the second part of August 1978. Several samples were taken from the whole column with the plankton tube near the deepest point of each fish pond and mixed in a vessel. From this bulked volume, subsamples for analysis of chlorophyll *a* concentration, biomass and composition of phytoplankton were taken. The resulting data were compared with the summer fish-stock (average between the original biomass of fish stock and the yield in autumn). The fish stock is expressed here, as in the case of Ruda fish pond, in the units of kg ha^{-1} m^{-1}. As the studied fish ponds had different average depths, this unit has been chosen to compensate for the different depths of the fish ponds considered. Data on the morphometry of fish ponds and biomass and composition of the fish stock were kindly provided by the State Fishery, Třeboň.

Data are also invoked from studies on the Římov Reservoir. This is a eutrophic, dimictic, canyon-type reservoir on the Malše river, in southern Bohemia. It serves as a drinking water supply (length 13.5 km, volume max. 34.5 × 10^6 m^3, max. depth 43 m, mean retention time 92 days). The sampling site was chosen near the dam. The reservoir was sampled in three-week intervals during the regular monitoring of the reservoir since its filling in 1979. The years 1985–1995 were chosen for the evaluation. Samples of phytoplankton were taken using a 4-m tube plugged with a stopper. We took 5–7 subsamples along the transversal line across the reservoir at the widest profile near the main sampling point, and mixed them in a barrel. Then the samples were taken for determination of chlorophyll-*a* content, biomass (FW) and qualitative composition of the phytoplankton assemblage. For the detail data on the fishstock management and zooplankton estimation see Sed'a & Kubečka (1997).

Finally, data from a tropical location are compared. The Catemaco Lake (Mexico Bay, Vera Cruz State, Mexico) is a eutrophic lake of volcanic origin, shallow and large (mean depth 7.6 m, about 80 km^2 of surface). It is polymictic and intensively fished. The main species in the lake were a small species of shad (*Dorosoma petenense*), an endemic characid *Bramocharax caballeroi* and an introduced *Oreochromis niloticus* (Tavera, 1996, Komárková & Tavera, 1996). All the samples of phytoplankton biomass were preserved with a Lugol solution and counted on the inverted microscope after sedimentation in the Utermöhl chambers (Lund, 1970; Javornický & Komárková, 1973). The biomass of algae (biovolume) was estimated as fresh mass using the PHPL – HBI computing program (Komárková & Hauser, unpublished).

The diversity of phytoplankton was expressed by the Shannon-Weaver index (H", bits) based on biomass of the species, expressed as logarithms to the base 2 (Odum, 1971; Padisák et al., 1993). The equitability of the assemblage (E) was calculated also according to Odum (1971). For demonstration of percentual shares of algae in phytoplankton of the Catemaco lake and in the intestines of *Dorosoma petenense*, the sedimented samples preserved by formaline were used. Chlorophyll *a* concentration was estimated according to the Lorenzen's method (Lorenzen, 1967) and, since 1990, according to the fluorescence method used at Římov Reservoir (modification by Vyhnálek et al., 1993).

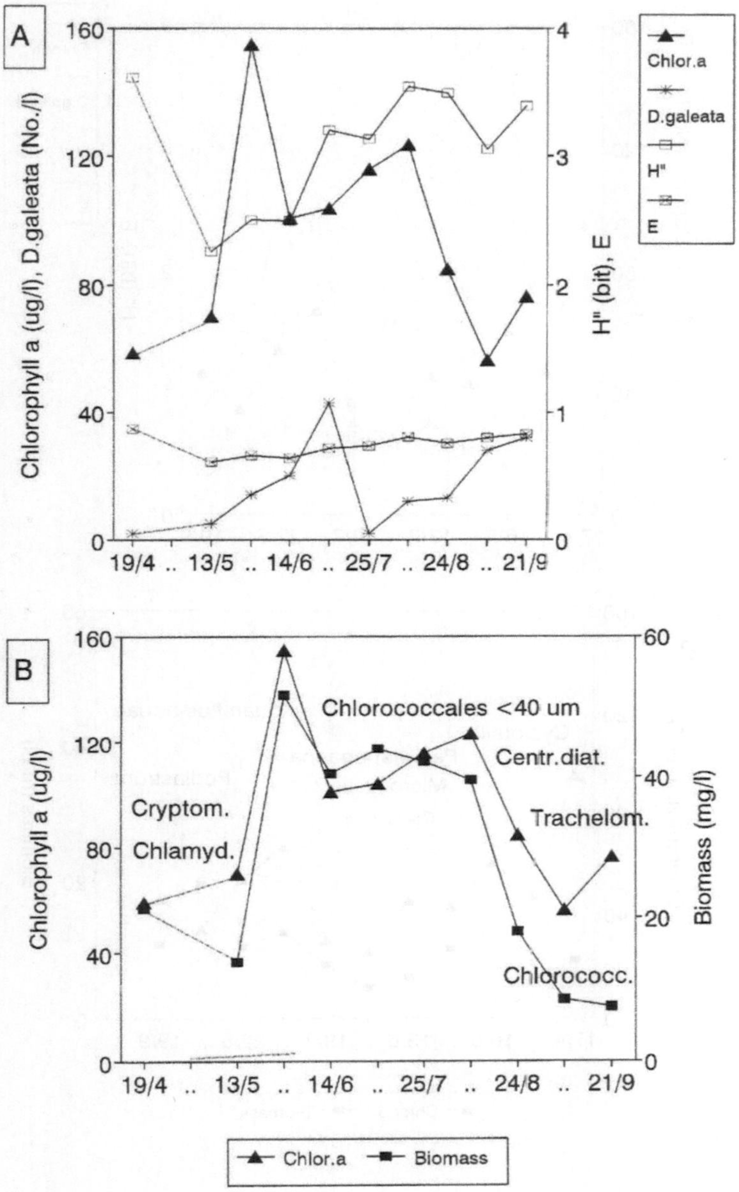

Figure 1. The fishpond Ruda, high fishstock: 2-year old carp, summer fishstock 365 kg ha^{-1} m^{-1}, i.e. 0.027 kg m^{-3}. Production of fish in autumn 413 kg ha^{-1}. A. Chlorophyll *a* concentration, numbers of *Daphnia galeata* and diversity indices. B. Phytoplankton biomass and composition.

Zooplankton samples were taken using the Apstein net (70, 140 and 700 μm), by hauling the water column from the bottom to the surface. The results are expressed as numbers per liter or the concentration of zooplankton N-protein per liter.

The main data on the localities, chemistry and biocoenoses are summarised in Table 1.

Results

The Ruda fishpond

Figures 1A, B and 2A, B show differences between the two years of different management of the fishpond. Figure 1 (A, B) shows data from 1977, when the fish

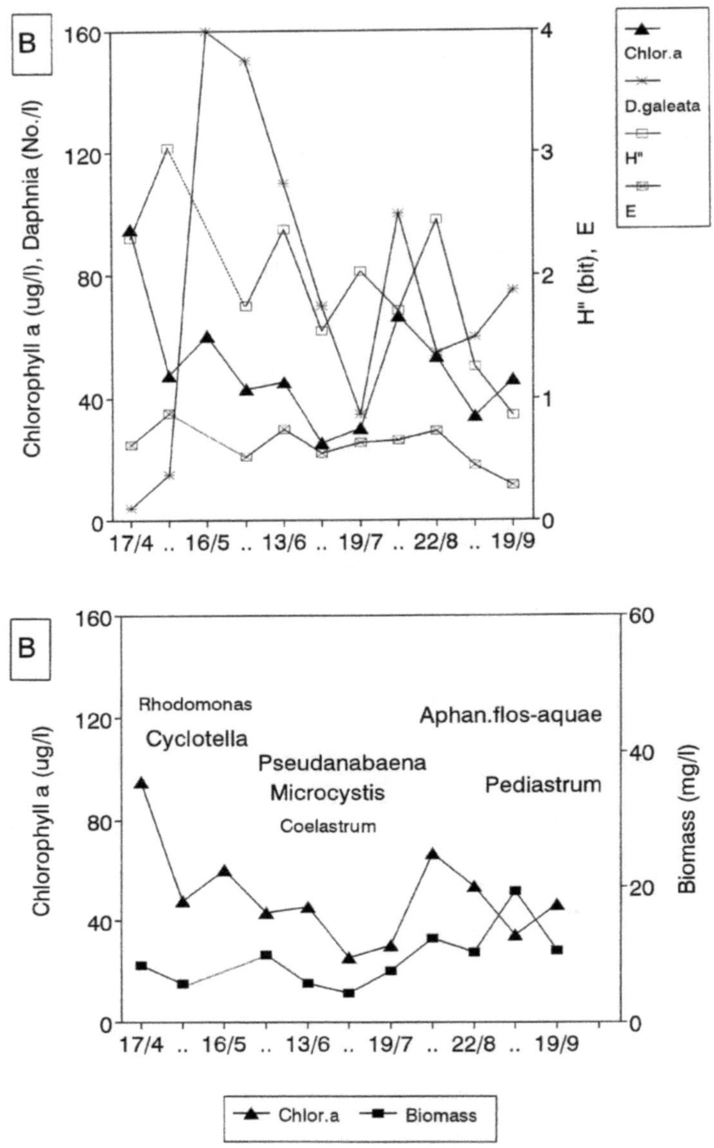

Figure 2. The fish pond Ruda, low fishstock: 1-year old carp, summer fishstock 227 kg ha^{-1} m^{-1}, i.e. 0.019 kg m^{-3}. Production of carp 116 kg. A. Chlorophyll *a* concentration, numbers of *Daphnia galeata* and diversity indices. B. Phytoplankton biomass and composition.

pond was overstocked, while Figure 2 (A, B) shows the data from 1978, when the fishstock was very low. In the productive year (1977), high biomass but small forms of phytoplankton developed (Figure 1B). The numbers of *Daphnia galeata* were small and they were able to exert only a light grazing effect on the phytoplankton (Figure 1A). Other species of crustaceans present (*Bosmina longirostris, Ceriodaphnia*) did not consume algae so much as bacteria and detritus. The phytoplankton developed enormously, reaching almost 160 μg l^{-1} of chlorophyll *a* at the beginning of June. The shift toward unicellular algae is evident (Figure 1B). The characteristics of phytoplankton diversity (H", E) are shown in Figure 1A: the Shannon-Weaver index reached a maximum of 3.61 bits/species with an average 3.11, even though the biomass concentration was extremely high. High values of H" were supported by high evenness (average 0.75, maximum 0.87). Both characteristics remained high, especially in August-September. From the point of intermediate disturbance theory (see Padisák et al., ed., 1993), this can be explained as a high frequency of light fluctuation with simultaneously high supply of nutrients, as the adult carps are feeding in the bottom and lifting the mud and sedimented organic matter to the water column. The diversity of the shallow littoral region and the exchange of water between the littoral part and the pelagial due to the wind can also contribute to the diversity of the phytoplankton.

In 1978 (Figure 2A, B, low fishstock), the assemblage of cladocera feeding on phytoplankton developed well in spring and were maintained in the biocoenoses throughout the season. Besides *Daphnia galeata* (Figure 2A), *D. pulicaria* also appeared (max. abundance 24 specimens/l): this is one of the most efficient of the filter feeders (Fott et al., 1980). Phytoplankton consisted of colonial and filamentous algae which were prominent despite the rapid consumption rates (Holm et al., 1983; Gliwicz & Pianowska, 1989). Moreover the 'canopy' species had a lower growth rate compared to the unicellular algae, so that the total biomass of algae was lower (Sommer, 1981) (Figure 2B). The Shannon-Weaver Index and the equitability decreased toward the end of the summer to very low values (below 1 bit and 0.2, respectivelly).

46 ponds in the Třeboň basin

The data presented in Figure 3A, B were measured during the 2-week sampling expedition of 46 fish ponds in the Třeboň basin in 1978. The relationship between the summer fishstock and chlorophyll *a* concentration is significant, positive and linear ($y = 0.64x–47.35$, $r = 0.826^{**}$, $p = 0.01$: Figure 3A). A similar relationship applies to the first part of the curve plotted in Figure 3B (relationship between the summer fishstock and percentage of unicellular and small coenobial algae $< 40\ \mu m$), that is, before the summer fish stock reaches 400 kg ha^{-1} m^{-1} (the curve was interpolated by eye). In the seventies, the net phytoplankton in understocked ponds was represented almost exclusivelly by the flakes of *Aphanizomenon flos-aquae* (L.) Ralfs (see the letter A to the right of the mark). In our dataset, *Aphanizomenon flos-aquae* formed the dominant population (at least 50%) while the summer fishstock was below 350 kg ha^{-1} m^{-1}. With an increase of the fish stock, the numbers of large filter feeders – *Daphnia pulicaria* and *D. galeata* – decreased and small forms of algae became dominant. The course of the relationship after the summer fishstock reached 400 kg ha^{-1} m^{-1} changed because no colonial or large algae were able to outcompete small species of chlorococal algae growing rapidly under conditions of increasing bioturbation.

Unfortunatelly we did not have the opportunity to study samples of zooplankton during the expedition. As all the fish ponds were managed in the same way and stocked with carp, we assumed the zooplankton to have occupied the same role as we had observed in the Ruda fish pond.

Fott et al. (1980) evaluated the relationship between the fishstock, zooplankton and phytoplankton in the fishponds near Blatná in South Bohemia. They concluded that *Aphanizomenon* occurrence is connected with the first year of the so called two-year management of the pond. It means that in the first year of management, the fishpond is stocked with sufficient numbers of 1-year old carp to reach the expected numbers of carp of a market size after the second year. The fishpond can be drained only every second year. From the sets of data in Figure 3A it may be deduced that the relationship between the appearance of *Aphanizomenon flos-aquae* was not connected with the age of the fish, only with the biomass of the fishstock (Komárková, 1983). The difference between the one- two- and three-year old carp is probably less important than is the intense consumption of large forms of *Daphnia* when the range of the fishstock in the studied set is so large.

Biomanipulation of the Římov Reservoir

Good results from the biomanipulation of shallow water-bodies led to more extensive projects which were successful, especially when biomanipulation was combined with other forms of restorative treatment (Björk, 1972; Cronberg, 1982; Benndorf, 1987; Walker, 1989). One of the biomanipulation experiments was carried out on the Římov Reservoir. The original high fishstock of bream (*Abramis brama*), perch (*Perca fluviatilis*) and roach (*Rutilus rutilus*) was lowered, using different ways of fish stock biomanipulation (Komárková et al., 1995) to achieve changes in the zooplankton and phytoplankton assemblages (Figure 4A). During 1984–1989, the fishstock of the reservoir was decreased from 600 to 100 kg ha^{-1}. The population of *Daphnia galeata* responded by producing much larger individuals (> 0.7 mm), which reached

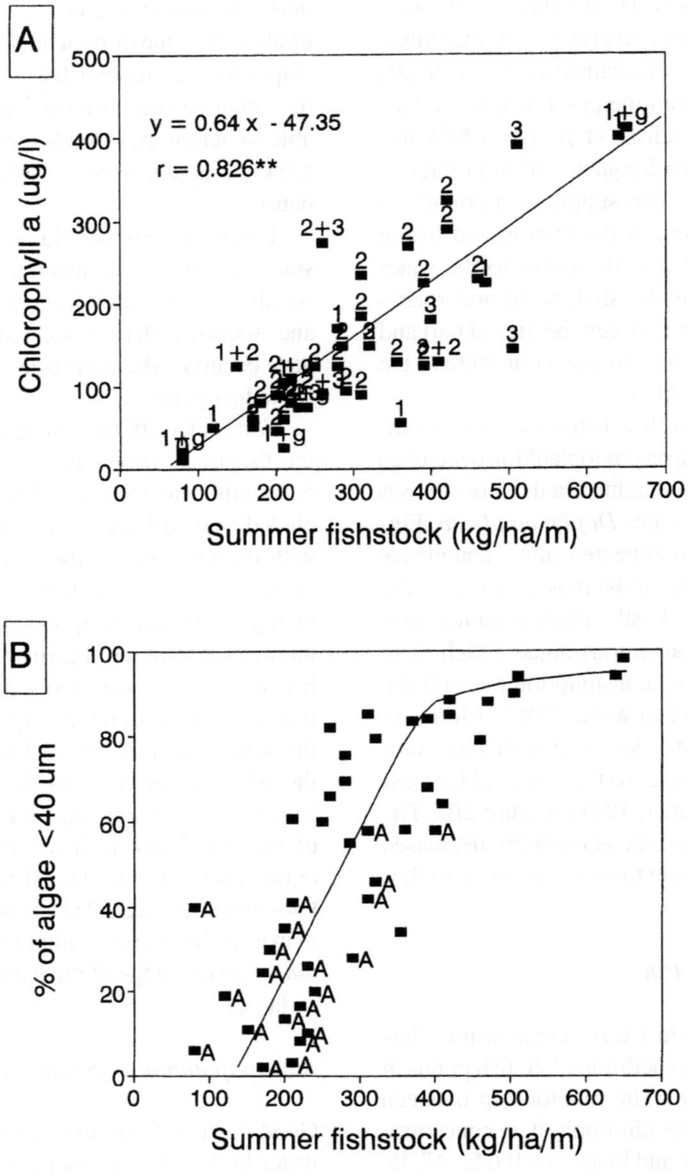

Figure 3. Results from the expedition sampling of 46 fish ponds in the Třeboň basin in August-September 1977. A – Relationship between the summer fishstock and chlorophyll *a* concentration. Numbers above the ticks mean age of the fish in years. B – Relationship between the summer fishstock and the percentage of nanophytoplankton, specimens smaller than 40 μm, mostly chlorococcal algae. The letter A means assemblages having more than 50% of *Aphanizomenon flos-aquae* in the biovolume. The curve is fitted by hand.

up to 8% of the total biomass of zooplankton (Figure 4A). The diminution in fishstock was not sufficient to permit the free establisment of *Daphnia pulicaria,* which would have been expected in a successful experiment. Neither the decrease of biomass of phytoplankton nor the change in the phytoplankton assemblage was explained, as other, more important factors regulated this reservoir (Komárková & Hejzlar, 1996). However, an evident change in phytoplankton composition was noted (Figure 4B). In 1986–1990, the flakes of *Aphanizomenon flos-aquae* reached 40–60% of the total biomass of phytoplankton. Later, this species was replaced by other large alge, *Ceratium hirundinella* Schrank and *Staurastrum planctonicum* Teil. It can be discerned, too, that, since 1988, the percentage of con-

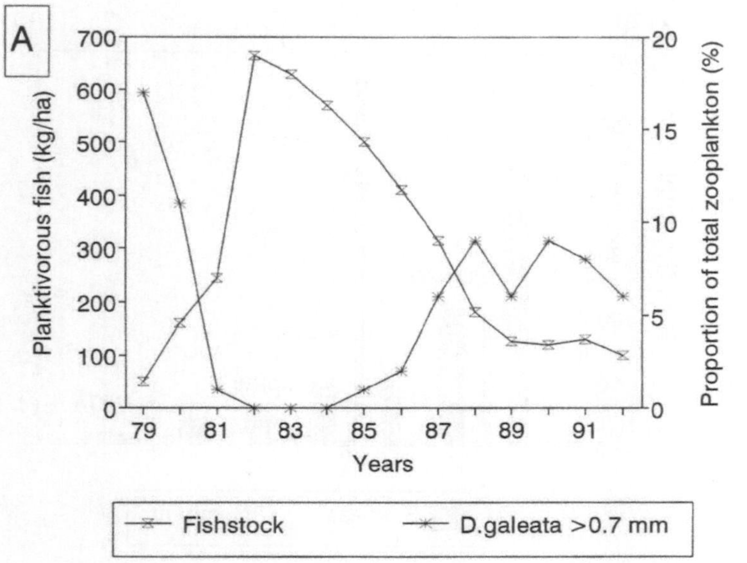

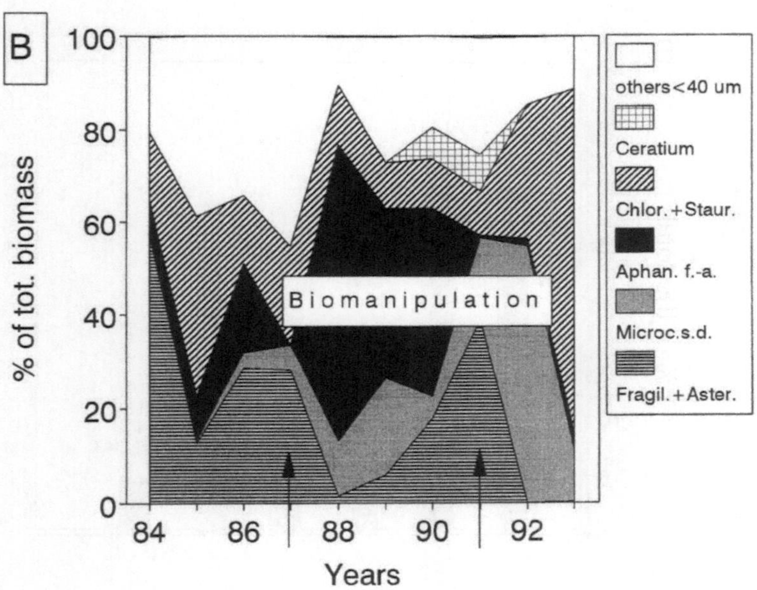

Figure 4. Course of the biomanipulation of the Římov reservoir. A – The curve of the fish population decreased rapidly after the start of the fishstock management. *Daphnia galeata*, large specimens, developed after the fishstock dropped below 300 kg ha^{-1} and reached more than 9% of the biomass of total zooplankton. B – Percentual shares of phytoplankton species in total biomass. The data for individual years represent means of three-week samplings during the main season (May to end of September) (n = 6). The arrows show the beginning of the appearance of large daphnids in 8% of the total zooplankton biomass.

sumable algae dropped to about 20% during the whole season and to 8% during high summer.

Catemaco lake, direct consumption of phytoplankton

A different situation frequently exists in tropical or subtropical lakes, where the niche of large zooplankton is occupied by herbivorous and omnivorous fish

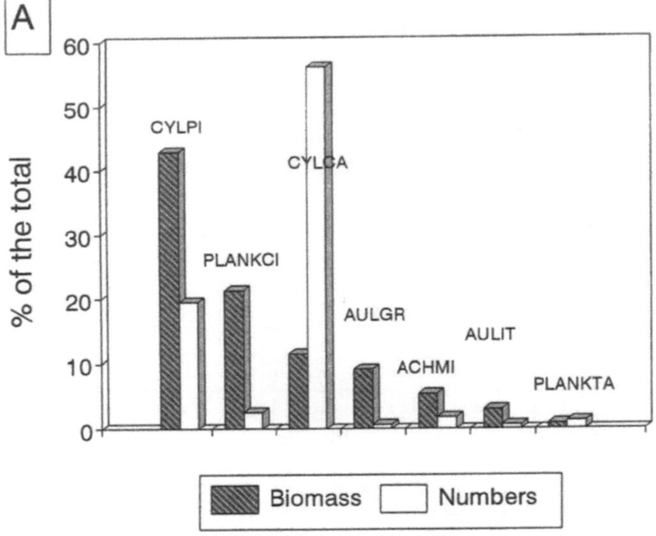

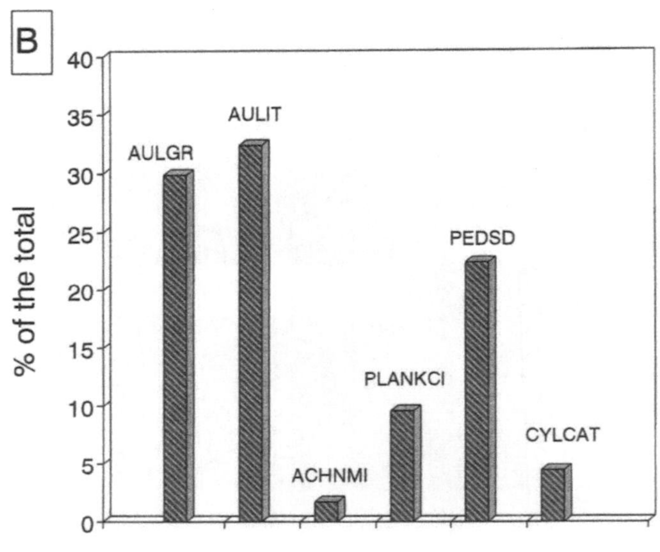

Figure 5. Comparison of percentual shares in the supply of the food-phytoplankton of the Catemaco Lake, and that in the gut content of the most important consumer, *Dorosoma petenense*. A: Composition of the phytoplankton biomass and numbers. B: Composition of the gut content expressed in the shares of the total biomass.
(CYCPI – *Cylindrospermopsis philippinensis*; PLANKCI – *Planktolyngbya circumcreta*; CYLCA – *Cylindrospermopsis catemaco*; AULGR – *Aulacoseira granulata*; ACHMI – *Achnanthes minutissima*; AULIT – *Aulacoseira italica*; PLANKTA – *Planktolyngbya tallingii*; PEDSD – *Pediastrum* spec. div. (*P. boryanum, kawrayskii, simplex*).

(*Dorosoma petenense, D. cepedianum, Bramocharax caballeroi, Oreochromis niloticus*). In Lake Catemaco, Mexico, we studied an interesting composition of phytoplankton (Komárková & Tavera, 1996). The most abundant species included two of *Cylindrospermopsis, C. catemaco* Kom.-Legn. et Tavera and *C. phillipinensis* (Taylor) Komárek, as well as loose populations of *Aulacoseira granulata* (Ehr.) Sim. and *A. italica* (Ehr.) Sim. The more abundant species of fish included a small herbivore, *Dorosoma petenense*. There was little cladoceran zooplankton, which was probably consumed together with the phytoplankton. The percentage composition of the phytoplankton is shown in Figure 5A. The comparison with the content of the guts (Figure 5B) shows the selection of food by the fishes. Much higher percentages of large diatoms are evident. A more detailed study will be published in future (Tavera, 1996).

Discussion

Even though the food web seems to be unimportant in lakes which are well stabilized and balanced in that respect, it becomes a driving variable in others. Direct consumption of phytoplankton by herbivorous or omnivorous fish exists mainly in the tropics and subtropics (Moriarty & Moriarty, 1984; Drenner et al., 1984; Dokulil, 1983). The mechanism of straining and selection of the food by fish depends on the structure of the gill rakers, which serve as mechanical filters, retaining particles greater than the spaces between them. The structure is specific for species and it changes during the life cycle of the specimen (Moriarty & Moriarty, 1976). In Lake Catemaco, large particles were removed from the phytoplankton by fish, the rest of the small cells (*Cylindrospermopsis catemaco*, 20×1.2 μm) could pass through the gill rakers, sustain in water and continue growing. Some portion of algae, mainly blue-greens, can pass through the intestines without being ingested (Dokulil, 1983). The cell content of nutrients could be enriched during passage through the intestines and their subsequent growth enhanced. The minute species *Cylindrospermopsis catemaco* has been abundant in the lake already for many years as it has been proved from the preserved samples from 1970 provided by E. Rott. It creates a background to other larger species, which are probably more available to consumption by fish. The reasons for the maintenance of small species in lake Catemaco and in the overstocked temperate fish pond, seems to be different (comp. Ruda, 1977, Figure 1A, 1B). While the fishpond conditions support diversification and production of small forms, in lake Catemaco the population of *Cylindrospermopsis* in Lake Catemaco seems to be eliminated from the consumption process; even if the high fishstock caused the increase in numbers of this cyanobacterium, it caused a decrease of phytoplankton diversity at the same time. The plagioclimactic state (Reynolds, 1994) of the lake is maintained by intensive daily cropping of the fish which keeps the population of fish growing and grazing.

The leading function of fish stock in the managed carp fish ponds can be demonstrated in Table 2. The Table is an extended modification of Czech version issued by Institute of Fisheries and Hydrobiology, Vodňany, Czech Republic, for the purposes of fish pond management. According to the Table, assistants controlling the development of the fish stock can obtain the relevant information from the zooplankton and phytoplankton composition. An objection may be raised against the unnatural status of fish ponds and reservoirs, which must be managed to be able to function. Yet the same regulations apply to the overstocked fishpond Ruda (Figure 1A), which was found to be hypertrophic as do to natural lakes. Rojo & Alvarez-Cobelas (1993) demonstrated that the higher diversity indices were related with the lower cell sizes in the case of small, shallow hypertrophic El Porcal Lake. Also Reynolds et al. (1993) concluded the same in the synthesis from the IAP Workshop on the Intermediate Disturbance Hypothesis, explaining the higher diversity by presence of fast growing small sized algae.

The indirect influence of fishstock on the phytoplankton (of eutrophic water bodies in temperate zones) leads to the following simplifications: The more planktivorous fish, the more unicellular algae, the higher biomass and chlorophyll concentration, the higher diversity of phytoplankton. Also these presumptions are valid only under specific circumstances (shallow, nutrient rich polymictic waters), however, their phytoplankton characteristics may not always fit to the expected trends in the eutrophication gradient. Some of the criteria concerning the evaluation of phytoplankton state for the identification of the trophic level do not seem to be very reliable especially with respect to eutrophy and hypertrophy. Almost all characteristics of phytoplankton, and even the concentration of nutrients and transparency, can be modified by shifts in the food chain (Rigler, 1961; Peters, 1975). When fishstocks are low, nutrients can remain in water, despite being available for phytoplankton (Hrbáček et al., 1978). Such systems are very sensitive and, unfortunatelly, not stable for a longer time under natural conditions (Benndorf, 1987). In contrast, an artificial balance can be maitained for a long time, as in the London Reservoirs (Duncan, 1990). Also the ranges of chlorophyll *a* concentrations are shifted towards lower values when the fishstock is low and the predation pressure on the phytoplankton increases. Another effect of the low fishstock in carp ponds used to be the appearance

Table 2. Biological features of four steps in fishpond management. Issued in Czech by Institute of Fisheries and Hydrobiology, Vodňany, Czech Republic, for the purposes of State Fishery. Extended and modified.

Step	Fishstock, mainly carp kg ha^{-1} m^{-1}	Zooplankton	Nannopl.	Net phytopl. Higher plants	Transp. cm
I	0–400	*Daphnia magna, D. pulicaria* + with eggs, large diaptomids	none, little greenish water	*Hydrodictyon, Chara,* submersed vegetation, *Aphanizom. f.-a.*	100–> than depth
II.	400–800	*D. pulicaria D. galeata* + with eggs Rotatoria little forms of *Diapt.*	green colour	*Aphanizom. klebahnii Microcystis aeruginosa M. wesenbergii*	40–100
III.	800–3 000	*Moina, Bosmina,* Rotatoria, Copepoda	green to brownish colour	*M. aeruginosa, Aphanocapsa incerta Aphanothece* sp. div., green algae	5–40
IV.	3 000–10 000	Chydoridae, *Bosmina,* Ciliata, Rotatoria,	middy, turbid water, green colour, bacteria, picopl.	'pea soup' *Aphanothece, Aphanocapsa incerta* minute green algae, *Euglena* sp. div.	5–20

Table 3. Phytoplankton diversity values based upon biomass in lakes of different trophic status. Data published in Padisák et al. (1993).

Author:	Lake	Type of lake (TG)	H"
Jacobsen & Simonsen	Godstrup	hypertrophic	0–4
Olrik & Nauwerck	Hjarbaek Fjord	hypertrophic	0.70–2.8
Eloranta	Vasikalampi	eutrophic	0.95–3.9
Moustaka-Gouni	Volvi	mesotrophic	0.95–3.9
Rojo & Alvarez-Cobelas	El Porcal	hypertrophic	0.15–2.9
Sommer	Plusssee	mesotrophic	0.20–3.8
Chorus & Schlag	Schlachtensee	hypertrophic	0.30–3
	Tegel See	eutrophic	0.00–3
Trifonova	Lobadzu	eutrophic	1.20–4
	Ruduschku	eutrophic	1.80–3.8

of monospecific water bloom of *Aphanizomenon flos-aquae.* Apart from the abundant zooplankton consisting of large daphnids, the development of species was supported by superphosphate fertilization (Hrbáček, 1964) and probably also by liming (Shapiro, 1990). The species does not form such populations any more, as the management of the ponds has changed substantially. The fertilization with superphosphate was stopped, the concentration of nitrogen increased and the ponds are stocked with a higher fishstock (Pechar, 1995).

From much evidence, both from our results and literature, it can be deduced that indices of diversity are weak indicators of trophic status. Holzmann (1993) analysed phytoplankton of three lakes from a nature reserve in upper Bavaria. In her paper, she introduces also values of H" based on the biomass from other lakes of the reserve. Mesotrophic lakes are characterized by mean values of 1.3-2.2, eutrophic 1.8, highly eutrophic

1.8 and dystrophic 0.7-1.2. Except for the dystrophic level, which is extreme, all the other values can be found at any other level, including lower stages of hypertrophy. When comparing values of diversity given by some authors, we can see that, far from eutrophic phytoplankton having a low diversity, phytoplankton can reach also the highest diversity under specific circumstances (high values of the Shannon-Weaver Index in phytoplankton were found in the eutrophic Smyslov fish pond-4.5, Fott 1975). The diversity indices can well characterize differences between the assemblages and associations. However, the factors which influence diversity are seldom governed by trophic state. Any disturbance in the environment or any shift inside the food chain pattern can cause marked changes in the composition of phytoplankton and in the values of Shannon-Weaver Index. Thus, it is almost impossible to state the ranges for values of the index which would be characteristic for the particular levels of the trophic gradient.

Acknowledgements

Some data on the Římov Reservoir chemistry and concentration of chlorophyll *a* used in this publication were taken from the database of the Hydrobiological Institute ASCR. They were ready because of financial support from Grant Agency of Czech Republic No. 204/94/1672. I want to thank very much the referee of my paper, who devoted much attention to the manuscript and suggested many important improvements. I am also obliged to Eugen Rott for samples of phytoplankton collected in the lake Catemaco in 1970 and to Manuel Alvarez for his care for the manuscript.

References

Alvarez-Cobelas, M. & B. A. Jacobsen, 1992. Hypertrophic phytoplankton: An overview. Freshwat. Forum 2: 184–199.

Benndorf, J., 1987. Food web manipulation without nutrient control: A useful strategy in lake restoration? Schweiz. Z. Hydrol. 49: 237–248.

Björk, S., 1972. Ecosystem studies in connection with the restoration of lakes. Verh. int. Ver. Limnol. 18: 379–387.

Chorus, I. & G. Schlag, 1993. Importance of intermediate disturbances for the species composition and diversity of phytoplankton in two very different Berlin lakes. In Padisák, J., C. S.Reynolds & U. Sommer (eds), 1993. Intermediate Disturbance Hypothesis in Plankton Ecology. Dev. Hydrobiol. 81, Kluwer Acad. Publ., Dordrecht, 249: 67–92.

Cronberg, G., 1982. Phytoplankton changes in Lake Trummen induced by restoration. Long-term whole-lake studies and food-web experiments. Folia Limnol. Scand. 18: 119 pp.

Dokulil, M., 1983. Aspects of gut passage of algal cells in *Sarotherodon mosambicus* (Peters), (Pisces, Cichlidae). In Schiemer, F. (ed), Limnology of Parakrama Saamudra: 165–170.

Drenner, R. W., J. P. Mummert, deNoyelles jr. & D. Kettle, 1984. Selective particle ingestion by a filter-feeding fish and its impact on phytoplankton community structure. Limnol.Oceanogr. 29: 941–948.

Duncan, A., 1990. A review: Limnological management and biomanipulation in the London reservoirs: Species composition, density and biomass. Hydrobiologia 200/201: 541–548.

Eloranta, P., 1993. Diversity and succession of the phytoplankton in a small lake over two-year period. In Padisák, J., C. S. Reynolds & U. Sommer (eds), 1993. Intermediate Disturbance Hypothesis in Plankton Ecology. Dev. Hydrobiol. 81, Kluwer Acad. Publ., Dordrecht: 25–32.

Fott, J. 1975. Seasonal succession of phytoplankton in the fish pond Smyslov near Blatná, Czechoslovakia. Arch. Hydrobiol. Suppl. 46: 259–279.

Fott, J., L. Pechar & M. Pražáková, 1980. Fish as a factor controlling water quality in ponds. In Barica J. & L. R. Mur (ed), Hypertrophic Ecosystems. Dev. Hydrobiol. 2: 255–262.

Gliwicz, M. Z. & J. Pianowska, 1989. The Role of predation in Zooplankton Succession. Sommer, U. (ed), Plankton Ecology, Springer Verl.: 253–296.

Holm, N. P., G. G. Ganf & J. Shapiro, 1983. Feeding and assimilation rates for *Daphnia pulex* fed *Aphanizomenon flos-aquae*. Limnol. Oceanogr. 28: 622–687.

Holzmann, R., 1993. Seasonal fluctuations in the diversity and compositional stability of phytoplankton communities in small lakes in upper Bavaria. Hydrobiologia 249: 101–109.

Hrbáček, J., 1964. Contribution to the ecology of water-bloom-forming blue green algae *Aphanizomenon flos-aquae* and *Microcystis aeruginosa*. Verh. int. Ver. Limnol. 15: 837–846.

Hrbáček, J., 1994. Food web relations. In Eiseltová, M. (ed), Restoration of Lake Ecosystems, IWBR Publ. 32: 44–58.

Hrbáček, J., O. Albertová, B. Desortová, V. Gottwaldová & J. Popovský, 1986. Relation of the zooplankton biomass and share of large cladocerans to the concentration of total phosphorus, chlorophyll-*a* and transparency in Hubenov and Vrchlice Reservoir. Limnologica 17: 301–308.

Hrbáček, J., B. Desortová, & J. Popovský, 1978. The influence of the fishstock on the phosphorus-chlorophyll ratio. Verh. int. Ver. Limnol. 20: 1624–1628.

Jacobsen, B. A. & P. Simonsen, 1993. Disturbance events affecting phytoplankton biomass, composition and species diversity in a shallow, eutrophic, temperate lake. In Padisák, J., C. S. Reynolds & U. Sommer (eds), 1993. Intermediate Disturbance Hypothesis in Plankton Ecology. Dev. Hydrobiol. 81, Kluwer Acad. Publ., Dordrecht: 9–14.

Javornický, P. & J. Komárková, 1973. The changes in several parameters of phytoplankton primary productivity in Slapy Reservoir 1960–1967, their mutual correlations and correlations with the main ecological factors. Hydrobiol. Stud. 2, Acad. Praha: 155–211.

Jeppesen, E., J. P. Jensen, P. Kristensen, M. Sondergaard, E. Mortensen, O. Sortkjaer & K. Olrik, 1990. Fish manipulation as a lake restoration tool in shallow, eutrophic, temperate lakes 2: treshold levels, long-term stability and conclusions. Hydrobiologia 200–201: 219–227.

Komárková, J., 1983. Factors influencing the development of *Aphanizomenon flos-aquae* bloom in Czechoslovak eutrophic fish ponds. Schweiz. Z. Hydrol. 45: 301–306.

Komárková, J., R. Faina & J. Pařízek, 1986. Influence of the watershed and fishstock upon the fish pond biocenoses. Limnologica 17: 335–354.

Komárková, J., V. Vyhnálek & J. Kubečka, 1995. Impact of fishstock manipulation on the composition of net phytoplankton in the Římov Reservoir (Czech Republic). Wat. Sci. Technol. 32: 207–216.

Komárková, J. & J. Hejzlar, 1996. Summer maxima of phytoplankton in the Římov reservoir in relations to hydrologic parameters and phosphorus loading. Arch. Hydrobiol. 136: 217–236.

Komárková, J. & R.L.Tavera, 1996. Cyanoprocaryota (Cyanobacteria) in the phytoplankton of the lake Catemaco (Veracruz, Mexico). Arch. Hydrobiol., Algol. Stud. 83: 403–422.

Lampert, W., W. Fleckner, H. Rai. & B. B. Taylor, 1986. Phytoplankton control by grazing zooplankton: A study on the spring clear water phase. Limnol. Oceanogr. 31: 478–490.

Lorenzen, C. J., 1967. Determination of chloro- and pheo-pigments: Spectrofotometric equations. Limnol. Oceanogr. 12: 343–346.

Lund, J. W. G., 1970. Primary production. Water Treatment and Examination 19: 332–358.

Matěna, J., & V. Vyhnálek, 1994. Food web management in reservoirs. In Eiseltová, M. (ed), Restoration of Lake Ecosystems, IWBR Publ. 32: 169–173.

Moriarty, C. M. & D. J. W. Moriarty, 1973. Quantitative estimation of the daily ingestion of phytoplankton by *Tilapia nilotica* and *Haplochromis nigripinnis* in Lake George, Uganda. J. Zool. 171: 15–23.

Moustaka-Gouni, M., 1993. Phytoplankton succession and diversity in a warm monomictic, relatively shallow lake: Lake Volvi, Macedonia, Greece. In Padisák, J., C. S. Reynolds & U. Sommer (eds), 1993. Intermediate Disturbance Hypothesis in Plankton Ecology. Dev. Hydrobiol. 81, Kluwer Acad. Publ., Dordrecht, 249: 33–42.

Odum E. P., 1971. Fundamentals of Ecology. Saunders, Philadelphia.

OECD, 1982: Eutrophication of waters. Monitoring, assessment and control. OECD. Paris.

Olrik, K. & A. Nauwerck, 1993. Stress and disturbance in the phytoplankton community of a shallow, hypertrophic lake. In Padisák, J., C. S. Reynolds & U. Sommer (eds), 1993. Intermediate Disturbance Hypothesis in Plankton Ecology. Dev. Hydrobiol. 81, Kluwer Acad. Publ., Dordrecht: 15–24.

Padisák, J., C. S. Reynolds & U. Sommer, (eds) 1993. Intermediate Disturbance Hypothesis in Plankton Ecology. Dev. Hydrobiol. 81, Kluwer Acad. Publ., Dordrecht.

Pechar, L., 1995. Long-term changes in fish pond management as an 'unplanned ecosystem experiment': Importance of zooplankton structure, nutrients and light for species composition of cyanobacterial blooms. Wat. Sci. Technol. 32: 187–196.

Persson, A. & S. F. Hamrin, 1994. Effects of cyprinids on the release of phosphorus from lake sediment. Verh. int. Ver. Limnol. 25: 2124–2127.

Peters, R. H. 1975. Phosphorus regeneration by natural populations of limnetic zooplankton. Int. Ver. Theor. Angewandt. Limnol., Verhandl. 19: 273–279.

Porter, K. G., 1973. Selective grazing and differential digestion of algae by zooplankton. Nature 244: 179–180.

Porter, K. G., 1976. Enhancement of algal growth and productivity by grazing zooplankton. Science 192: 1332–1334.

Reynolds, C. S., 1984: The ecology of freshwater phytoplankton. – Cambr. Univ. Press: 384 pp.

Reynolds, C. S., J. Padisák & U. Sommer, 1993. Intermediate disturbance in the ecology of phytoplankton and the maintenance of species diversity. In Padisák, J., C. S. Reynolds & U. Sommer (eds), Intermediate Disturbance Hypothesis in Plankton Ecology. Dev. Hydrobiol. 81, Kluwer Acad. Publ., Dordrecht: 183–188.

Reynolds, C. S., J. M. Thompson, A. J. Ferguson & S. W. Wiseman, 1982. Loss processes in the population dynamics of phytoplankton maintained in closed stystems. J. Plankt Res. 4: 561–600.

Reynolds, C. S., 1994. Ecosystem exploitation, sustainability and biodiversity: Are they compatible? Freshwat. Forum 4: 202.

Rigler, F. H., 1961. The uptake and release of inorganic phosphorus by *Daphnia magna* Straus. Limnol. Oceanogr. 6: 165–174.

Rojo, C. & M. Alvarez-Cobelas, 1993. Hypertrophic phytoplankton and the Intermediate Disturbance Hypothesis. In Padisák, J., C. S. Reynolds & U. Sommer (eds), 1993. Intermediate Disturbance Hypothesis in Plankton Ecology. Dev. Hydrobiol. 81, Kluwer Acad. Publ., Dordrecht: 43–57.

Seďa, J. & J. Kubečka, 1997. Long-term biomanipulation of Římov Reservoir (Czech Republic). Hydrobiologia 345: 94–108.

Shapiro, J., 1990. Current beliefs regarding dominance by bluegreens: The case for the importance of CO_2 and pH. Ver. int. Ver. Limnol. 24: 38–54.

Sommer, U., 1981. The role of r- and K- selection in the succession of phytoplankton in Lake Constance. Acta Oecologica 2: 327–342.

Sommer, U., 1988. Phytoplankton succession in microcosm experiments under simultaneous grazing pressure and resource limitation. Limnol. Oceanogr. 33: 1037–1054.

Sommer, U., 1993. Disturbance-diversity relationships in two lakes of similar nutrient chemistry but contrasting disturbance regimes. In Padisák, J., C. S. Reynolds & U. Sommer (eds), 1993. Intermediate Disturbance Hypothesis in Plankton Ecology. Dev. Hydrobiol. 81, Kluwer Acad. Publ., Dordrecht: 59–66.

Sommer, U., Z. M. Gliwicz, W. Lampert, & A. Duncan, 1986. The PEG-model of seasonal succession of planktonic events in fresh waters. Arch. Hydrobiol. 106: 433–471.

Sterner, R. W., 1989. The Role of Grazers in Phytoplankton Succession. In Sommer (ed), Plankton Ecology. Springer Verl.: 107–170.

Tavera, R. L., 1996. Phytoplankton of the tropical Lake Catemaco. PhD. Thesis. University of South Bohemia, Fac. of Biological Sciences, 62 pp.

Trifonova, I., 1993. Seasonal succession of phytoplankton and its diversity in two highly eutrophic lakes with different conditions and stratification. In Padisák, J., C. S. Reynolds & U. Sommer (eds), 1993. Intermediate Disturbance Hypothesis in Plankton Ecology. Dev. Hydrobiol. 81, Kluwer Acad. Publ., Dordrecht, 68–93.

Vyhnálek, V., J. Komárková, J. Sed'a, Z. Brandl, K. Šimek & N. Johanisová, 1991. Clear water phase in the Římov Reservoir (South Bohemia). Controlling factors. Verh. int. Ver. Limnol. 24: 1335–1339.

Vyhnálek, V., Z. Fišar, A. Fišarová & J. Komárková, 1993. In vivo fluorescence of chlorophyll *a*: Estimation of phytoplankton biomass and activity in Římov Reservoir (Czech Republic). Wat. Sci. Technol. 28: 29–33.

Walker, P. A., 1989. Feasibility of lake restoration through biomanipulation – a literature study. Aquasense Report No. 89012, Amsterdam, 60 pp.

Washington, H. G., 1979. Diversity, biotic and similarity indices. A review with special relevance to aquatic ecosystems. Water Res. 18: 653–694.

Wetzel, R. G., 1975. Limnology. – W.B.Saunders Comp. Philadelphia – London – Toronto, 743 pp.

Hydrobiologia **369/370:** 153–162, 1998.
M. Alvarez-Cobelas, C. S. Reynolds, P. Sanchez-Castillo & J. Kristiansen (eds), Phytoplankton and Trophic Gradients.

Seasonality of phytoplankton as an indicator of trophic status of the large perialpine 'Lago di Garda'

Nanna Büsing
Arbeitsgruppe Hydrobotanik, Institut für Botanik, Sternwartestr. 15, A-6020 Innsbruck, Austria

Key words: seasonality, phytoplankton, Lake Garda, eutrophication, trophic index

Abstract

The seasonality of phytoplankton and its vertical distribution in Lake Garda, Italy, was studied on the basis of monthly sampling during 1994. Chlorophyll a concentration, phytoplankton biovolumes and species composition are used for the trophic characterization of the lake, supplemented by water chemistry data (nutrients and total phosphorus) and Secchi depth. Horizontal differences in trophic status are investigated by comparing data from three sampling stations, whereas the long term changes in trophic status are based on a comparison of the recent findings with earlier studies of the phytoplankton.

Introduction

Lago di Garda (hereinafter, Lake Garda) is the largest lake in northern Italy and one of its most popular holiday resorts, attracting short-term and long-term visitors from central and northern Europe (16 million overnight lodgings per year). In spite of this high tourist pressure and the long water renewal time (27 years), the lake maintained a reasonable water quality until quite recently. The first symptoms of eutrophication – increasing phosphorus concentrations, oxygen depletion and waterblooms – have been observed in the last few years. Recognising the economic importance of Lake Garda, monitoring by the public health department of Verona in cooperation with the department of Biology of the University of Padova began 1987 and, since 1989, by the Istituto Agrario di San Michele all'Adige (Province of Trento). This monitoring consists in monthly analyses of Secchi depth, water chemistry, chlorophyll a and counts of the most abundant plankton species.

The first scientific description of the morphometry, bathymetry and hydrology of Lake Garda was given by Marsili in the eighteenth century (ex: Ambrosetti, 1992) but this remained undiscovered until 1930 (Regia Accademia delle Scienze, 1930). The early limnological studies of Garbini on Lake Garda, dating back to 1897, were concerned with the physical conditions of the lake, including its colour, transparency and temperature. The species composition of planktonic and benthic algae was also first described a century ago (Garbini, 1898; Kirchner, 1899; Forti, 1902). Apart from Marchesoni (1952), who counted only the net plankton species, more fundamental investigations on the quantity of phytoplankton as related to abiotic variables are available for several years since 1957 (Merlo & Mozzi, 1963; Gerletti, 1974). Since the early 1980s, research has covered various aspects such as mixing, the main chemical and biological characteristics, and the trophic status of the lake (Ambrosetti et al., 1983; Ruggiu & Mosello, 1984; Chiaudani & Premazzi, 1990; Cordella et al., 1992; Ambrosetti et al., 1992; Cordella & Salmaso, 1993; Salmaso & Cordella, 1994; Salmaso et al., 1996).

Besides the routine monitoring of water chemistry and phytoplankton quantities, the present investigation, carried out in 1994, focused on the seasonality of the phytoplankton community, its species composition, its vertical and horizontal distribution. I compare the results with those of earlier studies and review the trophic status of the lake on the basis of qualitative phytoplankton criteria (see Rott, 1984). Here, I concentrate on phytoplankton seasonality and on differences among the three sampling stations. Detailed

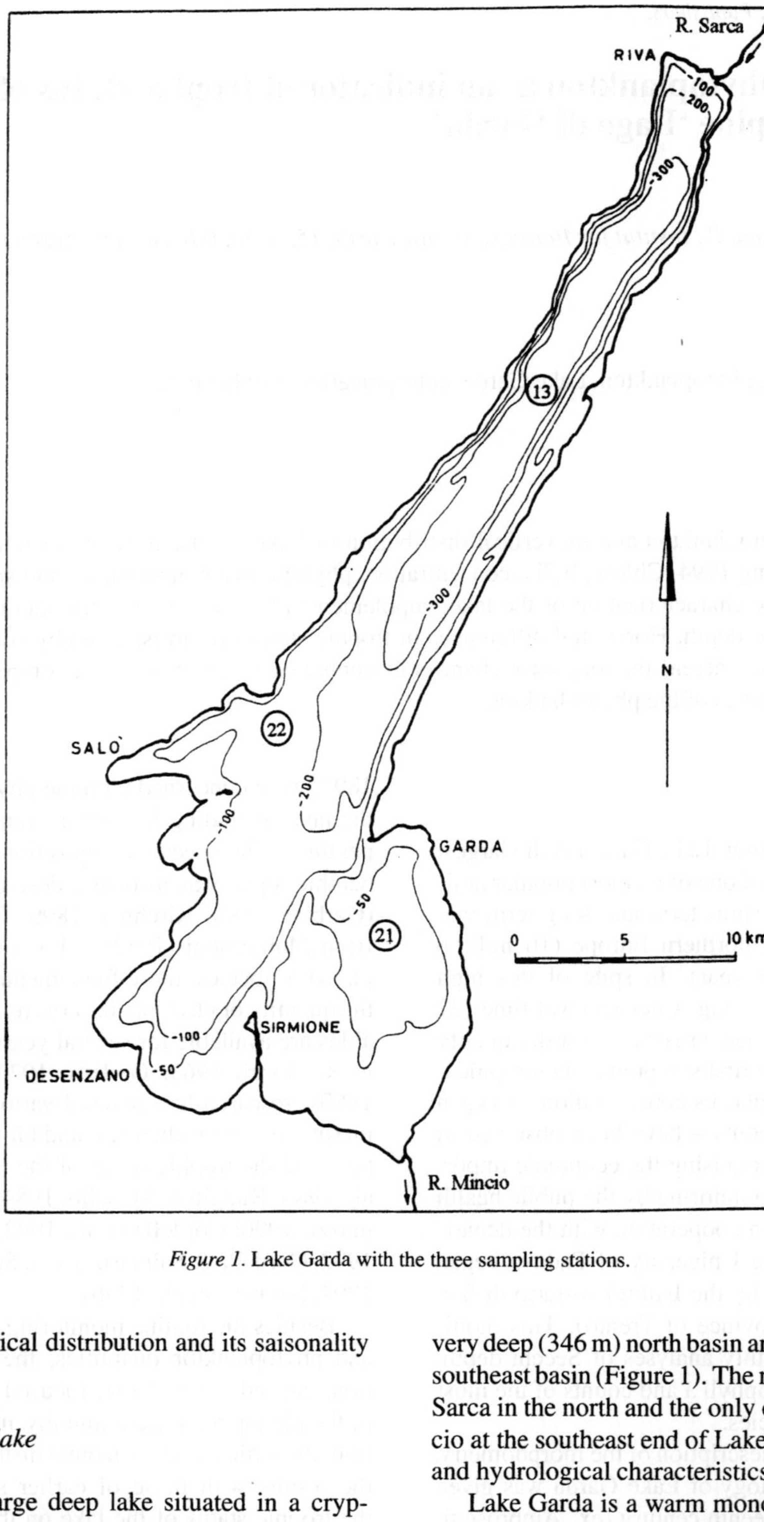

Figure 1. Lake Garda with the three sampling stations.

analysis of the vertical distribution and its saisonality will be given later.

Description of the lake

Lake Garda is a large deep lake situated in a cryptodepression (Figure 1). Among the world's largest lakes, Lake Garda is ranked 37th in depth (Herdendorf, 1987). The basin of Lake Garda is subdivided into the very deep (346 m) north basin and the shallower (70 m) southeast basin (Figure 1). The main inflow is the River Sarca in the north and the only outflow the River Mincio at the southeast end of Lake Garda. Morphometric and hydrological characteristics are given in Table 1.

Lake Garda is a warm monomictic lake characterized by a winter circulation (from December to March) and a distinct thermal gradient in summer (from May to October). In the period of summer stratification, mul-

Table 1. Morphometric and hydrological characteristics of Lake Garda

Location	Lat. 45°53′04"
	Long. 1°37′38"
Altitude [m]	65
Ice cover period	none
Area [km^2]	368
Mean depth [m]	133
Max. depth [m]	350
Volume [km^3]	50
Total catchment area [km^2]	2260
Residence time [a]	27
Mixing type	warm monomictic

Table 2. Water chemistry data of Lake Garda during the investigation period (February–October 1994). All averages are volume weighted mean values (0-15 m).

Variable	Ranges	Average
Secchi depth [m]	4.5–19.5	8.8
pH	7.3–8.8	8.0
Oxygen [%] epilimnion	77–131	99
Oxygen [%] hypolimnion	51–96	72
Conductivity	197–228	220
TP [μg l^{-1}]	6–36	17
SRP [μg l^{-1}]	<2–34	
NH_4-N [μg l^{-1}]	3–77	17
NO_3-N [μg l^{-1}]	30–562	366
Biovolume [mm^3 m^{-3}]	100 - 1700	606
Chlorophyll *a* [mg m^{-3}]	0.3–6.3	2

tiple thermoclines can be observed. They are caused by periodical heating and cooling periods in summer, the strong exposure to the wind (reaching maxima of 35 m s^{-1}) and the formation of internal seiches. Complete mixing of the water column is evident only in the southeast basin. The deep basin is meromictic insofar as the winter mixing reaches only to 150–200 m. Overturn of the entire water column of the large Insubrian lakes (not only Lake Garda) is convectional and coincides with particularly cold winters (Ambrosetti, 1992). The oxygen concentration in the hypolimnion at the end of summer stratification did not fall below 9 mg l^{-1} for many years, but results from routine monitoring of the last six years (Istituto Agrario di San Michele, 1989–1995; Ambrosetti, 1992) show a decrease to a minimum of 6 mg l^{-1}. Given the monimolimnion temperatures of $\sim$ 8 °C throughout the year, these concentrations correspond to saturation values of 77% and 51% respectively. The concentrations of total phosphorus in the hypolimnion have risen from less than 7 μg l^{-1} in 1959–1961 (Merlo & Mozzi, 1963) to 17 μg l^{-1} (annual average in the hypolimnion: this study)

Although the sewage system has almost been completed since 1988 (Chiaudani & Premazzi, 1990), additional diffuse sources in the near shore areas are evident. In spite of sanitation measures taken in the catchment area, the oligosaprobic to β-mesosaprobic quality waters of the main inflow (River Sarca) still cause significant nutrient enrichment to the lake (Istituto Agrario di San Michele, 1989–1995). Due to the relatively small catchment area of Lake Garda and the low precipitation in comparison to the other perialpine lakes, the run-off of phosphorus from agricultural origin is limited to 50 t phosphorus per annum (Chiaudani & Premazzi, 1990).

Materials and methods

From February to October 1994, samples for chemical analysis and phytoplankton were taken monthly at three sampling stations: P13 at the deepest point of the lake (350 m), P21 in the shallower southeast basin (70 m) and P22 (140 m) in the southwest of the lake (see Figure 1). Temperature, pH, oxygen and conductivity were measured in situ with a multiparameter probe (Idronaut 401). Meteorological data and results of chemical variables analysed in the laboratory were kindly supplied by the Hydrobiological Station of the Istituto Agrario di San Michele (Riva). For the qualitative examination of phytoplankton, net samples down to 20 m were taken with a net of 10 μm meshsize and fixed with formalin. Samples for the quantitative evaluation of the phytoplankton were taken at 0 m, 2.5 m, 5 m, 7.5 m, 10 m, 12.5 m, 15 m, 20 m, 25 m and 30 m depth with a Van Dorn water-sampler. These phytoplankton samples were fixed with Lugol's iodine and counted according to Utermöhl (1958). For the sedimentation, 26-ml sedimentation chambers were used. Usually the whole chamber was counted at the lowest magnification (6.3×), four transverses across the chamber at 16×, two at 25× and one at 63×, always depending on the number of individuals. Algae with cell diameters < 2 μm were not counted. For the calculation of species-specific biovolumes, all relevant dimensions of at least 30 individuals were measured, and of > 130 individuals in the case of conspicuously variable forms, such as *Cryptomonas*. The calculations of biovolume were made observing the recommendations of Rott (1981). All averages shown in Table 2 are based on the integration of the depth samples, volume-weighted according to depth. The diversity of phytoplankton was calculated using the Shannon function based on biomass (Shannon, 1948). For this calculation, two discrimination thresholds were used: (1) relative biovolume > 0.5% as proposed in the Workshop guidelines and (2) species with biovolume > 10 mm^3 m^{-3}, a threshold that seemed more relevant to my dataset. The six stages of succession during the year were discriminated by percentage similarity with subsequent sampling dates. As a measure of light exposure, the ratio between mixing depth and euphotic zone was calculated taking into account the bathymetry of the lake, as suggested by Reynolds (1984).

Results

Abiotic variables

The high transparency of Lake Garda was recognised in the earliest limnological studies, to the extent that Garbini (1897) added to the colours of the Forel-scale the one of Lake Garda, being especially blue. In this investigation, Secchi-depth measurements ranged from 4.5 m at the end of July to 19.5 m in February (Table 2). The measurements are clearly correlated with the phytoplankton biovolumes (Figure 3). The comparison of Secchi-disk readings were compared to parallel measurements made with a Li-Cor Quantum-sensor to derive a factor of 2.6 for the conversion of Secchi depth to euphotic depth. Thus, the theoretical depth of the euphotic zone of Lake Garda during my study varied between 11.7 m and 50.7 m.

The vertical profiles of temperature from February to October 1994 showed the typical multiple thermoclines observed in previous years (Merlo & Mozzi, 1963). The temperature at winter circulation remains above 8 °C. In the study year mixing depth in the meromictic north basin reached 180 m.

The ratio of z_{mix}:z_{eu} (Figure 3) responds to the prolonged mixing situation prior to April with a decreasing euphotic zone due to the enhanced growth of phytoplankton assemblages from February. The high values of this ratio in the early spring indicate that phytoplankters need to be adapted to spend a long time in darkness especially at station 13. Low values of z_{mix}:z_{eu} characterise the stratification period from April to October.

The oxygen saturation in the epilimnion reached maxima of more than 130% in July 1994 (Table 2). This oversaturation is linked to high photosynthetic activity. Undersaturation has been observed at the very end of the summer stratification (end of October), when the density of phytoplankton was also low. Despite meromixis, Lake Garda still has high minimum oxygen concentrations > 6.1 mg l^{-1} in the monimolimnion. However, these are significantly lower than those found in the fifties (Marchesoni, 1952: 10.4 mg l^{-1} at 340 m depth). At the end of November 1994, an isolated minimum of oxygen concentrations (< 70%) observed in the metalimnion (from 50 m to 60 m) was taken as a sign of enhanced bacterial activity. This indicates that organic matter (decaying plankton) is decomposed while still sedimenting through the water column.

The vertical distribution of phosphorus is positive clinograde with a distinct hypolimnetic maximum. The lowest concentration of total phosphorus (6 μg l^{-1}) was found at the surface in June; the highest at 300 m depth in November 1993. Comparison of hypolimnetic TP concentrations with those from earlier studies (Marchesoni, 1952; Gerletti, 1974; Ambrosetti, 1992) showes a strongly increasing trend. The late autumn concentration at 340 m depth has increased from 12 μg l^{-1} in 1951 (Marchesoni, 1952) to 36 μg l^{-1} in 1994. In the epilimnion the phosphorus concentrations are constantly low with minor variations from 10 μg l^{-1} at the end of summer stratification to 13 μg l^{-1} during circulation.

The concentrations of SRP in the epilimnion are usually below the limit of detection (2μg l^{-1}), whereas, in the hypolimnion, the annual average SRP content accounts for 78% of the total. This means that Lake Garda now stores large amounts of phosphorous at greater depth (140–350 m), which could be mobilised by a complete overturn.

Phytoplankton

The species composition of Lake Garda revealed 79 taxa, with Chlorophyceae being best represented (26 species) followed by Conjugatophyceae (12). Bacillariophyceae and Cyanophyceae were moderatly represented. Less well-represented were Chlamydophyceae, Chrysophyceae, Cryptophyceae and Dinophyceae.

Quantitatively, the most important taxonomic groups were diatoms, followed by Conjugatophyceae and blue-greens (Figure 2). Diatoms peaked in Feb-

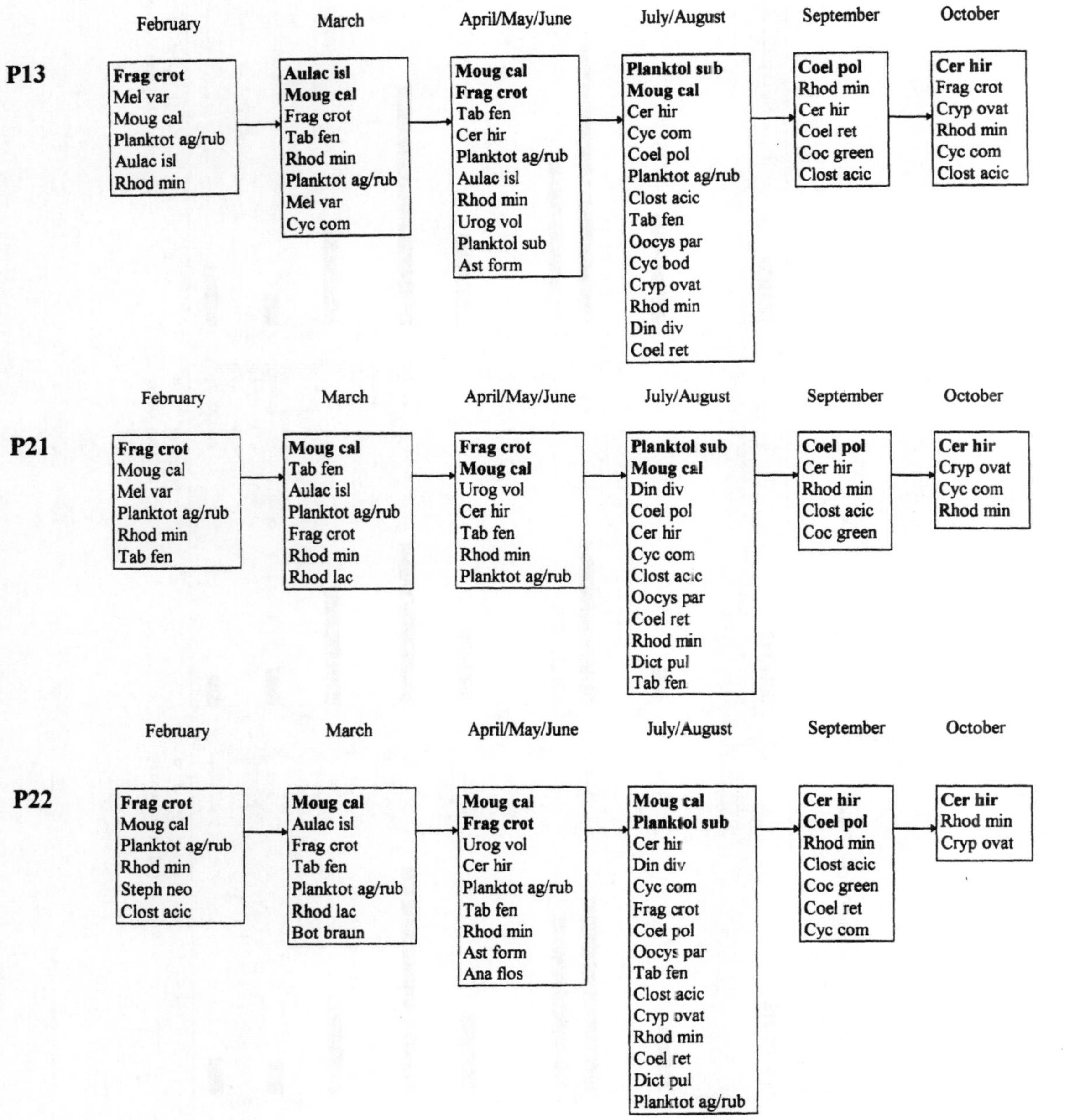

Abbreviations: Ana flos, *Anabaena flos-aquae*; Ast form, *Asterionella formosa;* Aulac isl, *Aulacoseira islandica ssp. helvetica;* Bot braun, *Botryococcus braunii;* Cer hir, *Ceratium hirundinella*; Clost acic, *Closterium aciculare*; Coel pol, *Coelastrum polychordum*; Coel ret, *Coelastrum reticulatum*; Coc green, coccal green algae; Cryp ovat, *Cryptomonas ovata*; Cyc com, *Cyclotella comensis*; Dict pul, *Dictyosphaerium pulchellum*; Din div, *Dinobryon divergens*; Frag crot, *Fragilaria crotonensis*; Mel var, *Melosira varians*; Moug cal, *Mougeotia cf. calcarea*; Oocys par; *Oocystis parva*; Planktol sub, *Planktolyngbya subtilis*; Planktot ag/rub, *Planktothrix agardhii/rubescens*; Rhod lac, *Rhodomonas lacustris*; Rhod min, *Rhodomonas minuta*; Steph neo, *Stephanodiscus neoastrea*; Tab fen, *Tabellaria fenestrata*; Urog vol, *Uroglena volvox*.

Figure 2. Successional phytoplankton stages at the three sampling stations. Only species contributing with more than 10 mm^3 m^{-3} have been taken in consideration. Dominant species (> 20% of the 'seasonal' biomass) are indicated by bold characters.

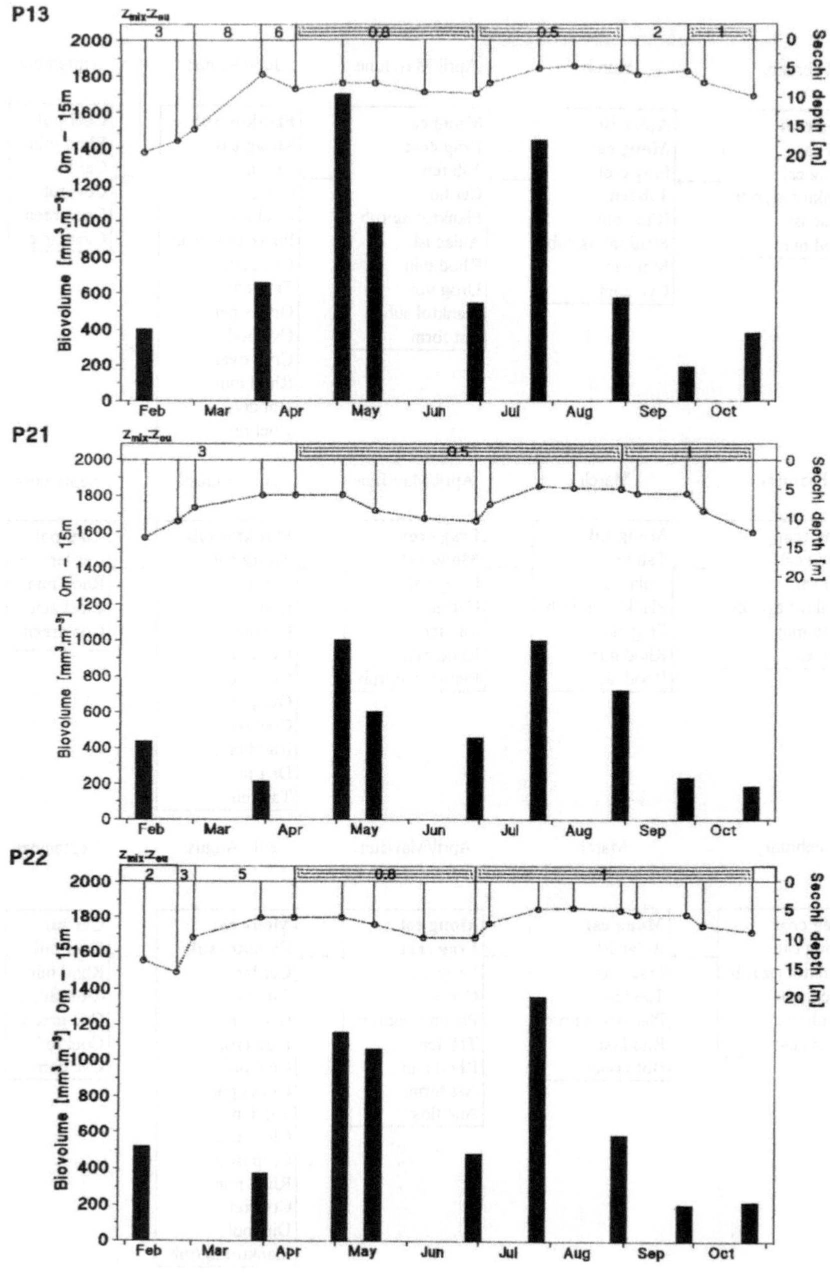

Figure 3. Seasonal distribution of the phytoplankton biovolume, Secchi depth and z_{mix}:z_{eu} ratio from February until October 1994.

ruary – forming more than 80% of the total biomass – and spring, mainly with *Fragilaria crotonensis* (blunt tipped type). Green algae showed two maxima: the first in May and the second in July both dominated by *Mougeotia cf. calcarea*. The blue-green algae peak in July with *Planktolyngbya subtilis* (former *Lyngbya limnetica*) being dominant, but *Planktothrix agardhii/rubescens* was found at all sampling stations in moderate quantities throughout the year. Minor peaks of other taxonomic groups as (dinoflagellates: *Ceratium hirundinella*; chlorococcal green algae: *Coelastrum polychordum, Coelas-*

trum reticulatum) and desmids: *Closterium aciculare, Closterium acutum, Staurastrum pingue)* were observed during August and September.

In the study year, six successional stages were distinguished (Figure 2):

1. In winter *Fragilaria crotonensis* predominated (> 80%) at each sampling station, but the accompanying species differed.
2. The early spring situation represents a transitional stage between the winter association and the spring maximum. *Fragilaria crotonensis* is replaced by *Mougeotia cf.calcarea* (at P13, also by *Aulacoseira islandica*) as first ranking species.
3. The spring association was characterized by the dominance of *Fragilaria crotonensis* and *Mougeotia cf. calcarea* with *Uroglena volvox* codominating at the southern sampling stations.
4. The summer association was dominated by *Planktolyngbya subtilis* and (still) *Mougeotia cf. calcarea* with *Ceratium hirundinella, Dinobryon divergens, Cyclotella comensis* and *Coelastrum polychordum* as accompanying species.
5. The late summer assemblage consisted of high quantites of *Coelastrum polychordum* and/or *Ceratium hirundinella* combined with moderate quantities of various green algae and *Rhodomonas minuta*.
6. The autumn community was dominated by *Ceratium hirundinella* with *Cryptomonas ovata* and *Rhodomonas minuta* as accompanying species.

The best defined assemblages are those founded on *Fragilaria crotonensis* and *Mougeotia cf. calcarea* dominance (each species > 20%) and the summer grouping having *Planktolyngbya subtilis* and *Mougeotia cf. calcarea* as dominants. The highest number of accompanying species was found in the summer. In autumn, when mixing depth is greater, a pulse of diatom growth together with remnants of the stratified period dominates the plankton (e.g. *Closterium aciculare, Ceratium hirundinella*).

During the mixing period *Fragilaria crotonensis* is dominant at all three sampling stations, but it decreases with the onset of stratification, as noted also by Cordella & Salmaso (1993). The species was recorded in all the prevous studies (Garbini, 1898; Kirchner, 1899; Marchesoni, 1952; Andreoli, 1977), although it had been more characteristic of the late of the stratification period in early autumn (Ruggiu, 1983). *Aulacoseira islandica ssp. helvetica* is typical for all south alpine lakes and for Lake Garda it has been recorded in moderate quantities since 1978 (Ruggiu, 1983). The continous presence of *Planktothrix* is also noted in the recent literature (Cordella & Salmaso, 1993), whereas in 1951 it was not recorded (Marchesoni, 1952) and found only in very low densities in 1970–1972 (Gerletti, 1974). The appearance of *Planktolyngbya subtilis* dates back only to 1991 (Cordella & Salmaso, 1993), when it was observed during the whole year and dominating in July and August, often in depth maxima with *Planktothrix agardhii/rubescens*. In recent studies (Cordella & Salmaso, 1993) of the southeast basin, *Mougeotia cf. calcarea* has been found to be a 'perennial' with scarcely any regularity, but in the present study as well as in earlier investigations (Gerletti, 1974), it was found to fluctuate regularly. The species occurs in Lake Garda from early spring until the end of summer, whereas the *Mougeotia* species in Lake Constance (Sommer, 1984) develops in the early autumn, in response to the richer nutrient supply provided by increased vertical mixing.

Mean phytoplankton biovolume ranged from 120 $mm^3\ m^{-3}$ in winter to 1700 $mm^3\ m^{-3}$ in spring. The seasonal distribution is bimodal with one peak in late spring and one in the summer (Figure 3). Maxima are highest at P13, where seasonal variation is also most clearly expressed. Biovolume maxima occur during the stratification and suggest that, for most phytoplankters, net growth is achievable only when they can spend most of the time in the euphotic zone.

Species diversities as shown in Figure 4 are high, when compared to literature values (Rojo & Alvarez Cobelas, 1993), especially in the case of the minima, which never fall below 1. This is due to the high species numbers and to the even community structure. Minimal diversity at all three sampling stations coincided with the late-spring biomass peak, while the greatest diversity accompanied the attrition of the summer peak in August and September. Diversities based on all species with a relative biovolume > 0.5% are, in many cases, higher than those based on the threshold of an absolute biovolume of > 10 $mm^3\ m^{-3}$ (method 2). Figure 4 shows clearly that the two indices give the greatest discrepancies at low total biovolume, when a portion of 0.5% corresponds to very few individuals in the sedimentation chamber. Species of a biovolume > 10 $mm^3\ m^{-3}$ can be assumed to be 'active' members of the population.

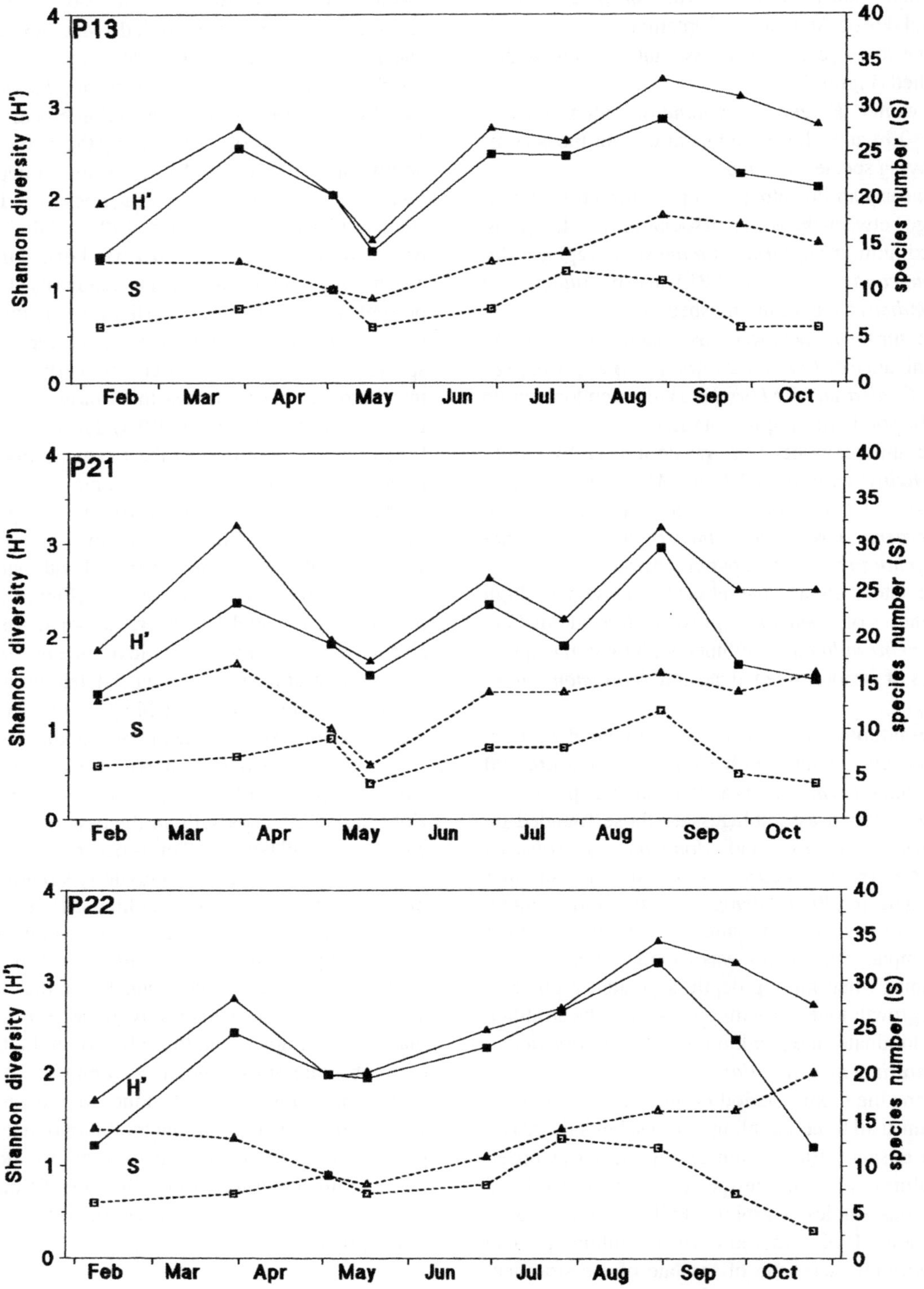

Figure 4. Biomass based Shannon diversity and species number in Lake Garda in 1994 at the three sampling stations. Triangle for > 0.5% and square for > 10 $mm^3 m^{-3}$.

Discussion

From the phytoplankton data, the tendency towards a more eutrophic assemblage is plainly evident, more so perhaps than from water chemistry, which indicates only a slight eutrophication. Some variables such as the annual average and maximum of Secchi depth, total nitrogen and chlorophyll ranking the lake on the boundary of oligotrophy, according to the OECD (1982) criteria, although the total phosphorus concentration and its clinograde distribution with depth are more associated with the mesotrophic category. Lake Garda still has a high transparency and a blue water colour. Compared to data from 1970–1972 (Gerletti, 1974), there has been a decline in Secchi-disk transparency in summer (from 11–15 m, in 1970–1972 to 7 m 1994).

Comparing the phytoplankton with the literature records (Ruggiu, 1983), not only an increase of quantities but also shifts in species composition are evident. For example, *Planktolyngbya subtilis*, which was recorded only in eutrophic Lake Lugano among the subalpine lakes during the 1980s (Ruggiu, 1983), is now found in large quantities in Lake Garda. The large number of species and the greater quantities of chloroccal green algae present in summer (especially of *Coelastrum polychordum*) are also indicative of eutrophic ecosystems (Rott, 1984). The establishment of the diatom-desmid assemblage typical for mesotrophic ecosystems (Reynolds, 1980) at station P13 (in this case with *Fragilaria crotonensis, Cyclotella comensis* and *Closterium aciculare*) in October, is favoured by the increased wind-mixing and epilimnetic enlargement. *Closterium* species occuring in large lakes are adapted to nutrient-rich waters (Rosén, 1981). *Dinobryon divergens* fits very well in this category of oligo- to mesotrophic ecosystem since it is typical for lakes with average TP concentrations of 12 μg l^{-1} (Rosén, 1981). *Planktothrix agardhii/rubescens* is the most famous example of a species typically restricted to a medium P-range (Sommer, 1986) and adaptable to low light intensities and therefore able to reach maxima at greater depth in the stratification period (in this case from 12.5 m to 30 m). The development of *Ceratium hirundinella* at the end of the annual succession in Lake Garda is typical for mesotrophic systems (Reynolds, 1980; Rott, 1984).

A tendency towards increasing eutrophy can also be inferred from the number of successional stages (Sommer, 1986), comparing nine large lakes in the Alps. In the most oligotrophic cases, only one phytoplankton peak in spring is found, typically starting after the onset of thermal stratification; in shallower lakes, the vernal bloom commonly starts immediately after ice-break (Sommer, 1986). The occurrence of two maxima of phytoplankton biovolume, now more pronounced even than in 1990 (Chiaudani & Premazzi, 1990) is consistent with mild eutrophication. The present study detects a shortening of the time between these peaks. In addition the average phytoplankton biovolume of more than 600 mm^3 m^3 and the peak values in spring and summer of 1700 mm^3 m^{-3} now encountered are fully consistent with mesotrophic descriptors (see also Rott, 1984).

In spite of the shallower depth and higher average water temperatures of the south and south-eastern basin of the lake, no great differences in either water chemistry or phytoplankton biomass were detected among the three central sampling stations. This is matched by similarities in the seasonally variation in diversity. However, the stronger exposure to wind action is likely to have caused the slightly higher diversity at station 13.

Conclusions: the trophic status of Lake Garda

Phytoplankton species composition, average- and peak biovolume clearly indicate mesotrophic conditions. This is also evident from the oxygen depletion during stratification below 150 m with a parallel increase of phosphorus concentration to a maximum of 36 μg l^{-1}. The fact that, in spite of a water renewal time of 27 y, Lake Garda is subject to slow eutrophication seems to be a consequence of its meromixis and of its high buffering capacity, favouring phosphorus binding in the epilimnion. Besides local waterblooms (Salmaso et al., 1992), a trend toward eutrophication can be inferred from the increased amounts of *Planktothrix rubescens* compared to investigations in 1971 (Gerletti, 1974). It is a matter of fact that the local effects of eutrophication are mainly restricted to near-shore areas. Not least for the economic interest invested in Lake Garda, extensive limnological studies must continue. To understand, resist and reverse eutrophication is an urgent priority.

Acknowledgements

I am grateful to Doz. Dr. E. Rott for his guidance and constructive criticism during this investigation. Thanks are due to Dr. E. Pipp for her kind help and her valuable comments on the manuscript. I am grateful to the Istituto Agrario di San Michele, Italy, for making their water chemistry data available and, especially, to Dott. F. Marchetti for his help taking the samples.

References

Ambrosetti, W., L. Barbanti, R. Mosello, A. Rolla & D. Ruggiu, 1983. Mescolamento, caratteristiche chimiche, fitoplancton e situazione trofica nei laghi profondi sudalpini. Prog. Finalizz. 'Promozione della Qualità dell'Ambiente'. C.N.R. AQ/2/20: 151pp.

Ambrosetti, W., L. Barbanti, R. Mosello & A. Pugnetti, 1992. Limnological studies on the southern alpine lakes Maggiore, Lugano, Como, Iseo and Garda. In: Guilizzoni P., G. Tartari & G. Giussani (eds), Limnolgy in Italy. Mem. ist. Ital. Idrobiol. 50: 117–146.

Andreoli, C., 1977. Il Lago di Garda – Distribuzione qualitativa e quantitativa del fitoplancton in un ciclo annuale (Settembre 1972 – Agosto 1973). Riv. Idrobiol. XVI Fasc. 3: 269-294.

Chiaudani, G. & G. Premazzi, 1990. Il Lago di Garda. Evoluzione trofica e condizioni ambientali attuali. Commissione delle Communità Europee. EUR 12925, 193 pp.

Chiaudani, G., G. Borsani, A. Negri, P. Poltronieri, G. Premazzi & E. Rodari, 1993. Lake Garda: Trophic evolution and present quality conditions. In Giussani G. & C. Callieri (eds.), Strategies for lake ecosystems beyond 2000, Proceedings. 5th International conference on the conservation and management of lakes: 70-73.

Cordella, P., N. Salmaso & E. Castellani, 1992. Ricerche limnologiche sulla sponda orientale del Lago di Garda. Ambiente Risorse Salute 4: 15–21.

Cordella, P. & N. Salmaso, 1993. Phytoplankton communities in the southern basin of Lake Garda, Italy. Verh. int. Ver. Limnol. 25: 754–761.

Forti, A., 1902. Contributo 4° alla conoscenza della Florula Ficologica Veronese. Nova Notarisia, Serie XIII(II): 49–68 p. Serie XIII(III): 97–124.

Garbini, A., 1897. Alcune notizie fisiche sulle acque del Benaco. Riv. Geogr. It. IV: 23–29.

Garbini, A., 1898. Alghe neritiche del Lago di Garda. Nova Notarisia X: 3–20.

Gerletti, M., 1974. Indagini sul Lago di Garda. I.R.S.A. 18: 186–205.

Herdendorf, C. E., 1987. Distribution of the World's Large Lakes. In Tilzer M. M. & C. Serruya (eds), Large Lakes: ecological structures and functions. Brock/Springer series in contemporary bioscience: 3–38.

Istituto Agrario di S.Michele a/Adige: Annali dei dati idrobiologici 1989–1995. Provincia Autonoma di Trento.

Kirchner, O., 1899. Florula Phycologica Benacensis. XXXVI Pubblicazione fatta per cura del Civico Museo di Rovereto, 37 pp.

Marchesoni, V., 1952. Ricerche orientative sulla microflora pelagica del Garda. Studi Trent. Sc. Nat. 29: 85-109.

Merlo, S. & C. Mozzi, 1963. Ricerche limnologiche sul Lago di Garda. Arch. Oceanogr. Limnol. 13: 1–125.

OECD, 1982. Eutrophication of waters. Monitoring, assessment and control. OECD, Paris, 154 pp.

Regia Accademia delle Scienze dell'Istituto di Bologna, 1930. Memoria intorno a Luigi Ferdinando Marsili. Zanichelli, Bologna, 509 pp.

Reynolds, C. S., 1980. Phytoplankton assemblages and their periodicity in stratifying lake systems. Holarctic Ecol. 3: 141–159.

Reynolds, C. S., 1984. The ecology of freshwater phytoplankton. Cambridge University Press, 384 pp.

Rojo, C. & M. Alvarez Cobelas, 1993. Hypertrophic phytoplankton and the Intermediate Disturbance Hypothesis. In Padisák J., C. S. Reynolds & U. Sommer (eds), Intermediate Disturbance Hypothesis in Phytoplankton Ecology. Dev. Hydrobiol. 81. Kluwer Academic Publishers, Dordrecht: 43–57.

Rosén, G., 1981. Phytoplankton indicators and their relations to certain chemical and physical factors. Limnologica 13/2: 263–290.

Rott, E., 1981. Some results from phytoplankton counting intercalibrations. Schweiz. Z. Hydrobiol. 43/1: 34–62.

Ruggiu, D., 1983. Caratteristiche e comportamento del fitoplancton nei laghi profondi sudalpini. In Ambrosetti, W., L. Barbanti, R. Mosello, A. Rolla & D. Ruggiu. Mescolamento, caratteristiche chimiche, fitoplancton e situazione trofica nei laghi profondi sudalpini. Prog. Finalizz. 'Promozione della Qualità dell'Ambiente'. C.N.R. AQ/2/20: 105–146.

Ruggiu, D. & R. Mosello, 1984. Nutrient levels and phytoplankton characteristics in the deep southern alpine lakes. Verh. int. Ver. Limnol. 22: 1106–1112.

Salmaso, N., F. Cavolo & P. Cordella, 1994. Fioriture di Anabaena e Microcystis nel Lago di Garda. Eventi rilevati e caratterizzazione dei periodi di sviluppo. Acqua Aria 1: 16–28.

Salmaso, N. & P. Cordella, 1994. Indagini limnologiche sul Lago di Garda. In Cordella P. & N. Salmaso (eds), Indagini limnologiche sui principali laghi della Regione del Veneto (1987–1992). Regione del Veneto: 79-124.

Salmaso, N., F. Decet, M. Manfrin, P. Cordella, 1996. Studio delle caratteristiche chimiche e della struttura della communità fitoplanctonica per una valutazione delle condizioni trofiche attuali. In Ecologia. Atti del VII Congresso Nazionale della Società Italiana di Ecologia. S.It.E. – Atti 17: 623–626.

Shannon, C. E., 1948. A mathematical theory of communication. Bell. Syst. Tech. J. 27: 623–656.

Sommer, U., 1984. Factors controlling the seasonal variation in phytoplankton species composition – A case study for a deep, nutrient rich lake (Lake Constance). Habilitationsschrift, Universität Konstanz, 200 pp.

Sommer, U., 1986. The periodicity of phytoplankton in Lake Constance (Bodensee) in comparison to other deep lakes of central Europe. In Munawar, M. & J. F. Talling (eds), Seasonality of freshwater phytoplankton. Dev. Hydrobiol.: 33, 236 pp.

Utermöhl, H., 1958. Zur Vervollkommnung der quantitativen Phytoplanktonmethodik. Mitt. Int. Ver. Limnol. 9: 1–38.

Hydrobiologia **369/370:** 163–178, 1998.

M. Alvarez-Cobelas, C. S. Reynolds, P. Sanchez-Castillo & J. Kristiansen (eds), Phytoplankton and Trophic Gradients.

Phytoplankton dynamics in two reservoirs with different trophic state (Lake Rosamarina and Lake Arancio, Sicily, Italy)

Luigi Naselli Flores & Rossella Barone
Dipartimento di Scienze Botaniche, Via Archirafi, 38, 90123 Palermo, Italy

Key words: seasonality, trophic gradients, man-made lakes, multivariate analysis, CCA

Abstract

Two man-made Sicilian lakes of differing trophic status were sampled weekly for a year, in order to compare their phytoplankton dynamics and to identify their driving factors. The water bodies store comparable volumes but the eutrophic Lake Arancio may be considered to be a shallow lake whereas the mesotrophic Lake Rosamarina is deep. The dynamics of their principal physical, chemical and biological features were analysed using, among others, Canonical Community Ordination. The results obtained suggest that increased nutrient availability causes an increase of phytoplankton biomass but no community change nor any alteration to the typical assemblages supported. The major influence in the more eutrophic lake was exerted by physical variates and, in particular, by the underwater light climate regime, whereas a significant influence of nutrients was observed in the mesotrophic one. Moreover, an assemblage of species able to regulate their buoyancy, or adapted to frequent dark/light cycles was characteristic of the eutrophic lake. We hypothesize that the physical characteristics of a water body and secondary modifications induced by increased nutrient availability, e.g. higher phytoplankton biomass, contraction of the euphotic depth, selective grazing, may drive the seasonal succession of phytoplankton in the upper part of the trophic spectrum.

Introduction

Eutrophication processes are promoted by nutrient enrichment; the evidence of such addition is directly manifest in increased average photosynthetic biomass. It follows that, since the development of the so called 'Vollenweider's models' which related nutrient inputs to P loading and to chlorophyll *a* concentration, high phytoplankton biomass continues to be one of the most common descriptors of eutrophic systems. Many models adopting this central assumption have been validated (Harper, 1992; Watson et al., 1992).

Only seldom has it been recognised that the increase in biomass generates complex changes in the physical environment, such as modifications in the underwater light climate (Sommer, 1988). How the shift in species composition of phytoplankton assemblages is due to these secondary modifications induced by nutrient increase is, at present, poorly understood (Straskraba et al., 1993). In addition, some environments, such as man-made lakes, are characterized by peculiar hydraulic regimes that may strongly influence the annual and interannual variability in abundance and composition of phytoplankton (Harris & Baxter, 1996).

A previous investigation on the man-made Lake Arancio suggested that the secondary effects induced by increased nutrient availability, along with the specific morphology and hydrology of the water body, have to be considered as important factors in selecting species (Barone & Naselli Flores, 1994).

Here, we show how physical factors may become important in reservoirs as nutrient availability increases. The work was carried out using data collected on two man-made lakes of different trophic state (Lake Arancio and Lake Rosamarina). In these environments the physical constraints are generally enhanced by the management applied and this offers a good opportunity for investigating their role in phytoplankton dynamics.

Table 1. Main characteristics of the studied lakes. A refers to maximum project value, B to the maximum value attained in the study years. A: Area; V: Volume; Z: mean depth; Zmax: Max depth; SD: Secchi disk transparency depth; Alk: Alkalinity; Cond: Conductivity at 18 °C.

Lakes	A [km²]		V [10⁶ m³]		Z [m]		Z_{max} [m]		pH	SD [m]	Total P [μg l^{-1}]	SRP [μg l^{-1}]	Alk [meq l^{-1}]	Cond [μS cm^{-1}]	Total N [mg l^{-1}]	DIN [μg l^{-1}]
	A	B	A	B	A	B	A	B								
Arancio	3.6	2.6	32	21.7	8.3	9.2	30	24	8.2–8.8	0.35–1.40	40–107	18–87	2.0–4.5	550–1130	1.8–6.0	60–900
Rosamarina	5.2	1.9	100	25.5	19.2	13.2	61	38	8.2–8.4	1.20–5.70	25–110	5–27	1.3–2.7	920–1410	0.6–1.4	55–680

Materials and methods

A detailed description of Lake Arancio is given in Barone & Naselli Flores (1994). The year of the survey was characterized, compared to the former investigation, by a major inflow, which permitted a doubling of the volume of water in store.

Lake Rosamarina was created in 1993 by damming the valley of the San Leonardo torrent, 3 km from the North coast of Sicily, to retain water for irrigation and drinking purposes. The catchment area covers a surface of 500 km^2 and is geologically characterized by carbonate rocks, clays and marls with some outcrops of Messinian evaporites; the climate is Mediterranean with an average annual rainfall of 650 mm falling on about 40 rain days in winter. The reservoir has a narrow, elongated shape with its principal axis along its North-South axis. The characteristics of the lakes are summarized in Table 1.

Samples were collected at weekly intervals, from January to December 1993 in Lake Arancio and from February to December 1995 in Lake Rosamarina, at a fixed station in the open water of both reservoirs.

Methods for measuring water temperature, pH, dissolved oxygen, conductivity at 18 °C, transparency, depth of the euphotic zone and for calculation of the attenuation coefficient were given in Barone & Naselli Flores (1994).

Water samples for nutrients, total phosphorus, total nitrogen and alkalinity were taken at the surface, from the middle of the water column, and near the bottom. Total alkalinity was immediately measured using a mixed bromocresol-green/methyl-red indicator and adding standard HCl up to the equivalence point. Nutrients, total phosphorus and total nitrogen analyses were performed as detailed in Barone & Naselli Flores (1994). Values were expressed as weighted averages for the water column of either lake. In addition, depth-weighted averages in the mixing zone were calculated for Lake Rosamarina.

Water samples for phytoplankton and chlorophyll *a* were collected at intervals of 1 m from the surface to the bottom in Lake Arancio, whereas they were taken at 1, 2, 5, 10, 20 meters of depth in Lake Rosamarina. Analytical methods were as described in Barone & Naselli Flores (1994). Total phytoplankton biomass and chlorophyll *a* concentrations are expressed as weighted averages the euphotic zone.

The taxonomic nomenclature follows that proposed by Van den Hoek et al. (1995)

Species were grouped in four strategic groups, following Arauzo & Alvarez Cobelas (1994). In particular, our classification of strategies in a r-K continuum used time span of occurrence, average growth and loss rates, and biovolume. Time span of occurrence, measured as percentage of weeks on an annual basis in which a given species was continuously recorded, is generally low for r-strategists, whereas a high time span of occurrence is typical of K-strategists; thus, species with a time span of occurrence less than 50%, average growth and loss rate above 0.10 d^{-1} and biovolume less than 3000 μm^3, were classified as r-strategists. Two additional strategic groups (r* and K*) were adopted for those species which showed time span of occurrence characteristic of r- or K, but with other features typical of the other group. Species diversity was calculated using Shannon's index.

Methods for zooplankton analysis are those described in Barone et al. (1991)

The data set used for statistical analyses consisted of weekly species biomass values and correspondent environmental variables. Only the taxa that were present with a relative abundance $>0.5\%$ in a minimum of two samples were included in the data analyses. For each lake, 21 limnological variables were monitored: 9 chemical (N-NH_4, N-NO_3, N-NO_2, SRP, SRSi, TP, TN, alkalinity, pH), 6 physical (temperature, surfacial irradiance, extinction coefficient, mixing and euphotic depth, conductivity), 3 morphological (lake volume, lake surface, average depth) and 2 biological (herbivorous zooplankton biomass and chlorophyll *a* concentration). Weighted averages in the mixing zone were applied to the chemical variables; physical variables monitored in the water column (temperature, conductivity) were averaged over the epilimnetic zone.

All ordinations were performed using CANOCO 3.1 (ter Braak, 1990). Principal component analysis (PCA) was used to detect the major environmental variables in either lake. Canonical correspondence analysis (CCA), a multivariate method to elucidate the relationships between biological assemblages of species and their environment (ter Braak & Verdonschot, 1995), was undertaken with $log(x+1)$ transformation. The significance with which environmental variables explain the variance of species data was tested using Monte Carlo methods, testing with 99 unrestricted permutations. Variables were considered to be significant when $p < 0.05$.

Results

Trophic state parameters

Figure 1 depicts fluctuations in Secchi-disk transparency, phytoplankton biomass and chlorophyll *a* values averaged through the euphotic zone in the two lakes. In accordance with OCDE trophic classification (OCDE, 1982), Lake Arancio would be classified as eutrophic within respect to its Secchi-disk transparency (values typical of high trophic state, generally ranging from 0.8 m to 1.3 m; values < 0.4 m were recorded at the end of October due to the high solid transport caused by the rain). Chlorophyll *a* concentration averaged 22 μg l^{-1} over the year with a maximum value of 48 $\mu g\ l^{-1}$, recorded at the end of May. Phytoplankton biomass, after an isolated peak at the beginning of April, started to increase in the middle of May, up to maximal values (> 20 mg l^{-1}) by the beginning of September. Following Rott (1984) and based on average phytoplankton biomass, the lake lies in the upper part of the trophic spectrum. The concensus of the principal indicators is that Lake Arancio is eutrophic.

In Lake Rosamarina, Secchi disk transparency values were rarely lower than 2.0 m. Higher values (> 3 m) were recorded in spring and in autumn. Phytoplankton biomass and chlorophyll *a* concentrations were one order of magnitude lower than in Lake Arancio. Moreover, the respective maximum values of both variables, 2.4 mg l^{-1} and 8 $\mu g\ l^{-1}$, were recorded in summer. According to various classifications, the lake should fall within the mesotrophic part of the trophic gradient. Total phosphorus concentrations ranged within an interval (Table 1) indicative of a higher trophic state. On the basis of these variables, Lake Rosamarina is considered mesotrophic. This deduction needs to be tempered by the recognition that this is a relatively new water body and movement towards a higher trophic state by total phosphorus concentration is possible.

Physical and chemical features

The hydraulic balance in Lake Arancio led to volume fluctuations of between 22 × 106 m^3 in May and 8 × 106 m^3 in October. This corresponds to a vertical range of water level of about 7 m. Winter and spring represent the filling period of the reservoir, which, at the end of May, starts emptying due to the irrigation demand and evaporation losses. From the middle of October, at the end of agriculture needs, the reservoir fills up again. The volume variations in Lake Rosamarina follow a similar trend but are much more less extreme, as the water stored has not yet been exploited as a resource. Thus, the oscillation in volume from May to September is 3 × 106 m^3 which corresponds to a vertical movement in water level of about 2 m.

The hydraulic characteristics and the differing morphometries of the two reservoirs are reflected in their contrasted thermal structure. Although surficial temperatures were similar, ranging between 11 °C in winter and 27 °C in summer, mixing depth was often quite different (Figure 2A and B). In Lake Arancio, the thermocline was not stable and showed a continuous variability in depth until mid August when due to drawdown, the reservoir showed a premature full circulation. In addition, the difference in temperature between the epilimnion and the hypolimnion never exceeded 3 °C. Oxygen saturation generally ranged between values $> 100\%$ in the surface layer and $> 25\%$ in the hypolimnion. Only in late June did hypolimnetic anoxia develop and even this was abruptly overcome by the next full-circulation event.

In contrast, Lake Rosamarina had a stable thermocline which started developing in the middle of April and, after some oscillations, found a constant location at about 5.5 m of depth by mid May, where it persisted until early August. In this period, the surface temperature was about 15 °C higher than that at the bottom. Thereafter, the thermocline was progressively depressed until the onset of full circulation at the

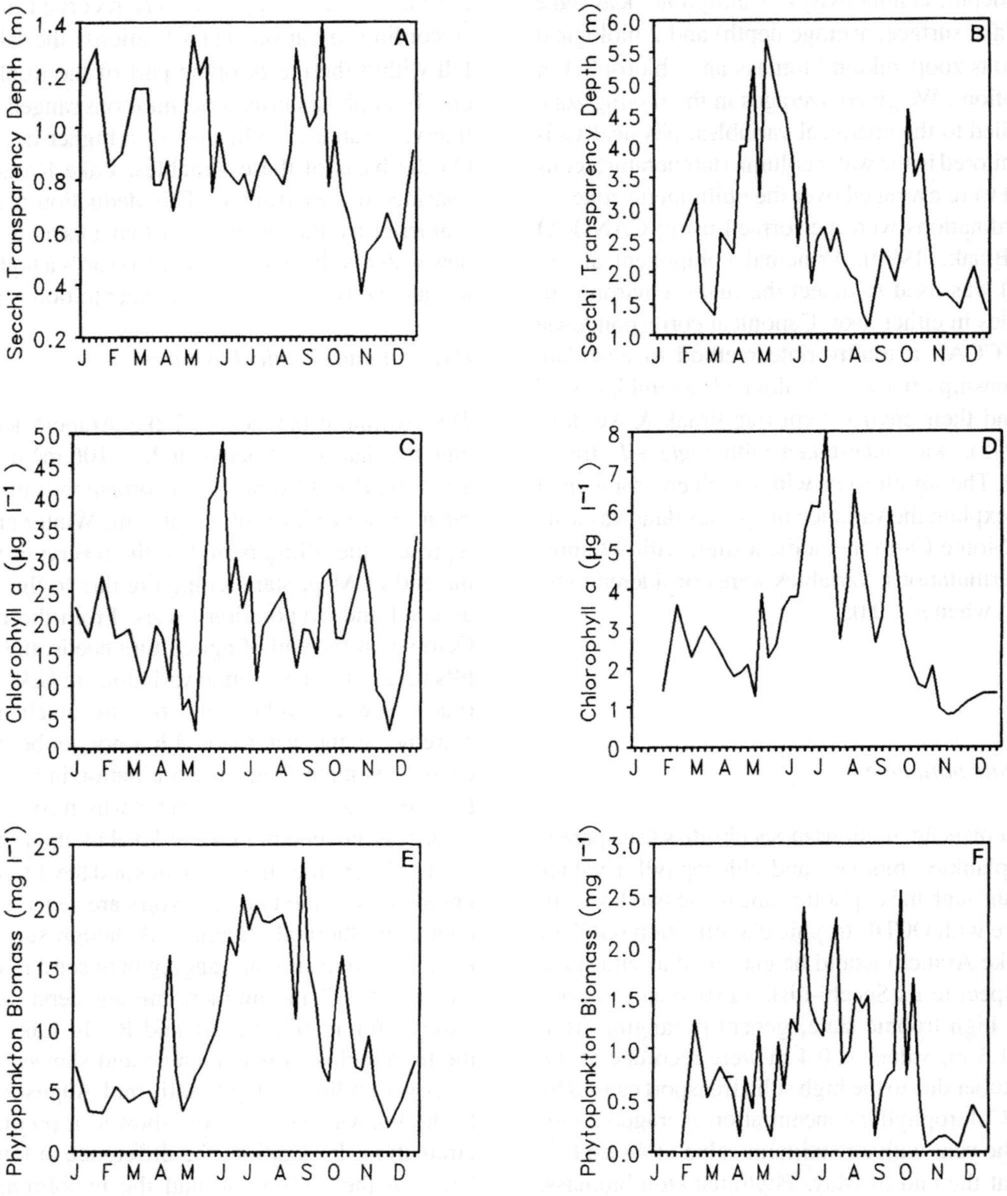

Figure 1. Seasonal fluctuations in the principal trophic state parameters in Lake Arancio (A, C, E) and in Lake Rosamarina (B, D, F).

beginning of November. During the summer months, oxygen saturation ranged between 75 and 100% in the epilimnetic layers whereas complete anoxia developed in the hypolimnion after July. In consequence, low pH values (< 7.00) occurred in the hypolimnion.

Euphotic depth co-varied with Secchi-disk transparency depth in both lakes. In Lake Arancio, the z_m/z_{eu} ratio was generally above 1.5. In contrast, Lake Rosamarina was characterized by a deep euphotic layer, which ranged between 8 and 14 m of depth, and the z_m/z_{eu} ratio was always less than 1.5.

Conductivity and alkalinity values highlighted a wide variation during the year in both lakes (Table 1).

Nutrients and total nitrogen

The nitrate concentrations of the two lakes showed comparable values and similar dynamics through time,

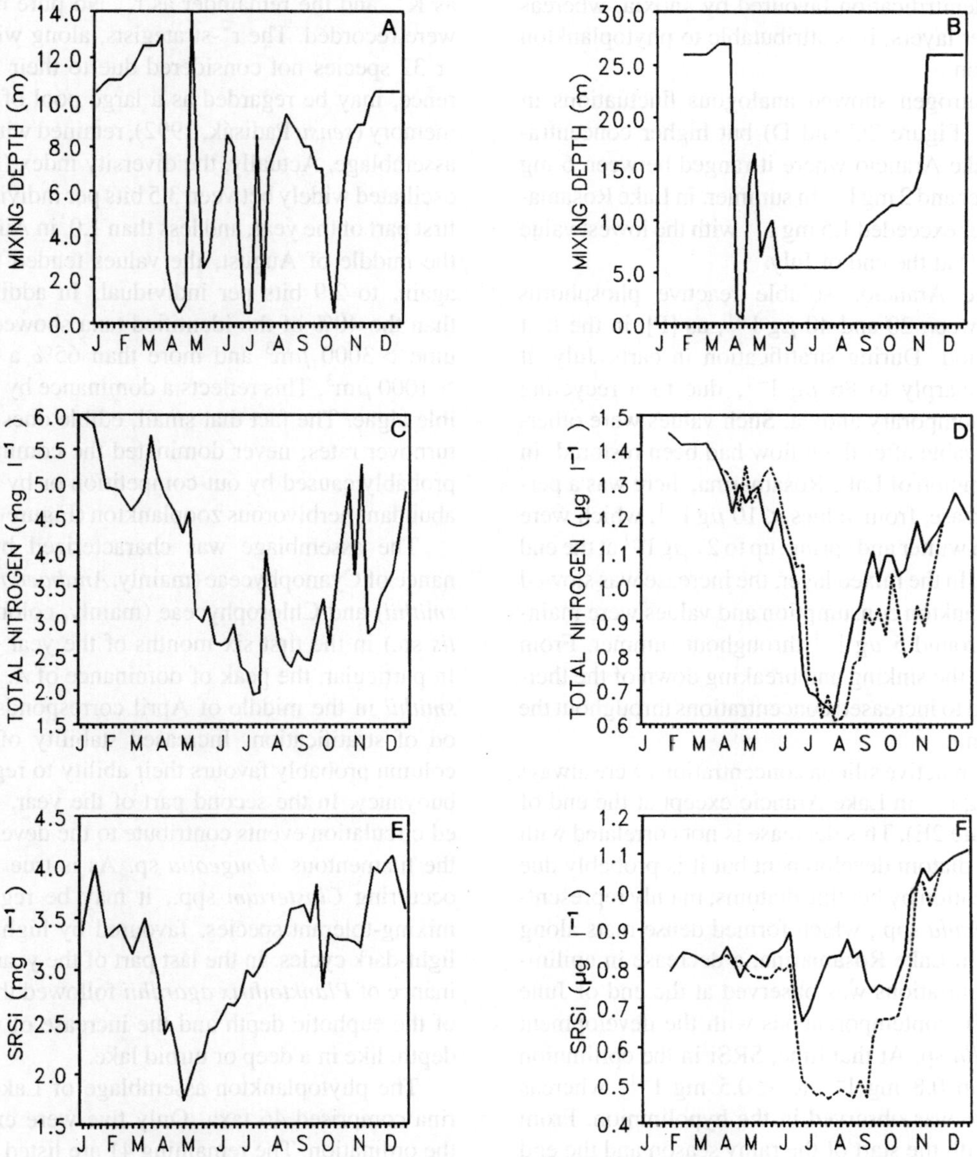

Figure 2. Mixing depth trends and seasonal variation in the concentrations of total nitrogen and soluble reactive silicon (SRSi) in Lake Arancio (A, C, E) and in Lake Rosamarina (B, D, F). The solid lines represent weighed averages in the water column whereas the dotted lines represent weighed averages in the mixing zone.

with maxima in the periods of inflow and minima in summer. They ranged between 50 $\mu g\ l^{-1}$ and 890 $\mu g\ l^{-1}$, as [N], in Lake Arancio and between 35 $\mu g\ l^{-1}$ and 670 $\mu g\ l^{-1}$ in Lake Rosamarina. In the latter, no significant differences were recorded between the concentrations in the mixed layer and in the whole water column despite the summer stratification.

In Lake Arancio, ammonia values are generally below 20 $\mu g\ l^{-1}$, as [N], with the minima in early spring (< 2 $\mu g\ l^{-1}$) and a peak (19 $\mu g\ l^{-1}$) immediately thereafter, probably due to spring zooplankton grazing. During summer, the concentrations showed little variation, with values between 5.0 and 8.5 $\mu g\ l^{-1}$. This emphasises the minor importance of stratification in early summer. In Lake Rosamarina, ammonia dominated over nitrate in late summer, and reached high values in the hypolimnion (> 300 $\mu g\ l^{-1}$), suggesting that the decrease in nitrate beyond the thermocline is due to denitrification favoured by anoxia, whereas in the upper layers, it is attributable to phytoplankton consumption.

Total nitrogen showed analogous fluctuations in both lakes (Figure 2C and D) but higher concentrations in Lake Arancio where it ranged between 6 mg l^{-1} in winter and 2 mg l^{-1} in summer. In Lake Rosamarina, it never exceeded 1.5 mg l^{-1} with the lowest value (0.4 mg l^{-1}) at the end of July.

In Lake Arancio, soluble reactive phosphorus ranged between 20 and 40 μg l^{-1}, as [P], in the first survey period. During stratification in early July, it increased sharply to 86 μg l^{-1}, due to a recycling caused by temporary anoxia. Such values were otherwise observable after the inflow had been restored. In the hypolimnion of Lake Rosamarina, there was a persistent increase, from values < 10 μg l^{-1}, which were recorded in winter and spring, up to 27 μg l^{-1} at the end of summer. In the mixed layer, the increase was slowed by phytoplankton consumption and values were maintained at around 5 μg l^{-1} throughout summer. From September, the sinking and breaking down of the thermocline led to increased concentrations throughout the water column.

Soluble reactive silicon concentrations were always above 2 mg l^{-1} in Lake Arancio except at the end of April (Figure 2E). This decrease is not correlated with planktonic diatom development but it is probably due to the utilization by benthic diatoms, mainly represented by *Navicula* spp., which formed dense mats along the shore. In Lake Rosamarina, a decrease in epilimnetic concentrations was observed at the end of June (Figure 2F), contemporaneous with the development of *Cyclotella* sp. At that time, SRSi in the epilimnion passed from 0.8 mg l^{-1} to <0.5 mg l^{-1}, whereas no decrease was observed in the hypolimnion. From October, with the start of the rainy season and the end of stratification, concentrations at both determinands increased to levels > 1 mg l^{-1}.

Phytoplankton biomass and species composition

Ninety-three taxa belonging to the phytoplankton assemblage of Lake Arancio were identified. Among these, the 61 with a biomass > 0.5% of the total in two consecutive samples are listed in Table 2, along with their time span of occurrence, average growth and loss rates, average biovolume and the strategic group to which we assigned them. Three species (*Anabaena solitaria* f. *smithii*, *Pediastrum simplex* and *Coelastrum astroideum*) are classifiable as K-strategists, 10 as K*, and the remainder as r*. No pure r-strategists were recorded. The r*-strategists, along with the other 32 species not considered due to their low occurrence, may be regarded as a large pool of ecological memory (*sensu* Padisák, 1992), retained within the full assemblage. Actually, the diversity index (Figure 3A) oscillated widely between 3.5 bits per individual, in the first part of the year, and less than 1.0, in August. From the middle of August, the values tended to increase again, to 2.9 bits per individual. In addition, more than the 40% of the identified taxa showed a biovolume > 3000 μm^3 and more than 65% a biovolume > 1000 μm^3. This reflects a dominance by large inedible algae. The fact that small, edible ones, with high turnover rates, never dominated the community, was probably caused by out-competition or by grazing by abundant herbivorous zooplankton (Figure 3C).

The assemblage was characterized by a dominance of Cyanophyceae (mainly, *Anabaena solitaria* f. *smithii*) and Chlorophyceae (mainly, colonial *Oocystis* sp.) in the first six months of the year (Figure 4). In particular, the peak of dominance of *A. solitaria* f. *smithii* in the middle of April corresponds to a period of stratification. Increased stability of the water column probably favours their ability to regulate their buoyancy. In the second part of the year, the repeated circulation events contribute to the development of the filamentous *Mougeotia* sp. As is true for the co-occurring *Closterium* spp., it may be regarded as a mixing-tolerant species, favoured by high frequency light-dark cycles. In the last part of the year, the dominance of *Planktothrix agardhii* followed the decrease of the euphotic depth and the increase of the mixing depth, like in a deep or turbid lake.

The phytoplankton assemblage of Lake Rosamarina comprised 46 taxa. Only five were excluded by the ordination. The remaining 41 are listed in Table 3. Among these, *Botryococcus braunii* and *Pediastrum duplex* were K-strategists, one (*Lagerheimia* cf. *genevensis*), was an r-strategist, 12 were K*-strategists and the remainder r*. About 60% of the organisms biovolumes of less than 1000 μm^3, even though the dominant species generally showed larger dimensions. Moreover, this assemblage tended towards higher growth rates than in Lake Arancio. In addition, a significant number of small species (*Scenedesmus* sp., *Oocystis* sp., *Planctonema lauterbornii*, *Cosmarium* sp., *Cyclotella* sp.), edible by zooplankton grazers, were classified as K*-strategist. The diversity index (Figure 3B) always stayed above 2.0 bits per individual and generally oscillated within narrow limits

Table 2. Phytoplankton taxonomic composition of Lake Arancio. AGR: Average growth rate; ALR: Average loss rate; TSAO: Time span of occurrence; BV: Biovolume; SG: Strategic group; sd: standard deviation.

	Abbreviations	AGR		ALR		TSAO	BV	SG
		d^{-1}	sd	d^{-1}	sd	%	μm^3	
Cyanophyceae								
Anabaena solitaria Klebahn f. *smithii* Komarék	ANA SOL	0.06	0.05	0.10	0.10	98	5860	K
Gomphosphaeria nägeliana (Unger) Lemmermann	GOM NAG	0.09	0.07	0.08	0.07	23	3800	r*
Microcystis aeruginosa (Kützing) Kützing	MIC AER	0.19	0.06	0.09	0.05	31	5200	r*
Microcystis wesenbergii Komarék	MIC WES	0.12		0.06	0.03	10	113	r*
Planktothrix agardhii (Gomont) Anagnostidis et Komarék	PLA AGA	0.11	0.10	0.11	0.10	48	1380	r*
Pseudanabaena sp.	PSE SPC	0.10	0.06	< 0.01		13	460	r*
Bacillariophyceae								
Cyclotella meneghiniana Kützing	CYC MEN	0.12	0.07	0.13	0.07	31	1809	r*
Fragilaria ulna Lange-Bertalot var. *acus* (Nitzsch) Lange-Bertalot	FRA ULN	0.08	0.09	0.06	0.04	67	384K	*
Navicula sp.	NAV SPC	0.05	0.07	0.09	0.06	8	324	r*
Nitzschia sp.	NIT SPC	0.09		0.08	0.02	8	480	r*
Nitzschia closterium (Ehrenberg) W. Smith	NIT CLO	0.22	0.13	0.12	0.06	23	980	r*
Stephanodiscus sp.	STE SPC	0.11	0.14	0.12	0.10	46	31	r*
Cryptophyceae								
Cryptomonas cf. *erosa* Ehrenberg	CRY ERO	0.06	0.05	0.07	0.07	100	424	K*
Cryptomonas cf. *marssonii* Skuja	CRY MAR	0.14	0.11	0.15	0.11	31	707	r*
Cryptomonas cf. *rostratiformis* Skuja	CRY ROS	0.10		0.25		10	4414	r*
Rhodomonas minuta Skuja var. *nannoplanctica* Skuja	RHO MIN	0.11	0.07	0.10	0.06	63	113	K*
Dinophyceae								
Peridinium sp.	PER SPC	0.10		0.07	0.06	23	28886	r*
Peridinium sp. 1	PER SP1	0.14	0.12	0.10	0.06	63	113	r*
Peridinium sp. 2	PER SP2	0.17	0.09	0.11	0.09	42	2574	r*
Peridinium sp. 3	PER SP3	0.09	0.06	0.16	0.19	25	6047	r*
Peridinium sp. 4	PER SP4	< 0.01		0.07		4	3820	r*
Euglenophyceae								
Euglena sp.	EUG SPC	< 0.01		0.13	0.01	13	7259	r*
Euglena sp. 1	EUG SP1	< 0.01		< 0.01		4	10788	r*
Euglena acus Ehrenberg	EUG ACU	0.10	0.03	0.10	0.04	23	7542	r*
Lepocinclis texta (Dujardin) Lemmermann	LEP TEX	< 0.01		0.10		4	31121	r*
Phacus sp.	PHA SPC	0.09	0.09	0.04	0.03	13	1060	r*
Phacus pyrum (Ehrenberg) Stein	PHA PYR	0.02	0.04	0.09	0.01	38	541	r*
Trachelomonas sp.	TRA SPC	0.02	0.01	0.09	0.04	25	1769	r*
Trachelomonas sp. 1	TRA SP1	0.13	0.07	0.29	0.11	13	1664	r*
Trachelomonas caudata (Ehrenberg) Stein	TRA CAU	0.08	0.10	0.10	0.03	17	3398	r*
Trachelomonas ovata Roll	TRA OVA	0.10	0.10	0.26	0.06	23	963	r*
Trachelomonas volvocina Ehrenberg	TRA VOL	0.21	0.02	0.10		12	905	r*
Chlorophyceae								
Carteria sp.	CAR SPC	0.03		0.12		6	1132	r*
Chlamydomonas sp.	CHL SPC	0.11	0.08	0.13	0.11	94	409	K*
Coelastrum astroideum De Notaris	COE AST	0.06	0.05	0.09	0.06	90	7888	K
Coelastrum microporum Nägeli	COE MIC	0.10	0.05	0.11	0.01	21	7888	r*
Coelastrum reticulatum (Dangeard) Senn	COE RET	< 0.01		0.10		4	7888	r*
Dictyosphaerium pulchellum Wood	DIC PUL	0.07	0.05	0.10		25	8450	r*
Elakatothrix gelatinosa Wille	ELA GEL	0.05	0.06	0.15	0.08	35	121	r*
Monoraphidium minutum (Nägeli) Komárková-Legnerová	MON MIN	0.04	0.05	0.10		46	40	r*
Oocystis sp.	OOC SPC	0.08	0.08	0.09	0.08	100	277	K*
Pediastrum boryanum (Turpin) Meneghini)	PED BOR	0.10		< 0.01		8	3888	r*

Table 2. Continued.

	Abbreviations	AGR		ALR		TSAO	BV	SG
		d^{-1}	sd	d^{-1}	sd	%	μm^3	
Pediastrum duplex Meyen	PED DUP	0.13	0.12	0.15	0.05	31	3888	r*
Pediastrum simplex Meyen	PED SIM	0.07	0.06	0.08	0.07	88	3384	K
Scenedesmus sp.	SCE SPC	0.11	0.08	0.10	0.04	58	1132	K*
Scenedesmus sempervirens Chodat	SCE SEM	0.04	0.05	0.07	0.05	58	228	K*
Scenedesmus magnus Meyen	SCE MAG	0.25		< 0.01		10	1584	r*
Scenedesmus quadricauda (Turpin) Brébisson *sensu* Chodat	SCE QUA	0.05	0.05	0.06	0.03	37	228	r*
Siderocystopsis fusca (Korsikov) Swale	SID FUS	0.13	0.08	0.16	0.08	48	2004	r*
Sphaerocystis schröteri Chodat	SPH SCH	< 0.01		0.17		6	1808	r*
Tetraedron minimum (A. Braun) Hansgirg	TET MIN	0.05	0.06	0.08	0.06	94	427	K*
Tetrastrum sp.	TET SPC	0.15	0.06	0.09		19	23	r*
Zygnematophyceae								
Closterium aciculare T. West	CLO ACI	0.10	0.07	0.15	0.15	90	1872	K*
Closterium limneticum Lemmermann	CLO LIM	0.15	0.15	0.12	0.02	33	1371	r*
Closterium limneticum Lemmermann var. *fallax* Ruzicka	CLO LIF	0.06	0.04	0.13	0.03	19	7738	r*
Closterium acutum Brébisson var. *variabile* (Lemmermann) W. Krieger	CLO VAR	0.05	0.09	0.11	0.02	33	452	r*
Cosmarium sp.	COS SPC	0.08	0.13	0.10	0.06	46	1073	r*
Cosmarium sp. 1	COS SP1	0.18		0.08		8	47	r*
Mougeotia sp.	MOU SPC	0.07	0.09	0.07	0.05	96	1774	K*
Staurastrum sp.	STA SPC	0.10		0.09		10	1924	r*
Staurastrum tetracerum (Kützing) Ralfs	STA TET	< 0.01		0.11		8	412	r*

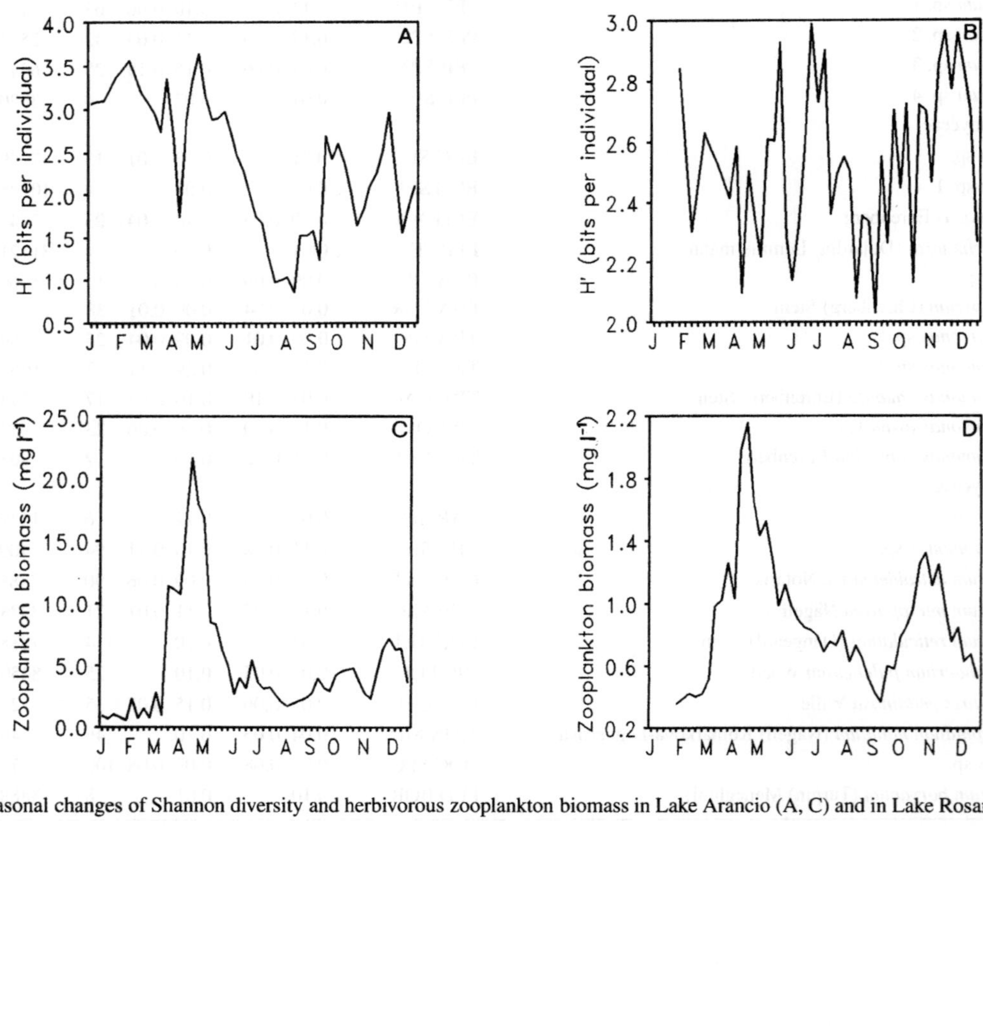

Figure 3. Seasonal changes of Shannon diversity and herbivorous zooplankton biomass in Lake Arancio (A, C) and in Lake Rosamarina (B, D).

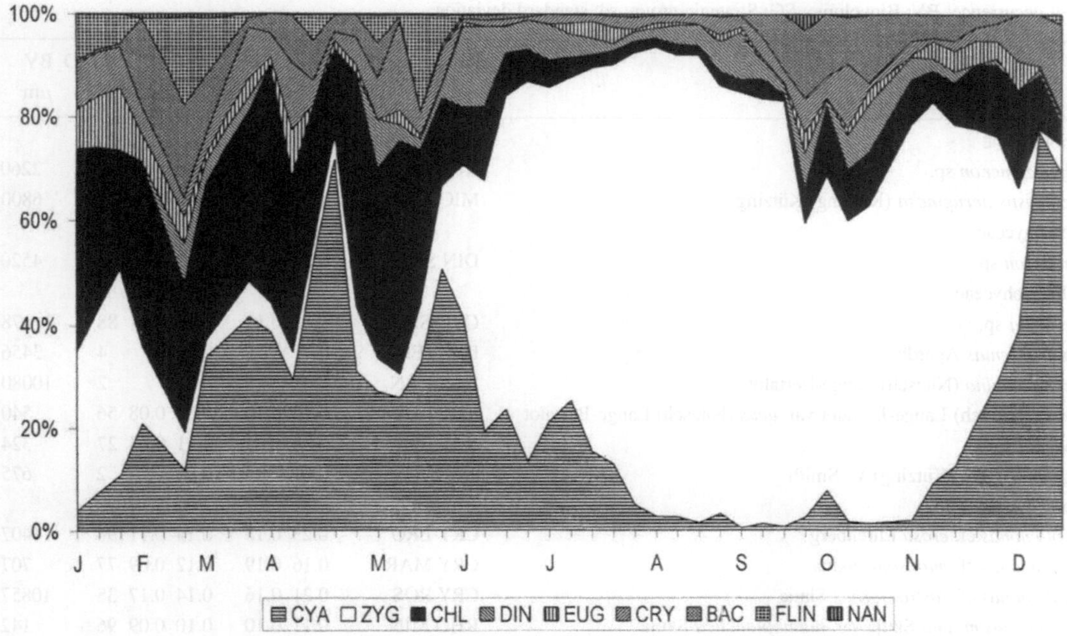

Figure 4. Relative abundance of taxonomic groups in Lake Arancio. CYA: Cyanophyceae; ZYG: Zygnema-tophyceae; CHL: Chlorophyceae; DIN: Dinophyceae; EUG: Euglenophyceae; CRY: Cryptophyceae; BAC: Bacillariophyceae; FLIN: Undetermined flagellates; NAN: Nannoplankton.

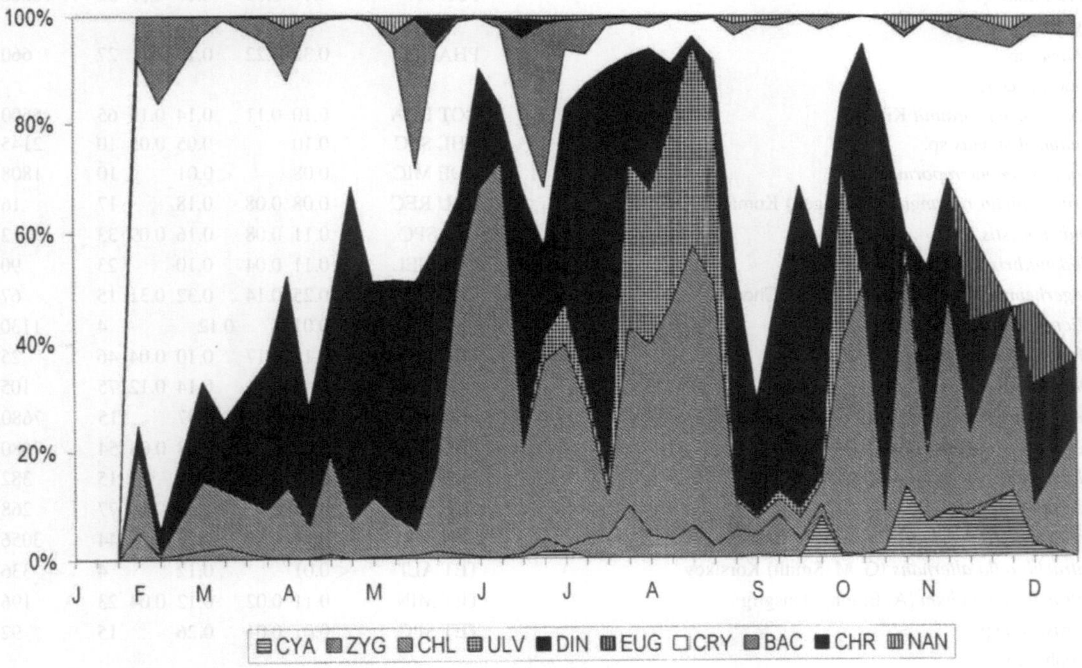

Figure 5. Relative abundance of taxonomic groups in Lake Rosamarina. CYA: Cyanophyceae; ZYG: Zygnematophyceae; CHL: Chlorophyceae; ULV: Ulvophyceae; DIN: Dinophyceae; EUG: Euglenophyceae; CRY: Cryptophyceae; BAC: Bacillariophyceae; CHR: Chrysophyceae; NAN: Nannoplankton.

Table 3. Phytoplankton taxonomic composition of Lake Rosamarina. AGR: Average growth rate; ALR: Average loss rate; TSAO: Time span of occurrence; BV: Biovolume; SG: Strategic group; sd: standard deviation.

	Abbreviations	AGR		ALR		TSAO	BV	SG
		d^{-1}	sd	d^{-1}	sd	%	μm^3	
Cyanophyceae								
Aphanizomenon sp.	APH SPC	0.10	0.07	0.10		19	2260	r*
Microcystis aeruginosa (Kützing) Kützing	MIC AER	0.02	0.04	0.05	0.03	25	6800	r*
Chrysophyceae								
Dinobryon sp.	DIN SPC	0.18	0.13	0.32	0.14	25	4520	r*
Bacillariophyceae								
Cyclotella sp.	CYC SPC	0.18	0.14	0.16	0.11	88	78	K*
Diatoma tenuis Agardh	DIA TEN	0.71		< 0.01		4	3456	r*
Fragilaria ulna (Nitzsch) Lange-Bertalot	FRA ULN	< 0.01		< 0.01		2	10080	r*
F. ulna (Nitzsch) Lange-Bertalot var. *acus* (Nitzsch) Lange-Bertalot	FRA ACU	0.10	0.10	0.07	0.08	56	540	r*
Navicula sp.	NAV SPC	0.05	0.05	0.11	0.03	27	324	r*
Nitzschia sigma (Kützing) W. Smith	NIT SIG	< 0.01		< 0.01		2	675	r*
Cryptophyceae								
Cryptomonas cf. *erosa* Ehrenberg	CRY ERO	0.23	0.18	0.14	0.11	94	1407	K*
Cryptomonas cf. *marssonii* Skuja	CRY MAR	0.16	0.19	0.12	0.09	77	707	K*
Cryptomonas cf. *rostratiformis* Skuja	CRY ROS	0.21	0.16	0.14	0.17	38	10857	r*
Rhodomonas minuta Skuja var. *nannoplanctica* Skuja	RHO MIN	0.11	0.10	0.10	0.09	96	142	K*
Dinophyceae								
Ceratium hirundinella (O. F. Müller) Dujardin	CER HIR	0.02	0.04	< 0.01		21	34712	r*
Gymnodinium sp.	GYM SPC	0.10		< 0.01		6	396	r*
Peridinium sp. 1	PER SP1	0.11	0.10	0.16	0.10	79	12441	K*
Peridinium sp. 2	PER SP2	0.18	0.10	0.11	0.09	25	50737	r*
Peridinium sp. 3	PER SP3	0.14	0.09	0.17	0.09	52	18868	K*
Euglenophyceae								
Phacus sp.	PHA SPC	0.32	0.22	0.10		27	660	r*
Chlorophyceae								
Botryococcus braunii Kützing	BOT BRA	0.10	0.11	0.14	0.15	65	5600	K
Chlamydomonas sp.	CHL SPC	0.10		0.05	0.05	10	2145	r*
Coelastrum microporum Nägeli	COE MIC	0.08		< 0.01		10	1808	r*
Crucigeniella rectangularis (Nägeli) Komárek	CRU REC	0.08	0.08	0.18		17	16	r*
Didymocystis sp.	DID SPC	0.11	0.08	0.16	0.09	33	2	r*
Elakatothrix gelatinosa Wille	ELA GEL	0.11	0.04	0.10		23	90	r*
Lagerheimia cf. *genevensis* (Chodat) Chodat	LAG GEN	0.25	0.14	0.32	0.31	15	67	r
Micractinium pusillum Fresenius	MIC PUS	< 0.01		0.12		4	1130	r*
Monoraphidium sp.	MON SPC	0.17	0.17	0.10	0.04	46	25	r*
Oocystis sp.	OOC SPC	0.15	0.11	0.14	0.12	75	105	K*
Pediastrum boryanum (Turpin) Meneghini	PED BOR	0.10	0.15	0.07		15	7680	r*
Pediastrum duplex Meyen	PED DUP	0.10	0.06	0.08	0.06	54	7680	K
Planktosphaeria gelatinosa G. M. Smith	PLA GEL	0.10	0.11	0.01		15	382	r*
Scenedesmus sp.	SCE SPC	0.11	0.13	0.14	0.11	77	268	K*
Sphaerocystis schröteri Chodat	SPH SCH	0.19	0.12	0.25	0.23	44	3056	r*
Tetrachlorella alternans (G. M. Smith) Korsikov	TET ALT	< 0.01		0.12		4	336	r*
Tetraedron minimum (A. Braun) Hansgirg	TET MIN	0.11	0.02	0.12	0.04	23	196	r*
Tetrastrum sp.	TET SPC	0.01	0.01	0.26		15	92	r*
Ulvophyceae								
Planctonema lauterbornii Schmidle	PLA LAU	0.13	0.12	0.23	0.30	56	500	K*
Zygnematophyceae								
Closterium acutum Brébisson var. *variabile* (Lemmermann) W. Krieger	CLO VAR	0.09	0.10	0.15	0.07	69	452	K*
Cosmarium sp.	COS SPC	0.18	0.12	0.27	0.25	54	170	K*
Staurastrum sp.	STA SPC	0.14	0.01	0.12	0.05	15	1846	K*

(2.1 < H' < 3.0), compared to Lake Arancio. This is probably due to the hydrological stability of the water body, whereas the high percentage of small organisms has to be sought in low herbivorous biomass (Figure 3D), which, in turn, reflects a low concentration of phytoplankton.

Cryptophyceae, mainly represented by *Rhodomonas minuta*, and Dinophyceae, with *Peridinium* spp., were dominant during circulation periods (Figure 5). All these were designated K*-strategists due to the long occurrence. During stratification a sequence of *Pediastrum duplex*, *Peridinium* sp. 2, *Planctonema lauterbornii* and *Botryococcus braunii* occurred. All these belonged to the K or K* groups, with the exception of *Peridinium* sp. 2 which was too short lived. With restored circulation, *Botryococcus braunii* maintained a relatively large part of the biomass (> 40%), co-dominating with *Rhodomonas minuta* and *Peridinium* spp. Cyanophyceae, mainly *Microcystis aeruginosa*, were of minor importance, and present mainly in late summer and autumn, with relative biomass values below 10%. Altogether, minor variability characterised the phytoplankton assemblage of Lake Rosamarina, owing probably to its hydrological stability.

Ordination

PCA of the environmental data of Lake Arancio highlighted volume variations, as well as other correlated parameters (mean and maximum depth and surface area), as most important in explaining variance (eigenvalue 1 = 0.533; eigenvalue 2 = 0.219; V correlation with axis 1: r = - 0.64).

CCA was performed, initially, on the whole environmental and species datasets. The eigenvalues for CCA axis 1 (0.47) and axis 2 (0.37) explained 43.9% of the variance in the species and the species-environment correlations for CCA axis 1 (0.99) and axis 2 (0.96) were high, indicating a strong relationship between the 21 environmental variables and the 61 taxa considered. Forward selection indicated that five of the twentyone variables made independent and significant contributions to the variance in the phytoplankton data. The most significant of these was volume, which explained 15.6% of the total variance followed by extinction coefficient (11.8%), conductivity (10.6%), mixing depth (9.8%) and, finally, temperature (8.7%).

The results of CCA based upon only these five variables are given in Table 4, and are illustrated in Figure 6. The eigenvalues of CCA axis 1 (0.44) and axis 2 (0.32), constrained to the five significant variables, account for 31.6% of the cumulative variance in the species data. The species-environment correlations of CCA axis 1 and 2 are high and the first two axes account for the 72.4% of the variance in the phytoplankton-environment relationship. These values are only slightly lower than those for the CCA using all twentyone environmental variables, indicating that the five variables provide a good representation of the major controlling gradients in the phytoplankton data of Lake Arancio.

The CCA restricted to these five variables shows the importance of volume variation (*r* = 0.76), attenuation coefficient (*r* = - 0.61) and temperature (*r* = - 0.47) in explaining variance. Axis 1 contrasts the low volume/low depth periods characteristic of summer-autumn seasons with a phytoplankton assemblage dominated by mixing-tolerant species (*Mougeotia* sp., *Closterium* spp.), favoured by high frequency light fluctuations. Axis 2 strongly correlates with conductivity (*r* = - 0.90) and contrasts the autumn and winter periods, characterized by water inflow and decrease in conductivity. These periods were dominated by filamentous Cyanophytes (mainly *Planktothrix agardhii*). Mixing depth was of minor importance in explaining phytoplankton distribution, but its relation to axis 1 (*r* = - 0.30) and to axis 4 (*r* = 0.59) indicate that its variation was more important during filling than during emptying.

PCA of the environmental data of Lake Rosamarina revealed nitrate concentrations as the most useful variable in explaining variance (eigenvalue 1 = 0.301; eigenvalue 2 = 0.244; N-NO_3 correlation with axis 1: r = 0.72).

CCA was performed, as for Lake Arancio, on the whole environmental and species datasets. The eigenvalues for CCA axis 1 (0.54) and axis 2 (0.35) explained 35.8% of the variance in the species. Also in this case, the species-environment correlations for CCA axis 1 (0.97) and axis 2 (0.93) were high, indicating a strong relationship between the 21 environmental variables and the 41 phytoplankton taxa considered. Forward selection indicated that seven of the twentyone variables made independent and significant contributions. The most significant of these was nitrate concentration, which explained 12.5% of the total variance, followed by nitrite concentration, (10.9%), soluble reactive silicon, (10.2%), soluble reactive phosphorus, (8.8%), herbivorous zooplankton biomass, (8.4%), extinction coefficient, (7.6%), and mixing depth, (6.3%).

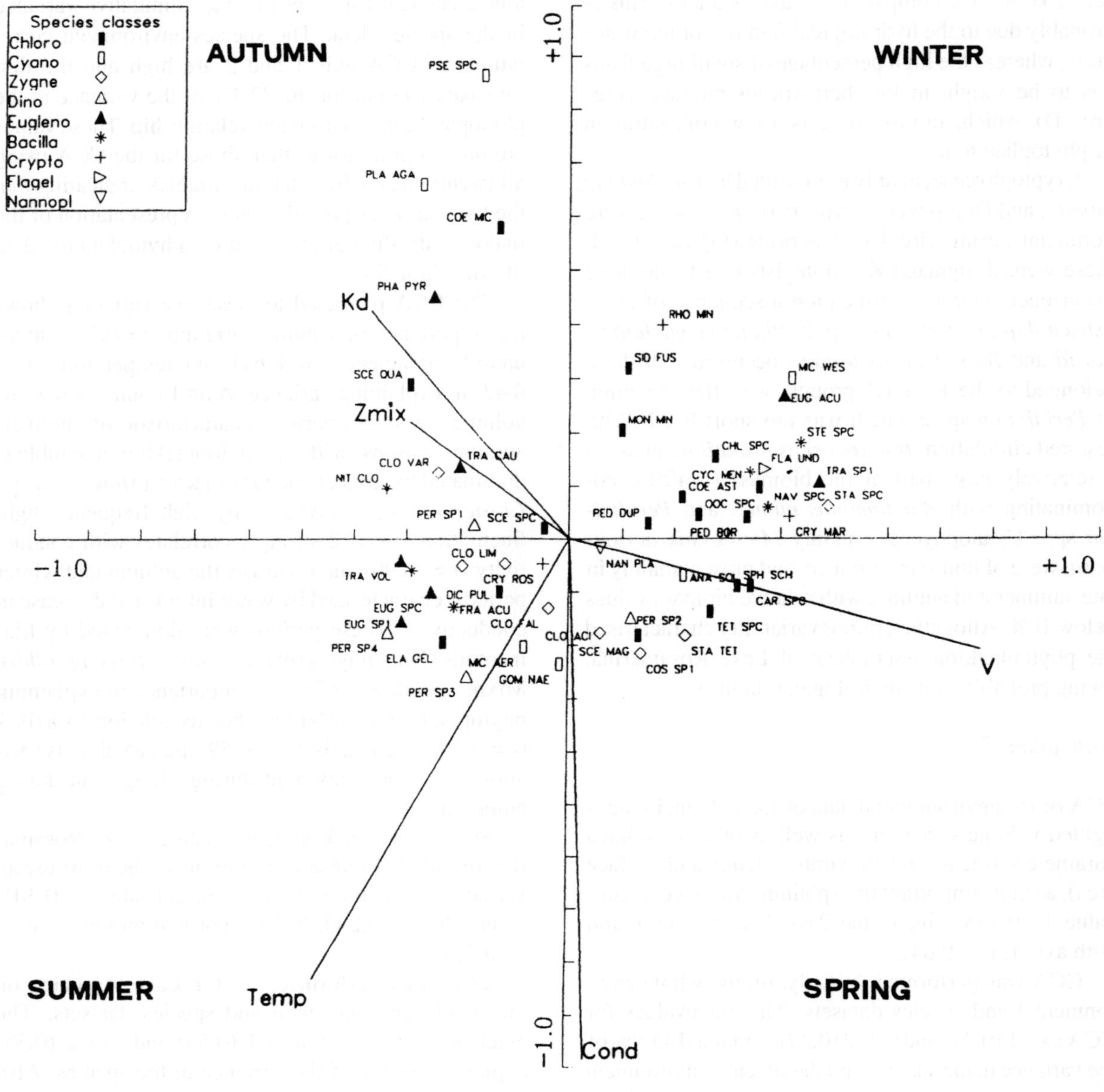

Figure 6. Ordination triplot of CCA performed on the species-environment relationships in Lake Arancio. To facilitate interpretation the sample dates were grouped into the corresponding season of the year and in this way reported in the corresponding semiplane. Chloro: Chlorophyceae; Cyano: Cyanophyceae; Zygne: Zygnematophyceae. Dino: Dinophyceae; Eugleno: Euglenophyceae; Bacilla: Bacillariophyceae; Crypto: Cryptophyceae; Flagel: Undetermined Flagellates; Nannopl: Nannoplankton; Kd: attenuation coefficient; Zmix: mixing depth; V: volume; Temp: surfacial temperature; Cond: conductivity. For species abbreviations see Table 1.

The results of CCA based on these seven variables are given in Table 4 and illustrated in Figure 7. The eigenvalues of CCA axis 1 (0.51) and axis 2 (0.34), constrained to the seven significant variables, account for 27.8% of the cumulative variance in the species data. The species-environment correlations of CCA axis 1 and 2 are high and the first two axes account for the 59.8% of the variance in phytoplankton-environment relationship. These seven variables were considered the principal factors regulating the lake's phytoplankton.

The CCA restricted to the seven variables stresses the importance of N-NO_3 ($r = -0.84$), and Kd ($r = 0.56$). Axis 1 contrasts high winter values of

Table 4. Summary statistics for the first four axes of CCA performed on Lake Arancio (AR) and on Lake Rosamarina (RM).

CCA axes	AR 1	RM 1	AR 2	RM 2	AR 3	RM 3	AR 4	RM 4
Eigenvalues	0.444	0.510	0.323	0.335	0.200	0.216	0.064	0.130
Species-environment correlations	0.973	0.950	0.914	0.913	0.905	0.833	0.658	0.767
Cumulative % variance:								
– of species data	18.3	16.3	31.6	27.6	39.8	33.9	42.5	38.1
– of species-environment relationship	41.9	36.1	72.4	59.8	91.2	75.1	97.3	84.3

Table 5. Correlation coefficients between species and nutrients concentrations and molar ratios in Lake Rosamarina. In brackets the significance levels are reported; *n*: number of observations; n.s.: not significant.

Species	*n*	$N\text{-}NO_3/N\text{-}NH_4$	SRSi/SRP	SRP	$N\text{-}NO_3$	$N\text{-}NH_4$
Planctonema lauterbornii	24	– 0.49 (< 0.02)	n.s.	– 0.60 (< 0.01)	n.s.	n.s.
Cosmarium sp.	24	– 0.54 (< 0.01)	n.s.	– 0.53 (< 0.01)	n.s.	n.s.
Rhodomonas minuta	39	n.s.	n.s.	n.s.	0.49 (< 0.01)	n.s.
Phacus sp	10	n.s.	n.s.	0.63 (<0.05)	n.s.	0.81 (<0.05)
Pediastrum duplex	22	n.s.	n.s.	0.45 (< 0.04)	-0.43 (< 0.05)	0.43 (< 0.05)
Cyclotella sp.	35	n.s.	0.35 (<0.05)	n.s.	n.s.	n.s.

nitrate and a phytoplankton assemblage of *Botryococcus braunii* and *Planctonema lauterbornii*. Axis 2 is strongly correlated with nitrite concentrations ($r=0.79$) and characterizes the transition period between spring and summer when it may represent an additional source of nitrogen for phytoplankton. This period is dominated by *Dinobryon* sp. Large motile dinophytes, represented by *Peridinium* spp. and *Ceratium hirundinella*, are mainly present in winter and early-spring, characterized by a deeper mixing depth and by higher transparency values. Finally, Z_{mix} and K_d are correlated with axis 3 ($r=0.53$ and -0.48 respectively).

Discussion

Even though the two lakes were comparable in volume during the study-year, their hydroclimates differed strongly. Actually, the hydrological forcing caused by the drawdown of Lake Arancio strongly influenced the seasonal pattern of temperature stratification and transformed a potentially monomictic lake into a polimictic, shallow one, in which the loss rates of organisms by sinking are reduced and nutrient recycling is enhanced by the easier transfer of shallow sediments into the water-column (Reynolds, 1996). Moreover, as Harris & Baxter (1996) showed, in deeper reservoirs in which the seasonal pattern of temperature stratification is strong but modulated by interannual changes in the strength of stratification, irregular hydrological events over-ride the seasonal cycle of phytoplankton biomass, yielding irregular patterns of variability. These may be explained as the effect of alternations between the characteristics of deep and shallow lakes within relatively short periods of time. This variability allows the persistence of a large number of rarer, fast-growing species, 'ready to develop' when the condition of the environment become suitable (Padisák, 1991).

In contrast, Lake Rosamarina, which had not yet been used in water supply, showed more regular hydrological and thermal patterns. Hydraulic stability of the

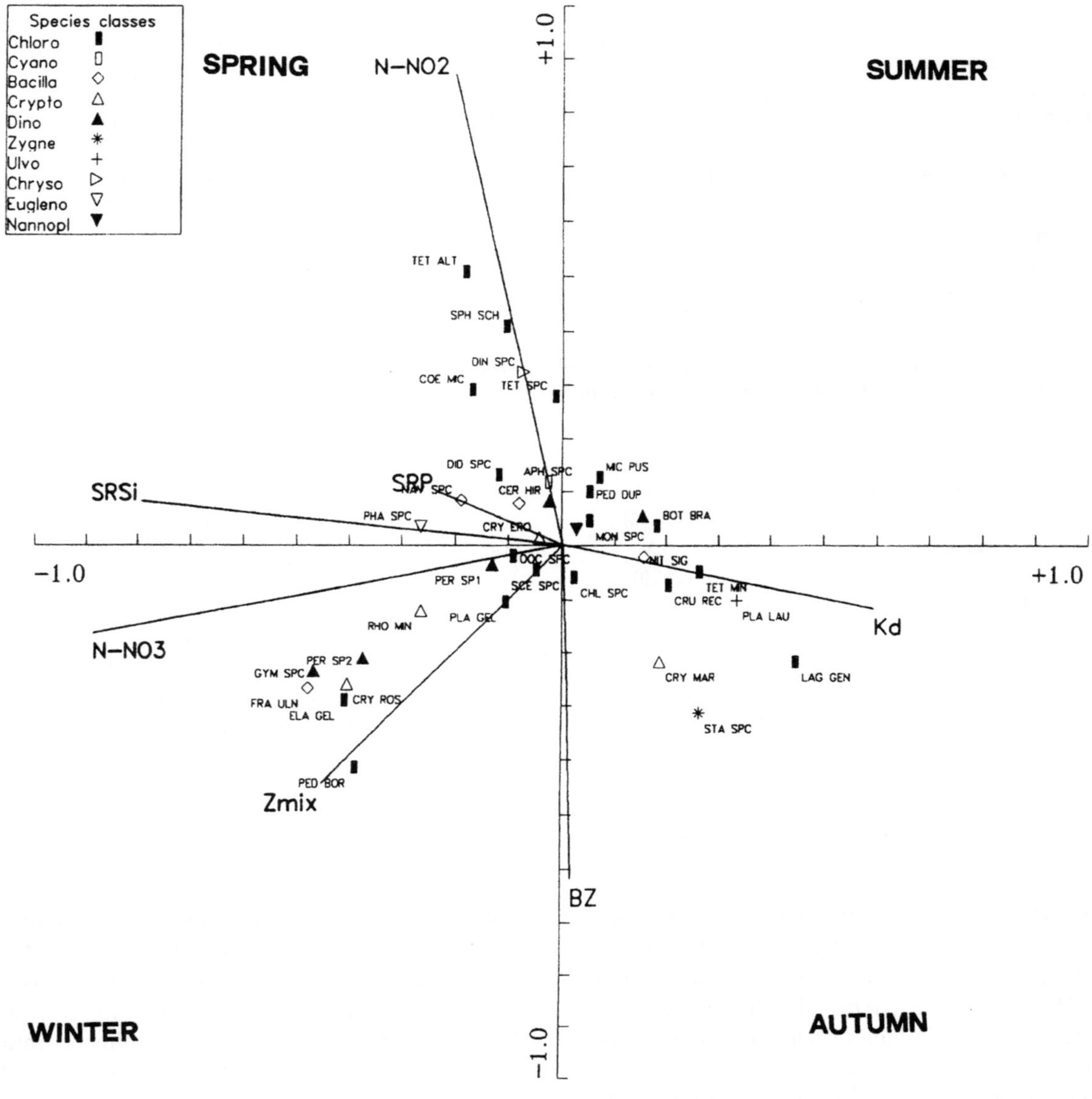

Figure 7. Ordination triplot of CCA performed on the species-environment relationships in Lake Rosamarina. To facilitate interpretation the sample dates were grouped into the corresponding season of the year and in this way reported in the corresponding semiplane. Chloro: Chlorophyceae; Cyano: Cyanophyceae; Zygne: Zygnematophyceae. Dino: Dinophyceae; Eugleno: Euglenophyceae; Bacilla: Bacillariophyceae; Crypto: Cryptophyceae; Ulvo: Ulvophyceae; Chryso: Chrysophyceae; Nannopl: Nannoplankton; Kd: attenuation coefficient; Zmix: mixing depth; BZ: herbivorous zooplankton biomass; SRP: soluble reactive phosphorous; SRSi: soluble reactive silicon. For species abbreviations see Table 2.

lacustrine environment suppresses irregularities in the seasonal cycle of phytoplankton which, consequently, becomes regulated most by nutrient availability. This environment supports a lower number of rare species and a generally lower diversity.

These differences owe to the different trophic states of the two water bodies and are only enhanced by the differences in lake morphology and physical constraints. Neither variability in the phytoplankton community in response to nutrient enrichment, nor lack of evidence for clear trends of community change rates across a trophic gradient are often cited in the literature (Havens et al., 1996). This can happen when the structure is not regulated by nutrient availability but by physical constraints to which the different water bodies are subjected directly (e.g. morphology and use of stored water) and indirectly (e.g. modifications in underwater light climate caused by increased biomass). In contrast, many relationship between the amount of phytoplankton biomass and trophic state have been developed (e.g. OCDE, 1982; Rott, 1984), and have become validated for temperate lakes showing uniform thermic and morphological features. Thus, the quantity of biomass seems to be the only predictable consequence of increased nutrient availability. Its quality, i.e. the structure of the assemblage, and why some species dominate over others is not clear. The answer has been sought in competitive mechanisms, though no clear paradigm has yet emerged.

In Sicilian reservoirs, subject to strong hydrological forcing, phytoplankton biomass was previously observed to rise with trophic state, but no clear alterations in community structure were identified (Naselli Flores & Barone, 1994). Thus, the mechanism that distinguishes among eutrophic, mesotrophic or oligotrophic phytoplankton communities probably lies in the modification of the physical environment experienced by their various communities.

The analysis of results suggests a slight decrease in the trophic state of Lake Arancio, from the hypertrophy observed in 1990 and 1991 (Barone & Naselli Flores, 1994) to the eutrophy of today, probably due to a dilution effect caused by increased inflow. Nevertheless, as previously noted, it is possible to deduce that nutrients do not fulfill any obvious role in the phytoplankton dynamics in this environment. On the contrary, physical constraints, enhanced by human activities, have been again demonstrated to play a key role in regulating phytoplankton dynamics. In particular, the annual variation of volume causes a continuous variation of the mixing zone and consequently of the mixing depth/euphotic depth ratio. Moreover, euphotic depth shows low values probably due to the high phytoplankton biomass ($r = -0.71$; $n = 48$; $p < 0.001$) and the phytoplankton species present need adaptative mechanisms to regulate their buoyancy or to tolerate frequent dark/light cycles. *Anabaena solitaria* f. *smithii* and *Oocystis* sp. showed a negative correlation with z_m/z_{eu} ratio ($r = -0.76$; $n = 28$; $p < 0.001$ and $r = -0.78$; $n = 21$; $p < 0.001$), to which *Mougeotia* sp. was positively correlated ($r = 0.54$; $n = 25$; $p < 0.001$).

Lake Rosamarina appeared to be classically mesotrophic during our study and its phytoplankton assemblage seemed to be primarly dependent on nutrient dynamics, as is usual for mesotrophic lakes (Makulla & Sommer, 1993; Blomqvist et al., 1994). Some of the more significant relationships are listed in Table 5. In particular, *Planctonema lauterbornii*, which was also positively correlated to the attenuation coefficient ($r = 0.57$; $n = 24$; $p < 0.01$), confirms its preference for environments with an low/intermediate trophic state (Jeddes, 1984; Padisák et al., 1990).

Moreover, from the results obtained by CCA analysis, it is possible to deduce that the secondary modification induced by an increase in nutrient availability, e.g. increased phytoplankton biomass, selective grazing, reduction of euphotic depth and increased z_m/z_{eu} ratio, promotes the development of a specific phytoplankton. From a qualitative point of view the process could be summarized as follows:

1. Increased nutrient loading favours the development of photosynthetic biomass.
2. This development is first supported by small fast growing, r-strategist algae.
3. Small species supply a food source to filter-feeding zooplankton and cause an increase in animal biomass.
4. Zooplankton grazing selects for larger algae and contributes to a faster recycling of nutrients in the mixing layer.
5. At the same time, the increase in phytoplankton biomass modifies the underwater light climate, causing a contraction of the euphotic layer.
6. Larger species, adapted to frequent light/dark cycles dominate the assemblage.

However, this qualitative scheme is not independent of the morphological, biological and hydrological characteristics of the water bodies in question. The length and structure of trophic webs is one such factor (Hansson & Carpenter, 1993), because the presence of predatory fishes may cause a cascade effect in favour of the development of phytoplankton biomass. These bottom-up or top-down relationships may therefore alter the scheme proposed.

Acknowledgements

The authors are grateful to Miguel Alvarez Cobelas for his comments which sensibly improved an earlier version of the manuscript. We are also indebted to Colin Reynolds for his valuable suggestions and help with the English.

References

Arauzo, M. & M. Alvarez Cobelas, 1994. Phytoplankton strategies and time scales in a eutrophic reservoir. Hydrobiologia 291: 1–9

Barone, R. & L. Naselli Flores, 1994. Phytoplankton dynamics in a shallow, hypertrophic reservoir (Lake Arancio, Sicily). Hydrobiologia 289: 199–214.

Barone, R., L. Naselli Flores & S. Calvo, 1991. Plankton communities in the artificial lakes of Sicily (Italy). Verh. int. Ver. Limnol. 24: 1409–1414.

Blomqvist, P., A. Pettersson & P. Hyenstrand, 1994. Ammonium-nitrogen: A key regulatory factor causing dominance of non-nitrogen-fixing cyanobacteria in aquatic systems. Arch. Hydrobiol. 132: 141–164.

Hansson, L.-A. & S. R. Carpenter, 1993. Relative importance of nutrient availability and food chain for size and community composition in phytoplankton. Oikos 67: 257–263.

Harper, D., 1992. Eutrophication of Freshwaters. Chapman & Hall, London, 336 pp.

Harris, G. P. & G. Baxter, 1996. Interannual variability in phytoplankton biomass and species composition in a subtropical reservoir. Freshwat. Biol. 35: 545–560.

Havens, K. E., T. L. East, R. H. Meeker, W. P. Davis & A. D. Steinman, 1996. Phytoplankton and periphyton responses to in situ experimental nutrient enrichment in a shallow subtropical lake. J. Plankton Res. 18: 551–566.

Jeddes, M. C., 1984. Limnology of Lake Alexandrina, River Murray, South-Australia, and the effect of nutrients and light on the phytoplankton. Aust. J. Mar. Freshwat. Res. 35: 399–415.

Makulla, A. & U. Sommer, 1993. Relationships between resource ratios and phytoplankton species composition during spring in five north German lakes. Limnol. Oceanogr. 38: 846–856.

OCDE, 1982. Eutrophisation des eaux. Méthodes de surveillance, d'évaluation et de lutte. OCDE, Paris, 164 pp.

Naselli Flores L. & R. Barone, 1994. Relationships between trophic state and plankton community structure in 21 Sicilian dam reservoirs. Hydrobiologia 275: 197–205.

Padisák, J., 1991. Relative frequency, seasonal pattern and possible role of species rare in phytoplankton in a large shallow lake (Lake Balaton, Hungary). Verh. int. Ver. Limnol. 24: 989–992.

Padisák, J., 1992. Spatial and temporal scales in phytoplankton ecology. Abstracta Bot. 16: 15–23.

Padisák, J., G. Toth & M. Rajczy, 1990. Stir-up effect of wind on a more-or-less stratified shallow lake phytoplankton community, Lake Balaton, Hungary. In Biro P. & J. F. Talling (eds), Trophic Relationships in Inland waters. 191: 249–254.

Reynolds, C. S., 1996. Phosphorus recycling in lakes: evidence from large limnetic enclosures for the importance of shallow sediments. Freshwat. Biol. 35: 623–645.

Rott, H. E., 1984. Phytoplankton as biological parameter for the trophic characterization of lakes. Verh. int. Ver. Limnol. 22: 1078–1085.

Sommer, U., 1988. Growth and survival strategies of planktonic diatoms. In Sandgren C. D. (ed), Growth and Reproductive Strategies of Freshwater Phytoplankton. Cambridge University Press, Cambridge: 227–260.

Straškraba, M., J.G: Tundisi & A. Duncan, 1993. State of-the-art of reservoir limnology and water quality management. Dev. Hydrobiol. 77: 213–288.

ter Braak, C.J.F., 1990. Update notes: CANOCO version 3.1, Agricultural Mathematics Group, Wageningen, 35 pp.

ter Braak C.J.F. & P.F.M. Verdonschot, 1995. Canonical correspondence analysis and related multivariate methods in aquatic ecology. aquat. Sci. 57/3: 255–289.

Van den Hoek, C., D. Mann & H. M. Jahns, 1995. Algae: An introduction to phycology. Cambridge University Press, Cambridge, 637 pp.

Watson, S., E. McCauley & J. A. Downing, 1992. Sigmoid Relationship between Phosphorus, Algal Biomass, and Algal Community Structure. Can. J. Fish. aquat. Sci. 49: 2605–2610.

Hydrobiologia **369/370:** 179–197, 1998.
M. Alvarez-Cobelas, C. S. Reynolds, P. Sanchez-Castillo & J. Kristiansen (eds), Phytoplankton and Trophic Gradients.

Phytoplankton succession in the oligotrophic Lake Stechlin (Germany) in 1994 and 1995

Judit Padisák[1], Lothar Krienitz[2], Wolfram Scheffler[2], Rainer Koschel[2], Jørgen Kristiansen[3] & István Grigorszky[4]
[1] *Balaton Limnological Institute of the Hungarian Academy of Science, H-8237 Tihany, Hungary*
[2] *Institut für Gewässerökologie und Binnenfischerei im Forschungsverbund Berlin e. V., Abteilung Limnologie Geschichteter Seen, Alte Fischerhütte 2, D-16775 Neuglobsow, Germany*
[3] *Department of Mycology & Phycology, Botanical Institute, University of Copenhagen, Ø Farimagsgade 2 D, DK-1353 Copenhagen K, Denmark*
[4] *Ecological Department of the Kossuth University, H-4010 Debrecen, Hungary*

Key words: phytoplankton succession, oligotrophic lakes, picophytoplankton, winter growth, diversity

Abstract

Phytoplankton samples were taken weekly from January to December of 1994 (epilimnion) and 1995 (0–25 m, euphotic zone) from the deep, stratified, alkaline, oligotrophic Lake Stechlin, Baltic Lake District, Germany. The purpose of the study was to gain detailed information about phytoplankton changes including those of picophytoplankton, to relate these changes to stratification patterns and nutrient chemistry of the lake and to compare them to results from other lakes of similar character.

During 1994–1995, a total of 142 phytoplankton taxa was encountered in quantitative samples, most being common in deep, oligotrophic lakes. Seasonal development of phytoplankton is characterized by a definite spring peak followed by a moderate summer peak. Autotrophic picophytoplankton made the largest contribution to the annual total biomass. This is probably true for other, temperate, non-acidic, oligotrophic lakes.

Development of the spring assemblage (autotrophic picophytoplankton and centric diatoms) starts in February–March and is terminated by the onset of stratification when diatoms sink to the hypolimnion. Picophytoplankton, especially *Synechococcus* sp., assembles in a narrow deep-layer maximum in the upper hypolimnion.

Our data show that neither deep circulations nor decreased incident radiation under winter ice and snow cover prevent the development of some specially adapted low-light – high-nutrient species. Our views about the length of vegetation period for phytoplankton need to be revised with respect to winter and isothermal conditions.

Introduction

Lake Stechlin is one of the most interesting waters of Northern Germany, owing to its morphometry, appearance, trophic state and history. It is a relatively pristine, deep, oligotrophic, glacial lake, save that it was subject to an effluent of cooling water from a nuclear power plant during the 1960s, 1970s and 1980s (Koschel et al., 1985; Koschel & Casper, 1986; Koschel, 1995).

The phytoplankton of Lake Stechlin has been studied more or less regularly since 1959. Results for 1973 and 1975 were published by the original investigator (Küchler, 1981, 1982), while data for periods 1959–1962, 1963–1964 and 1969–1972 appeared in the syntheses of Casper (1985a, b). The latter contained a checklist of the algal flora of Lake Stechlin. Data for 1992 have been published by Krienitz & Scheffler (1994). Extensive measurement of chemical variables and primary production of phytoplankton started in 1970 (Koschel, 1981, 1985, 1995; Mothes et al., 1985; Casper & Koschel, 1995).

This paper reports on the phytoplankton succession observed in 1994 and 1995. In order to gain the highest possible resolution, samples were taken weekly

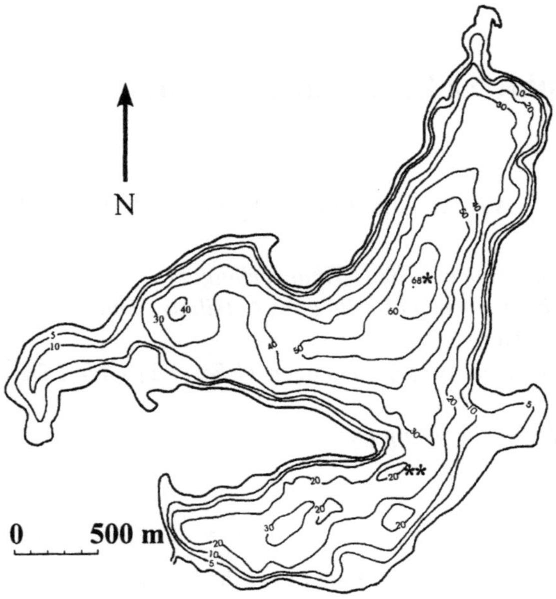

Figure 1. Location of sampling stations. * sampling station 1; ** sampling station 2.

Table 1. Annual averages of some physical and chemical variables in Lake Stechlin in 1994 and 1995. pH and conductivity data refer to the whole water column (0–65m); *: annual average 0–10 m; **: annual average 40–65 m.

	1994	1995
Secchi transparency [m]	8.20	7.50
Oxygen saturation [%]	111*; 83**	104*; 83**
pH	8.2	8.1
Conductivity [μS cm^{-1}]	278	272
PO_4-P [μg l^{-1}]	2*; 8**	2*; 6**
TP [μg l^{-1}]	16*; 19**	17*; 19**
NO_3-N+NO_2-N [μg l^{-1}]	18*; 94**	22*; 71**
NH_4-N [μg l^{-1}]	32*; 78**	25*; 25**
TN [μg l^{-1}]	587*; 584**	444*; 422**
silica [μg l^{-1}]	620*; 1610**	660*; 1390**

from January to December. Picoplankton counts were also performed and the species composition of centric diatoms was determined in separate preparations. In the discussion, special care is taken to evaluate the results in the context of the community composition along trophic gradients.

Description of study site

Lake Stechlin is a deep oligotrophic lake in the Baltic Lake District (53°10′ N, 13°02′ E, 59.9 m a.s.l.). The surface area is 4.25 km^2, the maximal depth is 68 m, the mean depth is 23 m, and the total lake volume is calculated as 97 10^6 m^3. The catchment area amounts of 12.4 km^2 of which 80% is covered by forests. The lake stratifies during April–June, the most stable thermocline is usually observed in late July. Inverse stratification occurs in winters with prolonged ice cover. Therefore, the lake fluctuates between being mono- or dimictic. Ice cover did not develop on Lake Stechlin in the winter of 1994/1995 while in 1995/1996 it lasted from early December to early April. In the winter of 1993/1994 ice cover lasted for a period of two weeks only. Some important chemical and physical features of the lake water are given in Table 1.

At the beginning of the 1960s, a nuclear power plant was built on the shore of the lake. Cooling water was drawn from the the Stechlin-Nehmitz lake system (Casper, 1985a). This operation led to significant thermal pollution and, moreover, reduced the mean retention time of water in the lake basin from 335 days to 202 days (Koschel & Casper, 1986). The operation of the nuclear power plant ceased in 1989.

Sampling and methods

Sampling

(i) Epilimnetic samples were taken from 3 to 5 depths each week between mid January and late December 1994 at a buoy installed at Sampling Station 2, approximately 400 m north of the Institute für Gewässerökologie und Binnenfischerei (water depth: 26 m: Figure 1). These samples were mixed for analyses of phytoplankton composition, biomass and autotrophic picophytoplankton numbers.

(ii) Between January and December 1995 samples from 10 depths within the 0–25 m layer were taken weekly but at sampling Station 1, located at the maximum depth of the lake. Picoplankton numbers, phytoplankton composition and biomass were then established from integrated samples.

Phytoplankton composition and abundance

Phytoplankton species were identified using the most up-to date phycological manuals and literature. A mini-

Table 2. List of algal taxa, their average contribution to annual average biomass and their number of occurrences in Lake Stechlin in 1994 an 1995. A total of 88 samples were analyzed during the two years of which 46 in 1994 and 42 in 1995.

	1994		1995	
	Average biomass %	Number of occurrences	Average biomass %	Number of occurrences
Cyanophyta				
Aphanocapsa elachista W. & G. S. West	0.33	13	–	–
Aphanocapsa grevillei (Hassall) Rabenhorst	0.04	5	0.01	2
Aphanothece sp.	0.01	4	–	–
Chroococcus limneticus Lemmermann	0.15	8		
Chroococcus minutus (Kützing) Nägeli	< 0.01	2	< 0.01	3
Chroococcus cf. *turgidus* (Kützing) Nägeli	1.03	26	0.38	14
Chroococcus sp.	0.56	15	1.53	24
Coelospharium kuetzingianum Nägeli	0.37	21	0.23	7
Eucapsis alpina Clements & Schantz	< 0.01	1	–	–
Merismopedia tenuissima Lemmermann	< 0.01	3	0.01	3
Microcystis wesenbergii Komárek	< 0.01	1	–	
Radiocystis geminata Skuja	0.17	12	0.14	11
other Chroococcales	0.08	5	–	–
Synechococcus sp. (picoalgal size)	15.17	46	33.91	42
colonial cyanobacteria of picoalgal cell sizes	2.83	46	0.75	42
Anabaena lemmermannii P. Richter	0.17	9	1.14	17
Anabaena sp.	0.35	15	–	–
Limnothrix planctonica (Woloszynska) Meffert	< 0.01	2	–	–
Planktolyngbya circumcreta (G. S. West) Anagnostidis & Komárek.	< 0.01	1	–	–
Planktothrix rubescens (DC. ex Gomont) Anagnostidis & Komárek	0.04	6	0.07	8
Oscillatoria chlorina Kützing ex. Gomont	0.13	2	–	–
Pseudanabaena catenata Lauterborn	0.01	3	–	–
Chlorophyta				
Carteria sp.	–	–	0.03	5
Carteria sp. big	–	–	0.01	1
Chlamydomonas sp.	0.04	8	–	–
Chlamydomonas sp. small	0.01	11	0.01	6
Eudorina elegans Ehrenberg	0.08	5	–	–
Pandorina morum (O. F. Müller) Bory	0.19	14	0.08	4
Phacotus lenticularis (Ehrenberg) Stein	0.16	3	0.58	12
other Phytomonadina	0.02	5	0.03	3
Pseudosphaerocystis lacustris (Lemmermann) Nováková	0.20	20	0.50	21
Ankistrodesmus fusiformis Corda	< 0.01	1	–	–
Botryococcus braunii Kützing and *B. terribilis* Komárek & Marvan	0.04	3	0.38	9
Coelastrum spp.	0.03	4	0.20	6
Coenochloris polycocca (Korschikov) Hindák	0.15	21	0.13	19
Crucigenia quadrata Morren	–	–	0.01	8
Dictyosphaerium sp.	–	–	< 0.01	1
Neocystis polycocca Hindák, *Pseudodictyosphaerium jurisii* (Hindák) Hindák and *Choricystis minor* (Skuja) Fott; (green algae of picoalgal size)	1.02	26	2.69	29
Oocystis lacustris Chodat	0.40	41	0.63	37
Oocystis solitaria Wittrock	0.07	17	0.06	3
Pediastrum boryanum (Turpin) Meneghini	< 0.01	1	0.02	6
Pediastrum duplex Meyen	0.01	1	0.02	2

Table 2. Continued.

	1994		1995	
	Average biomass %	Number of occurrences	Average biomass %	Number of occurrences
Planktosphaeria gelatinosa G. M. Smith	0.02	2	–	–
Pseudococcomyxa simplex (Mainx) Fott	< 0.01	1	–	–
Quadrigula pfitzeri (Schröder) G. M. Smith	0.08	15	0.05	6
Scenedesmus armatus Chodat	0.06	7		
Scenedesmus costato-granulatus Skuja	0.33	19	0.60	40
Scenedesmus obliquus (Turpin) Kützing	< 0.01	1		
Scenedesmus quadricauda (Turpin) Brébisson	0.10	5	< 0.01	1
Scenedesmus spp.	–	–	0.01	4
unidentified Selenstraceae spp.	< 0.01	3	< 0.01	2
Tetraedron caudatum (Corda) Hansgirg	–	–	< 0.01	1
Tetraedron minimum (A. Braun) Hansgirg	0.09	14	0.18	18
Willea vilhelmii (Fott) Komárek	0.09	18	0.20	10
other coccal green algae	0.07	9	< 0.01	1
Elakatothrix gelatinosa Wille	0.10	40	0.07	26
Koliella longiseta (Vischer) Hindák	< 0.01	2	–	–
Planktonema lauterbornii Schmidle	0.01	7	< 0.01	1
Ulothrix sp., filamentous	0.04	9	0.21	8
unicellular Ulotrichales sp.	0.09	13	0.14	14
Closterium acutum var. *variabile* (Lemmermann) W. Krieger	0.02	19	0.08	36
Cosmarium bioculatum Brébisson	0.16	13	0.21	13
Cosmarium phaseolus Brébisson ex. Ralfs	0.65	30	0.41	25
Cosmarium sp.	–	–	< 0.01	1
Staurastrum sp.	0.07	4	0.20	12
Staurodesmus cf. *incus* (Brébisson) Teiling	0.02	3	0.01	1
Spirogyra sp.	0.01	2	< 0.01	0
Euglenophyta				
Euglena sp.	–	–	0.02	1
Heterokontophyta				
Xanthophyceae				
Monallanthus sp.	–	–	0.03	4
Trachydiscus sexangulatus Ettl	0.37	15	0.48	27
cf. *Tetraedriella jovetii* (Bourrelly) Bourrelly	0.96	27	0.34	32
Chrysophyceae				
Bitrichia chodatii (Reverdin) Chodat	0.17	14	0.03	7
Chrysamoeba sp.	0.09	6	< 0.01	1
Chrysochromulina parva Lackey	1.24	44	0.09	18
Chrysococcus sp.	1.22	15	0.16	31
Chrysolykos planktonicus Mack	0.11	6	0.02	5
Dinobryon bavaricum Imhof	0.56	16	0.09	12
Dinobryon acuminatum Ruttner	0.21	17	< 0.01	1
Dinobryon divergens Imhof	0.03	9	–	–
Dinobryon pediforme (Lemmermann) Steinecke	< 0.01	1	0.23	11
Dinobryon sociale Ehrenberg	5.10	25	4.87	13
Dinobryon cysts	0.16	8	–	–
Kephyrion spp.	0.09	9	0.05	22
Mallomonas spp.	0.82	35	0.27	13
Ochromonas sp.	0.61	19	0.01	1
Paraphysomonas vestita (Stokes) de Saedeler	0.04	4	–	–

Table 2. Continued.

	1994		1995	
	Average biomass %	Number of occurrences	Average biomass %	Number of occurrences
Pseudokephyrion ellipsoideum (Pascher) Schmid	0.12	17	< 0.01	2
Spiniferomonas sp.	0.02	8	–	–
Stichogloea doerderleinii (Schmidle) Wille	0.29	21	0.49	25
Synura sp.	0.03	5	–	–
small chrysoflagellates	< 0.01	1	0.26	28
Chrysophyceae cysts	0.12	6	0.12	8
Bacillariophyceae				
Cyclotella sp. and *C. pseudocomensis* Scheffler	6.21	46	12.07	42
C. krammeri Håkansson	1.36	23	3.21	39
C. ocellata Pantocsek	< 0.01	1	< 0.01	2
C. radiosa (Grunow) Lemmermann	5.00	37	4.87	42
C. tripartita Håkansson	12.42	32	2.70	42
Stephanodiscus alpinus Hustedt	1.14	11	0.8	19
St. minutulus (Kützing) Cleve & Möller	0.23	13	0.07	22
St. neoastraea Håkansson & Hickel – group	0.24	6	1.46	16
Stephanocostis chantaicus Genkal & Kuzmina	0.07	17	0.12	36
Stephanodiscus hantzschii Grunow	0.01	1	< 0.01	1
Aulacoseira granulata (Ehrenberg) Simonsen	–	–	0.02	2
Aulacoseira islandica (O. Müller) Simonsen	–	–	0.01	2
Achnanthes cf. *exigua* Grunow	< 0.01	2	< 0.01	1
A. minutissima Kützing	0.32	10	0.04	7
Achnanthes sp.	–	–	< 0.01	1
Amphora pediculus (Kützing) Grunow	0.03	6	0.01	1
Asterionella formosa Hassall	3.82	37	0.29	15
Cocconeis placentula Ehrenberg	0.01	2	–	–
Cymbella spp.	0.02	5	0.01	1
Diatoma tenuis Agardh	0.17	13	0.04	2
D. vulgaris Bory	0.02	1	–	–
Epithemia sp.	0.01	2	0.04	2
Fragilaria capucina Desmazières	0.06	8	0.17	4
F. construens (Ehrenberg) Grunow	0.11	9	–	–
F. crotonensis Kitton	0.01	2	0.21	11
Fragilaria sp.	0.01	4	0.14	12
Gomphonema sp.	0.02	6	–	–
Navicula cryptocephala Kützing	0.01	5	–	–
N. pupula (Kützing)	0.01	1	–	–
N. cf. *radiosa* Kützing	0.01	2	–	–
Navicula sp.	0.04	5	–	–
Nitzschia spp.	0.10	13	0.03	6
Rhopalodia gibba (Ehrenberg) O. Müller	–	–	0.04	1
Fragilaria ulna (Nitzsch) Lange-Bertalot *acus*-Sippen	0.53	30	0.16	16
Fragilaria ulna (Nitzsch) Lange-Bertalot *angustissima* Sippen	0.39	16	0.03	3
Fragilaria ulna (Nitzsch) Lange-Bertalot *ulna*-Sippen	0.48	14	0.03	3
Fragilaria sp.	0.04	10	0.02	5
Tabellaria fenestrata (Lyngbye) Kützing	0.04	1	–	–
Tabellaria flocculosa (Roth) Kützing	0.03	1	0.02	2

Table 2. Continued.

	1994		1995	
	Average biomass %	Number of occurrences	Average biomass %	Number of occurrences
Cryptophyta				
Chroomonas nordstedtii Hansgirg	7.27	40	8.46	39
Cryptomonas erosa Ehrenberg and *C. ovata* Ehrenberg	1.70	42	1.02	20
Katablepharis ovalis Skuja	0.46	28	0.33	24
Rhodomonas lacustris Pascher & Ruttner	10.32	46	5.72	42
Dinophyta				
Ceratium hirundinella (O. F. Müller) Schrank	3.26	26	1.69	17
Gymnodinium helveticum Penard	1.37	29	0.88	25
Gymnodinium spp. + gymnoid forms	0.48	28	0.60	18
Peridiniopsis penardiforme (Lindemann) Bourrelly	2.08	16	0.57	2
Peridinium inconspicuum Lemmermann	0.75	10	0.17	2
Peridinium umbonatum Stein	1.05	12	0.19	11
Peridinium sp.			0.23	5

	Biomass $\mu g\ l^{-1}$	Biomass $\mu g\ l^{-1}$
Coccal blue-green	96.60	120.39
Heterocystinae	2.40	3.71
Homocystinae	0.91	0.24
Phytomonadina	2.35	2.46
Tetrasporales	0.94	1.63
Chlorococcales	11.98	16.89
Desmidiales	4.27	2.99
Ulotrichales	1.10	1.36
other filamentous	0.04	–
Euglenophyta	–	0.05
Chrysophyceae	51.35	21.79
Xanthophyceae	6.20	2.78
Cryptophyceae	91.90	50.59
Dinoflagellata	41.82	14.14
Centrales	124.13	82.57
pennatae diatoms	29.29	4.15
Total biomass	465.29	325.73
Maximum/average biomass	3.15	2.82
Diversity, species number, evenness		
Species level		
H"	4.67	3.89
Species number	129	109
H"max	7.01	6.77
J	67%	57%
Genus level		
H"	4.07	3.43
Number of genera	90	78
H"max	6.49	6.29
J	63%	55%
Group level		
H"	2.77	2.54
Number of categories	15	15
H"max	3.91	3.91
J	71%	65%

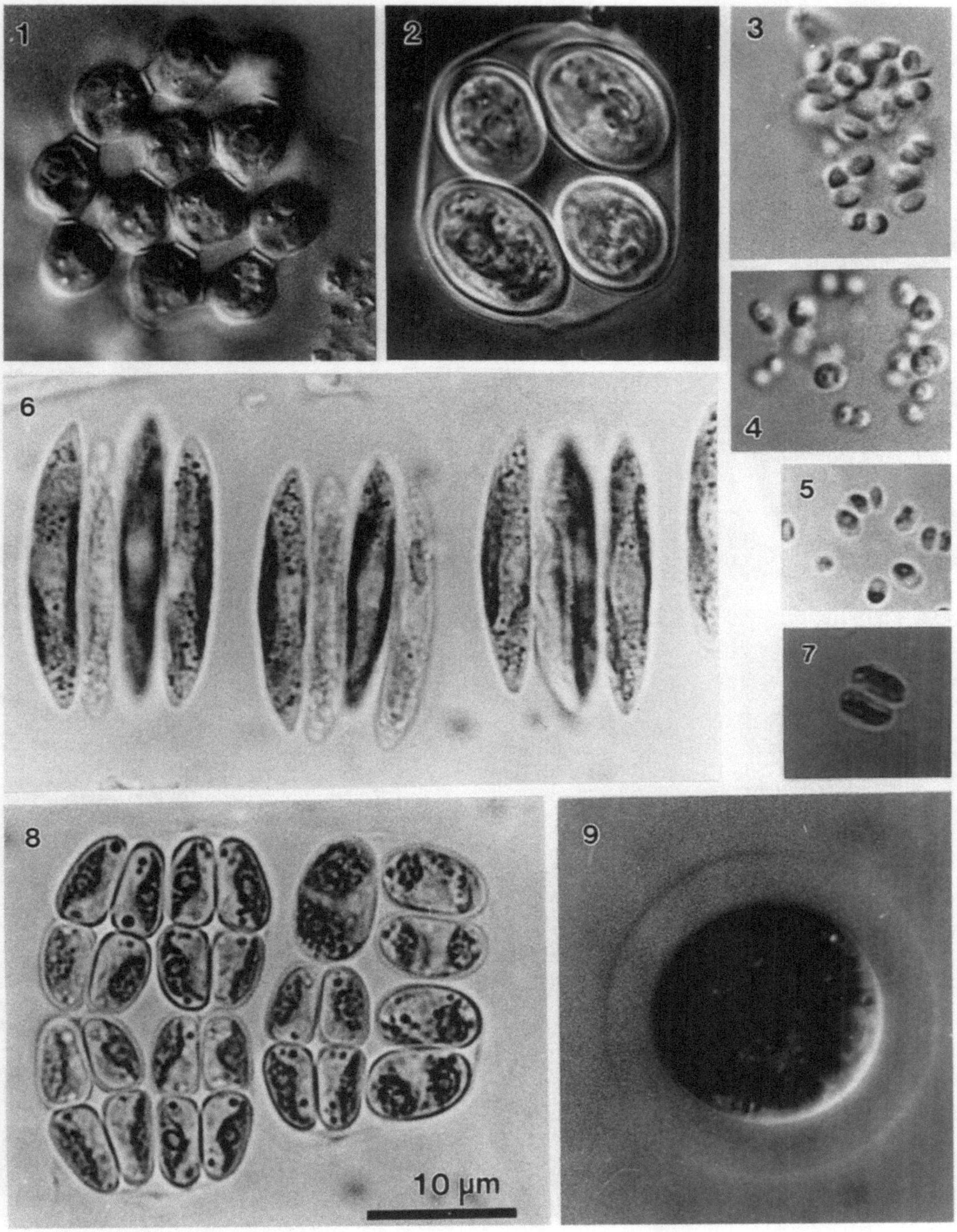

Figure 2. Coccal green algae from Lake Stechlin. 1: *Coelastrum indicum*; 2: *Oocystis lacustris*; 3: *Neocystis diplococca*; 4: *Pseudodictyosphaerium jurisii*; 5: *Choricystis minor*; 6: *Quadrigula pfitzerii*; 7: *Scenedesmus costato-granulatus*; 8: *Willea vilhelmii*; 9: *Planktosphaeria gelatinosa*.

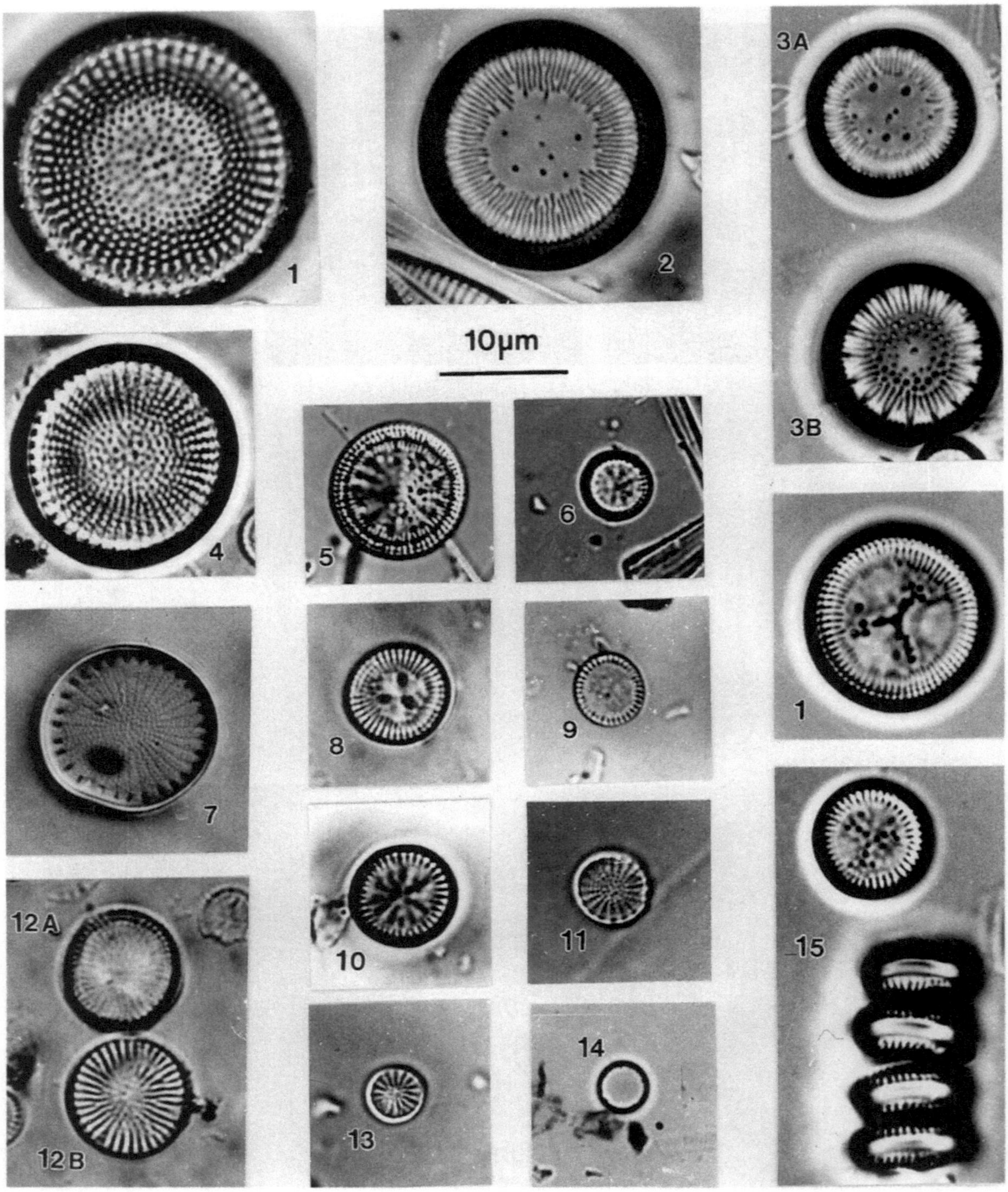

Figure 3. Centric diatoms from Lake Stechlin. 1: *Stephanodiscus* sp. from the *S. neoastraea*-group; 2, 3A: *Cyclotella krammeri*; 3B: C. *radiosa*; 4: *Stephanodiscus alpinus*; 5, 6: *C. pseudocomensis*; 7: *S. hantzschii* morphotype II; 8: *Cyclotella ocellata*; 9: *C.* cf. *distinguenda* var. *unipunctata*; 10. *C.* cf. *comensis*; 11: *Stephanodiscus minutulus*; 12–13: *Stephanocostis chantaicus*, A: initial cell, B: post-initial cell; 14. *Cyclotella* sp., 15–16: *C. tripartita*.

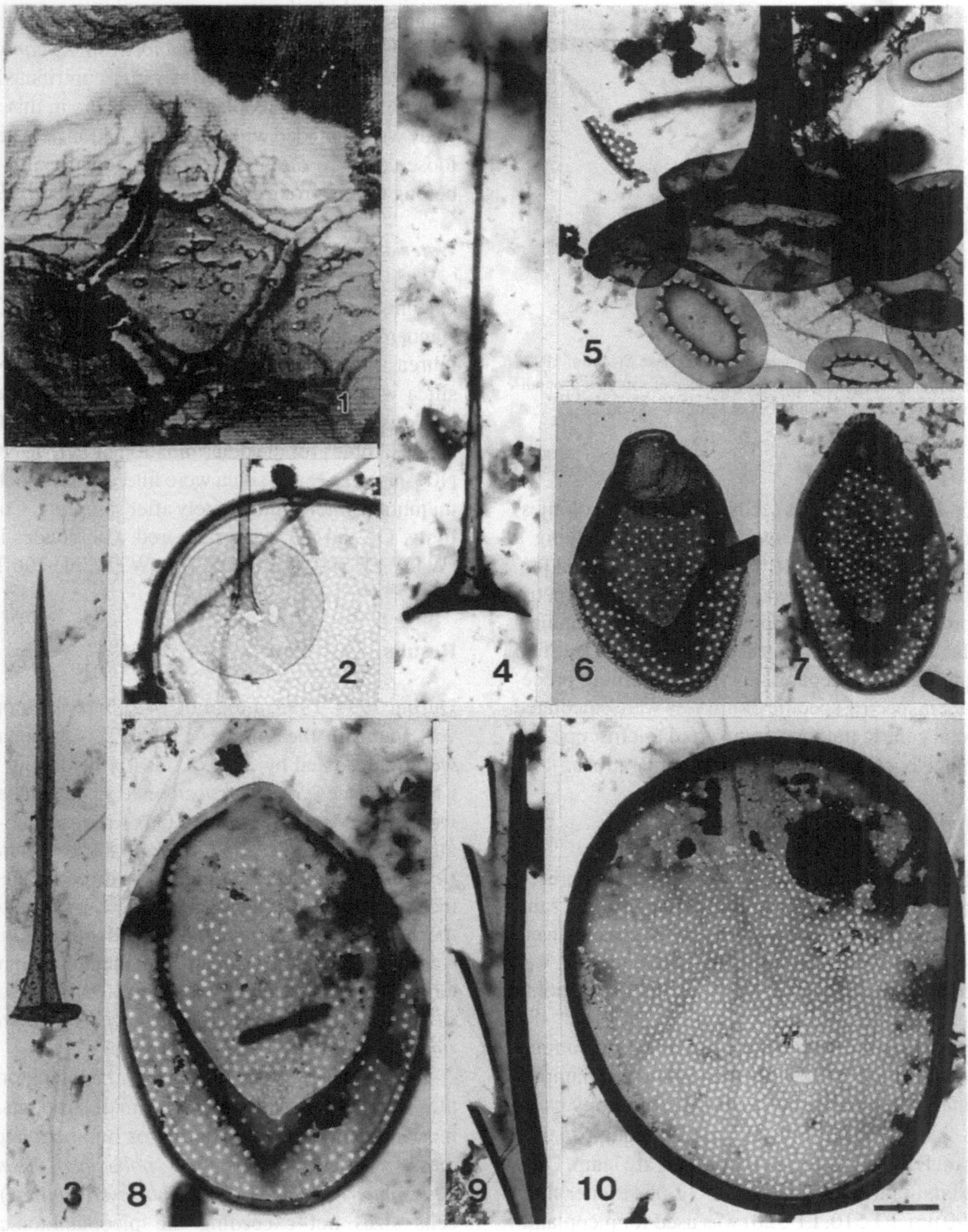

Figure 4. Dinoflagellates and silica-scaled chrysophytes from Lake Stechlin. 1: The plate arrangement (4' 2a 7") of *Peridinium umbonatum*, pore and the two intercalary plates; 2: *Paraphysomonas vestita*, basal part of the spine; 3: *Spiniferomonas trioralis*, spine scale; 4: *S. bourrellyi*, spine scale; 5: *Chrysosphaerella coronacircumspina*, plate scales and basal part of the spine scale; 6: *Mallomonas alpina*, domebearing scale: 7:*M. alpina* domeless scale; 8: *M. elongata*, domeless scale; 9: *M. caudata*, bristle tip; 10: *M. caudata*, scale. Panels 2–10: TEM, 10,000x, scale bar 1 μm.

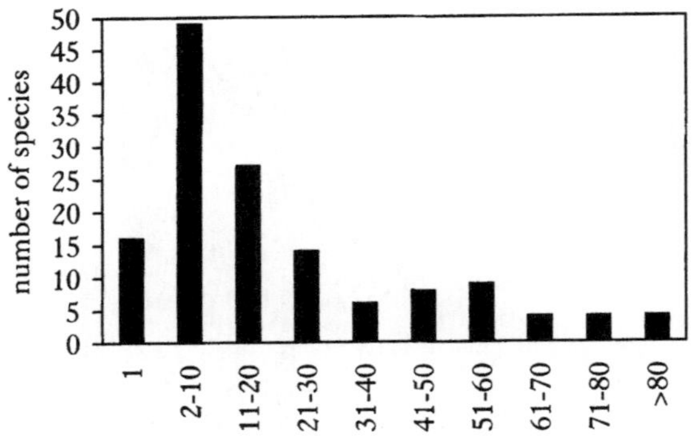

Figure 5. Distribution of algal species found (Lake Stechlin, 1994–1995) in different frequency classes. Number of samples are indicated on the x axis, the total number of samples was 88.

mum of 400 settling units (cells, filaments or colonies) were counted in each Lugol-fixed sample, giving a counting accuracy of ± 10% for total phytoplankton. Phytoplankton biomass was estimated by geometrical approximations using Hamilton's (1990) computerized plankton counter. Centric diatoms were identified in parallel diatom preparations (gentle burning procedure), their species-specific biomasses were calculated from total centric numbers in inverted microscope and percentage contributions in diatom preparations.

Picoplankton numbers

Autotrophic picoplankton (APP) was counted preferably immediately after sampling in unpreserved samples. If it was not possible, unpreserved samples were deep-frozen within 1 hr after sampling and APP was counted in melted samples no later than one month after sampling.

APP cells were concentrated on black membrane then embedded in ∼30% glycerine solution. Samples were analyzed with epifluorescence microscopy using a Zeiss-Axiovert 35 inverted microscope equipped with an Osram HBO 50W/AC mercury short arc lamp. Only blue excitation (Zeiss filter set 45 17 66, exciter filter 450–490 nm, FT 510, LP 520) was used for picoplankton counting because in previous trials, no significant difference had been found between counting with blue and green excitation and green eukaryotic picoalgal cells that occassionally occur in Lake Stechlin were easier to distinguish with blue excitation. According to the original definition picoplankton comprises cells having diameters from 0.2 to 2 μm (Sieburth et al., 1978). Picophytoplankton was considered here as each organisms, unicellular or colonial, having cell sizes of < 2 μm. Colonial forms (*Aphanocapsa, Aphanothece, Pseudodictyosphaerium, Neocystis*) contributed only about 4% to total picoplankton numbers in this study. They were included with the picoplankton here because loose cells are often present and, functionally, these belong to this group of unicells (Søndergaard, 1991).

Water chemistry

Chemical data were analyzed according to the OECD standards and standards of Deutsche Einheitsverfahren. PO_4-P, TP, NO_3-N, NO_2-N, NH_4-N, TN, and silica were measured with the Perstop flow injection analysis system TECATOR FIA STAR 5010/5030. The water samples for the determination of PO_4-P, NO_3-N, NO_2-N, NH_4-N and silica were filtered through 0.6 μm membrane filters immediately after sampling. Conductivity, O_2 and pH were measured with sondes in situ (WTW LF 196, WTW OX 196, WTW, pH 196 T).

Results

During 1994–1995, a total of 142 phytoplankton taxa was found in the samples (Table 2). Cyanobacteria were represented by 22 taxa, mostly belonging to the Chroococcales. *Synechococcus*-like picocyanobacteria were common in all seasons. Other coccal blue-green algae: *Chroococcus* spp., *Radiocystis geminata* and *Coelosphaerium kuetzingianum* occurred in late summer. Heterocytic cyanobacteria were represented by *Anabaena lemmermannii* and an unidentified (akinetes were not seen) *Anabaena* sp. Filamentous cyanobacteria, other than heterocytic, were found only occasionally. Notable is the reoccurrence of *Planktothrix rubescens*.

Green algae were regularly found in plankton samples, though, usually only in small amounts. The most numerous, both in terms of numbers and biomass, were those (*Neocystis polycocca, Pseudodictyosphaerium jurisii* and *Choricystis minor*) with cells sizes not exceeding the picoplanktonic limit (< 2 μm). As they were counted with epifluorescence method, species-specific abundance data are not known. Very common green algal species are *Scenedesmus costato-granulatus* and *Cosmarium phaseolus*. *Phacotus lenticularis* occurred in larger quantities only in 1995. Some characteristic forms are documented on Figure 2. We found species of the Sele-

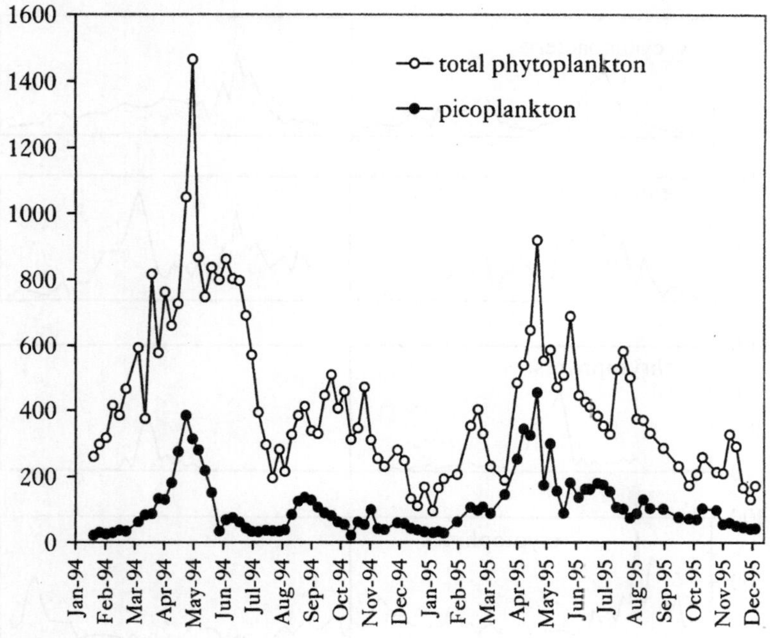

Figure 6. Biomass (μg fw l^{-1}) of phytoplankton and autotrophic picophytoplankton in Lake Stechlin in 1994–1995. Data for 1994: epilimnion, for 1995: euphotic zone.

nastraceae and Coelastraceae which are common in eutrophic waters, however in Lake Stechlin they develop only very low cell numbers. *Coelastrum indicum* Turner (Figure 2/1), well known from the tropics, is also common in oligotrophic, eutrophic and polytrophic waters of northern Germany.

Euglenophyta are insignificant in Lake Stechlin. Three species of Xanthophyceae were recorded in the plankton, *Trachydiscus sexangulatus* and *Tetraedriella jovetii* occurs regularly. The former can be easily confused with small centric diatoms if magnification is smaller than 1000×.

Among centric diatoms (Figure 3), *Cyclotella tripartita, C. pseudocomensis* and an unidentified small *Cyclotella* sp. were the most abundant species present. The first two taxa have been the subject of separate papers (Scheffler, 1994; Scheffler & Padisák, 1997a, b). *C. pseudocomensis* and *Cyclotella sp.* are treated together, because small individuals of *C. pseudocomensis* are impossible to separate from the other small sized *Cyclotella* sp. in LM. Beside the species listed in Table 2, *Ellerbeckia arenaria* (Moore) Crawford was found regularly in net samples. However, this species was so rare during the study period that no specimen was found in the quantitive material.

During these studies, 19 taxa of chrysophytes were identified under the inverted microscope (Table 2). Of the silica-scaled species, only *Paraphysomonas vestita* (Stokes) de Saedeleer could be identified to species. Previous LM records of silica-scaled species include *Mallomonas acaroides* Perty ('var. *striatula* Asmund'), *M. akrokomos* Ruttner in Pascher, *M. crassisquama* (Asmund) Fott, *M. caudata* Iwanov em. Krieger ('*M. fastigiata* Zacharias'), *M. heterospina* Lund and *M. tonsurata* Teiling em. Krieger (Casper, 1985 b). EM studies verified some of these identifications and revealed some further species: *Paraphysomonas vestita*, *Spiniferomonas trioralis* Takahashi, *S. bourrellyi* Takahashi, *Chrysosphaerella coronacircumspina* Wujek & Kristiansen, *Mallomonas alpina* Ruttner in Pascher, *M. elongata* Reverdin, and *M. caudata* (Figure 4). *Dinobryon sociale* occurred in relatively high amounts, quantities of other species seem to be quite unpredictable. For example *Chrysochromulina parva*, *Mallomonas* spp. and a *Chrysococcus* sp. were abundant only in 1994.

The four species of cryptophytes occurred regularly and in large quantities in both years.

Peridiniopsis penardii occurred mostly in spring. *Peridinium umbonatum* (Figure 4/1) and *P. inconspicuum* together with *Ceratium hirundinella* built up the

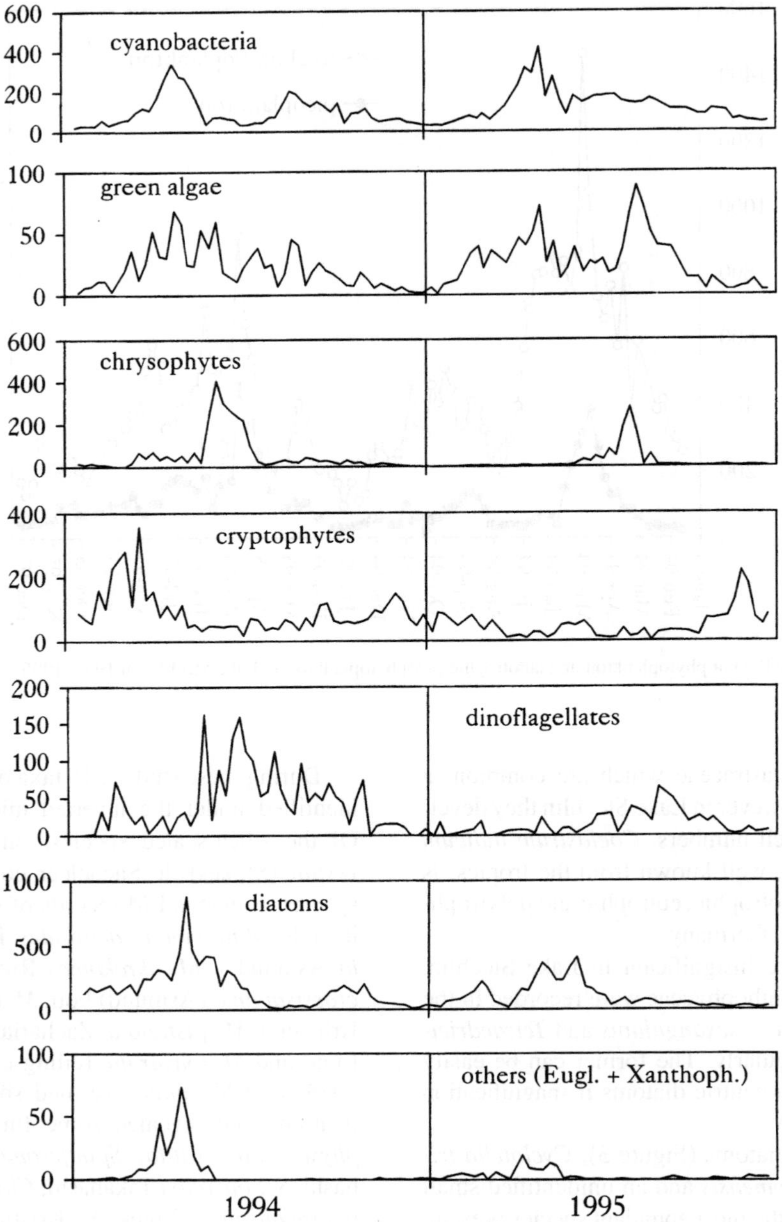

Figure 7. Biomass (μg fw l^{-1}) of different algal groups in Lake Stechlin in 1994–1995. Cyanobacteria biomass includes picophytoplankton.

higher summer dinoflagellate biomass. *Gymnodinium helveticum* occurred regularly in small amounts throughout the study period.

Of the 142 taxa that we found in quantitative analyses of a total of 88 samples, only 16 taxa were found only once. A large number of species occurred in most of the samples. Distribution in occurrence classes is shown in Figure 5.

Phytoplankton biomass in Lake Stechlin produces a definite spring maximum and a moderate second maximum in late summer (Figure 6, 7). In the non-stratified period (September–March), cryptophytes (*Rhodomonas, Chroomonas*) and centric diatoms dominate together with autotrophic picoplankton. The spring maximum begins to develop as early as March when APP and one or other of the centric diatoms

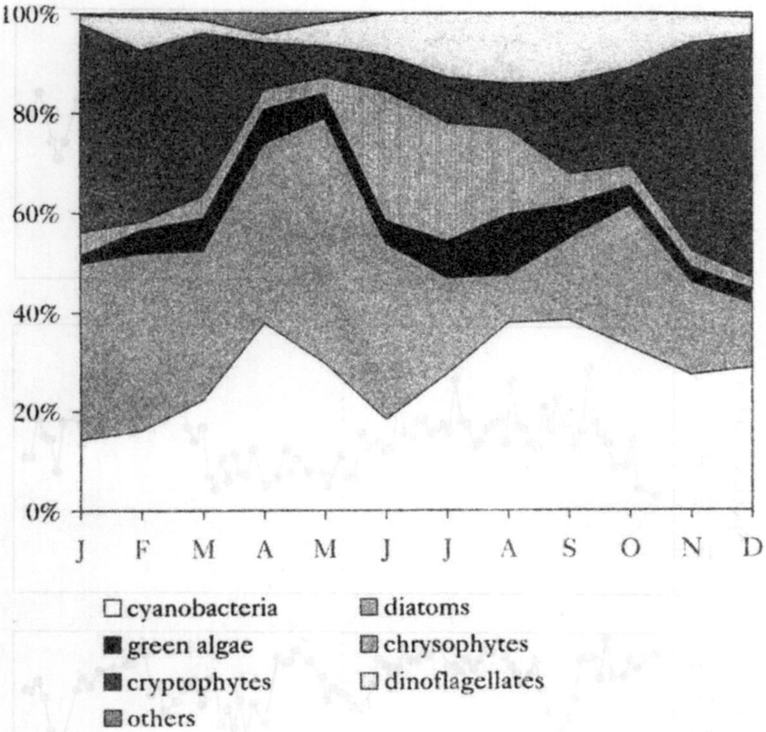

Figure 8. Percentage biomass contribution of different algal groups in Lake Stechlin in 1994–1995. Cyanobacteria contribution includes picophytoplankton. Data are monthly averages.

(*Cyclotella tripartita* in 1994 and *C. pseudocomensis* + an unidentified small *Cyclotella* sp. in 1995) increases slowly. Net increase rate was 0.126 d^{-1} (19 April–27 April) for *Cyclotella tripartita* in 1994 which corresponds to one doubling every 5.5 days. Parallel estimates for 1995 are difficult because of taxonomic problems and the erratic nature of the spring diatom development. The highest net growth rates of the unicellular blue-green APP populations were calculated as 0.13 d^{-1} and 0.12 d^{-1} (3–11 March and 12–19 April, 1994; upper 10 m) and 0.13 d^{-1} and 0.10 d^{-1} (1–7 February, 14–21 March, 1995; upper 25 m). The spring maximum peaks early in May, after which, the biomass rapidly decreases. Diatoms sink to the hypolimnion but autotrophic picoplankton forms a persistent deep layer maximum in the upper hypolimnion (data are shown in Scheffler & Padisák, 1997a, b, and Padisák et al., 1997). A clear-water phase lasts usually until late July. At the beginning of this period chrysophytes, especially *Dinobryon* spp. dominate. Phytoplankton biomass reaches its lowest values in the end of this period, species of the late summer phase are characteristic. In the period from August to October, a diverse phytoplankton assemblage develops with most groups roughly equally represented (Figure 8).

Phytoplankton diversity shows two annual minima: the first occurs after the spring biomass maximum, the second can be observed after the complete breakdown of stratification (Figure 9). Over the study period, diversity was found to correlate closely with evenness ($r = 0.93$, $n = 88$, $P < 0.1\%$). Its correlation with species number was also significant ($r = 0.53$, $n = 88$, $P < 0.1\%$).

Concentrations of PO_4-P in the upper 10-m water layer was below the level ($< 5\ \mu g\ l^{-1}$) that is considered analytically detectable (Figure 10). That was also the case in the water layer deeper than 40 m except the late autumn and winter months (November–January) when maxima of 16.5 $\mu g\ l^{-1}$ (1994) and 10 $\mu g\ l^{-1}$ (1995) were measured. On an annual basis, PO_4-P concentrations in the deeper layer exceeded those in upper layers by a factor of 2.3 in both years.

Concentrations of inorganic N-forms (= NO_2-N + NO_3-N + NH_4-N) in the upper 10-m water layer were below 100 μg N l^{-1} except a short period in March–April 1994. In water layers deeper than 40 m, higher

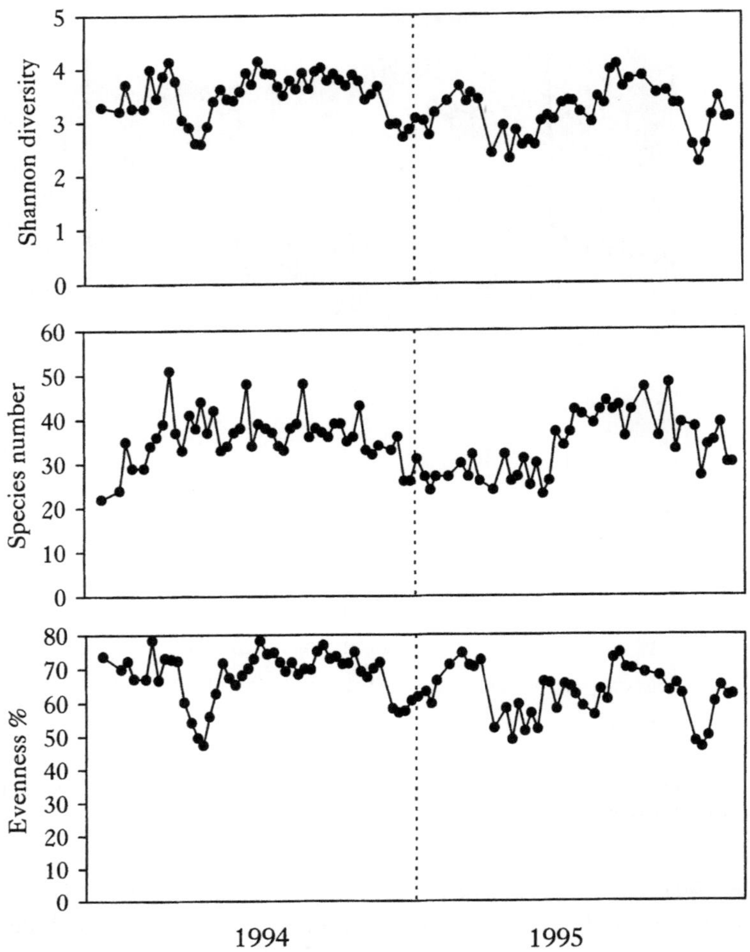

Figure 9. Shannon diversity (base of logarithm: 2), species number per sample and evenness% of phytoplankton in Lake Stechlin in 1994–1995.

inorganic nitrogen values were found during (1994) or by the end (1995) of the stratified period. Inorganic nitrogen concentrations were higher by factors of 2.7 (1994) and 2.0 (1995) in the lower than in the upper layers at annual levels. NO_2-N + NO_3-N exhibited opposite trends in the epi- and hypolimnion.

The average of inorganic N/P ratio was 18.5 (upper layer) and 21.7 (deep layer) in 1994 while 21.7 (upper layer) and 19.2 (deep layer) in 1995. Dissolved silica concentrations exhibited opposite trends in the upper and deeper layers, however, they rarely dropped below 0.5 mg l^{-1}.

In 1994 the thermocline began to develop in middle April and it stabilized within two weeks (Figure 11a). Over the long-lasting calm weather from mid-July to mid-August, a secondary thermocline developed within the epilimnion that eroded quickly following a sharp temperature drop in late August. In 1995 the thermocline began to develop in late April. As a consequence of a cold and wind in mid-May it was quickly eroded and began to redevelop only late in May. The autumnal breakdown of stratification began in August; full circulation occurred at about the end of October in both years.

Discussion

According to previous studies on phytoplankton succession in Lake Stechlin (Busse, unpubl. data, Küchler, 1981, 1982, Casper, 1985 b), the annual development of the phytoplankton was altered by the operation of

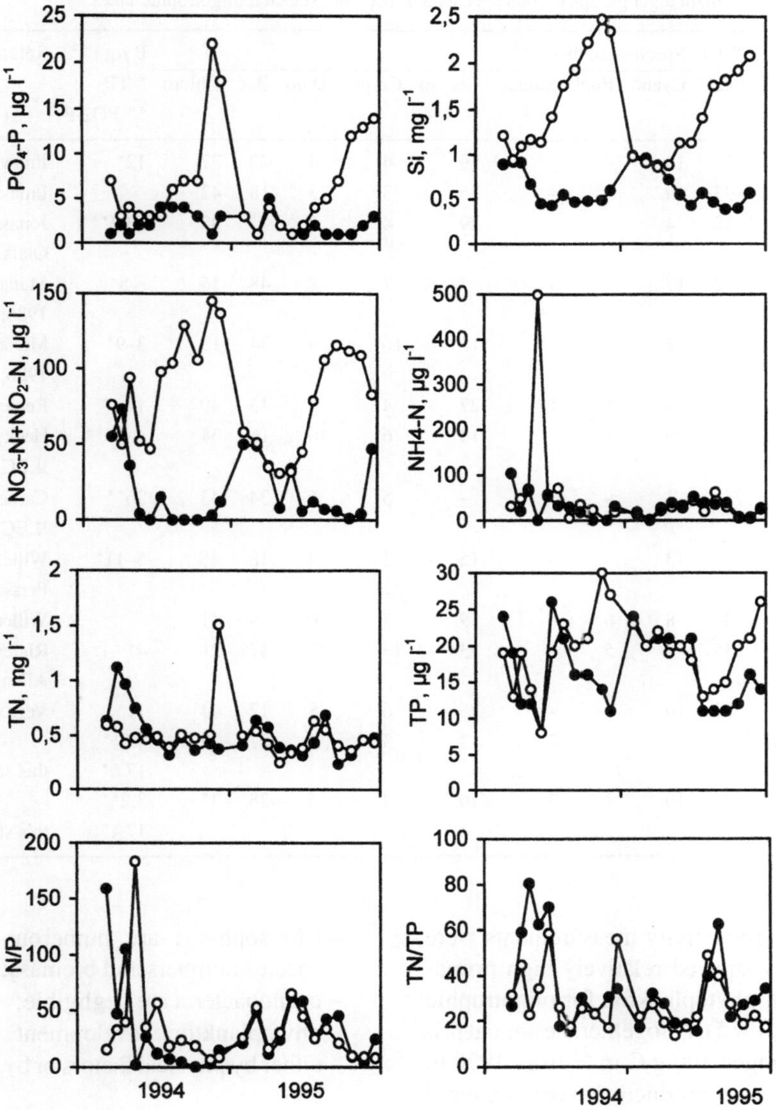

Figure 10. Concentrations of PO_4-P, Si, NO_3-N + NO_2-N, NH_4-N, TN and TP and the dissolved and TN/TP ratios (by weight) in Lake Stechlin in 1994 and 1995. Closed circles: averages of the 0–10 m water layer; open circles: averages of the 40–65 m water layer.

the nuclear power station (1966–1989): biomass might peak sharply either in spring or summer, population changes of dominant algal groups would occur rapidly, and annual biomass maxima and averages varied within a wide range. Our attempts to compare the present results with the previous observations run into difficulties because different methodological approaches and quantifications. If allowance is made for distortions forced by the original protocols, it is not obvious that the phytoplankton populations have changed much since the 1960s. However, some temporal compositional responses of the phytoplankton to thermal and nutrient loading from the nuclear power plant are evident. For example, *Planktothrix rubescens*, a species difficult not to notice, occurred in net samples taken from Lake Stechlin in 1926 (Krieger, 1927), was subsequently found to relatively frequent (in the period 1963–64) but, then, it disappeared completely. Ours is the first record for almost 30 years. Summer development of pennate diatoms e.g. *Tabellaria fenestrata* var. *actinastroides* in 1969 and *Fragilaria crotonensis* (Casper, 1985 b) may also be a result of changed condition.

Table 3. Contributions of main algal groups to total species number from selected oligotrophic lakes.

Lake	Total	Species number								P $\mu g\ l^{-1}$	Reference
	total	Cyano	Eugl	Xantho	Chryso	Crypto	Dino	Bac	Chloro	* TP ** PO_4-P	
Lake Pääjärvi	70	14	–	–	19	4	4	22	37	12*	Ilmavirta, 1975, 1981
Lake Pääjärvi	93	10	–	–	25	3	3	18	42		Ilmavirta, 1974
Thingvallavatn	48	4	–	–	29	4	8	17	38	13.7**	Jónasson et al., 1992 Ólafsson, 1992
Lake Huron	72	17			7	7	8	48	16	~5*	Munawar & Munawar, 1986; ILEC, 1995
Lake Superior	69	12			16	16	9	34	13	3–9*	Munawar & Munawar, 1986; ILEC, 1995
Langvatn	147	5	5	1	27	4	5	13	40	14.1*	Reinertsen, 1982
Tanganyika	103	19	1		14	6	10	16	34	4–10**	Hecky et al. 1978 ILEC, 1987–89
Titicaca	75	1	–	1	–	5	6	34	43	26**	Carney et al. 1987 ILEC, 1995
Vättern	177	13	–	–	15	5	4	18	45	5–11*	Willén, 1976 Persson et al., 1989
Hundsjön	86	8	6	–	19	7	10	9	41		Willén, 1980
Char	155	8	5	–	23	13	12	17	21	4*	Rigler, 1974 Allan et al., 1994
Sanabria	52	10	–	–	15	6	8	17	44	11**	Vega et al., 1992
Lake Stechlin, 1994	129	17	–	2	17	3	5	27	29	4.3** 17.0*	this study
Lake Stechlin, 1995	109	10	1	3	16	4	5	28	35	3.2** 17.4*	this study

Regular primary productivity measurements were started in 1970. They showed relatively high primary production rates of phytoplankton for oligotrophic lakes (Koschel, 1981, 1985). Altogether the annual primary production averaged 100 g C m^{-2} from 1970 to 1995. The highest primary production were measured at beginning of the 1970s and at present. The primary production of phytoplankton showed a more or less regular seasonal periodicity, with a high spring and/or summer maximum and a smaller autumn maximum. However, the seasonal variation of primary production was considerable during the discontinuous operation of the nuclear power plant and was said to be influenced by the increased nutrient level and changed physical conditions (Koschel & Casper, 1986; Koschel, 1995). This seasonal dynamic of primary production did not reflect the contemporaneous dynamic (Koschel, 1981).

Interpretations of phytoplankton composition, development, dominant groups and background factors driving seasonal planktonic events in oligotrophic lakes are conditioned by several widely accepted suppositions. Among them are that:

- Chrysophytes are numerous both in terms of species numbers and biomass;
- cyanobacteria are negligible;
- phytoplankton development is strongly limited either by physical factors or by the scarcity of nutrients.

This study on phytoplankton succession in Lake Stechlin partly supports the above statements but it also suggests that a good deal of our understanding needs to be reviewed.

In Table 3 the species representation of the main algal groups in some selected oligotrophic lakes is compared (data are exclusively from quantitative studies). Few Euglenophyte species are found in such waters. Xanthophyceae also seem to be also very rare but they may have been confused with other groups. For example, *Trachydiscus sexangulatus*, a constant element of phytoplankton in Lake Stechlin can be differentiated from small centric diatoms only if 1000x magnification and oil immersion is used during countings. Other species can be confused with chlorophytes. On average, 10% of the species belong to Cyanobac-

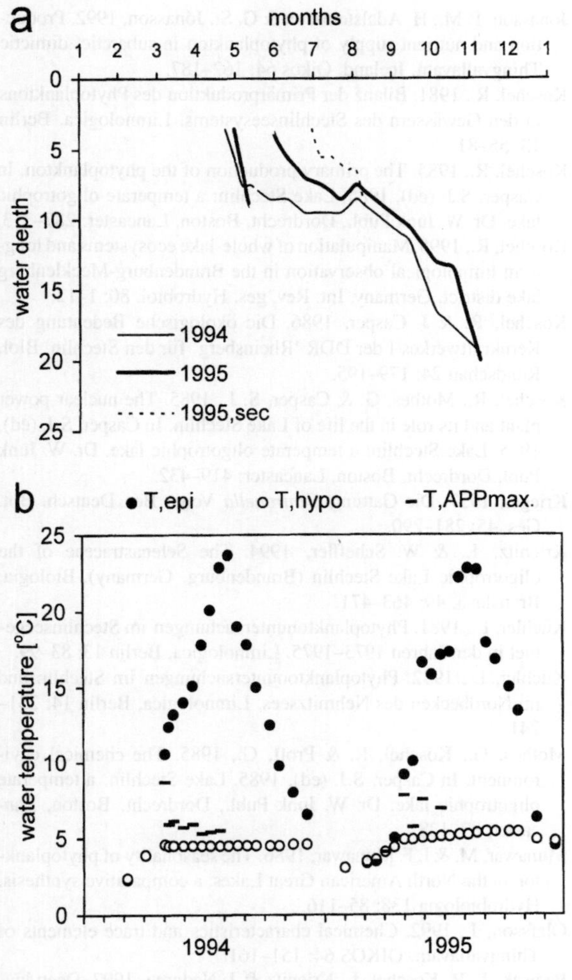

Figure 11. The depth of thermocline defined as dt/dz max in Lake Stechlin in 1994 and 1995; b: Average temperatures in the epi- (T, epi) and hypolimnion (T, hypo), and at the depth of the APP maximum (T, APPmax) in Lake Stechlin in 1994 and 1995. Tickmarks on x axis indicate months.

teria, the contribution of dinoflagellates and cryptophytes remains under 10%. The most species-rich groups are chrysophytes, diatoms and green algae. Qualitative composition of phytoplankton of Lake Stechlin shows close similarities with that of other oligotrophic lakes, as a matter of fact, the Stechlin data are very close to the average of the other lakes listed here.

While it is relatively easy to find comparable species lists, there are very few studies with data with which to compare annual average contributions of different taxonomic groups to total biomass. In general, contribution of blue-green algae to total biomass in temperate oligotrophic lakes is very low (Willén, 1976; Fängström & Willén, 1987), higher values, for example 22% in Lake Tanganyika (Hecky et al., 1978) or 60% in the Kariba Reservoir (Ramberg, 1987) have been found only in tropical lakes. This finding is certainly valid for temperate lakes with acidic pH since blue-green algae are not favoured in these ones. Recent studies (Stockner, 1991) have shown that contribution either to total biomass or to net primary production of autotrophic picophytoplankton increases with decreasing trophic state. Our results show that low contribution of blue-greens to total phytoplankton in temperate, non-acidic lakes cannot be claimed unless additional picophytoplankton counting is performed. In a tropical-temperate comparision, the difference pertains only in taxonomic level: in temperate lakes Chroococcales, in tropical ones various filamentous species contribute mostly to blue-green biomass (Carney et al., 1987, Ramberg, 1987). According to Blomquist et al. (1995), differences in distribution and dynamics of N-forms can lay behind this difference.

Another outcome of our studies in Lake Stechlin challenges the accepted view about the starting points of annual phytoplankton development in stratified lakes. Current knowledge says, that during winter phytoplankton growth is restricted by insufficient light (low incoming radiation, light extinction by ice and snow, deep water circulation in lakes without ice-cover). Release from this physical control by ice-break (lakes of moderate depth) or by onset of stratification (deep lakes) permits the start of seasonal succession (Sommer, 1991). In winter 1993/1994, ice cover was thin and lasted for a 2-weeks period in February, while in 1994/1995, failed to develop at all. In both years, deep circulations occurred from October to early May. The spring dominants (picoplankton and *Cyclotella tripartita* in 1994, picoplankton, *C. pseudocomensis* and a small *Cyclotella sp.* in 1995) started to grow as early as February–March. A constant, slow growth occurred until the end of April and, by that time, inorganic nutrients were exhausted from the water column. The onset of the stratification in early May did not initiate the growth of the spring assemblage, on the contrary it terminated the growth. Both centric and autotrophic picophytoplankton populations seem to be well adapted to growth in cold, isothermal, low light – high nutrient environment. However, the fate of the populations during and after the onset of stratification differ: diatoms sink rapidly to the hypolimnion (see data in Scheffler & Padisák, 1997a, b) while picoalgae, especially *Synechococcus* sp. accumulated in a narrow deep-layer maximum in the upper hypolimnion (see data in Padisák et al., 1997).

The 1995/1996 winter was very cold, ice cover lasted from early December to mid-April. Under 45 cm ice + 20 cm snow cover in February considerable population increase of the rare diatom, *Stephanocostis chantaicus* was observed (Scheffler & Padisák, in press, b) indicating that neither deep circulations nor increased light extinction by ice and snow cover prevents growth of specially adapted species. These records together with other, quite recent ones (Wiedner & Nixdorf, this volume) call the attention to the importance of phytoplankton studies in winter periods. In general, our view of the 'vegetation period' in phytoplankton may need to be modified.

References

Allan, R. J., M. Dickmann, C. B. Gray & V. Cromie, 1994. The book of Canadian lakes. The Canadian Association on Water Quality, Monogr. ser. 3: 563–570.

Blomquist, P., A. Pettersson & P. Hyenstrand, 1995. Ammonium-nitrogen: a key regulatory factor causing dominance of non-nitrogen-fixing cyanobacteria in aquatic systems. Arch. Hydrobiol. 132: 141–164.

Carney, H. J., P. J. Richerson & P. Eloranta, 1987. Lake Titicaca (Peru/Bolivia) phytoplankton: Species composition and structural comparison with other tropical and temparate lakes. Arch. Hydrobiol. 110: 365–385.

Casper, S. J. (ed.), 1985a. Lake Stechlin. A temperate oligotrophic lake. Dr. W. Junk Publ., Dordrecht, Boston, Lancaster.

Casper, S. J., 1985b. The phytoplankton. In Casper, S.J. (ed), 1985a. Lake Stechlin. A temperate oligotrophic lake. Dr. W. Junk Publ., Dordrecht, Boston, Lancaster: 157–195.

Casper, P. & R. Koschel, 1995. Description of Lake Stechlin. Limnologica, Berlin 25: 281–284.

Fängström, I. & E. Willén, 1987. Clustering and canonical correspondence analysis of phytoplankton and environmental variables. Vegetatio 71: 87–95.

Hamilton, P. B. 1990. The revised edition of a computerized plankton counter for plankton, periphyton and sediment analyses. Hydrobiologia 194: 23–30.

Hecky, R. E., E. J. Fee, H. Kling & J. W. Rudd, 1978. Studies on the planktonic ecology of Lake Tanganyika. Can. Fish. mar. Serv. tech. Rep. 816, 51 pp.

ILEC, 1987–1989. Data book of world lake environments – a survey of the state of world lakes. ILEC and UNEP, Otsu.

ILEC, 1995. Data book of world lake environments – a survey of the state of world lakes – 3. The Americas. ILEC and UNEP, Kusatsu.

Ilmavirta, V., 1974. Diel periodicity in the phytoplankton community of the oligotrophic lake Pääjärvi, southern Finland. I. Phytoplanktonic primary production and related factors. Ann. Bot. Fenn. 11: 136–177.

Ilmavirta, V., 1975. Diel periodicity of phytoplankton community of the oligotrophic Lake Pääjärvi, southern Finland. II. Late summer phytoplanktonic biomass. Ann. Bot. Fenn. 12: 37–44.

Ilmavirta, V., 1981. The ecosystem of the oligotrophic Lake Pääjärvi 1. Lake basin and primary production. Verh. int. Ver. Limnol. 21: 442–447.

Jónasson, P. M., H. Adalsteinsson & G. St. Jónasson, 1992. Production and nutrient supply of phytoplankton in subarctic, dimictic Thingvallavatn, Iceland. Oikos 64: 162–187.

Koschel, R., 1981. Bilanz der Primärproduktion des Phytoplanktons in den Gewässern des Stechlinseesystems. Limnologica, Berlin 13: 55–81.

Koschel, R., 1985. The primary production of the phytoplankton. In Casper, S.J. (ed), 1985. Lake Stechlin: a temperate oligotrophic lake. Dr. W. Junk Publ., Dordrecht, Boston, Lancaster: 287–313.

Koschel, R., 1995. Manipulation of whole-lake ecosystems and long-term limnological observation in the Brandenburg-Mecklenburg lake district, Germany. Int. Rev. ges. Hydrobiol. 80: 1–12.

Koschel, R. & J. Casper, 1986. Die ökologische Bedeutung des Kernkraftwerkes I der DDR 'Rheinsberg' für den Stechlin. Biol. Rundschau 24: 179–195.

Koschel, R., Mothes, G. & Casper, S. J., 1985. The nuclear power plant and its role in the life of Lake Stechlin. In Casper, S.J. (ed), 1985. Lake Stechlin: a temperate oligotrophic lake. Dr. W. Junk Publ.,Dordrecht, Boston, Lancaster: 419–432.

Krieger, 1927. Die Gattung *Centronella* Voigt. Ber. Deutsch. Bot. Ges. 45: 281–290.

Krienitz, L. & W. Scheffler, 1994. The Selenastraceae of the oligotrophic Lake Stechlin (Brandenburg, Germany). Biologia, Bratislava, 49: 463–471.

Küchler, L., 1981. Phytoplanktonuntersuchungen im Stechlinseegebiet in den Jahren 1973–1975. Limnologica, Berlin 13: 83–99.

Küchler, L., 1982. Phytoplanktonuntersuchungen im Stechlin und im Nordbecken des Nehmitzsees. Limnologica, Berlin 14: 231–241.

Mothes, G., Koschel, R. & Proft, G., 1985. The chemical environment. In Casper, S.J. (ed), 1985. Lake Stechlin: a temperate oligotrophic lake. Dr. W. Junk Publ., Dordrecht, Boston, Lancaster, 87–128.

Munawar, M. & I. F. Munawar, 1986. The seasonality of phytoplankton in the North American Great Lakes, a comparative synthesis. Hydrobiologia 138: 85–116.

Ólafsson, J., 1992. Chemical characteristics and trace elements of Thingvallavatn. OIKOS 64: 151–161.

Padisák, J., R. Koschel, L. Krienitz & J. Nedoma, 1997. Deep layer autotrophic picophytoplankton maximum in the oligotrophic Stechlinsee, Germany: origin, activity, development and erosion. European J. Phycol. 32: 403–416.

Persson, G., H. Olsson, T. Wiederholm & E. Willén, 1989. Lake Vättern, Sweden: A 20-year perspective. Ambio 18: 208–218.

Ramberg, L., 1987. Phytoplankton succession in the Sanyati basin, Lake Kariba. Hydrobiologia 153: 193–202.

Reinertsen, H., 1982. The effect of nutrient addition on the phytoplankton community of an oligotrophic lake. Holarctic Ecol. 5: 225–252.

Rigler, 1974. Char Lake Project. PF-2. Final Report. Can. Comm. Int. Biol. Progr., Toronto, 96 pp.

Scheffler, W., 1994. *Cyclotella pseudocomensis* nov. spec. (Bacillariophyceae) aus norddeutschen Seen. Diatom Res. 9: 355–369.

Scheffler, W. & J. Padisák, 1997a. *Cyclotella tripartita* (Bacillariophyceae), a dominant species in the oligotrophic Stechlinsee (Germany). Nova Hedw. 65: 221–232.

Scheffler, W. & J. Padisák, 1997b. *Cyclotella tripartita*, eine für Deutschland neue Diatomee des Planktons. DGL Jahrestagung, Wissenschaftliche Kurzreferaten, Schwedt/Oder, Germany, 16–20 September, 1996.

Sieburth, J. M., V. Smetacek & J. Lenz. 1978. Pelagic ecosystem structure: heterotrophic compartments of the plankton and their relationship to plankton size fractionation. Limnol. Oceanogr. 23: 1256–1263.

Sommer, U., 1991. Phytoplankton: directional succession and forced cycles. In Remmert, H. (ed), The Mosaic-Cycle Concept of Ecosystems, Springer Verlag, Berlin: 132–146.

Søndergaard, M., 1991. Phototropic picoplankton in temperate lakes: seasonal abundance and importance along trophic gradient. Int. Rev. ges. Hydrobiol. 76: 502–522.

Stockner, J. C., 1991. Autotrophic picoplankton in freshwater environments: the view from the summit. Int. Rev. ges. Hydrobiol. 76: 483–492.

Vega, J. C., C. de Hoyos & J. J. Aldasoro, 1992. The Sanabria lake. The largest natural freshwater lake in Spain. Limnetica 8: 49–57.

Wiedner, C. & B. Nixdorf, 1997. Success of chrysophytes, cryptophytes and dinoflagellates over blue-greens (cyanobacteria) during an extreme winter (1995/96) in eutrophic shallow lakes. Hydrobiologia 369/370: 229–235.

Willén, E., 1976. Phytoplankton in L. Hjälmaren. Acta Univ. Upsal. 378: 1–18.

Willén, T., 1980. Phytoplankton from Swedish lakes, 3. Lake Hundsjön and other kettle lakes of central Sweden. Arch. Hydrobiol. 89: 135–159.

Hydrobiologia **369:** 199–216, 1998.
M. Alvarez-Cobelas, C. S. Reynolds, P. Sanchez-Castillo & J. Kristiansen (eds), Phytoplankton and Trophic Gradients.

Grazing by large river zooplankton: a key to summer potamoplankton decline? The case of the Meuse and Moselle rivers in 1994 and 1995

Véronique Gosselain[1], Jean-Pierre Descy[1], Laurent Viroux[1], Célia Joaquim-Justo[2], Astrid Hammer[3], Arnaud Métens[1] & Sabine Schweitzer[4]
[1] *Laboratory of Freshwater Ecology (LFE), Facultés Universitaires Notre-Dame de la Paix, rue de Bruxelles, 61, 5000 Namur, Belgium*
[2] *Laboratoire d'Ecotoxicologie, Institut de Zoologie, Université de Liège, Quai Van Beneden, 22, 4020 Liège, Belgium*
[3] *Erasmus student at LFE, from the University of Rostock, Germany*
[4] *Centre de Recherches Ecologiques de l'Université de Metz, France*

Key words: potamoplankton, plankton interactions, *in situ* grazing, European rivers

Abstract

To explain summer declines in phytoplankton biomass in large rivers, we compared the effect of zooplankton grazing on the planktonic algae of two large European rivers, the Meuse and the Moselle. *In situ* grazing was measured during two years (1994 and 1995), using the Haney method. Total zooplankton community filtration rates recorded in the river Meuse ranged between 1 and 32% of the water volume filtered per day. A drastic algal decline was observed early July both years and may be explained by high densities of a rotifer-dominated zooplankton community (500–700 ind. l^{-1}) with more than 75% of *Brachionus calyciflorus*. During the summer period in 1994, when grazing was over 20%, edible algal biomass was controlled by a diversified rotifer community (up to 2500 ind. l^{-1}), while a non-edible algal assemblage developed. In contrast, phytoplankton biomass remained comparatively low in the Moselle throughout the low-flow period, as did zooplankton numbers during most of this time (fewer than 200 ind. l^{-1} during the summer period). The proportion of crustaceans in this zooplankton was rather higher than in the Meuse, and they dominated at times, in biomass as well as in numbers. Nevertheless, measured *in situ* grazing rates (1–15%) could not explain the low summer algal biomass, even if low filtration rates may at times represent a significant carbon loss for phytoplankton, when and where net algal production was low. As a conclusion, the role of phytoplankton – zooplankton interactions in controlling algal biomass in large rivers is discussed.

Introduction

Summer declines of potamoplankton have been reported from several large rivers, including the Meuse (Gosselain et al., 1994; Gosselain et al., 1998), the Moselle (Descy & Willems, 1991; Descy, 1993), the Rhine (de Ruyter van Steveninck et al., 1992; Admiraal et al., 1994), the Spree (Köhler, 1995) and the Seine (Garnier et al., 1995). These declines may be of short duration, as in the River Meuse, or span many weeks and affect long river sectors (as in the Moselle). Longitudinal (downstream) declines may also be superimposed, as in the Rhine (de Ruyter van Steveninck et al., 1992; Admiraal et al., 1994), the Spree (Köhler, 1995) and the Seine (Garnier et al., 1995). In some cases, a low summer phytoplankton biomass may result in water quality problems, as large rivers usually have an important organic matter loading; along polluted stretches, the balance of the oxygen is so delicate that a decrease of planktonic production may fail to offset critical oxygen deficits.

As stated by Reynolds & Descy (1996) in their recent synthesis on phytoplankton in large rivers, longitudinal declines of producer biomass can result from changes in physical conditions along the river course. Net algal production is related to the ratio between the river depth and its euphotic layer, so that a decrease in chlorophyll **a** is expected in the downstream sections of large rivers, as a consequence of increasing turbidity and depth. This is in agreement with general concepts on river function (e.g. Vannote et al., 1980) which predict a trend toward heterotrophy for the high order sections of long watercourses. Furthermore, in relation with temporal summer declines, Reynolds & Descy (1996) pointed out that conditions apparently optimal (summer temperatures, more sunlight over a longer day, less turbidity, slower downstream travel) do not necessarily coincide with the highest net algal increase. In fact, *'these are the conditions under which the potamoplankton dynamics become very sensitive to cell-specific loss processes, particularly where zooplankton development is favoured or where shallow water and accelerated sinking rates contribute to enhanced losses to sedimentation'* (Reynolds & Descy, op. cit.).

A second approach to the investigation of algal declines in large rivers is to consider biotic interactions, namely zooplankton feeding on algal biomass. In a previous publication dealing with observations in the River Meuse in 1992 (Gosselain et al., 1994), we hypothesised that this process may have a significant influence on the potamoplankton community. According to our deliberations on zooplankton and mathematical simulations, the phytoplankton summer variations could be explained as a combination of low growth rates and losses from zooplankton grazing and sedimentation. Combining grazing measurements from the Meuse, assembled over three years (1994–1996; Gosselain et al., 1998), with measured grazing rates by the rotifer-dominated communities, we found loss rates between 1 and 113%. However, this grazing pressure resulted from small zooplankton feeding, affecting algae only in the size range < 20 μm; this seems to explain why larger phytoplankton units became dominant in the summer assemblage. In one of the few studies conducted on plankton interactions in rivers, de Ruyter van Steveninck et al. (1992) demonstrated the possible role of biotic interactions within the planktonic community in the Dutch section of the River Rhine. They estimated zooplankton filtration rates in the range 0.07–0.18 l $[ZooC]^{-1}$ h^{-1} or 0.01–0.18 d^{-1}. However, depending on the season, only 2.3 to 28% of the algal biomass loss could be explained by this level of grazing. Garnier et al. (1995) calculated grazing fluxes in the River Seine to be in the range 200–450 mg C m^{-2} d^{-1}, which was estimated to represent between 20 and 50% of phytoplankton growth (and occasionally up to 75% for diatoms or chlorophytes). In the River Spree, biotic interactions have also been strongly implicated but, going downriver, zooplankton rapidly declined and could only account for few algal losses (Köhler, 1995).

Besides a possible effect of zooplankton grazing on phytoplankton biomass, many reports emphasise the consequences of filtration by benthic filter feeders, especially the zebra mussel *Dreissena polymorpha* Pallas. In the River Rhine, increasing populations of the zebra mussel have been suspected to have modified food web structure, along with the amphipod *Corophium curvispinum* Sars (Admiraal et al., 1994). Similarly, large numbers of molluscs (mainly *Unio tumidus* Phil. and *Anodonta anatina* L.) have been recorded in parts of the lower Spree, arguing for an effective grazing by benthic filter feeders (Köhler, 1995). Some impact of zoobenthos on phytoplankton summer declines have also been suggested by Garnier et al. (1995) as a possible explanation for enhanced phytoplankton mortality in the River Seine.

In this respect, it is interesting to report the impact of new settled zoobenthos, in particular of the zebra mussel *D. polymorpha*, in some American large rivers and lakes. In the River Seneca, dramatic changes in water quality, particularly on the oxygen resources, have been brought about by zebra-mussel infestation (Effler et al., 1996). Recorded filtering rates for settled bivalves range from 100 to 400 ml $ind.^{-1}$ h^{-1} (Sprung, 1995) for large individuals (20 mm shell length) and higher values have also been reported (e.g. MacIsaac et al., 1992). Filtering rates are biomass-dependent and a strong temperature effect has been reported, with a maximum recorded at 20 °C (Aldridge et al., 1995). According to Sprung & Rose (1988), adult zebra mussels prefer food in the 15-40 μm size range but they stressed also that the inhaling siphon aperture size seems to enable the mollusc to ingest all algal size units. Furthermore, MacIsaac et al. (1995) reported ingestion of microzooplankton, probably at a rate approaching values for phytoplankton food. Effler et al. (1996) also showed that low dissolved oxygen was related to a high respiration demand from the zoobenthos. Böhme (1994) suggested that oxygen consumption in large rivers might be partly dependent on benthic community respiration.

The purpose of this paper is to assess the effect of zooplankton grazing on the planktonic algae of two large rivers, the Meuse and Moselle rivers, mainly through *in situ* zooplankton grazing measurements. We

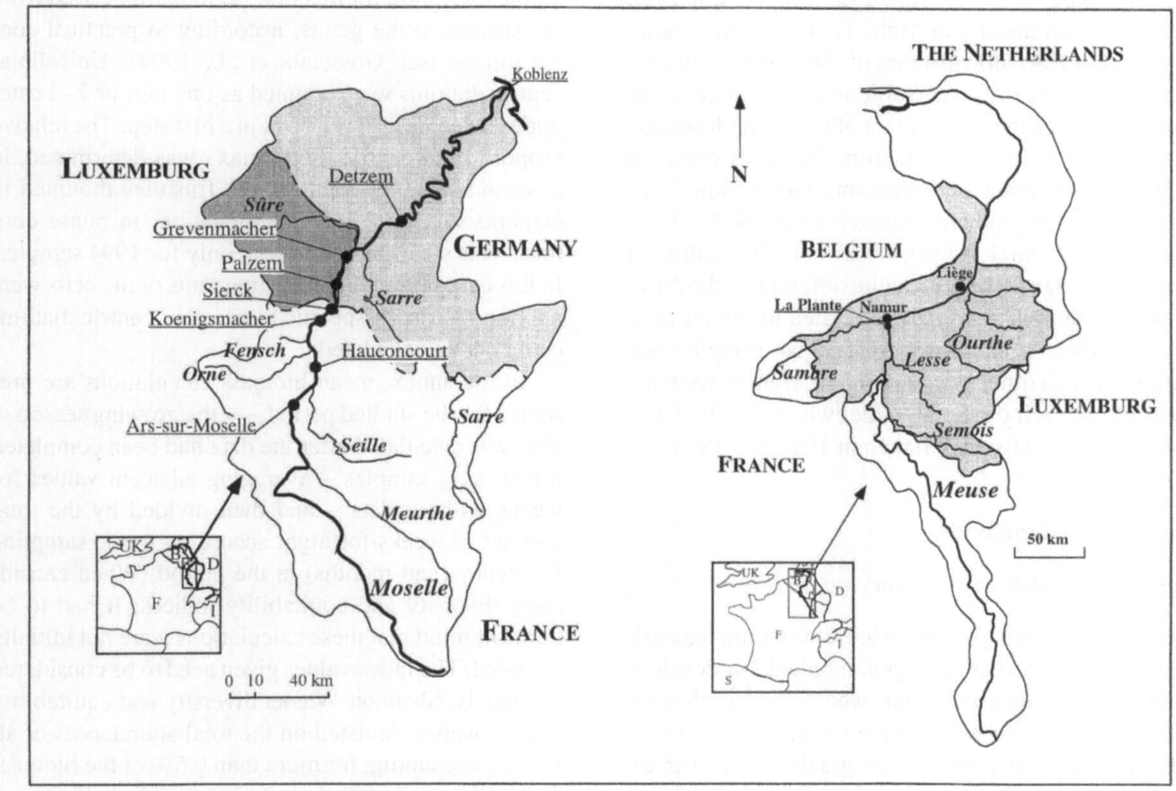

Figure 1. Map of the Meuse and Moselle rivers basins. Main tributaries are reported in italic letters and studied stations are underlined.

present hereafter the data from two years of investigations, 1994 and 1995, in both rivers.

Description of the studied sites

The Meuse and Moselle rivers both rise in north-east of France (Figure 1). The Meuse flows through Belgium and The Netherlands, where it meets the lower Rhine, forming the Dutch Delta, which opens in the North Sea. Its total length is 885 km and its catchment area is about 36,000 km^2, 40% of it being within Belgian territory and where the river has been regulated for navigation with weirs and locks. The Moselle is a major tributary of the Rhine: its total length is 535 km and its catchment area about 28,000 km^2. The river has been partly canalised but some natural stretches have been maintained, so that the river flows alternately through deep, slow-flowing reaches and wild, shallow, sometimes meandering reaches.

Both are nutrient-rich, eutrophic rivers (see annex). Nevertheless, the River Moselle has more alkaline and more electrically conductive waters (Descy & Willems, 1991).

Phytoplankton composition is similar in the two rivers, with centric diatoms dominating throughout the plankton growth season. Nevertheless, the Moselle carries some typically halophilous or euryhaline taxa, originating from the Meurthe and Seille rivers (Descy & Willems, 1991), among them *Skeletonema potamos* (Weber) Hasle and *S. subsalsum* (Cleve-Euler) Bethge. It should be noted that *S. potamos* has been recorded recently in the Meuse as well, where it now sometimes accounts for half the phytoplankton numbers. Green algae seem to achieve higher biomass in the Meuse. The zooplankton is dominated by rotifers in both rivers but microcrustacean biomass sometimes exceeds rotifer biomass in the Moselle, during the summer months.

The main study site on the Meuse, 'La Plante', is located at 537 km from the source. At this site, the mean depth is 3.95 m and the mean width is 100 m. Seven stations along the Moselle, spanning 140 km, were sampled, viz.: Ars-sur-Moselle (river km 245), upstream from the city of Metz; Hauconcourt (Argancy dam, km 269), downstream of the city of Metz, the confluence with the river Seille and the outlet of the cooling basin of the power plant of La Maxe; Koenigsmacker (km 297), upstream from the outlet from the nuclear power plant of Cattenom; Sierck (km 310); Palzem (km 323) and Grevenmacher (km 340) in Luxembourg; Detzem (km 386), in Germany. The different sampling designs owe to the quite different study objectives. The sampling strategy is related to the purpose of the studies: in the River Meuse, we sampled one station through time; longitudinal differences were the focus of research on the Moselle, where *in situ* determinations were also undertaken in 1994 and 1995.

Materials and methods

Sampling and data acquisition

So far as it was possible, the Meuse was sampled each week during the season of potamoplankton development; grazing measurements were made each fortnight. The Moselle was sampled at fortnightly intervals; in situ measurements of grazing and primary production were made on five occasions (1–4 August 1994; 30 May–1 June, 18–20 July, 2–4 August and 22–24 August 1995). Various environmental parameters were measured throughout the study: temperature, discharge, surface irradiance, vertical light attenuation, dissolved nutrients (N, P, Si), particulate carbon and nitrogen, chlorophyll *a* [Chl *a*, by the standard spectrophotometric technique developed by Péchar (1987) or by HPLC analysis (Descy & Métens, 1996)] and other general measurements, using standard methods detailed elsewhere (see Descy & Gosselain, 1994).

Phytoplankton was collected from the surface with a 3-l opaque Van Dorn bottle; 1 l was immediately preserved with Lugol's solution. After sedimentation, phytoplankton was examined, under a standard microscope (Leitz Laborlux D:12.5 × 40 magnification) in subsamples mounted in a Bürker cell. Algal units from all samples, save those from the Moselle in 1994, were counted and their dimensions were measured. Data were computed using Hamilton's software (Hamilton, 1990), which records numbers and calculates unit biovolumes from the dimensions entered – except for the samples from the Moselle in 1994. Conversion to carbon (pg C) was then done by means of the Eppley equations or, for the river Moselle in 1994, the Strathmann equations (Smayda, 1978). The level of taxonomic ascription for this first examination varied from the species to the genus, according to practical considerations (see Gosselain et al., 1994). Unicellular centric diatoms were counted as one unit or 2–3 categories, according to cell size, in a first step. The relative proportion of centric diatom taxa was determined, in a second step, by counting the frustules mounted in Naphrax at 12.5 × 100 magnification, in phase contrast. This step was carried out only for 1994 samples. In this case, individual measurements of the cells were not done so taxon-specific biomass of centric diatoms could not be calculated.

In the annex, mean biomass calculations are presented for the studied period – ~ the growing season –. This was calculated after the data had been completed for missing samples – averaging adjacent values for weeks without data – and then divided by the total number of weeks/fortnight according to the sampling frequency (and months) in the period. When considering diversity and equitability indices, it had to be borne in mind that these calculations were not initially intended. The index values given need to be considered cautiously. Shannon-Wiener diversity and equitability indices were calculated on the total abundances of all species accounting for more than 0.5% of the biomass during the studied period. As biomass per taxon was not calculated for centric diatoms from the Meuse data, abundances of all centric diatom taxa were taken into account for their calculation. As the 1993–94 data from the Moselle did not cover the whole plankton growing season, we include also data from 1993. As for the 1994 data from the Moselle, units were not measured during counting but a single value of biovolume per taxon was applied.

Primary production was measured in short (between $^{1}/_{2}$ and 2 h) surface incubations of river water added with $NaH^{14}CO_3$; subsamples were divided among nine 100 ml glass bottles and exposed to relative irradiances in the range 0–100% of the sunlight, in a field incubator immersed just beneath the water surface. Sunlight was monitored simultaneously; information was stored by a data logger. The photosynthesis-light relationship was established after measurement of the radioactivity (dpm) of the algae collected on Millipore HA 0.45 μm filters; subsequently, daily production was calculated as in Descy et al. (1987).

Subsurface zooplankton was sampled using the same 3-l Van Dorn bottle and collected on 37 or 63 μm-mesh Nitex plankton net; a minimum of 15 l was collected. Qualitative and quantitative analysis of zooplankton samples were performed under a dissecting microscope at 35 X magnification, either on fresh material or on acidifiied Lugol-fixed samples (Leakey et al., 1994), and after concentration by a factor ranging from 250 to 1000. Rotifers were determined to genus or species level while crustaceans were counted as either cladocerans or copepods, with no further level of identification. Details can be found elsewhere (Viroux, 1997).

Method for grazing experiments

Grazing measurements were carried out using a 6.5-l Haney (1971) chamber, the design of which followed Gawler & Chapuis (1987). Cultures of *Dictyosphaerium ehrenbergianum* Näg. (which was wholly dominated by *Chlorella*-like unicells of 4.4 μm mean diameter) were used as tracer food. Exponential-phase algae were incubated for at least 24 h with 250 μCi $NaH^{14}CO_3$ per 250 ml culture. Flasks containing 15 ml of labelled algae were then prepared so as not to exceed 10% of the phytoplankton biomass present in the grazing chamber. The carbon/Chl *a* ratio used for the River Meuse was 37 (Descy & Gosselain, 1994) and 30 for the River Moselle (based on our own measurements); the carbon content of the cultured algae was assessed by several measurements using a Carlo Erba carbon-nitrogen analyser NA 1500. Incubations in the river lasted between 6 and 11 min from the closure of the chamber to the end of the sieving of its content through a 37 μm-mesh (or rarely a 63 μm-mesh) Nitex screen. This range of time has been chosen as it is shorter than the known gut passage time for most zooplankters (Downing & Rigler, 1984). The animals were then narcotised in soda water. Before freezing in liquid nitrogen, they were either suspended in a small amount of water (samples for subsequent sorting) or collected on 25-mm Whatman GF/C filters (samples for total community measurements); all samples were stored in 20-ml scintillation vials. A small volume of the water collected from the sieving through the Nitex screen was preserved for determining the radioactivity of the algal suspension. Background corrections were made by running control incubations without any labelled algae: the radioactivity measured on the zooplankton and phytoplankton samples from these controls were subtracted from the counts obtained for the zooplankton and for the algal suspension, respectively. Zooplankton samples collected on GF/C filters were treated for the measurement of radioactivity by dissolving animals in 1 ml Lumasolve (Lumac LSC). Community filtering rates (CFR) were calculated according to Haney (1971) and expressed in% d^{-1}, i.e. the percentage of the volume filtered per day. Zooplankton ingestion rate could then be calculated by multiplying the daily filtering rate by the algal biomass in the river water (Chl *a*, converted to carbon). In 1994, maximum ingestion rate in the Moselle was calculated, assuming the same impact on the whole algal biomass. In 1995, ingestion rate was corrected for the proportion of edible units, assessed from the phytoplankton counts and size measurements, considering only algal units of greatest axial linear dimension (GALD) < 20 μm (Reynolds, 1986). Details of the method will be given elsewhere (Gosselain et al., 1998).

The samples preserved for measuring individual filtering rates (IFR) were sorted and a known number of hand-picked individuals were placed in 6 ml scintillation vials. From individual filtering rates, resulting from these measurements, an assessment of the population filtering rate (SFR, sub-community filtering rate) was made by multiplying the IFR by the corresponding number of individuals in the river water (for details, see Gosselain et al., 1996).

For the purpose of this paper and because IFR were rarely available for all taxa, their contribution was also back-calculated (BCG, back-calculated grazing), by multiplying specific abundances by a species-specific individual filtering rate for each taxon. For rotifers, IFR of *Brachionus calyciflorus* Pallas and *Keratella cochlearis* Gosse were used and applied to other taxa, according to size and diet. Global values were applied respectively to nauplii and copepodites. For cladocerans, only *Bosmina* was excepted from the application of a single estimate. It must be borne in mind that this calculation does not take into account either the algal biomass or influence of the temperature. Nevertheless we assume that it approaches the actual grazing activity when the temperature is greater or equal to 20 °C.

Results

The Meuse

In the Meuse, variation in environmental conditions, discharge and temperature during 1994 and 1995 (Figure 2) were similar and quite typical for this river. After a high-flow period, discharges declined in the spring (from mid-April in 1994, early May in 1995) towards the low-flows of summer, which persisted until late September. Temperature followed an opposite pattern.

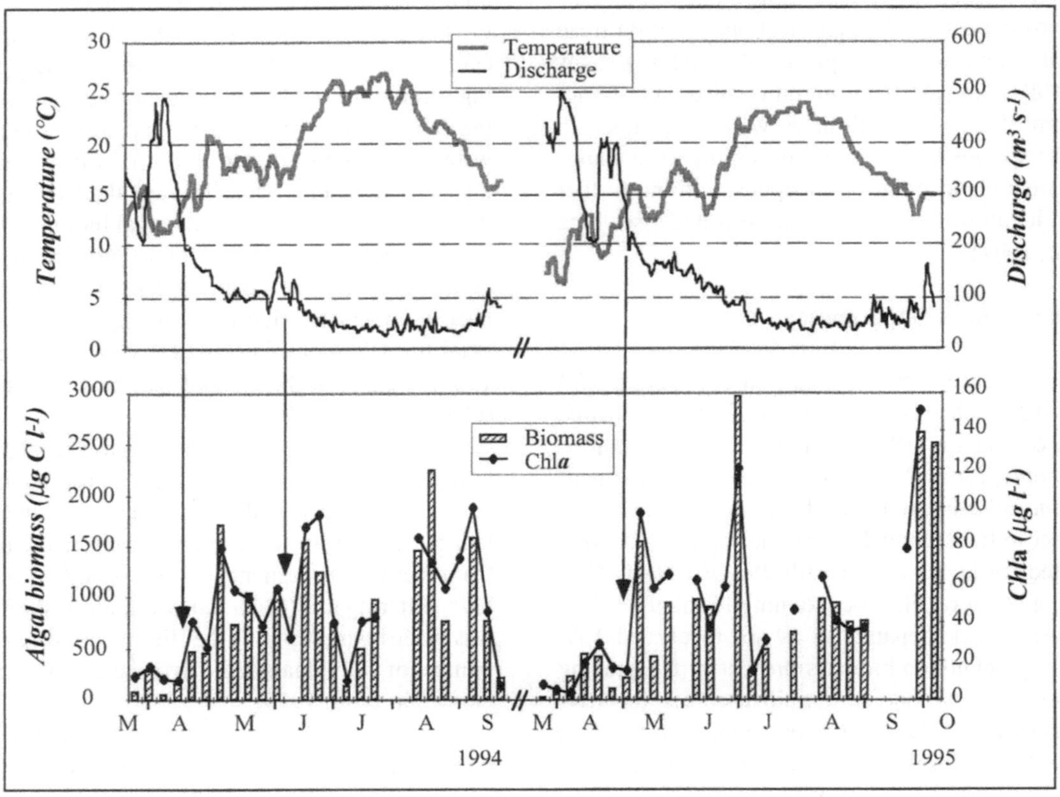

Figure 2. Top – Water discharge ($m^3 s^{-1}$) and temperature (°C) in the River Meuse during the years 1994 and 1995. Data from the CIBE (Compagnie Intercommunale Bruxelloise des Eaux) at Tailfer, close to the study site. *Bottom* – Variations of phytoplankton at 'La Plante', River Meuse, during the growing season in 1994 and 1995. Bars correspond to biomass (μg C l^{-1}), calculated from cell counting, and the dashed line to the measured Chl *a* (μg Chl *a* l^{-1}). Arrows indicate some discharge events influencing phytoplankton.

The main nutrients (N, P, Si) exhibited some variations but remained always well above concentrations limiting algal growth (see annex).

In both years, a first algal bloom followed the spring decrease in water discharge. Algal biomass then declined, perhaps because of lower temperature. A second bloom built up, with a maximum at the end of June (reaching the maximum Chl *a* concentration of the year in 1995, close to 160 $\mu g\ l^{-1}$), at a time when temperature and discharge were optimal for plankton growth. At the beginning of July, in both years, a sharp ten-fold decrease in phytoplankton was then observed, followed by a quick recovery. Maximum Chl *a* concentration in 1994 was observed during the late summer period (around 100 $\mu g\ l^{-1}$). In 1995, the high biomass recorded at the end of September was probably related to changes in flow rate and retention times resulting from river regulation operations.

In both years, centric diatoms were dominant all throughout the growing season (Figure 3). Green algae began to develop in May, probably responding to higher temperatures and improved light income. Algae of other phylogenetic groups appeared at this time but remained at relatively low levels. In 1994, green algae played a more important role in the second bloom (end of June) than in 1995, but *Peridinium* sp. reached a high biomass. During this first part of the algal development, unit sizes were in the range 5–15 μm GALD (yet with a significant biomass in the 20–25 μm GALD range in 1995). As far as the zooplankton is concerned, a spring 'bloom' of rotifers developed in May (253 ind. l^{-1} on 23 May 1994 and 677 ind. l^{-1} on 8 June 1995; see Figure 3); *Euchlanis* sp., along with *B. calyciflorus* in 1994 and with *K. cochlearis* and Synchaetidae in 1995, were the dominant taxa of this episode (Figure 4; bottom). In 1994, this phase was followed by an intense pulse of *D. polymorpha* veligers. In 1995, on

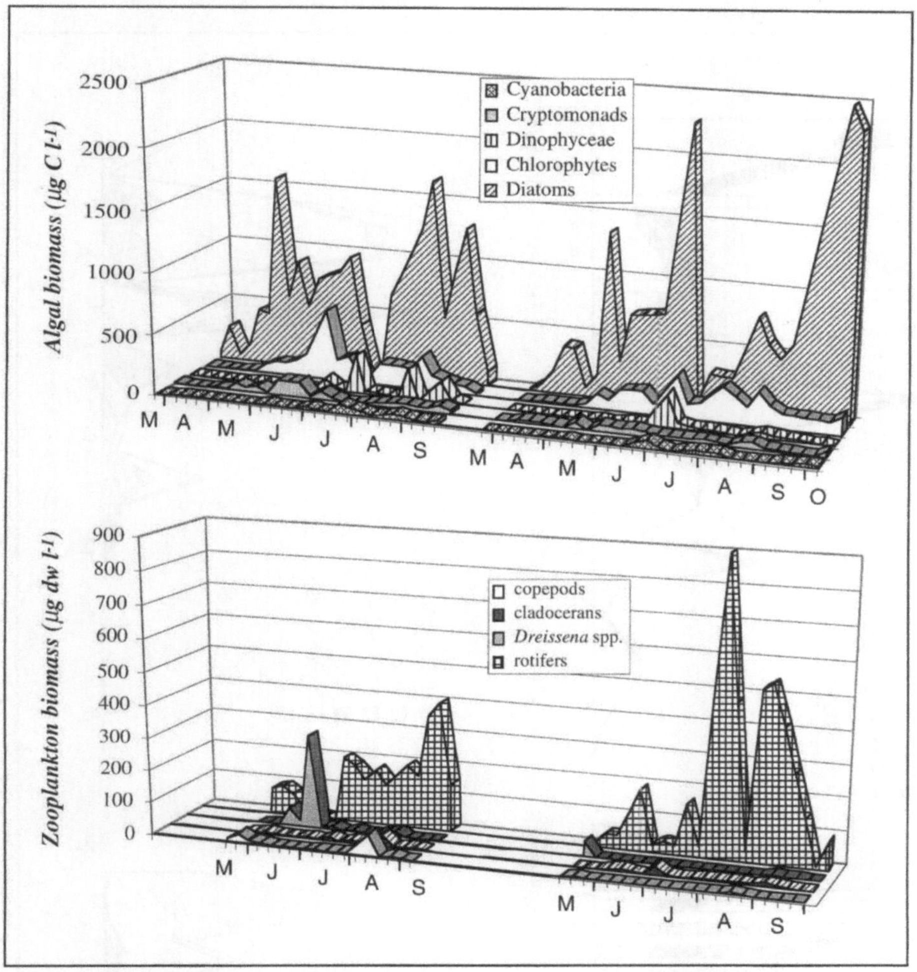

Figure 3. Variations of the main algal groups expressed in biomass (μg C l^{-1}; *top*) and variations of zooplankton expressed in biomass (μg dw l^{-1}; *bottom*) at 'La Plante', River Meuse, during the growing season in 1994 and 1995.

the other hand, *Dreissena* veligers were observed only once, in May, with 91 ind. l^{-1}. Nevertheless, a peak may have occurred before sampling started.

At the beginning of July, a sharp, ten-fold algal decrease occurred: populations of all algal groups diminished, particularly of species with GALD in the range 5–15 μm (and, notably, those in the 15–20 μm GALD range in 1995). At this time, rotifers increased in numbers and biomass, more especially so in 1994.

A slow recovery in algal numbers followed and filamentous centrics (*Aulacoseira* spp. with a few *Skeletonema potamos*) became more prominent and were the dominant centrics for about 3 weeks at the end of August in 1994, and in one sample in 1995 (August 25). In this summer period, a certain broadening of the distribution of algal size was observed in both years.The effect was more noticeable in 1994 than in 1995: units larger than 25 μm GALD were never significant in 1995 but the 130–180 μm GALD was well represented in the second half of August 1994.

Zooplankton biomass during mid-summer remained relatively high in both years (reaching 200 μg dw l^{-1} – around 1500 ind. l^{-1} – in 1994, and close to 500 μg dw l^{-1} – around 2000 ind. l^{-1} – in 1995). The dominant species were nevertheless quite different from one year to the other. In 1994, *B. calyciflorus* and *B. angularis* Gosse shared dominance, while in 1995, *K. cochlearis* dominated over all other rotifer species. Significant

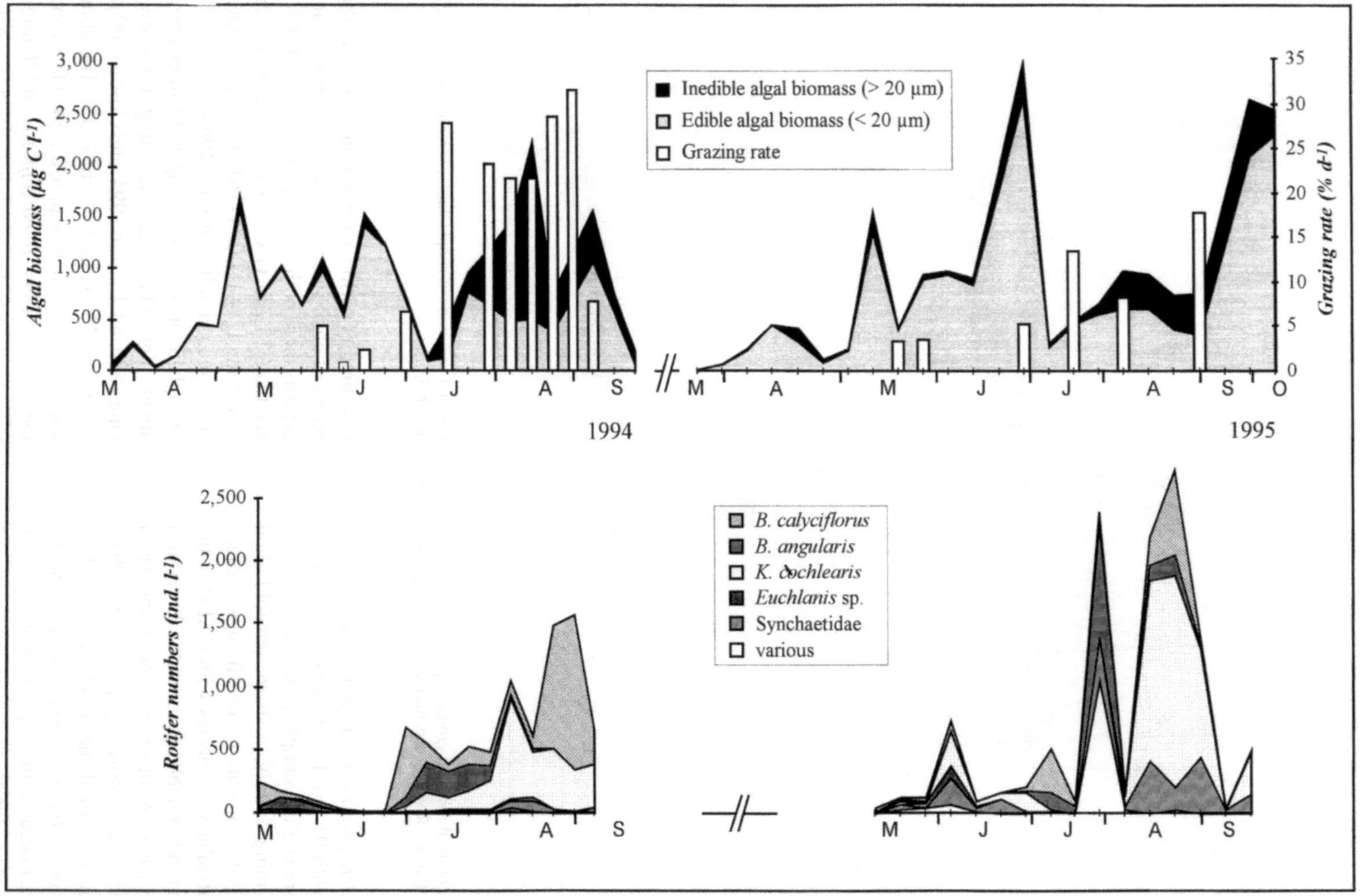

Figure 4. Top – Total community grazing rate (CFR; columns) overlaid on phytoplankton biomass (area) expressed as edible ($< 20\ \mu$m GALD) and inedible ($> 20\ \mu$m GALD) units at 'La Plante', River Meuse, during the growing season in 1994 and 1995. *Bottom* – Detail of the rotifer taxa expressed in numbers (ind l^{-1}) over the growing season at 'La Plante', River Meuse, in 1994.

numbers of small cladocerans appeared in August 1994 (6.5 ind. l^{-1} or 32.3 μg dw l^{-1}), with *Bosmina* spp. making up about 15% of them. They reappeared in late June 1995 (12.2 ind. l^{-1} or 27.2 μg dw l^{-1}), when *Bosmina* spp. represented 98% of them. Copepodites reached maximum densities in August (6.7 ind. l^{-1} in 1994 and 2.1 ind. l^{-1} in 1995).

In Figure 4, measured grazing rates are superimposed on the phytoplankton biomass changes within two broad size categories: less than 20 μm GALD, considered as edible by the rotifer-dominated zooplankton, and greater than 20 μm GALD, assumed to be inedible; the lower diagrams show the parallel dynamics of the rotifers. Grazing activity in 1994, expressed as a daily rate, ranged between 1 and 32% of the total volume, and remained over 20% d^{-1} throughout July and August. The five 1995 measurements done in 1995 yielded daily rates of between 3.4 and 17.9% d^{-1}. These diagrams strongly suggest, for both years of measurement, a link between a relatively well diversified rotifer population, relatively high grazing rates and a change in size structure of the phytoplankton community, i.e. in a wide range of sizes contrasting with the narrow spring GALD range. In summer 1994, the clear shift to larger algae was seemingly influenced by the grazing pressure applied to the smaller algae. Had the edible biomass in summer 1995 been in the same range as in 1994, the biomass of larger algae would perhaps have remained as low as in 1994.

Supposing the zooplankton to have been feeding only on the edible algal fraction (< 20 μm GALD), grazing losses in 1994 comprised between 69 and 301 μg C l^{-1} d^{-1} (which, on an areal basis, is equivalent to 274-1188 mg C m^{-2} d^{-1}) and, in 1995, between 70 and 211 μg C l^{-1} d^{-1} (276-834 mg C m^{-2} d^{-1}), with the highest values corresponding to the algal bloom on June 29.

Rotifer sub-community filtering rates (SFR), calculated for 1994, showed a rotifer contribution to total grazing (CFR) ranging between 40 and 134% (see Gosselain et al., 1996). Back-calculated filtering rates (BCG) indicate that rotifers represent the major contributor to total grazing. Most of the time, the rotifer BCG was ten times higher than that of the crustacean contribution; significant consumption by micro-crustaceans, principally by *Bosmina*, was found only once (12.5% of total BCG on June 29, 1995).

The Moselle

The phytoplankton of the Moselle was similar in composition to that of the Meuse: in the two years, centric diatoms largely dominated at all stations, except for occasional, local developments of other types of algae, including Dinophyceae in July 1994 (Figure 5), or green algae and Cryptophyceae in 1995 (not shown). A first spring bloom occurred in April and a second bloom reached its maximum at the end of June; it was composed of various small centric diatoms. It is worth noticing that this bloom declined downstream. The zooplankton community (Figures 5 & 6), made of rotifers and, at times, *Dreissena* larvae, bloomed at the same periods as phytoplankton. In June 1994, total zooplankton numbers rose to 4400–6500 ind. l^{-1} (1300-1600 μg dw l^{-1}), depending on the station. In 1995, spring peaks were less marked, reaching only 750 ind. l^{-1} (180 μg dw l^{-1}), except at Hauconcourt (1300 ind. l^{-1} or 250 μg dw l^{-1} for the second spring bloom). Dominant taxa were *B. calyciflorus*, *B. angularis* and *K. cochlearis* in 1994 and *B. calyciflorus*, *Synchaeta* sp. and *K. cochlearis* in 1995.

A summer algal biomass decline, affecting all groups in a similar way, was observed from July downstream of the French border and somewhat later in the upstream sector. At this time, rotifer numbers were already low and the low algal biomass level was observed during the entire second part of summer. Detailed analyses of unit sizes, carried out in 1995 for the date of experiments, indicated a progressive shift from small (6–19 μm GALD) to large units (> 20 μm GALD) among unicellular centrics. A longitudinal decrease of filamentous diatoms is also to be noticed. From early August on, zooplankton decreased, as did the phytoplankton, and remained at low levels (Figure 6). During this period, crustaceans were dominant on a biomass basis and also, most of the time, in terms of abundance. In 1995, they persisted for some weeks. *Dreissena* larvae dominated the zooplankton over the whole summer.

Grazing and primary production were measured in early August in 1994 (Figure 5), at a time when zooplankton numbers were between 100 and 350 ind. l^{-1} (which was 110–313 μg dw l^{-1} without taking into account *Dreissena* veligers). Grazing rates were approximated to be between 8 and 15% d^{-1}. In 1994, grazing was expressed as maximum ingestion over the water column and compared to algal net production. It is striking that consumption by grazers, in the range 30.2–149.0 μg C l^{-1} d^{-1} (which is 121.0–521.6 mg

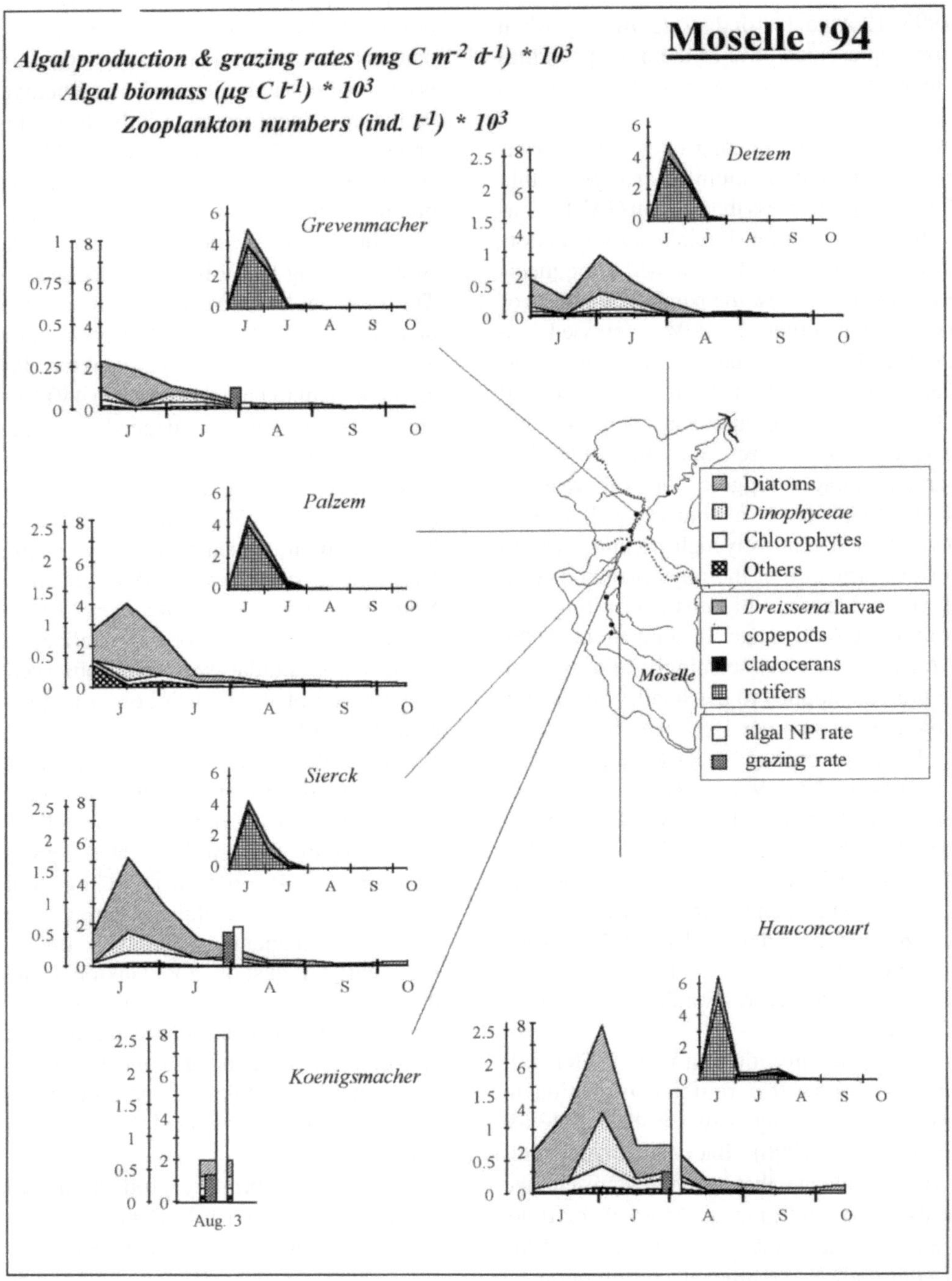

Figure 5. Variations of the main algal groups, expressed as biomass ($\mu g\ C\ l^{-1}$; inner axes), and variations of zooplankton, expressed in numbers (ind l^{-1}) at different stations in the River Moselle from June 1994 and till the end of the growing season. Phytoplankton net production and losses by grazing are overlaid as bars and are expressed as carbon areal values ($mg\ C\ m^{-2}\ d^{-1}$; outer axes).

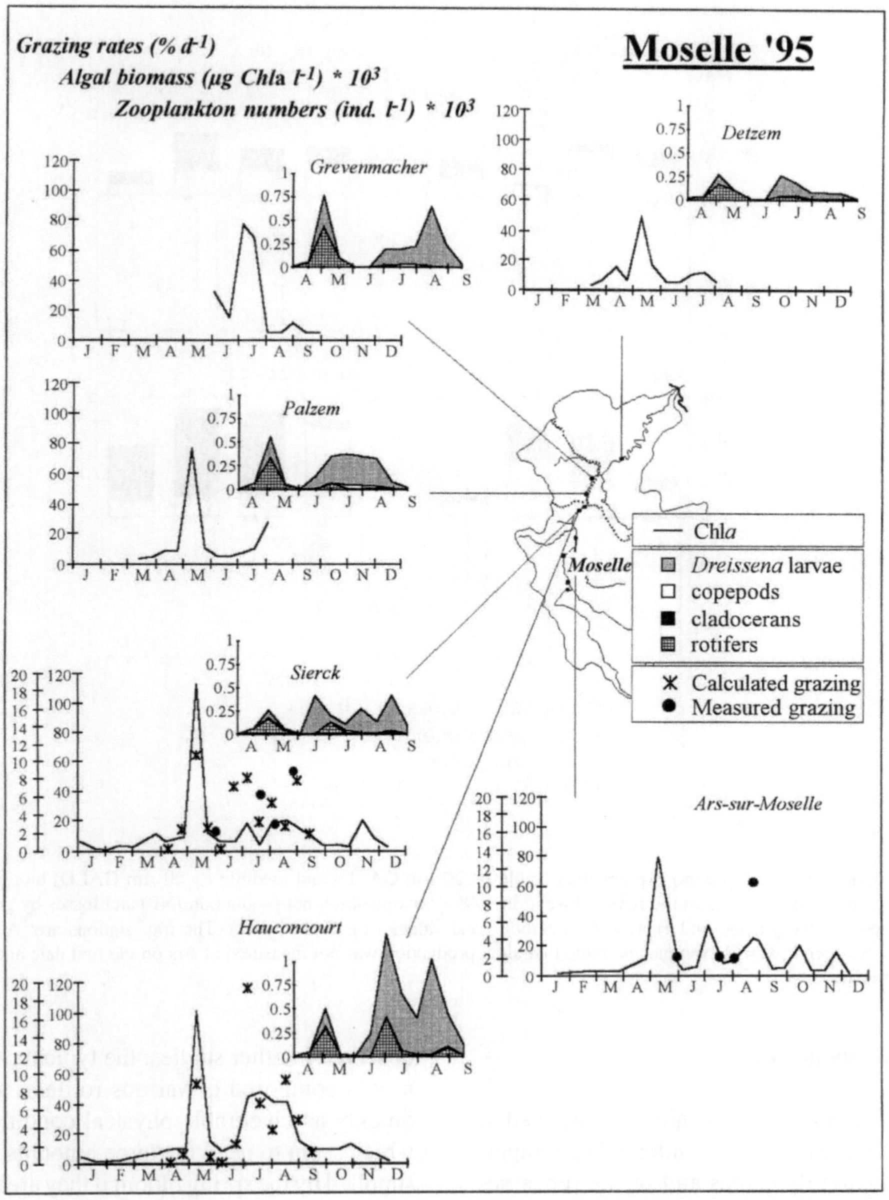

Figure 6. Variations of Chl *a* (μg Chl *a* l^{-1}; data from the IRH, Nancy, France, from the DEAL, Luxembourg and from the LfW, Germany), and variations of zooplankton, expressed in numbers (ind l^{-1}) at the different stations in the River Moselle during the growing season in 1995. Measured grazing (CFR) and back-calculated grazing (BCG) overlaid on Chl *a* curves for 3 of the sampling stations.

C m^{-2} d^{-1}), may exceed phytoplankton production. Indeed, algal production may be very low in the deepest sectors, as in Grevenmacher. In 1995, as could be expected from the low zooplankton numbers, grazing rates were also relatively low, between 1 and 11% d^{-1} (Figure 6). Considering the edible phytoplankton fraction, losses through grazing (2.9–245.5 mg C m^{-2} d^{-1}; Figure 7) represented up to 22% of algal net production (< 0–1690 mg C m^{-2} d^{-1}; Figure 7). The back-calculated filtering rates (BCG; Figure 6) for 1995 show, by contrast with the River Meuse, that the crustacean contribution to community grazing may approach or exceed that of rotifers. They are in agreement with total community filtering rates.

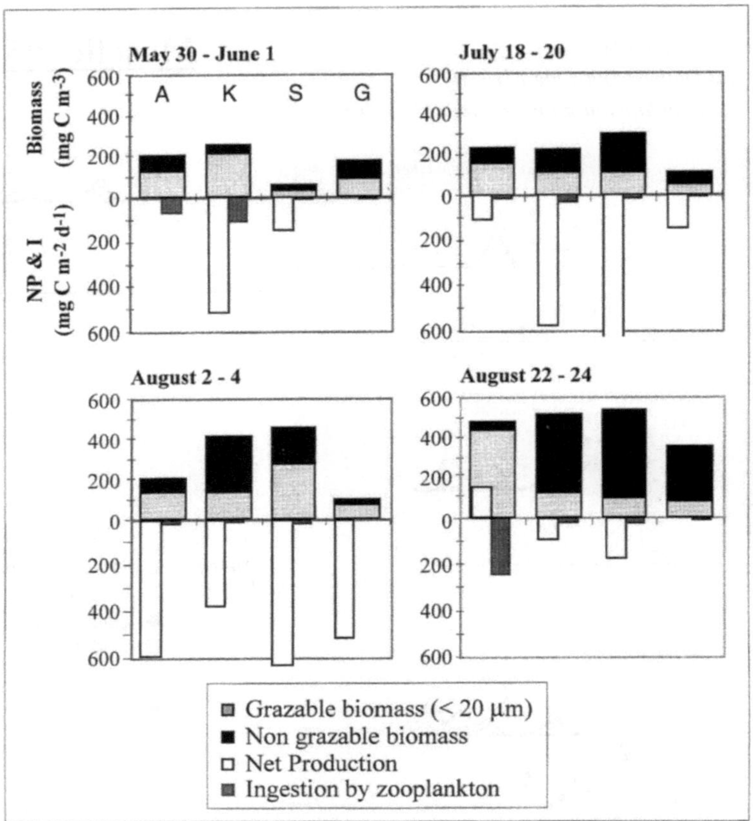

Figure 7. Phytoplankton biomass (top bars) expressed as edible (< 20 μm GALD) and inedible (> 20 μm GALD) biomass at the different stations and for the four sampling series in the River Moselle in 1995. Phytoplankton net production (NP) and losses by grazing (I) are given in parallel (bottom part of the graphs) and expressed as carbon areal values (mg C m^{-2} d^{-1}). The four stations are: A = Ars-sur-Moselle; K = Koenigsmacher; S = Sierck; G = Grevenmacher. Note that algal production was not measured at Ars on the first date and at Grevenmacher at the first and fourth dates.

Discussion and conclusion

Although different sampling strategies were applied to the two rivers, the datasets permit interesting comparisons of the plankton dynamics and of the processes involved. Both rivers exhibit a similar phytoplankton, which responded to discharge, light and temperature in predictable ways (see e.g. Descy et al., 1987). Notable differences between them surround the events in summer, after the onset of zooplankton development in 1994 and 1995: the phytoplankton decrease was of short duration in the Meuse, while a low algal biomass persisted through the whole summer in the Moselle. In both rivers, the proportion of large units of phytoplankton increased in the mid-summer algal assemblage.

The sequence of zooplankton development was similar in both rivers and conforms to what has been observed in other studies: the typical riverine community is composed of various rotifers, able to develop quickly as favourable physical conditions occur, and which seem to need the large amounts of edible algae supplied by the spring bloom if they are to achieve maximal development (Marneffe et al., 1996; de Ruyter van Steveninck et al., 1992). The difference between the two rivers lies in the low summer biomass of zooplankton in the Moselle, at a time when, in the Meuse, zooplankton achieve the largest numbers and biomass. It is also worth noticing the greater contribution by micro-crustaceans in the Moselle. The rivers also differ in the abundance of *Dreissena* veligers present, which reached much higher numbers in the Moselle (this study and Viroux, 1997).

When it comes to zooplankton grazing, the results from the Meuse clearly point to a significant influence on the phytoplankton community. The key observations are the following:

(i) the buildup of the rotifer community in spring corresponds to the decline of the algal spring bloom; this does not exclude other losses, for instance, through enhanced sedimentation or mortality by parasitism (Gosselain et al., 1994);

(ii) the dense rotifer assemblage was able to graze the algae at rates exceeding 20% d^{-1} throughout the summer; the measured grazing rates reached 32% d^{-1}, which can be compared to the values calculated for the River Rhine by de Ruyter van Steveninck et al. (1992) from the rotifer populations developed in the river. Values resulting from simulations with mathematical models (Gosselain et al., 1994; Garnier et al., 1995) are in the same range;

(iii) a distinct change in the size structure of the algal community occurred in summer; it may be due to highly size-selective grazing pressure by small zooplankton; their grazing may at times exceed primary production rates and thus control the biomass of small algal units edible by rotifers (Gosselain et al., 1998).

The grazing rates measured in either river clearly reflect rotifer abundance and this observation validates calculations of grazing losses based on zooplankton numbers. However, there seems to be a significant influence of community composition: for instance, rotifer assemblages dominated by *Brachionus* spp. typically graze at higher rates than those dominated by smaller taxa, as *Keratella* spp.. Community grazing rates may also increase when crustaceans, even small taxa, represent a large share of the zooplankton biomass, as in the Moselle in summer.

However, the summer decline in the Moselle is still far from being fully explained. In spite of favourable physical conditions, phytoplankton and zooplankton biomass remained at low levels from mid-summer on and measurements of zooplankton grazing were in a lower range than in the Meuse. Nevertheless, it is to be emphasized that all *in situ* grazing assays were performed in summer, after the phytoplankton and zooplankton blooms. When considering the spring rotifer numbers, it appears that the decrease of the spring algal bloom was likely due to grazing by this abundant zooplankton. For instance, back-calculated grazing (BCG) for June 22, 1994 gives values ranging between 93 and 165% d^{-1} depending on to the station but reached only 13% d^{-1} at the plankton spring bloom in 1995.

Further, more crustaceans have been recorded in summer in the Moselle than in the Meuse. The role of crustaceans is still unclear, owing to the probable patchiness of these organisms. Their grazing activity may not have been correctly evaluated, as indicated by the number of animals retrieved from the Haney chamber, compared to the abundance estimates from larger sampling volumes of river water.

Whatever the uncertainties remaining about zooplankton grazing in the River Moselle, it appears that the long-lasting summer decline of the plankton of this river requires another explanation. Various data collected during the period of study (Bachmann et al., 1997) indicate that high densities of benthic filter-feeders exist in the River Moselle. Although the mean densities are difficult to evaluate due to extreme patchiness of the bottom substrata, areal densities (ind. m^{-2}) between 700 and 7000 *D. polymorpha* adults, 17 to 8800 *Corbicula* sp. and 12 to 97500 *Corophium curvispinum* have been reported in the studied river section. As mentioned in the introduction, several studies point to a very dense incidence of benthic filter-feeders in rivers where they can develop high populations. In the Moselle, predation by these animals, which are able to ingest particles over a wide size spectrum, might explain the summer depression of both phytoplankton and zooplankton.

We conclude from this comparison between two eutrophic lowland rivers that significant biotic interactions may develop within the potamoplankton. In contrast to other studies dealing with the interaction between zoo- and phytoplankton in rivers (de Ruyter van Steveninck, 1992; Basu & Pick, 1997), we based our approach on direct measurements of zooplankton grazing. In this way, daily grazing rates are directly comparable with phytoplankton production rates, as well as to other loss rates (sedimentation, mortality,...). Our results point to a significant zooplankton control over phytoplankton, provided that the zooplankton community be well-developed and diversified. In the Meuse and Moselle rivers, zooplankton biomass can reach 500 μg dw l^{-1} (Marneffe et al., 1996) and 1600 μg dw l^{-1} (this study), respectively. Consequently, potential grazing rates may be even higher than those measured in our study. One of the most important features of the phytoplankon – zooplankton interaction in rivers is its strong dependency upon flow rate, which must be slow enough for a long period of time (several weeks) to allow zooplankton development along the course of the river. The zooplankton assemblage, dominated by r-strategists, is inherently unstable and may fluctuate, so that the grazing pressure on planktonic algae is highly variable. For instance, there is a large within-year and year-to-year variability of zooplankton in the Meuse and Moselle rivers related to meteorological and hydrological fluctuations. However, another cause of zooplankton variability may be predation, for instance by benthic filter-feeders, as it seems to be the case in the River Moselle.

References

Admiraal, W., L. Breebart, G. M. J. Tubbing, B. van Zanten, E. D. de Ruyter van Steveninck & R. Bijkerk, 1994. Seasonal variation in composition and production of planktonic communities in the lower River Rhine. Freshwat. Biol. 32: 519–531.

Aldridge, C. W., B. S. Payne & A. C. Miller, 1995. Oxygen consumption, nitrogenous excretion, and filtration rates of *Dreissena polymorpha* at acclimation temperatures between 20 and 32 °C. Can. J. Fish. aquat. Sci. 52: 1761–1767.

Bachmann, V., E. Cegielka, P. Wagner, P. Usseglio-Poltera & J. C. Moreteau, 1997. Installation de l'amphipode *Corophium curvispinum* et de la palourde asiatique *Corbicula* sp. dans le cours français de la Moselle – Establishment of the Amphipod *Corophium curvispinum* and the Asiatic Clam *Corbicula* sp. in the French part of the Mosel river. Hydroécol. Appl. 7: 185–191.

Basu, B. K. & F. R. Pick, 1997. Phytoplankton and zooplankton development in a lowland, temperate river. J. Plankton Res. 19: 237–253.

Böhme, M., 1994. Release and consumption of oxygen by a phytoplankton dominated community of a eutrophic lowland river. Verh. int. Ver. Limnol. 25: 1585–1589.

de Ruyter van Steveninck, E. D., W. Admiraal, L. Breebaart, G. M. J. Tubbing & B. van Zanten, 1992. Plankton in the River Rhine: structural and functional changes observed during downstream transport. J. Plankton Res. 14: 1351–1368.

Descy, J.-P., 1993. Ecology of the phytoplankton of the River Moselle: effects of disturbances on community structure and diversity. Hydrobiologia 249: 111–116.

Descy, J.-P. & C. Willems, 1991. Contribution à la connaissance du phytoplancton de la Moselle (France). Cryptogamie, Algol. 12: 87–100.

Descy, J.-P. & V. Gosselain, 1994. Ecological importance and dynamics of phytoplankton in a large lowland river (River Meuse, Belgium). Hydrobiologia 289: 139–155.

Descy, J.-P. & A. Métens, 1996. Biomass-pigment relationships in potamoplankton. J. Plankton Res. 18: 1557–1566.

Descy, J.-P., P. Servais, J. S. Smith, G. Billen & E. Everbecq, 1987. Phytoplankton biomass and production in the River Meuse (Belgium). Wat. Res. 21: 1557–1566.

Downing, J. A. & F. H. Rigler, 1984. A manual on methods for the assessment of secondary productivity in fresh waters. IBP Handbook 17; Blackwell scientific publications, Oxford.

Effler, S. W., C. M. Brooks, K. Whitehead, B. Wagner, S. M. Derr, M. Perkins, C. A. Siegfried, L. Walrath & R. P. Canale, 1996. Impact of zebra mussel invasion on river water-quality. Wat. Env. Res. 68: 205–214.

Garnier, J., G. Billen & M. Coste, 1995. Seasonal succession of diatoms and chlorophyceae in the drainage network of the River Seine: Observations and modelling. Limnol. Oceanogr. 40: 750–765.

Gawler, M. & R. Chapuis, 1987. An improved version of the Haney grazing chamber. Freshwat. Biol. 18: 1–4.

Gosselain, V., J.-P. Descy & E. Everbecq, 1994. The phytoplankton community of the River Meuse, Belgium: seasonal dynamics (year 1992) and the possible incidence of zooplankton grazing. Hydrobiologia 289: 179–191.

Gosselain, V., L. Viroux, & J.-P. Descy, 1998. Can a community small-bodied grazers control phytoplankton in rivers? Freshwat. Biol. 39: 9–24.

Gosselain, V., C. Joaquim-Justo, L. Viroux, M. Mena, A. Métens, J.-P. Descy & J.-P. Thomé, 1996. Laboratory and *in situ* grazing rates of freshwater rotifers and their contribution to community grazing rates. Arch. Hydrobiol. Suppl. 113: 353–361.

Hamilton, P., 1990. The revised edition of a computerized plankton counter for plankton, periphyton and sediment diatom analyses. Hydrobiologia 194: 23–30.

Haney, J. F., 1971. An *in situ* method for the measurement of zooplankton grazing rates. Limnol. Oceanogr. 16: 971–977.

Köhler, J., 1995. Growth, production and losses of phytoplankton in the lowland River Spree: carbon balance. Freshwat. Biol. 34: 501–512.

MacIssac, H. J., C. J. Lonnee & J. H. Leach, 1995. Suppression of microzooplankton by zebra mussels: importance of mussel size. Freshwat. Biol. 34: 379–387.

MacIsaac, H. J., W. G. Sprules, O. E. Johannsson & J. H. Leach, 1992. Filtering impacts of larval and sessile zebra mussels (*Dreissena polymorpha*) in western Lake Erie. Oecologia 92: 30–39.

Marneffe, Y., J.-P. Descy & J.-P. Thomé, 1996. The zooplankton of the lower River Meuse, Belgium: seasonal changes and impact of industrial and municipal discharges. Hydrobiologia 317: 1–16.

Reynolds, C. S., 1986. Experimental manipulations of the phytoplankton periodicity in large limnetic enclosures in Blelham Tarn, English Lake District. Hydrobiologia 138: 43–64.

Reynolds, C. S. & J.-P. Descy, 1996. The production, biomass and structure of phytoplankton in large rivers. Arch. Hydrobiol. Suppl. 113: 161–187.

RIWA, 1996a. Rapport annuel 1994 – Tome B: la Meuse, RIWA, Amsterdam, septembre 1996, 144 pp.

RIWA, 1996b. Rapport annuel 1995 – Tome B: la Meuse, RIWA, Amsterdam, février 1996, 128 pp.

Smayda, T. J., 1978. From phytoplankters to biomass. In Sournia, A. (ed), Phytoplankton Manual. UNESCO, Paris: 273–279.

Sprung, M., 1995. Physiological energetics of the zebra mussel *Dreissena polymorpha* in lakes. II. Food uptake and gross growth efficiency. Hydrobiologia 304: 133–146.

Vannote, R. L., G. W. Minshall, K. W. Cummins, J. R. Sedell & C. E. Cushing, 1980. The river continuum concept. Can. J. Fish. aquat. Sci. 37: 130–137.

Viroux, L., 1997. Zooplankton development in two large lowland rivers, the Moselle (France) and the Meuse (Belgium), in 1993. J. Plankton Res. 19: 1743–1762.

Physical & chemical data

Site		" La Plante "
Location	latitude	50° 26' 45'' N
	longitude	4° 50' 40'' E
Mean depth		3.95 m
Mean width		100 m
Residence time		~ 4 days for a 30 km sector in summer conditions
Mixing frequency		always mixed and homogenous for phytoplankton
Ice cover		none

year **1994**

		min.	Max.	Mean	n
Total P	µg P l^{-1}	62.7	187	132.8	16
Total N	mg N l^{-1}	2.3	16.3	4.3	(12)[a]
SRP	µg P l^{-1}	5	210	71	50[a]
DIN	mg N l^{-1}	1.4	3.6	2.4	15
pH		7.4	8.8	8.2	24
Alkalinity	meq l^{-1}	2.2	4.0	3.3	21
Suspended matter	mg l^{-1}	2.6	106	17.5	52[a]
Extinction coefficient	m^{-1}	1.2	4.7	2.4	25

year **1995**

		min.	Max.	Mean	n
Total P	µg P l^{-1}	59	296	169.8	9
Total N	mg N l^{-1}	3.3	3.8	3.6	(9)[a]
SRP	µg P l^{-1}	2	297	55	52[a]
DIN	mg N l^{-1}	1.7	4.1	2.6	27
pH		7.3	8.8	8.1	16
Alkalinity	meq l^{-1}	1.4	3.8	2.9	52[a]
Suspended matter	mg l^{-1}	3.0	204	26.3	50[a]
Extinction coefficient	m^{-1}	1.7	4.3	2.8	15

[a] values from RIWA (1996a, 1996b) at Tailfer, close to the sampling station. Numbers of samples (n) are in brackets when the values are calculated from monthly means.

Phytoplankton data (see material & methods chapter for more explanations)

year		1994				
Period of sampling		March 23 - September 25 (6 months)				
Frequency of sampling		once a week n = 23				
Characteristic sequences of species abundance		1) *Stephanodiscus hantzschii* group 2) *S. hantzschii* + small *Cyclotella* sp. 3) *S. hantzschii* + *Chlorella*-like (+ *C.meneghiniana*) 4) *Chlorella*-like + *Scenedesmus* spp. 5) *Chlorella*-like + *Scenedesmus* spp. + *C. meneghiniana* 6) *Chlorella*-like + *Scenedesmus* spp. + *S. hantzschii*				
		Biomass according to Eppley			***Total biovolume***	
Number of species accounting for ~ of the annual biomass	> 0.5 %	17			13	
	> 5 %	2 (AU SPP & CE UNI)			2	
	> 25 %	1 (CE UNI)			1	
Number of biomass phases of dominance per year		1 Diatoms always dominant				
Equilibrium phases ?		(*)				
Number of distinguishable peaks of biomass per year		4				
Maximal biomass of dominant species at those peaks		1) 1.292 mg C m^{-3}	5.103 mg C m^{-2}	(CE UNI)	13.497 mm^3 m^{-3}	53.313 mm^3 m^{-2}
		2) 950 mg C m^{-3}	3.753 mg C m^{-2}	(CE UNI)	11.474 mm^3 m^{-3}	45.322 mm^3 m^{-2}
		3) 1.372 mg C m^{-3}	5.419 mg C m^{-2}	(AU SPP)	17.503 mm^3 m^{-3}	69.137 mm^3 m^{-2}
		4) 972 mg C m^{-3}	3.839 mg C m^{-2}	(CE UNI)	12.269 mm^3 m^{-3}	48.462 mm^3 m^{-2}
Mean biomass	per week	884 mg C m^{-3}	3.491 mg C m^{-2}		8.947 mm^3 m^{-3}	35.343 mm^3 m^{-2}
(for the period March-Sept.)	per month	3.977 mg C m^{-3}	15.709 mg C m^{-2}		40.264 mm^3 m^{-3}	159.044 mm^3 m^{-2}
Peak/mean biomass ratio		1) 1,46			2,05	
		2) 1,08			1,47	
		3) 1,55			2,73	
		4) 1,1			1,96	
Shannon-Wiener diversity (H'')		3,64			3,23	
The equitability		0,42			0,37	

year		1995				
Period of sampling		March 23 - October 10 (6.5 months)				
Frequency of sampling		once a week n = 22				
Characteristic sequences of species abundance		*Unicellular centric diatoms not yet examined*				
		Biomass according to Eppley			***Total biovolume***	
Number of species accounting for ~ of the annual biomass	> 0.5 %	14			8	
	> 5 %	1 (CE UNI)			2 (CE UNI & AU SPP)	
	> 25 %	1 (CE UNI)			1 (CE UNI)	
Number of biomass phases of dominance per year		1 Diatoms always dominant				
Equilibrium phases ?		(*)				
Number of distinguishable peaks of biomass per year		3 *not taking into account atypical peak of Sept.*				
Maximal biomass of dominant species at those peaks		1) 1.157 mg C m^{-3}	4.570 mg C m^{-2}	(CE UNI)	12.066 mm^3 m^{-3}	47.660 mm^3 m^{-2}
		2) 2.256 mg C m^{-3}	8.911 mg C m^{-2}	(CE UNI)	28.128 mm^3 m^{-3}	111.105 mm^3 m^{-2}
		3) 634 mg C m^{-3}	2.504 mg C m^{-2}	(CE UNI)	9.884 mm^3 m^{-3}	39.042 mm^3 m^{-2}
Mean biomass	per week	983 mg C m^{-3}	3.884 mg C m^{-2}		10.463 mm^3 m^{-3}	41.330 mm^3 m^{-2}
(for the period March-Oct.)	per month	4.388 mg C m^{-3}	17.333 mg C m^{-2}		46.682 mm^3 m^{-3}	184.395 mm^3 m^{-2}
Peak/mean biomass ratio		1) 1,57			1,58	
		2) 3,01			3,09	
		3) 0,99			1,23	
Shannon-Wiener diversity (H'')		*not calculated as unicellular centric diatoms have not yet been examined*				
The equitability		*id.*				

(*) As biomass was not calculated for each unicellular centric diatoms, we can not precisely distinguish equilibrium phases according to biomass. Nevertheless when *S. hantzschii* or *C. meneghiniana* are dominant in terms of abundance they are also in terms of biomass.
CE UNI = unicellular centric diatoms; AU SPP = Aulacoseira spp.

Annex 2 - The River Moselle

Physical & chemical data

Site		" Sierck "
Location	latitude	49° 26' 40'' N
	longitude	6° 22' E
Mean depth		3.00 - 3.50 m
Mean width		100 m
Residence time		~ 5 days for a 25 km sector in summer conditions
Mixing frequency		always mixed and homogenous for phytoplankton
Ice cover		none

year **1994**

		min.	Max.	Mean	n
Total P*	µg P l^{-1}	190	780	350	24
Total N*	mg N l^{-1}	2.2	7.0	3.9	24
SRP*	µg P l^{-1}	90	360	210	24
DIN*	mg N l^{-1}	1.4	5.3	3.3	24
NO_3^-*	mg N l^{-1}	1.4	5.0	3.0	24
pH*		7.0	8.8	7.7	24
Alkalinity	meq l^{-1}	-	-	-	-
Suspended matter*	mg l^{-1}	6.6	198.0	30.9	24
Extinction coefficient(1)	*m^{-1}*	*1.8*	*2.9*	*2.1*	*10*

year **1995**

		min.	Max.	Mean	n
Total P*	µg P l^{-1}	190	510	340	24
Total N*	mg N l^{-1}	1.5	6.2	3.6	24
SRP*	µg P l^{-1}	100	400	216	24
DIN*	mg N l^{-1}	1.0	4.8	3.1	24
NO_3^-*	mg N l^{-1}	0.5	4.5	2.8	24
pH*		6.9	8.7	7.7	24
Alkalinity	meq l^{-1}	2.2	3.0	2.6	4
Suspended matter*	mg l^{-1}	7.2	91.0	25.3	24
Extinction coefficient	m^{-1}	1.8	3.7	2.5	4

* Data from the "Agence de l'eau Rhin-Meuse"

(1) Values from Koenigsmacher, 1993

Phytoplankton data (see material & methods for more explanations)

As data from 1993 and 1994 did not cover the whole plankton growing season, we chose to present data from 1993.
Remember than units were not measure during countings but a single value of biovolume apply per taxon.
On the contrary to the river Meuse data, countings gave a taxinomic list of genus or species present in the river Moselle.

year			**1993**				
Period of sampling			March 29 - October 11 (6.5 months)				
Frequency of sampling			forthnitghly n = 14				
Characteristic sequences of species abundance		1)	*Stephanodiscus hantzschii* group (+ *Cyclostephanos dubius*)				
		2)	*Skeletonema potamos* + *Scenedesmus* div. sp. (+ *Dictyosphaerium* div. sp. + *Cyclotella meneghiniana*)				
		3)	*S. potamos* + *C. meneghiniana*				
		4)	*Dictyosphaerium* div. sp. + *Scenedesmus* div. sp. (+ *C. meneghiniana*)				
		5)	*S. potamos* + *S. subsalsum*				
		6)	*S. potamos* + *S. subsalsum* + *Scenedesmus* div. sp.				
		7)	*Synechocystis* sp. + *Scenedesmus* div. sp.				
		8)	*Scenedesmus* div. sp. + *S. potamos* + *S. subsalsum*				
		9)	*S. potamos* + *Aulacoseira* div. sp.				
			Biomass according to Eppley			***Total biovolume***	
Number of species accounting for	> 0.5 %		25			20	
~ of the annual biomass	> 5 %		5			3	
	> 25 %		*(CE UNI)*			1 (*Peridinium* div. sp.)	*(+ CE UNI)*
Number of biomass phases of dominance per year			1 Diatoms always dominant				
Equilibrium phases ?			-				
Number of distinguishable peaks of biomass per year			2				
Maximal biomass of dominant species at those peaks		1)	2.094 mg C m^{-3}	8.271 mg C m^{-2}	*S. hantzschii* group	5.218 mm^3 m^{-3}	20.611 mm^3 m^{-2}
		2)	2.506 mg C m^{-3}	9.899 mg C m^{-2}	*C. meneghiniana*	9.728 mm^3 m^{-3}	38.426 mm^3 m^{-2}
			1.904 mg C m^{-3}	7.521 mg C m^{-2}	*Peridinium* div. sp.	8.625 mm^3 m^{-3}	34.069 mm^3 m^{-2}
			423 mg C m^{-3}	1.671 mg C m^{-2}	*S. potamos*	938 mm^3 m^{-3}	3.705 mm^3 m^{-2}
Mean biomass	per week		1.824 mg C m^{-3}	7.205 mg C m^{-2}		5.048 mm^3 m^{-3}	19.940 mm^3 m^{-2}
(for the period March-Oct.)	per month		4.210 mg C m^{-3}	16.630 mg C m^{-2}		11.651 mm^3 m^{-3}	46.021 mm^3 m^{-2}
Peak/mean biomass ratio		1)	1,6			1,4	
		2)	3,7			4,8	
Shannon-Wiener diversity (H'')						2,69	
The equitability						0,29	

Hydrobiologia **369/370:** 217–227, 1998.
M. Alvarez-Cobelas, C. S. Reynolds, P. Sanchez-Castillo & J. Kristiansen (eds), Phytoplankton and Trophic Gradients.

The phytoplankton succession in the lower part of hypertrophic Nakdong River (Mulgum), South Korea

Kyong Ha, Hyun-Woo Kim & Gea-Jae Joo
Department of Biology, Pusan National University, Pusan, S. Korea, 609-735

Key words: regulated river, *Microcystis* bloom, discharge, temperature, community composition

Abstract

Investigations were carried out to determine the mechanism of phytoplankton succession in the lower part of Nakdong River. Intensive monitoring was conducted from April 1993 to April 1995 at weekly or biweekly intervals. This river is the main source of drinking water for more than 8 million residents living in Pusan and the Southeastern region of Korea, and it is also important for industrial purposes. Due to the rapid urbanization and industrialization, this river has become hypertrophic and it is heavily regulated (mean chl. a, 79 ± 232 μg l^{-1} n = 76; mean TN, 4.3 ± 2.0 mg l^{-1} n = 30; mean TP, 165 ± 108 μg l^{-1} n = 45). Bacillariophyceae were dominant year-round in 1993, with spring and late fall peaks. Dominant communities in 1994 were Bacillariophyceae in March and April, Chlorophyceae and Cryptophyceae in May, Cyanobacteria in July and August, Chlorophyceae and Cryptophyceae in October, and Bacillariophyceae in December. As drought persisted through the summer of 1994, elevated water temperature (over 30 °C) possibly triggered Cyanobacteria bloom (mainly *Microcystis aeruginosa*, maximum cell density, 1.6×10^7 cells ml^{-1}). The most common diatom, *Stephanodiscus hantzschii*, repeatedly dominated from late fall to next spring (mean and maximum cell density, 7.5×10^4, 1.2×10^5 cells ml^{-1}, respectively). This small centric diatom was favored by the low discharge and the cold water (water temperature, 3–10 °C). After the diatom bloom, this community shifted to peaks of colonial Chlorophyceae and motile Cryptophyceae, owing to the high rate of zooplankton grazing activity and increased water temperature. Overall, the phytoplankton periodicity was primarily governed by the hydrologic regime (discharge). Changes in silica concentration, water temperature and high zooplankton density might have played an important role in phytoplankton dynamics during the non-flooding periods. Compared to other large rivers, strong *Microcystis* bloom events in summer and *Stephanodiscus* bloom events in winter were noticeable in the dry year. The patterns of phytoplankton succession observed in this study may have a significance since most of the large rivers in Far Eastern Asian countries are subjected to eutrophication and regulation of discharge.

Introduction

Seasonal phytoplankton development and succession are governed by such biological factors as competition, grazing and parasitism, and also by such physical and chemical factors as climate, discharge, water temperature, light intensity, conductivity, pH, dissolved oxygen and nutrient availability (Reynolds, 1984a). Especially, the intensity and frequency of disturbances driven by precipitation and discharge are important factors regulating phytoplankton dynamics (Reynolds et al., 1988; Sommer, 1989; Reynolds et al., 1993).

Numerous studies on the seasonal dynamics of the phytoplankton community in aquatic ecosystem have been conducted (Wetzel, 1983; Harris, 1986; Descy, 1993; Garnier et al., 1995). However, the studies of phytoplankton periodicity in the lotic ecosystem have been less studied than in the lentic ecosystem. Recently, the studies of phytoplankton ecology in large rivers are becoming more and more popular (Paerl & Ustach, 1982; Gosselain et al., 1994; Reynolds & Descy, 1996). And even though more and more rivers are starting to behave like reservoir, our knowledge of biological

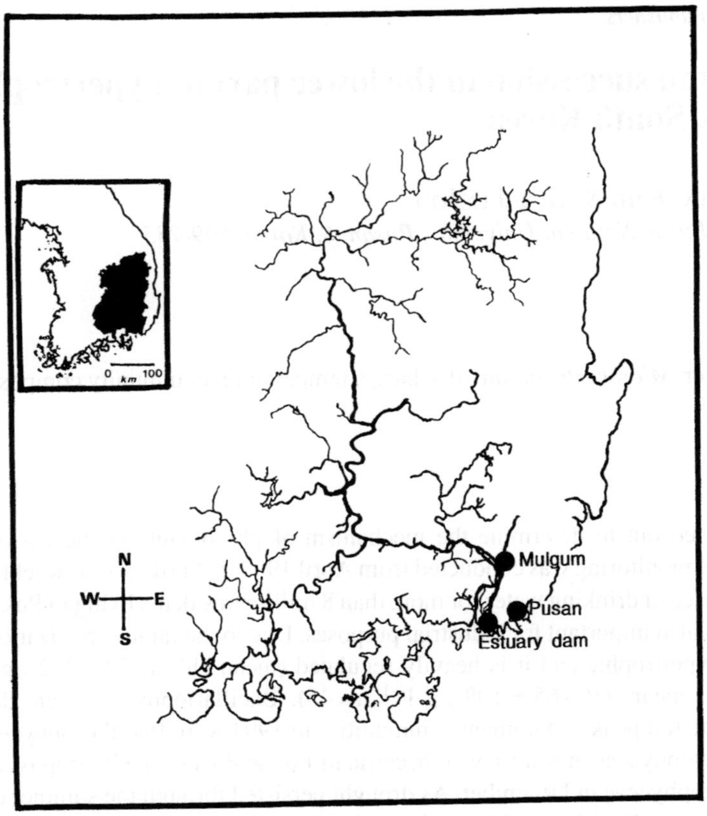

Figure 1. Map showing the basin of the Nakdong River, major tributaries and the study site; Mulgum (•).

community in regulated rivers is receiving little attention from scientists.

The study of the phytoplankton community in the Nakdong River has a history of about 30 years. Phytoplankton flora of this river has been relatively studied more than in others of South Korea. However, the focus was mostly on species composition and community structure, mainly on a monthly or seasonal basis (Joo, 1995). There are few, if any, description of phytoplankton seasonality and factors regulating dynamics of phytoplankton. The Nakdong River provides an excellent opportunity for studying phytoplankton periodicity in a hypertrophic river-reservoir system. Here, we present an investigation of variability in the structure and abundance of phytoplankton in the lower part of that River.

Description of the study site

The Nakdong River, with a main channel of approximately 526 km in length and the drainage area of 23,817 km^2, is the longest river in South Korea. The drainage area is situated in 35–37 °N and 124–131 °E (Figure 1). The river is a major source of drinking water, and it is the main source of the regions agricultural irrigation and industrial supply. Owing to the high demand of water, the river has been drastically modified during the last 3 decades, with 4 dams in the major tributaries and one estuary dam in the mouth to prevent salt water intrusion. These physical alterations and other human influences have caused dramatic changes in hydrologic regime, and they have accelerated eutrophication.

The mean annual precipitation on the central part of the basin is about 1100 mm and the coefficient of river regime is very high (260–372) due to the monsoon and typhoon. The rainfall during summer (June through September) provides about 60–70% of total annual precipitation, while fall and winter are dry with scanty precipitation. Annual discharge is typical of such systems in the monsoon region. Mean discharge at Jindong (rk, river km; 843) was 286 ± 624 m^3 sec^{-1} (1993–1995). In summer, fluctuations in discharge are

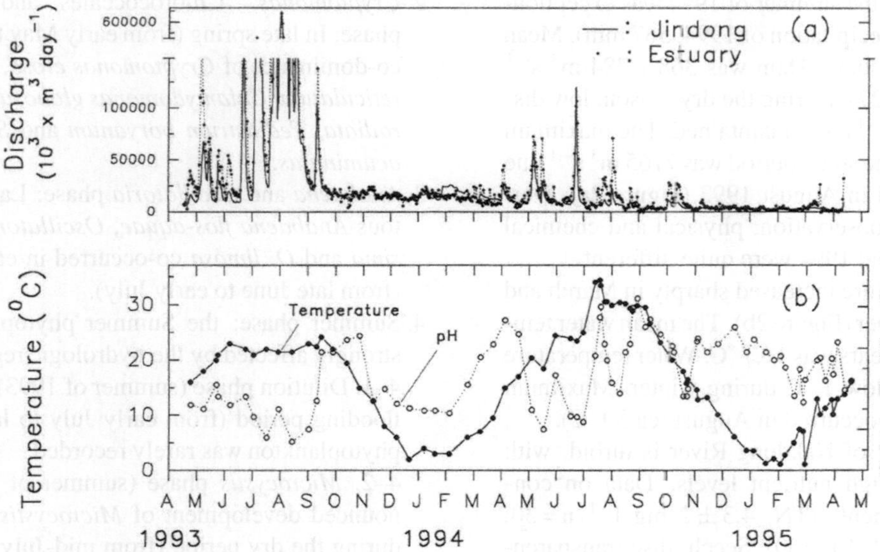

Figure 2. (a) daily discharge of the Nakdong River (– Jindong, rk 84.3; • estuary dam, rk 0) (b) temporal changes in water temperature (•) and pH (o).

highly variable depending on precipitation (*e.g.*, summer of 1993; 7765 m^3 sec^{-1}). From October to May, water level fluctuation is minimal at the lower part of the river (less than 50 cm). The hydrologic retention time here (rk 0–100) is longer than in other stretches due to the regulated discharge at the estuary dam (rk 0). Annual mean retention time for the stretch between rk 0 to rk 200 was 5–6 days. However, when the dry period extends for more than several months, retention time between rk 27 to rk 42 can increase to 1–3 days (Song, 1992).

Mulgum, where the study took place, is located 27 km from the estuary dam (maximum depth ca 10 m; mean depth 4–5 m; river width 250–350 m). Mixing is continuous through the entire water column. The gradient of this river is exceedingly low. Its water surface elevation at rk 200 is only 18.9 m above mean sea level.

Materials and methods

A weekly and/or biweekly sampling was conducted from April 1993 to April 1995. To determine the limnological characteristics of the study site, physical and chemical variables – Secchi disc transparency, water temperature, turbidity, water level, discharge, light intensity, conductivity, alkalinity, dissolved oxygen, chl. *a*, pH, TN, DIN, NO_3^--N, NH_4^+-N, TP, PO_4^{3-}-P and Si concentration – were investigated. Data on daily river discharge at Jindong and at the estuary dam were obtained from the Flood Control Center in Pusan. The daily flushing rate (% remaining water volume day^{-1}) was calculated in order to determine hydrologic retention at the lower part of the river. All calculations were based on the total volume of water between Jindong and the estuary dam. To classify the trophic state of the study site, the OECD (1982) criteria were used.

The phytoplankton samples were immediately preserved with Lugol's solution. Utermöhl's sedimentation method was used to identify and enumerate phytoplankton. Biovolume was estimated from mean cell dimensions and cellular shape. The Shannon diversity (H′, Shannon & Weaver, 1949) was based on cell numbers, even for colonial and filamentous forms, except for the nannoplankton.

Results

Physical and chemical variates

Heavy rainfall occurred at the end of June and until early August. Precipitation in Korea is heavily concentrated in summer. Over 60% of the annual total (1000–1200 mm) falls in summer (from June to September). The annual precipitation measured during the summer of 1993 (annual precipitation, 1181 mm) was higher than the annual average for the previous thirty years

(1036 mm), while the summer of 1994 was exceptionally dry (annual precipitation of 1994, 567 mm). Mean discharge at the Estuary Dam was 368 ± 784 m^3 s^{-1} (1993–1995, n = 727). During the dry season, low discharge (< 30 m^3 s^{-1}) was maintained. The maximum discharge during the study period was 7765 m^3 s^{-1} due to a heavy rainfall in August 1993 (Figure 2a). During the period of observation, physical and chemical features of 1993 and 1994 were quite different.

Water temperature increased sharply in March and decreased in October (Figure 2b). The mean water temperature for both years was 16.7 °C. Water temperature rarely dropped below 1 °C during winter. Maximum water temperature occurred in August (ca 30 °C).

The lower part of Nakdong River is turbid, with hard water and high nutrient levels. Data on concentration of nutrients (TN, 4.3 ± 2 mg l^{-1} n = 30; TP, 165 ± 108 μg l^{-1} n = 60), Secchi disc transparency (73 ± 24 cm; n = 69) and phytoplankton biomass (mean chl. *a*, 78.6 ± 232 μg l^{-1} n = 75; chl. *a* peak value, 231 ± 43 μg l^{-1} n = 5) showed a range for hypertrophic waters (Table 1). During flooding events (especially in August and September of 1993) transparency was low (< 25 cm) as a result of silt. During the dry year of 1994, a low transparency (< 40 cm) in summer was due to *Microcystis aeruginosa* blooms. Temporal changes in nutrient concentrations (TN, DIN, NO_3^--N, NH_4^+-N, TP, PO_4^{3-}-P) were less conspicuous than changes in silica concentration. We observed sharp drops in silica concentration (< 0.05 mg l^{-1}), several times during the low discharge season.

Species composition and temporal dynamics

During our study, 185 algal taxa were identified. Six phytoplankton groups -Chlorophyceae, Bacillariophyceae, Cryptophyceae, Cyanobacteria, Euglenophyceae and Dinophyceae- were represented. Bacillariophyceae were abundant year-round. The number of species accounting for more than 0.5% of the biannual biomass was 28 (Table 2). Only four of these contributed more than 5% to total biomass (as biovolume). *Stephanodiscus hantzschii* (25.8% of the total biovolume) was the dominant species and *Microcystis aeruginosa* (10.8%), *Cyclostephanos dubius* (6.2%) and *Synedra acus* (5.1%) followed.

The succession of species contributing more than 0.5% to total biovolume can be summarized as follows:

1. *Stephanodiscus hantzschii* was dominant (over 90% of total biovolume) in spring.
2. *Cryptomonas*, Chlorococcales and *Cyclotella* phase: In late spring (from early May to mid-June) co-dominance of *Cryptomonas erosa, Coelastrum reticulatum, Chlamydomonas globosa, Golenkinia radiata, Pediastrum boryanum* and *Scenedesmus acuminatus*.
3. *Anabaena* and *Oscillatoria* phase: Large filamentous *Anabaena flos-aquae, Oscillatoria angustissima* and *O. limosa* co-occurred in early summer (from late June to early July).
4. Summer phase: the Summer phytoplankton was strongly affected by the hydrologic regime.
   4–1. Dilution phase (summer of 1993): During the flooding period (from early July to late August), phytoplankton was rarely recorded.
   4–2. *Microcystis* phase (summer of 1994): Pronounced development of *Microcystis aeruginosa* during the dry period (from mid-July to late September).
5. Mixed Bacillariophyceae, Cryptophyceae and Chlorococcales phase: After the Cyanobacteria bloom, *Aulacoseira italica, A. distans, Cryptomonas erosa, Coelastrum* spp. and *Scenedesmus* spp. increased.
6. Large Bacillariophyceae phase: During late fall (from late October to mid-November), phytoplankton was dominated by *Melosira varians, Aulacoseira granulata, A. italica, Synedra acus* and *Cyclostephanos dubius, Fragilaria crotonensis*, which increased at the end of fall, continued into the winter.
7. Winter phase: Flushing rate now became a primary determinant.
   7–1. *Fragilaria* and *Cyclotella* phase: In the winter of the rainy year (from December 1993 to late February 1994), small *Cyclotella* were followed by *Fragilaria crotonensis*.
   7–2. *Stephanodiscus hantzschii* bloom phase: In the dry winter with low discharge (from mid-December 1994 to late January 1995), a large *Stephanodiscus hantzschii* bloom appeared.

Dynamics of the phytoplankton community

1. Temporal changes in phytoplankton community structure.

The temporal changes in the phytoplankton community of the Nakdong River were remarkable. Bacillariophyceae were dominant year-round in 1993, producing spring and late fall peaks. However, dominant communities in 1994 were Bacillariophyceae-

Table 1. Physical and chemical characteristics of the lower part of the Nakdong River. All samples were collected weekly and/or biweekly from April, 1993 to April, 1995.

Parameters	Unit	Mean	Standard deviation	Min.	Max.	n
Secchi transparency	cm	73	24	10	120	69
Temperature	°C	16.2	9.3	1.1	34.4	76
Turbidity	NTU	15	48	3	405	73
Discharge	$m^3\ sec^{-1}$	368	784	0	7665	727
pH		8.4	0.8	6.7	10.2	76
Dissolved oxygen	$mg\ l^{-1}$	10.5	3.7	3	19	75
Dissolved oxygen	% Saturation	106	38	37	237	75
Conductivity	$\mu S\ cm^{-1}$	327	121	95	640	76
Nitrate	$mg\ l^{-1}$	2.1	0.9	0.2	4.7	69
Ammonia	$mg\ l^{-1}$	0.6	0.7	0.01	3.72	68
SRP	$\mu g\ l^{-1}$	34	23	0	101	66
Total P	$\mu g\ l^{-1}$	165	108	44	460	36
DIN	$mg\ l^{-1}$	3.4	1.9	0.6	12.4	60
Total N	$mg\ l^{-1}$	4.3	2	1.3	8.5	30
Alkalinity	$mg\ l^{-1}$ as $CaCO_3$	57	16	22	92	74
Reactive Silica	$mg\ l^{-1}$	3.6	2.5	0.2	10.5	69
Chl. *a*	$\mu g\ l^{-1}$	78.6	232	1.4	1973	75

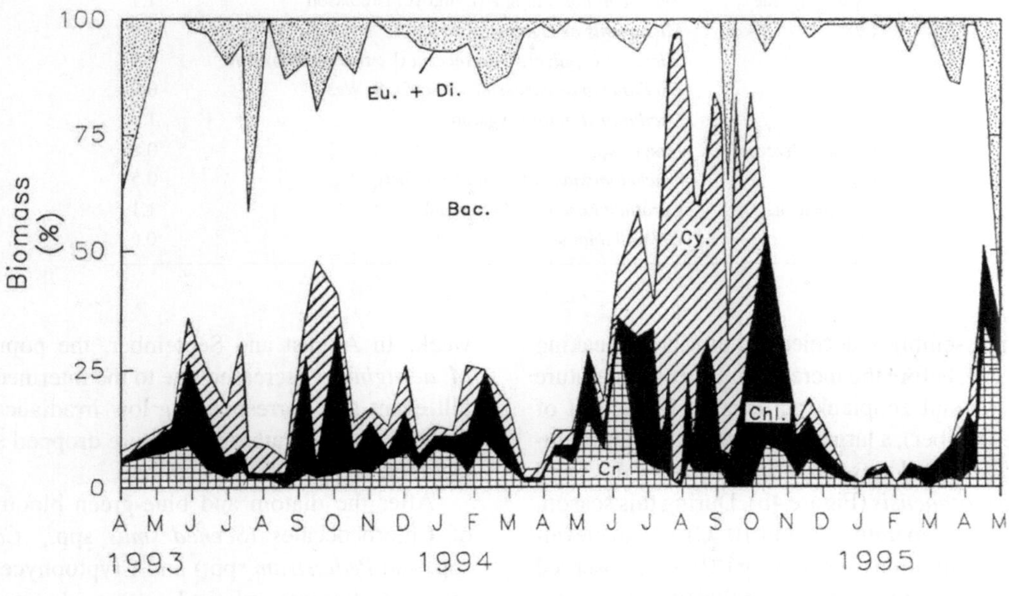

Figure 3. Percent total biovolume of major algal groups in the Nakdong River during the study period (Ch.: Chlorophyceae, Bac.: Bacillariophyceae, Cy.: Cyanobacteria, Eu. + Di.: Euglenophyceae + Dinophyceae).

Chlorococcales and Cryptophyceae-Cyanobacteria-Chlorococcales and Cryptophyceae-Bacillariophyceae (Figure 3). During the rainy year (1993), spring and fall peaks were observed. However, as drought persisted in 1994, the bloom of *Microcystis* was sustained during the summer, and a strong *Stephanodiscus* bloom was observed in winter.

In both seasons of diatom dominance, *Stephanodiscus hantzschii* was the main taxon (Figure 4a). A significant increase in cell number and biovolume of

Table 2. Percent biovolume of 28 dominant species over 0.5% of total biovolume in the Nakdong River during the study period. * indicates total number of species occurred from this study.

Chlorophyceae	*Chlamydomonas globosa* Snow	0.5
(96)*	*Closterium* sp.	2.9
	Coelastrum reticulatum (Dangeard) Senn	0.9
	Staurastrum tetracerum f. *trigona* Lundell	1.3
	Staurastrum tetracerum var.*subexcavatum* Gronbald	1.2
Chrysophyceae	*Cyclotella comta* (Ehrenberg) Kutzing	3.7
(56)*	*Cyclostephanos dubius* (Fricke) Round	6.2
	Cyclotella meneghiniana Kutzing	1.1
	Cyclotella stelligera Cleve and Grunow in Van Heurck	0.8
	Fragilaria crotonensis Kitton	2.5
	Aulacoseira granulata (Ehrenberg) Simonsen	2.9
	Aulacoseira italica (Ehrenberg) Simonsen	3.1
	Melosira varians Agardh	4.1
	Nitzchia palea (Kutzing) W. Smith	0.5
	Stephanodiscus hantzschii (Ehr) Grunow	25.8
	Surirella splendida (Ehrenberg) Kutzing	1.3
	Synedra acus Kutzing	5.1
	Synedra ulna Ehrenberg	1.1
Cryptophyceae (1)*	*Cryptomonas erosa* Ehrenberg	3.7
Cyanophyceae	*Anabaena flos – aquae* (Lyngbye) Brebisson	1.7
(20)*	*Microcystis aeruginosa* Kutzing	10.9
	Microcystis pulverea var.*incerta* (Lemmermann) Crow	2.1
	Oscillatoria angustissima W. et G. S. West	0.9
	Oscillatoria limosa Agardh	1.9
Euglenophyceae	*Phacus* sp.	0.8
(8)*	*Trachelomonas volvocina* Ehrenberg	0.5
Dinophyceae	*Ceratium hirundinella* Schrank	1.1
(4)*	*Glenodinium* sp.	0.6

the diatom assemblage accelerated in March, peaking in early April, before the increase in water temperature (over 16 °C) and zooplankton density. In the fall of 1993 (late October), a large diatom crop was represented primarily by *Aulacoseira italica, Synedra acus* and *Fragilaria crotonensis* (Figure 4b). During this season, a lower water temperature (about 10 °C) and an elevated silica concentration (over 8.6 mg l^{-1}) were observed (Figure 4c). Additional species included *Aulacoseira granulata, Melosira varians, Cyclostephanos dubius* and *Cyclotella comta.*

During the dry summer of 1994, a *Microcystis* bloom developed from mid-July to late October. The number and biovolume of *Microcystis aeruginosa* represented 80–98% of total phytoplankton biomass (Figure 4d). The highest peak appeared on July 21 (1.6×10^7 cells ml^{-1}) and a high level of cell density ($> \times 10^6$ cells ml^{-1}) was maintained for a further week. In August and September, the population of *M. aeruginosa* decreased due to the intermediate rainfall event and corresponding low irradiance in early August. As the water temperature dropped sharply in October, this population declined.

After the diatom and blue-green bloom, species of Chlorococcales (*Scenedesmus* spp., *Coelastrum* spp. and *Pediastrum* spp.) and Cryptophyceae (*Cryptomonas*) became relatively more abundant. Their dominance was observed several times following bloom events.

In the winter of 1994/1995, a diatom bloom (*Stephanodiscus hantzschii* > 90% of total biovolume) occurred. In spite of low water temperatures (less than 4 °C), this bloom was maintained for two months. A noticeable peak (1.1×10^5 cells ml^{-1}, over 80% of total biovolume) developed between late December and early January.

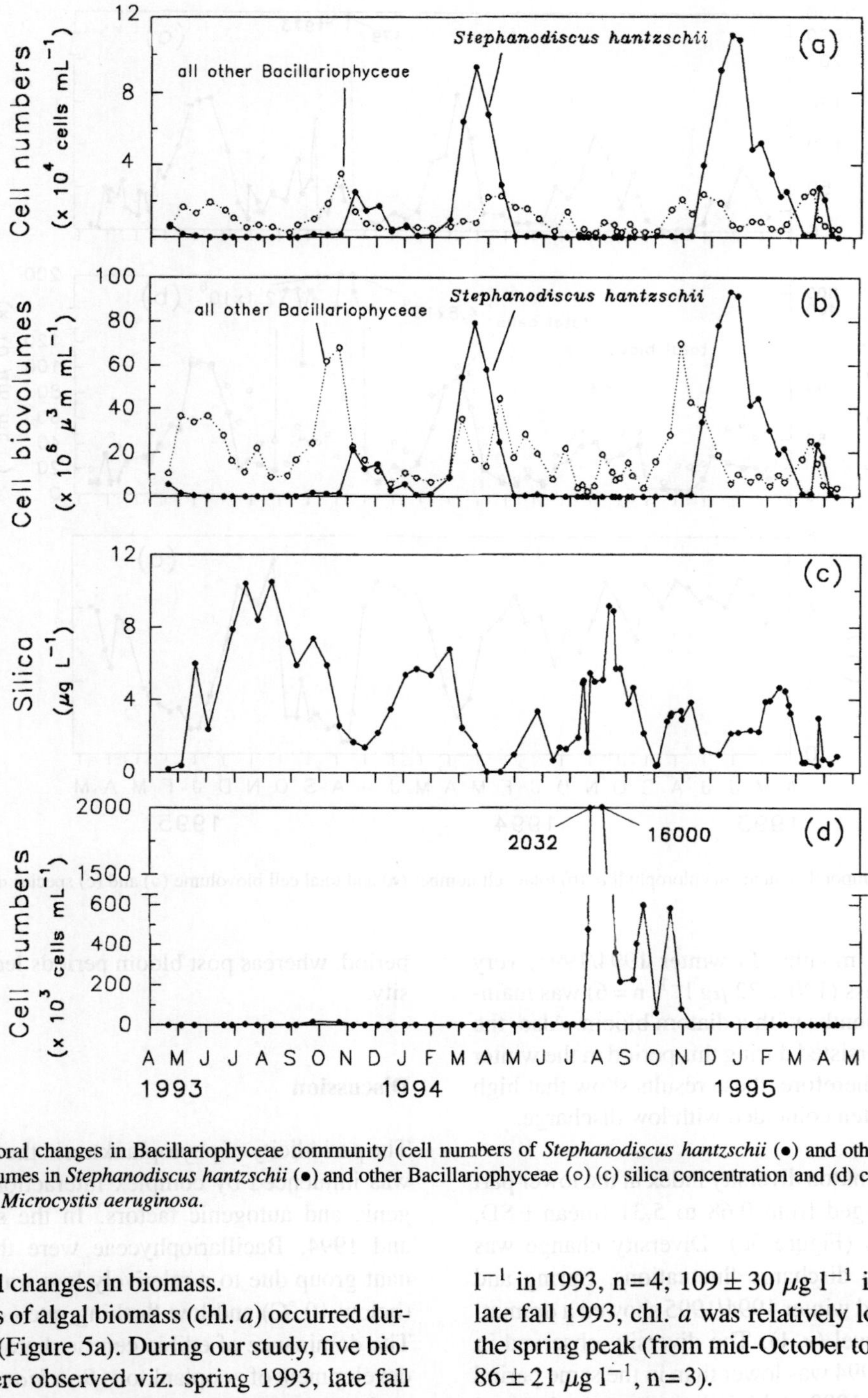

Figure 4. (a) temporal changes in Bacillariophyceae community (cell numbers of *Stephanodiscus hantzschii* (•) and other Bacillariophyceae (○)) (b) cell biovolumes in *Stephanodiscus hantzschii* (•) and other Bacillariophyceae (○) (c) silica concentration and (d) cell abundance (based on cell number) of *Microcystis aeruginosa*.

2. Temporal changes in biomass.

Many peaks of algal biomass (chl. *a*) occurred during the season (Figure 5a). During our study, five biomass peaks were observed viz. spring 1993, late fall 1993, spring and summer 1994 and winter 1994/1995, associated with low discharge. Temporal change in biovolume showed a similar pattern to that of chl. *a* (Figure 5b).

High concentrations of algal biomass were usually observed during the growing season of spring and fall. In spring 1993 and 1994, algal biomass was 94 ± 36 μg l^{-1} in 1993, n = 4; 109 ± 30 μg l^{-1} in 1994, n = 4. In late fall 1993, chl. *a* was relatively lower than during the spring peak (from mid-October to mid-November; 86 ± 21 μg l^{-1}, n = 3).

High flushing rates (> 50% per day) usually occurred in the rainy season, except in summer of 1994. During summer 1994, a high algal biomass was maintained (from late July to late August, 284 ± 578 μg l^{-1}; n = 10), when the *Microcystis* bloom occurred. Despite the slight reduction in algal biomass as a result of rainfall events in early-August, this bloom lasted

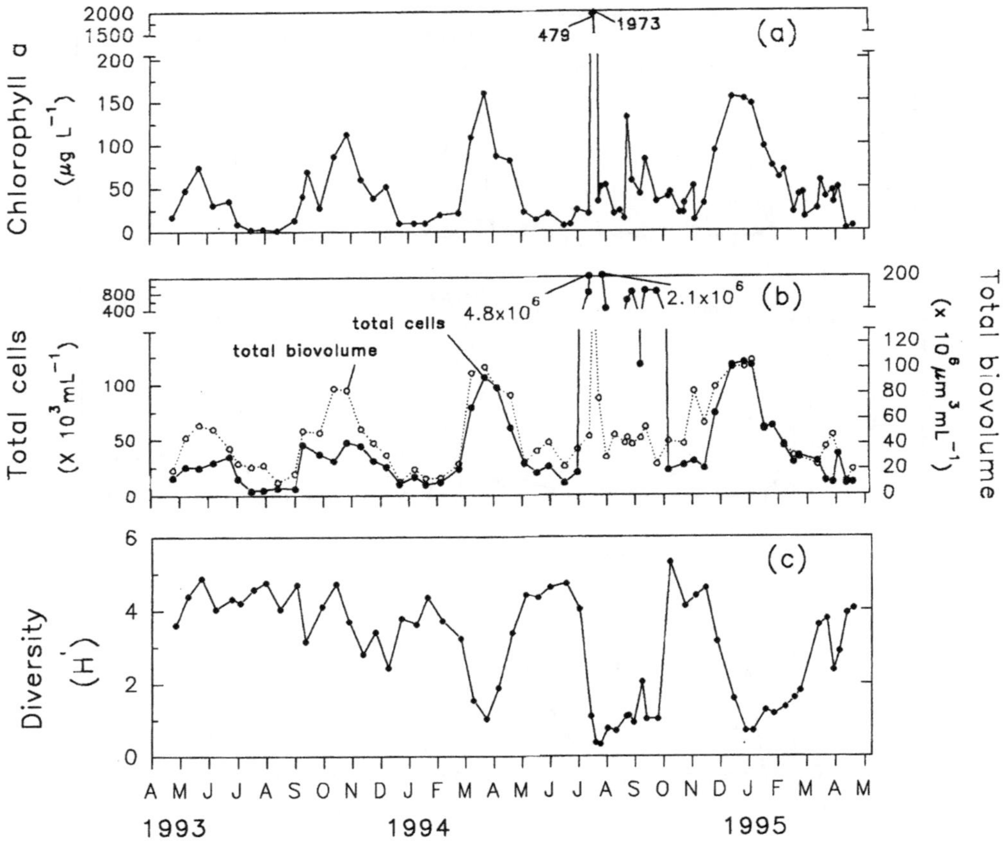

Figure 5. (a) temporal changes in chlorophyll *a*, (b) total cell number (•) and total cell biovolume (○) and (c) species diversity (H′).

for almost three months. In winter 1994/1995, very high algal biomass (120 ± 32 μg l^{-1}, n = 6) was maintained for two months with a diatom bloom. Also, the low discharge persisted during this period in the winter of 1994/1995. Therefore, these results show that high algal biomass often coincided with low discharge.

3. Diversity.

The phytoplankton diversity index at the lower part of the River ranged from 0.68 to 5.31 (mean ± SD, 2.9 ± 1.5 n = 64) (Figure 5c). Diversity change was clearly linked to discharge fluctuations. Spring and summer 1994 and winter 1994/1995 showed a dramatic decrease in level (< 1). The diversity observed in the summer of 1994 was lower than in the same period of the rainy year 1993. A high diversity was observed in the rainy year, going from high discharge to low flow condition. Diversity in the dry year was low as a result of near-monospecific blooms following long periods of water stability, low precipitation and low discharge. Low diversity corresponded to the bloom period, whereas post bloom periods led to high diversity.

Discussion

The periodicity of phytoplankton in the Nakdong River was influenced by complex interactions among allogenic and autogenic factors. In the spring of 1993 and 1994, Bacillariophyceae were the most dominant group due to a relatively low water temperature (below 10 °C) and low flushing rate (< 0.3% day^{-1}). The dominance of edible centric diatoms led to rapid development of zooplankton (*Brachionus* spp., *Bosmina longirostris, Cyclops* spp.). At Mulgum from 1993 to 1995, more than 80% of the zooplankton consisted of rotifers (Kim, 1996). *Brachionus calyciflorus, Keratella cochlearis, Polyarthra* spp,. *Asplanchna* spp. were most common. Cladocera and copepods were the minor constituents. The rotifers peaked in spring and fall at intermediate temperatures (10–25 °C) and abundant food (small centric diatom). In summer and winter, the density of major groups ranged from moderate to low at extreme temperatures (< 10 °C, > 30 °C), flooding events, discharge, and inedible phytoplankton.

The relatively low algal biomass in spring 1995 (from April to May, 2–20 $\mu g\ l^{-1}$, n = 6) coincided with the increase of zooplankton activity and frequent precipitation. It was somewhat similar to the 'clear-water phase' of stratified lakes (Lampert et al., 1986). During this phase, relatively high Secchi disc transparency (> 90 cm; annual mean, 73 cm) and a high zooplankton filtration (from 41.6 to 86.4 $ml\ h^{-1}$) were observed (Kim, 1996). In addition, Chlorococcales (*Coelastrum, Pediastrum, Scenedesmus*) and large motile *Cryptomonas* proliferated in the same fashion as sometimes seen in lakes during clearwater periods.

Summer phytoplankton community structure was strongly affected by the pattern of discharge. During the heavy rain period (summer of 1993), the phytoplankton density was low. However, in 1994, an elevated water temperature with high sunlight input may possibly have triggered the *Microcystis* bloom. After the last major rainfall event (mid-June), it took at least 2 weeks before the bloom started developing. During the first week, discharge sharply dropped, and water temperature and irradiance increased. During the second week, water temperature increased from 23 to 33 °C. We observed a sharp increase in *Microcystis* cell density (from $10–10^2$ cells ml^{-1} to $10^5–10^7$ cells ml^{-1}) in late July. As water temperature dropped between early and late October (from over 30 °C to 20 °C), *Microcystis* declined (late October, 1.6×10^3 cells ml^{-1}). There have been many attempts to explain Cyanobacteria bloom formation in natural communities. Elevated water temperature, CO_2/pH, low light, N/P ratio, zooplankton grazing and buoyancy hypotheses have been advocated to explain Cyanobacteria dominance (Shapiro, 1990; McQueen & Lean, 1987; Jones, 1994). Many workers reported that blue-green algae have capabilities that enable them to outcompete other communities at high pH or low CO_2 concentration (King, 1970; Shapiro et al., 1982; Paerl & Ustach, 1982). However, closer observation of the development of the *Microcystis* bloom in the Nakdong River revealed a slightly different fact. Sharp rises of pH in the middle of July 1994 occurred after the active reproduction of *Microcystis,* which lasted for almost one week. Evidence of selective grazing of zooplankton at the early stage of the bloom was not observed (Kim, 1996). When the summer blooms occurred, density of zooplankton was low, due to high water temperature. The concentrations of TN and TP were high in the lower part of the river (TN, 4.5 ± 1.1 $mg\ l^{-1}$; TP, 142 ± 76 $\mu g\ l^{-1}$; average from 1993 to 1995, TN, 4.1 ± 2.0 $mg\ l^{-1}$; TP, 151 ± 94.8 $mg\ l^{-1}$). In particular, high N and P concentration were maintained before and after the initial bloom formation. During this period, N_2-fixing blue-greens such as *Anabaena* spp., appeared at low density (< 100 cells l^{-1}). We therefore believe that high water temperatures along with other physical conditions, such as low flow and lack of rain, were major factor contributing to the *Microcystis* bloom.

Toxin production by *Microcystis* at Mulgum during the study period was high. Microcystins-LR and -RR at the surface were high in July 28, 1994 (171 $\mu g\ l^{-1}$) (H. D. Park, pers. comm.). The fate and effect of these toxins on the aquatic food web are unknown.

In fall, a peak by large diatoms occurred at low water temperature (about 10 °C) and elevated silica concentration. When the silica concentration was high in the fall, *Aulacoseira* spp., *Synedra acus* and *Fragilaria crotonensis* tended to occur. When the silica level was less than 1.5mg l^{-1} or at the detection limit (from March to April in 1994, 1995; from December of 1994 to January of 1995), *Stephanodiscus* increased. This was similar to the patterns of Bacillariophyceae succession in stratified lakes (Moed et al., 1976; Sell et al., 1984; Sommer, 1985). An enclosure experiment in the spring of 1995 to see the importance of silica in phytoplankton succession showed that *Stephanodiscus hantzschii* was dominant at low silica (1–1.5 mg l^{-1}), whereas *Fragilaria crotonensis* and *Synedra acus* had a competitive advantage in silica added enclosures (Ha, 1996). In late fall 1993, a decrease in the amount of light and the drop in water temperature caused a decline in phytoplankton biomass, which was at its minimum towards winter. However, in winter 1994/1995, an unusual *Stephanodiscus hantzschii* bloom was observed. This was a very surprising event compared to other river systems. Descy et al. (1987) showed for European eutrophic rivers (e.g., R. Meuse, R. Thames, R. Avon, R. Rhine, R. Neckar, R. Danube) that cold water (near 15 °C), low light and low discharge probably favor *Stephanodiscus*. But in the Nakdong River, this bloom developed at much lower temperatures (4–8 °C). It is assumed that the river became a stagnant reservoir during this low discharge period.

The phytoplankton periodicity observed in this study was somewhat different from that of well studied lakes and rivers. Succession of major phytoplankton groups in temperate lakes revealed that small Dinophyceae appear in winter, Bacillariophyceae in spring, colonial Chlorophyceae in late spring, Cyanobacteria in summer-early fall and Bacillariophyceae in fall (Reynolds, 1984b; Harris, 1986). In most large rivers with temperate hydrologic regime, Bacillariophyceae blooms tend to occur in spring and fall, whereas a mixture of Chlorococcales and Bacillariophyceae appear during summer (Kiss, 1987; Joo, 1990; Stoyneva & Draganov, 1991; Gosselain, 1994). *Stephanodiscus* blooms in the Nakdong River during low water temperature is a common phenomenon in other eutrophicated rivers as well. (Note that the same bloom, although at a different degree, was observed in several other eutrophic rivers in Europe). However, the *Microcystis* bloom observed in the Nakdong River is a rare phenomenon for eutrophicated river-reservoir systems, except for that reported by Paerl (1987). Overall, the phytoplankton periodicity in the Nakdong River was simultaneously affected by hydrologic regime and eutrophication.

Most other rivers in East Asia are also in a heavily eutrophicated and regulated condition, and have characteristics similar to the that of Nakdong River. Therefore, the succession patterns of phytoplankton observed in Nakdong River, provide a good baseline for further studies on the riverine phytoplankton in this region.

Acknowledgements

We thank Dr. C. S. Reynolds for improving the manuscript and for offering constructive criticisms on earlier drafts. We also thank Dr. M. Álvarez Cobelas and Dr. J. Padisák for inviting us to the IAP workshop. Two anonymous reviewers improved our manuscript. We also thank Park, S.B., J. S. Kim, H. S. Chang, Y. H. Joo, K. H. Chang, E. A Cho, K. S. Chung and G. I. Cho for their field assistance. And additional thanks to Mr. S. C. Charles Moon for proofreading. This research was funded by the Institute of Environmental Technology and Industry (RRC) of the Pusan National University (KSF-RRC 96-01-03).

References

Descy, J. P., P. Servais, J. S. Smitz, G. Billen & E. Everbecq, 1987. Phytoplankton biomass and production in the River Meuse (Belgium). Arch. Hydrobiol. Algol. Stud. 78: 225–245.

Descy, J. P., 1993. Ecology of the phytoplankton of the River Moselle: effects of disturbances on community structure and diversity. Hydrobiologia 249: 111–116.

Garnier, J., G. Billen & M. Coste, 1995. Seasonal succession of diatoms and Chlorophyceae in the drainage network of the Seine River: Observations and modeling. Limnol. Oceanogr. 40: 750–765.

Gosselain, V., J. P. Descy & E. Everbecq, 1994. The phytoplankton community of the River Meuse, Belgium: seasonal dynamics (year 1992) and the possible incidence of zooplankton grazing. Hydrobiologia 289: 179–191.

Ha, K., 1996. The mechanism of phytoplankton succession in the lower part of Nakdong River. M.S. Thesis. Pusan National University, Pusan, 86 pp.

Harris, G. P., 1986. Phytoplankton Ecology. Chapman and Hall, New York, 327 pp.

Jones, G. J. (ed), 1994. Cyanobacterial Research in Australia. CSIRO, Canberra, 193 pp.

Joo, G. J., 1990. Limnological studies of oxbow lakes in the Southeastern United States: Morphometry, physico-chemical characteristics and patterns of primary productivity. The University of Alabama, Tuscaloosa, PhD. dissertation, 122 pp.

Joo, G. J., 1995. Ecological Studies on the Nakdong River (1957–1994). Kumjung Press, Pusan, 178 pp (in Korean).

Kim, H. W., 1996. Characteristics of physico-chemical parameters and zooplankton dynamics in the Nakdong River. M.S. Thesis. Pusan National University, Pusan, 87 pp.

King, D. L., 1970. The role of carbon in eutrophication. J. Wat. Pollut. Cont. Fed. 42: 2035–2051.

Kiss, K. T., 1987. Phytoplankton studies in the Szigetkoz section of the Danube during 1981–1982. Arch. Hydrobiol. Suppl. 78: 247–273.

Lampert, W., Fleckner, W., Rai, H. & B. E. Taylor, 1986. Phytoplankton control by grazing zooplankton: A study on the spring clear-water phase. Limnol. Oceanogr. 31: 478–490.

McQueen, D. J. & D. R. S. Lean, 1987. Influence of water temperature and nitrogen to phosphorus ratio on the dominance of blue-green algae in Lake St. George, Ontario. Can. J. Fish. aquat. Sci. 44: 598–604.

Moed, J. R., H. L. Hoogveld & W. Apeldoorn, 1976. Dominant diatoms in Tjeukemeer (The Netherlands): II. Silicon depletion. Freshwat. Biol. 6: 355–362.

OECD., 1982. Eutrophication of waters: Monitoring, assessment and control. Paris, 154 pp.

Paerl, H. W. & J. F. Ustach, 1982. Blue-green algal scums: An explanation for their occurrence during freshwater blooms. Limnol. Oceanogr. 27: 212–217.

Paerl, H. W., 1987, Dynamics of Blue-green algal (*Microcystis aeruginosa*) blooms in the lower Neuse River, North Carolina: Causative factors and potential controls. Water Resources Research Institute of the University of North Carolina., 164 pp. UNC-WRRI-87-229.

Reynolds, C. S., 1984a. The ecology of freshwater phytoplankton. Cambridge Univ. Press, Cambridge and New York, 384 pp.

Reynolds, C. S., 1984b. Phytoplankton periodicity: the interactions of form, function and environmental variability. Freshwat. Biol. 14: 111–142.

Reynolds, C. S., 1988. The concept of ecological succession applied to seasonal periodicity of freshwater phytoplankton. Verh. int. Ver. Limnol. 23: 683–691.

Reynolds, C. S., J. Padisák & U. Sommer, 1993. Intermediate disturbance in the ecology of phytoplankton and the maintenance of species diversity: A synthesis. Hydrobiologia 249: 183–188.

Reynolds, C. S. & J. P. Descy, 1996. The production, biomass and structure ofj phytoplankton in large rivers. Arch. Hydrobiol. Suppl. 113, Large rivers 10 1–4: 161–187.

Sell, D. W., H. J. Carney & G. L. Fahnenstiel, 1984. Inferring competition between natural phytoplankton populations: The Lake Michigan example reexamined. Ecology 65: 325–328.

Shannon, C. E. & W. Weaver, 1949. The Mathematical Theory of Communication. Urbana, Univ. Illinois Press, 117 pp.

Shapiro, J., B. Forsberg, V. Lamarra, M. Lynch, E. Smeltzer & G. Zoto, 1982. Experiments and experiences in biomanipulation: studies of biological ways to reduce algal abundance and eliminate blue-greens. EPA 600/3-82-096, 251 pp.

Shapiro, J., 1990. Current beliefs regarding dominance by blue-greens: The case for the importance of CO_2 and pH. Verh. int. Ver. Limnol. 24: 38–54.

Sommer, U. (ed.), 1989. Plankton Ecology. Springer-Verlag, Berlin, 358 pp.

Sommer, U., 1985. Comparison between steady state and non-steady state competition: Experiments with natural phytoplankton. Limnol. Oceanogr. 30: 335–346.

Song, K. O., 1992. Water quality modeling in the Nakdong River. PhD. dissertation. National Fisheries, University of Pusan, Pusan, 134 pp.

Stoyneva, M. P. & S. J. Draganov, 1991. Green algae in the phytoplankton of the Danube (Bulgarian sector) – species composition, distribution, cell numbers and biomass. Arch. Protistenkd. 139: 243–260.

Wetzel, R. G., 1983. Limnology. 2nd Edition. Saunders College Publishing, Philadelphia, 860 pp.

Hydrobiologia **369/370:** 229–235, 1998.
M. Alvarez-Cobelas, C. S. Reynolds, P. Sanchez-Castillo & J. Kristiansen (eds), Phytoplankton and Trophic Gradients.

Success of chrysophytes, cryptophytes and dinoflagellates over blue-greens (cyanobacteria) during an extreme winter (1995/96) in eutrophic shallow lakes

Claudia Wiedner & Brigitte Nixdorf
Brandenburg Technical University Cottbus, Fac. Environ. Sciences, Water Protection, Seestr. 45, D-15526 Bad Saarow, Germany

Key words: phytoplankton succession, chrysophytes, winter, ice-cover, trophic gradient

Abstract

Phytoplankton succession in three shallow eutrophic and hypertrophic lakes located in the Scharmützelseeregion (East Brandenburg, Germany) is described for the extreme winter of 1995/96 before and during ice-cover. The lakes have been investigated since 1993 and pass along the tophic gradient from eutrophic, Petersdorfer See and Melangsee to hypertrophic, Langer See. All lakes are dominated by cyanobacteria from early spring to late autumn. Due to low water temperature and low light intensity before ice-cover a decline of cyanobaterial biovolume initiated which led to the total disappearance of cyanobacteria in Melangsee for the whole ice-covered period, whereas their portion in Langer See with at least 6% (3 mm^3 l^{-1}) and in Petersdorfer See with 25% (5 mm^3 l^{-1}) of total biovolume was still considerable. The cyanobacterial breakdown was linked to a switch in dominance to flagellates which built up together with diatoms a subsequent peak of biovolume during ice-cover.

Maximum phytoplankton biovolume under the ice peak reached 20 mm^3 l^{-1} with chl-*a* concentrations up to 100 μg l^{-1}. Oxygen saturation increased during ice-cover with a final oversaturation of more than 200% in the upper water column at the end of ice cover. These findings indicate that the winter period is also characterized by intensive photoautotrophic activities. Therefore, this period should be more worthy of note in investigations of phytoplankton succession and ecology.

Introduction

In comparison to the number of investigations of phytoplankton during summer, little attention has been paid to the winter, ice-covered periods. Most winter studies cover lakes located in polar and subpolar regions (e.g. Lizotte & Priscu, 1994) or in high mountains (e.g. Rott, 1988; Tilzer, 1973) with a trophic range from oligo- to mesotrophic.

The present paper presents data of phytoplankton succession in three shallow eutrophic to hypertrophic lakes of the temperate climate zone for the extreme cold winter 1995/96. The lakes have been investigated since 1993 and are dominated by cyanobacteria, either *Planktothrix agardhii* or *Limnothrix* spp., from early spring to late autumn (Rücker et al., 1996).

Materials and methods

The study area is located in the Scharmützelsee region, East Brandenburg (Germany). Investigated lakes are polymictic and very shallow in the sense of Nixdorf & Deneke (1996); relevant lake characteristics are given in Table 1. Sampling was carried out biweekly to monthly. At each sampling date, the thickness of the ice and snow cover was recorded. Oxygen saturation and temperature profiles were measured with a Hydrolab-H20 multiparameter probe with a range from 0% to 200% for oxygen. Incident and under-ice radiation was measured as PAR using spherical quantum sensors (Li-cor LI-193SA). Mixed samples from the water column were taken with 2,3 l-LIMNOS-sampler for the analysis of chlorophyll-*a*, dissolved nutrients and phytoplankton.

Table 1. Lake description. Mean and ranges of nutrients, Chl-*a* and Secchi depth, including data from 1995 and 1996.

Lake	Langer See	Melangsee	Petersdorfer See
Lake area (km^2)	1.55	0.11	0.23
Zmax (m)	3.80	3.30	4.00
Zmean (m)	2.10	1.60	2.10
TP ($\mu g\ l^{-1}$)			
mean	117.3	54.9	58.5
range	48–244.0	13.0–98.0	18.0–97.0
TN ($\mu g\ l^{-1}$)			
mean	2212.7	2060.0	3441.1
range	786.4–3780.0	1200.0–2820.0	1010.0–4990.0
DIP ($\mu g\ l^{-1}$)			
mean	14.3	4.8	6.4
range	2.0–61.6	1.8–12.5	1.0–18.1
NO_3-N ($\mu g\ l^{-1}$)			
mean	92.3	13.1	23.4
range	1.0–445.4	1.0–163.3	1.0–114.7
NH_4-N ($\mu g\ l^{-1}$)			
mean	197.6	147.3	498.4
range	10.1–1144.0	11.0–1020.0	34.4–2252.2
Chl-*a* ($\mu g\ l^{-1}$)			
mean	94.2	53.9	66.2
range	27.2–195.7	5.0–151.7	34.5–105.8
Secchi depth (m)			
mean	0.7	1.1	0.6
range	0.3–1.8	0.4–2.0	0.4–1.0

Phytoplankton samples were fixed with Lugol solution and counted by the method of Utermöhl (1958) as modified by Rott (1981). Chlorophyll-*a* and chemical parameters were determined according to the standard methods (DIN 38412-L16 for Chlorophyll-*a* and DIN 38405 D9, 10, 11, 21 and 38406 E5 for DIP, NO_3, NH_4).

Results

In all lakes there was a substantial decline in total phytoplankton biovolume due to a decrease in cyanobacteria just before ice cover. This was followed by a smaller peak built up by flagellates and diatoms during ice cover (Figure 1). Lakes differ both in the relative and absolute portion of cyanobacteria on the total phytoplankton biovolume and the groups of flagellates that became dominant during the ice-covered period. Cyanobacteria disappeared in Melangsee until the end of ice-cover, in Langer See they rested over the whole ice-covered period with at least 6% (3 to 5 $mm^3\ l^{-1}$) and in Petersdorfer See with at least 25% (5 $mm^3\ l^{-1}$) of total phytoplankton biovolume. Cyanobacterial species under ice were in both lakes the same: *planktothrix agardhii, Limnothrix* spp., *Pseudanabaena limnetica, Aphanizomenon flosaquae* and *Planktolyngbya limnetica*. Within flagellates cryptophytes did occur with high biovolume and dinophytes with less biovolume in all lakes. Chrysophytes did not occur in Larger See at all and in Petersdorfer See with a small portion on total phytoplankton biovolume. In Melangsee they were dominant during the whole ice-covered period with a succession of different taxa: *Uroglena volvox* followed by *Dinobryon* spp. followed by *Synura* sp. (Figure 2). The occurrence of the diatom *Stephanodiscus* sp. which built up high biovolumes for a short period in Melangsee and Langer See was associated with a decline in the share of cryptophytes in both lakes.

The physical and chemical conditions in the lakes developed similarly in each lake. Ice-cover started at the end of November 1995 and lasted until the end of April 1996. Maximum thickness of ice-cover reached

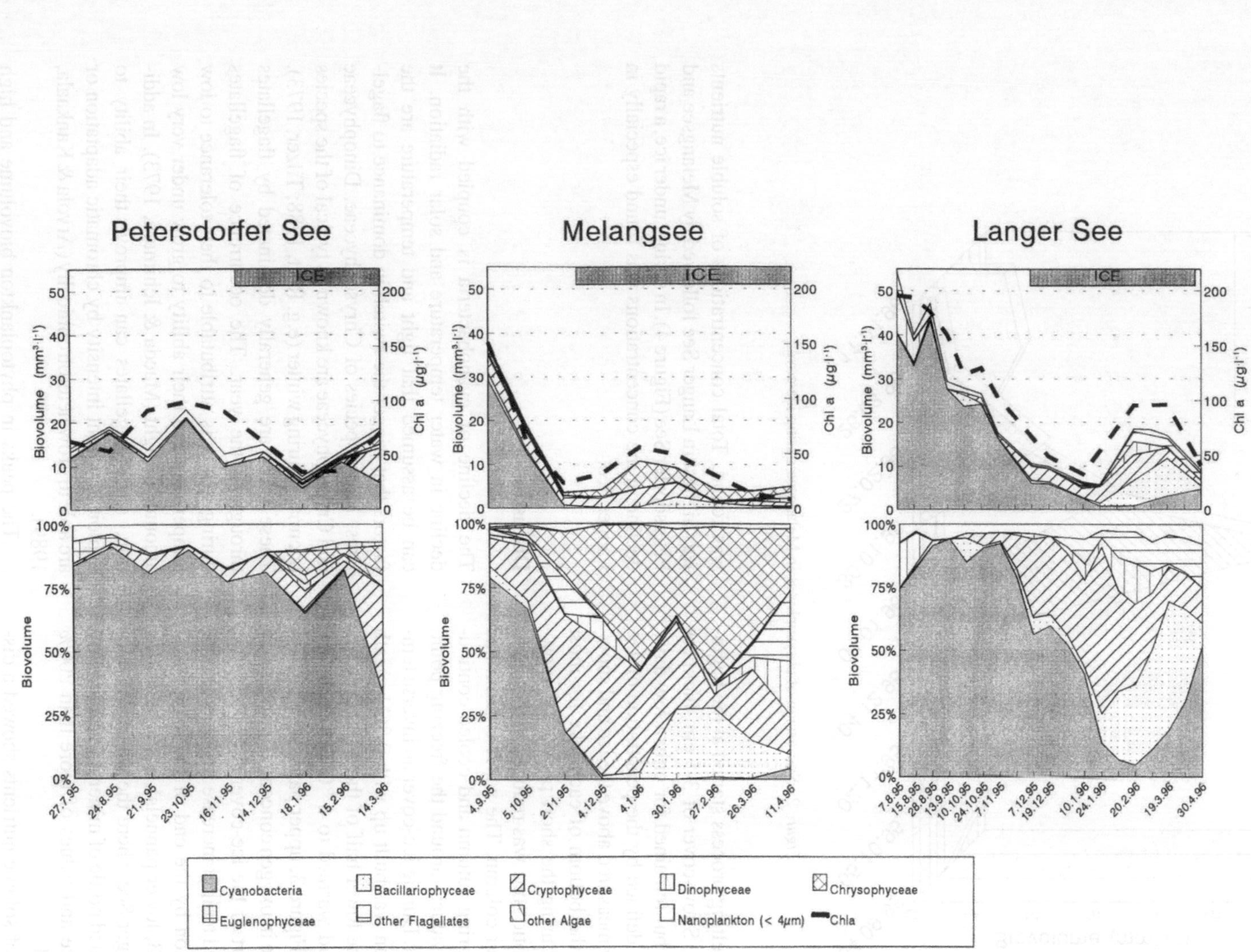

Figure 1. **Phytoplankton biovolume and chlorophyll-*a* concentration (upper pannels) and relative biovolume (lower pannels).**

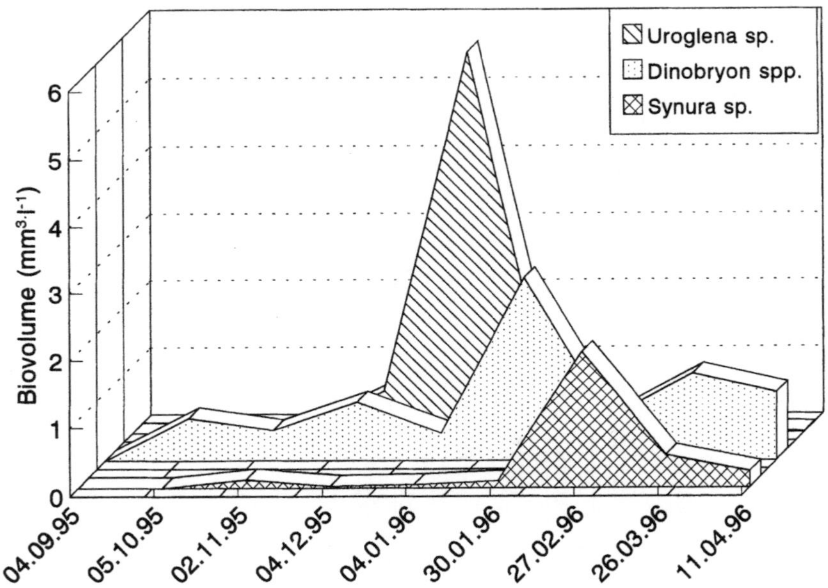

Figure 2. Succession of chrysophytes in Melangsee during ice-cover.

50 cm before the melting process started at the beginning of April 1996. Snow-cover of 3-5 cm thickness was recorded twice but remained for only a few days. The euphotic zone, defined by the penetration of 1% of the light intensity measured above the ice and snow-cover, extended to the bottom of each lake for most of the time. Only during the short periods of snow-cover was the euphotic zone was reduced to the upper 1–1.5 m of the water column. The water column had been well mixed during autumn and cooled continuously to its lowest values, around the freezing point, just before ice-cover. During ice-cover an inverse temperature stratification was built up with 0 °C beneath the ice to 4 °C in the lower half of the water column. Finally the lakes had warmed to 6 °C before the ice had finally melted (Figure 3, upper pannels).

An initial decline in oxygen concentration occurred during the early part of the ice-covered period, after which there followed a rapid increase in oxygen, leading to supersaturation by the end of ice-cover in all three lakes (Figure 3, lower pannels). Supersaturation was greatest in Langer See where the water was bubbling like a just opened bottle of mineral water directly after opening the ice and values of more than 200% have been measured.

Concentrations of soluble nutrients showed a distinct course in all lakes: Before ice-cover soluble nutrients were very low, often at the level of detection. During ice-cover an enormous increase in soluble nutrients was found. Total concentrations of soluble nutrients was highest in Langer See followed by Melangsee and Petersdorfer See (Figure 4). In spring, under ice, a rapid decline in the concentrations was found especially in Langer See.

Discussion

The decline of cyanobacteria is coupled with the decline in water temperature and solar radiation. It can be assumed that light and temperature are the factors that trigger the switch in dominance to flagellates. Many species of Chrysophyceae, Dinophyceae and Cryptophyceae are known as typical of the species occurring during winter (e.g. Rott, 1988; Tilzer, 1973). These lakes are generally dominated by flagellates throughout the year. The occurrence of flagellates during winter is attributable to their tolerance to low temperature and their ability to grow under very low amounts of light (Maeda & Ichimura, 1973). In addition, some flagellates can improve their ability to exploit low light intensity by chromatic adaptation or are able to grow heterotrophically (Arvola & Kankaala, 1989).

The peaks in phytoplankton biovolume and high chlorophyll-*a* concentrations as well as the increasing oxygen saturation during ice-cover indicate high photosynthetic activity in the lakes we investigated. Simi-

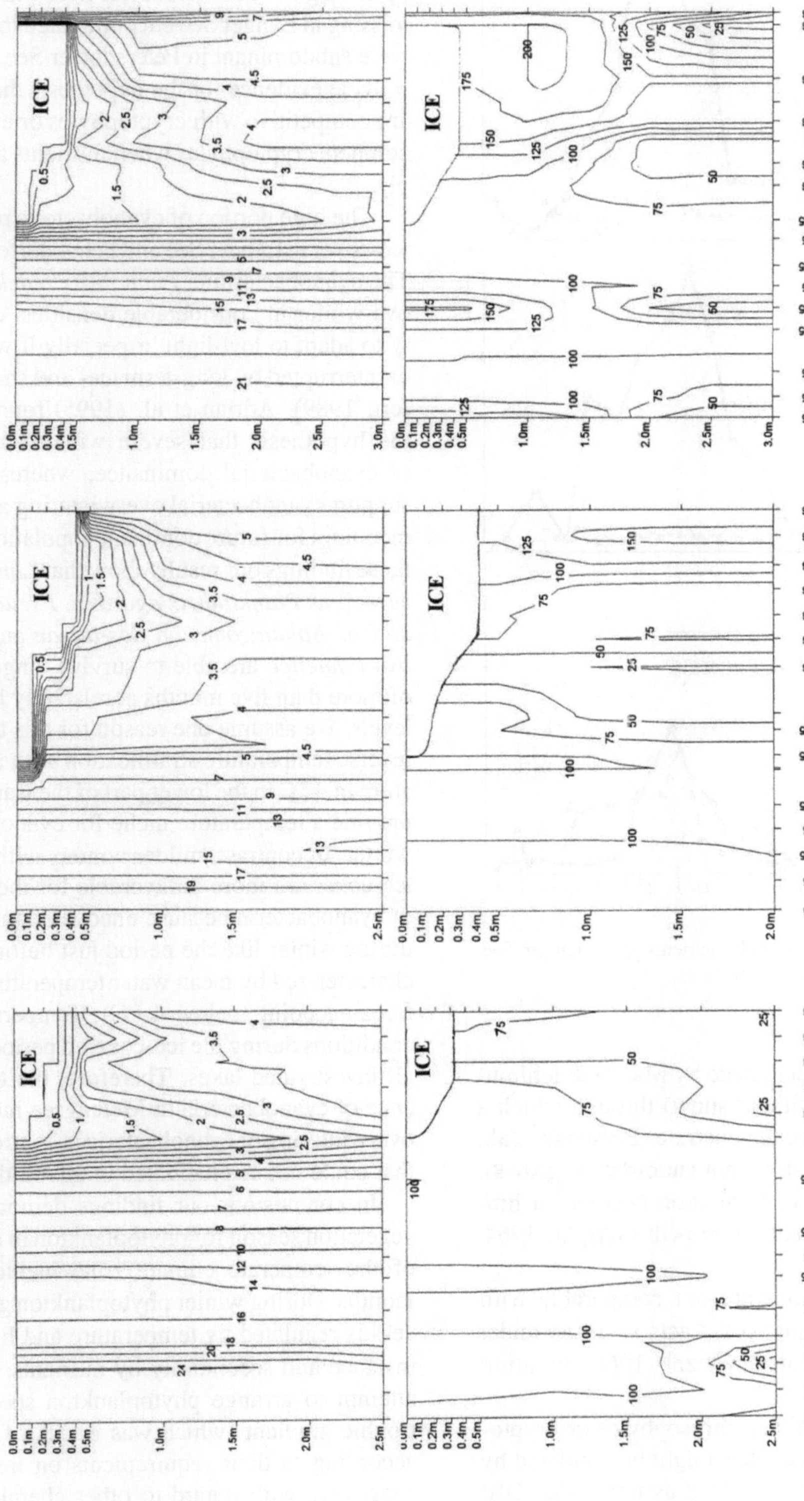

Figure 3. Isoplethes for temperature (upper pannels) and oxygen saturation (lower pannels).

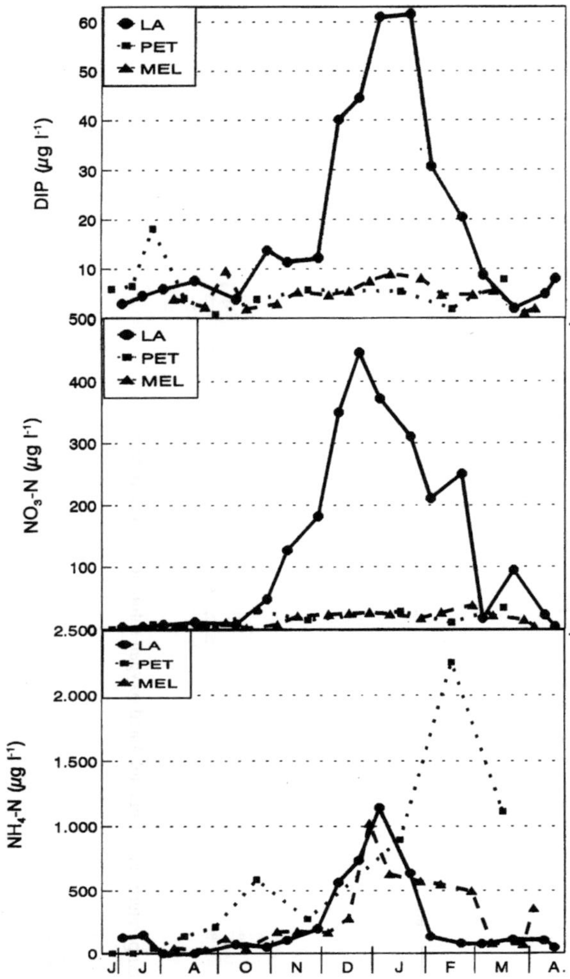

Figure 4. Concentration of soluble nutrients (LA = Langer See, PET = Petersdorfer See, MEL = Melangsee).

lar observations have been made by Maeda & Ichimura (1973). Clear ice (without snow) through which a high proportion of light can penetrate (Bolsenga et al., 1991) is a prerequisite for photoautotrophic growth. Under snow-covered ice, light soon becomes a limiting factor for phytoplankton growth (Wright, 1964; Tulonen et al., 1994).

Oxygen concentrations are not comparable with summer levels because ice-cover acts as a trap under which oxygen accumulates (Wetzel, 1966; Wharton et al., 1986).

The dominance of either chrysophytes or cryptophytes in the investigated lakes might be regulated by the different nutrient supply. It is assumed that DIP below a concentration of 10 μg l^{-1} may become a limiting factor. Concentrations of DIP reached up to 60 μg l^{-1} in Langer See, up to 9 μg l^{-1} in Melangsee and up to 6μg l^{-1} in Petersdorfer See. Chrysophytes were missing in Langer See but dominated in Melangsee and were subdominant in Petersdorfer See. These findings provide evidence for the hypothesis that chrysophytes are competitive with cryptophytes or even able to outcompete cryptophytes when nutrients are in short supply.

The high portion of cyanobacteria remaining under ice-cover in Langer See and Petersdorfer See is notable. The cyanobacterium, *Limnothrix redekei,* is known to overwinter in considerable densities, due to its ability to adapt to low-light, especially if winter growth is uninterrupted by long-term ice- and snow-cover (Meffert, 1989). Adrian et al. (1995) found evidence for the hypothesis that severe winters prevent the rise of cyanobacterial dominance, whereas mild winters support cyanobacterial overwintering and improve the inoculum for future dominant populations. Contrary to these findings our results show that *Limnothrix redekei* as well as *Planktothrix agardhii, Pseudanabaena limnetica, Aphanizomenon flos-aquae* and *Planktolyngbya limnetica* are able to survive long-term ice-cover of more than five months at relatively high biovolume levels. We assume one reason for this to be that stable inverse temperature stratification with a mean temperature of 4 °C in the lower part of the water column may provide a temperature niche for cyanobacteria during winter. In contrast, milder winters with discontinuous ice-cover are more unfavorable for the overwintering of cyanobacteria because uncovered turbulent periods during winter like the period just before ice-cover are characterized by mean water temperatures close to the freezing point (below 4 °C). Temperature and light conditions during the ice-covered period were equal in all investigated lakes. Therefore, the total disappearance of cyanobacteria in Melangsee indicates that the overwintering of cyanobacteria is controlled by factors that could not be elucidated in this study.

In conclusion, our findings demonstrate that the vegetation season for phytoplankton in eutrophic lakes of the temperate climate zone includes the winter months. During winter phytoplankton growth and success is regulated by temperature and light, in the first instance and secondarily by nutrients. Therefore, the attempt to arrange phytoplankton species along the trophic gradient (which was a task of the 10th IAP) according to their requirements on nutrients is only suggestive with regard to other chemical and physical parameters. Moreover, cyanobacteria are able to pass long, severe winters under certain circumstances and occur with a high inocula in early spring. Therefore, the general assumption that severe winters may prevent the rise of cyanobacterial dominance whereas mild winters support cyanobacterial overwintering and following dominance (Adrian et al., 1995) should be qualified.

Further investigations of lakes with high trophic levels under different conditions (with and without ice- and snow-cover) during winter and of the ability of cyanobacteria to survive such periods are necessary to a proper understanding of the succeeding development of phytoplankton populations in spring.

Acknowledgements

The authors are grateful to G. Lippert, T. Hanke, U. Abel and C. Tahedl for careful work in the laboratory, W. Terlinden and I. Hentschke for practical work in the field.

References

Adrian, R., R. Deneke, U. Mischke, R. Stellmacher & P. Lederer, 1995. A long-term study of the Heiligensee (1975–1992). Evidence for effects of climatic change on the dynamics of eutrophied lake ecosystems. Arch. Hydrobiol. 133: 315–337.

Arvola, L. & L. Kankaala, 1989. Winter and spring variability in phyto- and bacterioplankton in lakes with different water colour. Aqua Fennicia 19: 29–39.

Bolsenga, S. J., C. E. Herdendorf & D. C. Norton, 1991. Spectral transmittance of lake ice from 40–850 nm. Hydrobiologia 218: 15–25.

Lizotte, M. P. & J. C. Priscu, 1994. Natural fluorescence and quantum yields in vertically stationary phytoplankton from perennially ice-covered lakes. Limnol. Oceanogr. 39: 1399–1410.

Maeda, O. & S. Ichimura, 1973. On the high density of a phytoplankton population found in a lake under ice. Int. Revue ges. Hydrobiol. 58: 673–685.

Meffert, M.-E., 1989. Planktic unsheated filaments (Cyanophyceae) with polar and central gas vacuoles. II. Biology, population dynamics and biotopes of Limnothrix redekei (van Goor) Meffert. Arch. Hydrobiol. 116: 257–282.

Nixdorf, B. &. R. Deneke, 1997. Why very shallow lakes are more successful opposing reduced nutrient loads. Hydrobiologia (in press).

Rott, E., 1981. Some results from phytoplankton counting intercalibrations. Schweiz. Z. Hydrol. 43: 34–62.

Rott, E., 1988. Some aspects of the seasonal distribution of flagellates in mountain lakes. Hydrobiologia 161: 159–170.

Rücker, J., C. Wiedner & P. Zippel, 1997. Factors controlling the dominance of *Planktothrix agardhii* and *Limnothrix redekei* in shallow lakes. Hydrobiologia (in press).

Tilzer, M., 1973. Diurnal periodicity in the phytoplankton assemblage of high-mountain lakes. Limnol. Oceanogr. 18: 15–30.

Tulonen, T., P. Kankaala, A. Ojala & L. Arvola, 1994. Factors controlling production of phytoplankton and bacteria under ice in a humic, boreal lake. J. Plankton Res. 16: 1411–1432.

Utermöhl, H., 1958. Zur Vervollkommnung der quantitativen Phytoplanktonmethodik. Mitt. int. Ver. Limnol. 9: 1–38.

Wetzel, R. G., 1966. Variations in productivity of Goose and hypertrophic Sylvan lakes, India. Investigations of Indiana Lakes and Streams 7: 147–184.

Wharton, R. A. Jr., C. P. McKay & B. C. Parker, 1986. Oxygen budget of a perennially ice-covered Antarctic lake. Limnol. Oceanogr. 31: 437–443.

Wright, R. T., 1964. Dynamic of a phytoplankton community in an ice-covered lake. Limnol. Oceanogr. 9: 163–178.

Hydrobiologia **369/370:** 237–242, 1998.
M. Alvarez-Cobelas, C. S. Reynolds, P. Sanchez-Castillo & J. Kristiansen (eds), Phytoplankton and Trophic Gradients.

Structural changes during eutrophication of Lake Balaton, Hungary, as revealed by the Zipf-Mandelbrot model

Szilveszter Juhos & Lajos Vörös
Balaton Limnological Research Institute of the Hungarian Academy of Sciences, P.O. Box 35, H-8237 Tihany, Hungary

Key words: Zipf-Mandelbrot model, dominance-diversity curves, eutrophication, hierarchies, community structure

Abstract

In order to explore the changes in the community structure of Lake Balaton in response to progressive eutrophication, the historic plankton database (1965–1990) was analysed, using the Zipf-Mandelbrot (ZM) model of dominance and diversity. Our results confirm differences in community organisation between the less eutrophic eastern (Siófok) basin and the near hypertrophic western (Keszthely) basin: the ZM model shows there to be greater influence of the traditional limiting environmental factors in the less eutrophic basin. There is also a greater evenness in the total biomass in the less eutrophic basin, determined by the variability in the environmental control applied.

Introduction

The eutrophication of Lake Balaton may be traced back to the late 1960s (Herodek, 1984; Padisák, 1994). At that time, the maximal biomass in the western Keszthely basin was 5–8 mg l^{-1} but, by 1982, the biomass had reached 60 mg l^{-1}. Now, filamentous nitrogen-fixing cyanobacteria have become prominent.

Cylindrospermopsis raciborskii has accounted for over 95% of the phytoplankton biomass in some recently recorded blooms (G. Tóth & Padisák, 1986).

This profound change in composition of the phytoplankton is traced through a more or less continuous series of regular algal counts, assembled for either basin of the lake since 1965. The data also reveal the strong spatial gradient in the lake: there is a significant difference (in biomass and in the composition of planktonic communities) between the more eutrophic western (Keszthely) basin and relatively less eutrophic (Siófok) basin.

The objective of this work has been to explore the changes in the structure of the phytoplankton communities and to compare those of the eutrophic and mesotrophic basins. We based the work on the assembly of dominance-diversity graphs (DD, or rank-frequency diagrams). The graphs are constructed from counting data, after sorting the species in order of decreasing abundance. This ranked list provides the ordinate; the abundances are plotted on the abscissa. There have been few previous attempts to express mathematically the form of these graphs (Pielou, 1975; Engen, 1978). Application of the Zipf-Mandelbrot (ZM) model of Mandelbrot (1953), however, does provide a good fit to DD data (Aleya & Devaux, 1992) and its similar origins in information theory makes it appropriate to the current problem (Margalef, 1957; Frontier, 1987). According to the ZM model, the appearance of a species must satisfy some species-specific preconditions. These include, for example, appropriate thresholds for light and temperature, and the presence of one or more other species. In this way, we may infer that (say) the coincidence of warm water, optimal light conditions and depletion of the dissolved inorganic nitrogen occurs in only a short part of the whole year. Because these preconditions have a rather higher probability of fulfilment in the summer, then it is rather more likely that blooms of Cyanobacteria will occur then too. Furthermore, an event in the

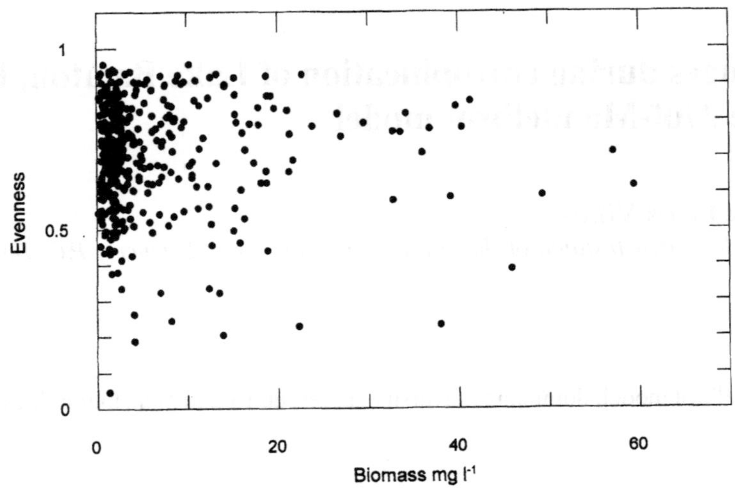

Figure 1. Relationship between the biomass and the evenness of the samples. There is no direct relationship, but *the maximal* available evenness is determined by the biomass.

model (i.e., the appearance of a species) can meet a precondition for another event. It is thus possible to recognise a chain of hierarchically organized events. This implies the occurrence of stable communities and the promotion of short- and long-term successional patterns.

In addition to these plausible conditions, the ZM model has only two other parameters and one prefactor in its analytical form:

$$F_r = F_0(r + \beta)^{-\gamma}$$

or, in logarithms:

$$log(F_r) = log(F_0) - \gamma log(r + \beta)$$

where F_r is the probability of the appearance (the abundance) of the species at the *r*th place in the DD graph and F_0 is a scale factor equal to the sum of frequencies. γ is connected to the evenness of the sample: it has higher values for 'simple' and lower values for 'complex' systems. More precisely $\gamma \propto 1/D$, where D is a fractal dimension reflecting the complexity of the community (Mandelbrot, 1982). The parameter β is related to the diversity of the environment. It serves as a measure of the complexity of the initial preconditions. As we will demonstrate, γ is related to evenness as defined by Frontier (1985): the two parameters depend on each other in a non-trivial manner.

Description of the sites studied

Lake Balaton (surface area of 596 km^2, mean depth 3.2 m) is the largest shallow lake in Central Europe. The mean residence time is 5 years. The lake is rather elongated. The more eutrophic of its two main basins, the western, Keszthely basin, supported peak chlorophyll-*a* concentrations of up to 200 $\mu g\ l^{-1}$ during the large summer bloom of 1994. The eastern, Siófok basin has clearer water: the chlorophyll concentration rarely exceeds 20 $\mu g\ l^{-1}$. The main inflow the lake is the Zala River, which drains about half of the total catchment area of the lake (5775 km^2), enters the enriched Keszthely basin. The only outflow from the lake, the Sió Canal, drains the Siófok basin. The lake is the most important recreational area in Hungary, with more than two million visitors every year.

Materials and methods

The data set used here has been assembled over a number of years (Tamás, 1974; Herodek & Tamás, 1976; Vörös & Nagy-Göde, 1993) and has been submitted to a previous analysis (Vörös & Nagy-Göde, 1993). Although this long-term data set is based on year-round sampling, the present study makes use only of information relating to the summer months (July–August–September of each year). The data represent a total of 390 phytoplankton samples (197 from the Keszthely basin, 193 from Siófok), the algae in each

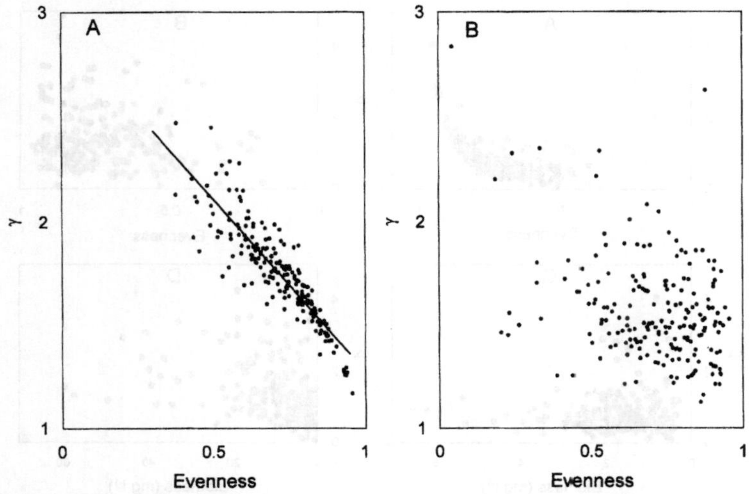

Figure 2. The dependence of the γ parameter on the evenness in the less eutrophic Siófok (A) and the more eutrophic Keszthely (B)basin. A clear inverse proportionality is observable in the Siófok basin (A: $r^2 = 0.79$) but there is no correlation in the Keszthely basin (B).

enumerated and recorded by class, species and specific biomass. The Graphics Language Editor (GLE) and its non-linear least squares fitting program (fitls) were used for the calculations. This program is available from the main Internet sites and uses the Levenberg-Marquart algorithm (Press et al., 1992). The program ran on an IBM-PC compatible computer running UNIX (Linux). Because the number of species were different in these sets, we used only the first 15 species in the ranked list. This usually accounted for more than 95% of the total biomass and total number of cells present. We did not use any other means to select species for analysis should be otherwise included or excluded. To determine whether the DD plots reflected ordered (invoking definitive interactions such as interspecific competition) or random (negligible interactions between species) assemblies, we generated some arbitrary communities, having differing predetermined distribution-density functions. These routines, written in C, were taken from the Numerical Recipes (Press et al., 1992) collection.

In the literature, species-abundance is usually expressed as a percentage of the total biomass or cell number. The abundance equivalent in the ZM model is the probability of appearance of the species. The equation,

$$\Sigma_{i=1}^{S} p_i = 1$$

must be true where p_i stands for the probability of a species and S is the number of species in the sample. This simple boundary condition means that there is at least one species with at least one individual in the sample and that the sum of probabilities cannot exceed 1. We normalized our data for every value in order to compare different assemblies with different total biomass

$$p_i = a_i / \Sigma_{i=1}^{s} a_i$$

where a_i is the abundance and p_i is the estimated probability of appearance. Providing we are using estimated probabilities instead of percentages or abundance values during the parameter fit, we need not estimate the F_0 scaling factor in the ZM model, because the sum of frequencies is equal to one.

The Shannon diversity, evenness and total biomass were derived from the database for each sampling date.

Results

Comparing the evenness and biomass values in the two basins, we found no direct mutual correlation, although these variables are, presumably, not independent. At higher biomass values, the *maximal* eveness is lower, as few species contribute to the dominance of blooms (see Figure 1). The evenness reflects the "complexity"or "randomness" of the sample, as γ does in the ZM model. Frontier observed strong correlation (inverse proportionality) between these parameters:

$$J \propto 1/\gamma$$

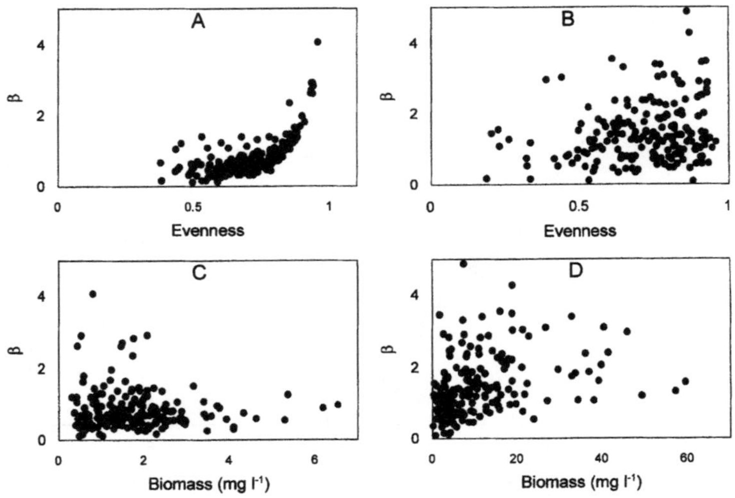

Figure 3. Relationship between the β parameter (diversity of the environment), the evenness and the biomass in the two basins with different trophic levels. A: In the less eutrophic Siófok basin, the evenness is correlated to the diversity of the environment. B: No direct correspondence between the evenness and these two values in the more eutrophic Keszthely basin. C: At high biomass values the diversity of the environment is low in the less eutrophic Siófok basin. D: A less clear dependence is found in the more eutrophic Keszthely basin.

According to our results, this is true only for the less eutrophic basin ($r^2 = 0.79$), there being no connection between these parameters in the highly eutrophic Keszthely basin ($r^2 = 0.11$, Figure 2).

The β parameter is related to the diversity of the environment or the richness of possibilities supported by the habitat (Barangé & Campos, 1991). Comparing this parameter to the evenness, we found a corresponding difference between the two basins. In the Siófok basin, evenness is correlated to β – a non-linear but clear correspondence is observed – but there is no correlation between them in the eutrophic Keszthely basin (Figures 3A, B). Comparing the biomass and the β values, however, high biomass occurs at low environmental diversity values in both basins. The effect is more striking in the less eutrophic basin (Figures 3C, D).

Although the two ZM parameters are independent from each other, the γ vs β plot nevertheless shows very interesting features. Just as there is a maximum evenness for a given biomass, so there is also a barrier for γ at any given value of β. The points are bounded by an envelope (Figure 4.). This is true for both the less eutrophic Siófok and the more eutrophic Keszthely basins. The analytical form of the ZM model does not prescribe the presence of this envelope curve; thus, we were able to generate other kinds of communities with alternative community structures (e.g., Poisson, Gaussian, exponential etc.). Fitting the ZM model to such generated 'communities', we get a point series lying exactly on this envelope curve independent of the generated sample distribution function.

Discussion

According to the ZM model, the appearance and extinction of species are stochastic processes governed by the dependence chains in the community and by the dependence chains related to the environment of that community. Such dependence chains represent succession. The relative importance of internal (i.e., competition) or external (meteorological and nutrient conditions) to the outcome depends on the composition of the community. The community acts as a mirror to the habitat; the sample contains plenty of information about its environment. This information can be calculated with the help of the β parameter; this reflects quantitatively the chance of the appearance of a new species. According to the Intermediate Disturbance Hypothesis (IDH), the probability of a community with high diversity is related to the frequency of the perturbations (Connell, 1978; Reynolds et al., 1993). A habitat with low level of perturbations has a low 'environmental diversity', and the result of very frequent perturbations is also a poor environment: other factors are suppressed or overridden.

In spite of the beautiful theories of the ZM model and IDH and against our expectations, the diversity of the environment is usually higher at the eu-

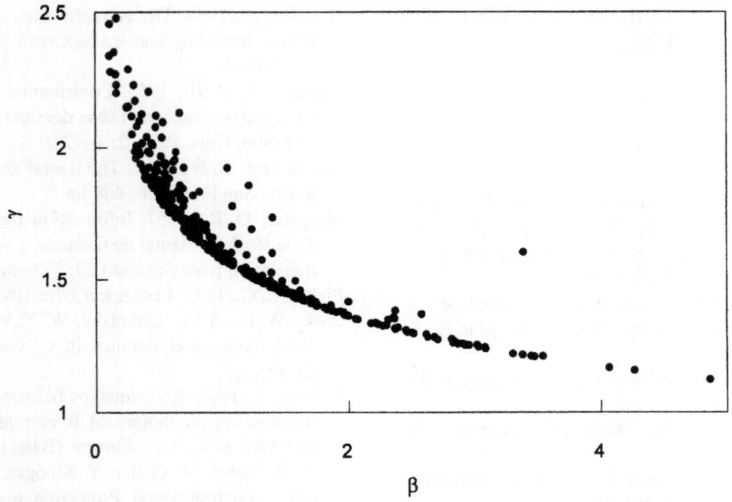

Figure 4. The correspondence between the two parameters in the Zipf-Mandelbrot model. The envelope of the points on the graph represents the unordered communities without interactions between the species. The majority of the points from the eutrophic basin lies near to this envelope.

trophic Keszthely basin than in the Siófok basin (Figures 3C, D). Perhaps the ZM model is less valid in this area, the conditions supposed by the ZM model being incompletely fulfilled. This is also supported by our other results: in the correlation between the evenness and γ, both values give measures of the higher complexity observable in the Siófok basin. With increasing diversity of the environment, the evenness also increases (Figure 3A) as is suggested by the IDH theory. Thus, the community in this basin more nearly complies to the conditions prescribed by the ZM model.

The eutrophic Keszthely basin is quite opposite in this respect. There is only a slight dependence between biomass and the β parameter (Figure 3D), and the evenness varies randomly as a function of the diversity of the environment (Figure 3B). The eveness is not proportional to γ, as it was in the Siófok basin and in other studies (Frontier, 1987).

The comparison of the two ZM constructs is also interesting. The data points are enveloped by a curve. This curve corresponds to the lowest value attainable by the community under a given diversity of the environment (Figure 4). The distance of a point from the curve is thus a measure of the randomness of the community: the computer-generated random samples align perfectly on this curve; data generated from samples from the eutrophic Keszthely basin come close to doing so. This fact suggests that the samples come from unordered random communities. As these samples are not, of course, purely random, it is safe to conclude only that the ZM model describes the data set from the more eutrophic Keszthely basin less satisfactorily.

Conclusion

An analysis, based on the Zipf-Mandelbrot model, of dominance-diversity relations in the two basins of the Lake Balaton has been considered. The model shows significant differences between the basins, although the validity of the model is challenged by the data for the highly eutrophic Keszthely basin. Our results suggest that the organization of the communities in the eutrophic and in the non-eutrophic basins are fundamentally different, and these differences can be partially explored on the basis of the ZM model. The remaining questions are whether there is a sharp transition between the two forms of the community organization and whether these differences are the most important signs of change in community organization during eutrophication. The analysis of more data from more lakes may provide answers to these questions, as well as more insights into the organization of phytoplankton communities.

Acknowledgements

We are grateful to Vera Istvánovics, Colin Reynolds and Chris Coulon for advice and comments on the manuscript and for their help with the English. This work was supported by the grant No. T 016348 of the Hungarian Research Fund (OTKA).

References

Aleya, L. & J. Devaux, 1992. The concept of seasonal succession theory applied to phytoplankton through the coupling use of diversity index and rank-frequency diagrams in an eutrophic system. Int. Revue ges. Hydrobiol. 77: 579–591.

Barangé, M. & B. Campos, 1991. Models of species abundance: a critique of and an alternative to the dynamics model. Mar. Ecol. Prog. Ser. 69: 293–298.

Connell, J., 1978. Diversity in tropical rain forests and coral reefs. Science 199: 1304–1310.

Engen, S., 1978. Stochastic Abundance Models, Chapman & Hall, London.

Frontier, S., 1985. Diversity and structure in aquatic ecosystems. Oceanogr. mar. Biol. ann. Rev. 23: 253–312.

Frontier, S., 1987. Application of fractal theory to ecology. In P. Legendre & L. Legendre (eds), Developments in Numerical Ecology. NATO ASI Series G., Springer Verlag, Berlin, 335–378.

G.-Tóth, L. G & J. Padisák, 1986. Meteorological factors affecting the bloom of *Anabaenopsis raciborskii* Wolosz. (Cyanophyta: Hormogonales) in the shallow Lake Balaton, Hungary. J. Plankton Res. 8: 353–363.

Herodek, S. & G. Tamás, 1976. Phytoplankton biomass, primary production and the eutrophication of Lake Balaton. Hidrológiai Közlöny 56: 219–228 (in Hungarian).

Herodek, S., 1984. The eutrophication of Lake Balaton: Measurements, modeling and management. Verh. int. Ver. Limnol. 22: 1087–1091.

Mandelbrot, B. B., 1953. Contribution a la théorie mathématique des communications. Thèse doctorat, Université de Paris, Publ. Inst. Stat. Univ. Paris. 2: 1–121.

Mandelbrot, B. B., 1982. The fractal geometry of nature. Freeman & Co., San Francisco, 468 pp.

Margalef, D. R., 1957. Information theory in ecology. Memorias de la Real Academia de Ciencias y Artes de Barcelona (English translation from Spanish) 23: 373–449.

Pielou, E. C., 1975. Ecological Diversity, Wiley & Sons, New York.

Press, W. H., S. A. Teukolsky, W. T. Vetterling & B. P. Flannery, 1992. Numerical Recipes in C, Cambridge University Press, Cambridge.

Padisák, J., 1994. Relationships between short-term and long-term responses of phytoplankton to eutrophication of the largest shallow lake in Central Europe (Balaton, Hungary). In H. Sund, H.-H. Stabel, W. Geller, Y. Xiaogan, Y. Kechang & S. Fenning (eds), Environmental Protection and Lake Ecosystem. China Science and Technology Press, Beijing, 419–438.

Reynolds, C. S., J. Padisák & U. Sommer, 1993. Intermediate disturbance in the ecology of phytoplankton and the maintenance of species diversity: a synthesis. In J. Padisák, C. S. Reynolds & U. Sommer (eds), Intermediate Disturbance Hypothesis in Phytoplankton Ecology. Developments in Hydrobiology/Hydrobiologia 249: 183–188.

Tamás, G., 1974. The biomass changes of phytoplankton during the 1960s. Ann. Biol. Tihany 41: 323–342.

Vörös, L. & P. Nagy-Göde, 1993. Long term changes of phytoplankton in Lake Balaton (Hungary). Verh. int. Ver. Limnol. 25: 682–686.

Hydrobiologia **369/370:** 243–257, 1998.
M. Alvarez-Cobelas, C. S. Reynolds, P. Sanchez-Castillo & J. Kristiansen (eds), Phytoplankton and Trophic Gradients.

Trends and relations among basic phytoplankton characteristics in the course of the long-term oligotrophication of Lake Maggiore (Italy)

D. Ruggiu, G. Morabito, P. Panzani & A. Pugnetti
CNR – Istituto Italiano di Idrobiologia, 28922 Verbania Pallanza, Italy

Key words: long-term studies, phytoplankton, lake recovery

Abstract

Lake Maggiore underwent a process of eutrophication in the course of the 1960s and 1970s which caused the lake status to change from oligotrophy to meso-eutrophy, sewage phosphorus being the main responsible factor. Starting from the late 1970s the P load has been gradually reduced, so that total phosphorus in-lake concentration is now below 10 $\mu g\ l^{-1}$ at winter mixing, compared to a maximum value of 30 $\mu g\ l^{-1}$ in 1978. Nearly two decades after the start of this reversal in trophic conditions, long-term analyses become fruitful, not least because of a scarcity of published long-term data on lake re-oligotrophication, and consequently of agreed predictions on specific points. This paper presents such an analysis, regarding some phytoplankton characteristics (numbers, biovolume, cell size, chlorophyll *a*, species richness and composition) and their mutual relations in the time series 1981-1995, as well as recent primary production compared with early observations. Notwithstanding phosphorus reduction, the phytoplankton remained quite stable for many years. Only after 1987, when TP concentration fell to about 15 $\mu g\ l^{-1}$ at mixing, definite changes in species diversity and composition appeared. At the same time, average cell size decreased abruptly and persistently, together with chlorophyll and total biovolume while cell numbers were unchanged or rising. Annual primary production was halved in comparison to a past reference period and the efficiency of the photosynthetic process was apparently enhanced.

Introduction

Lake Maggiore, one of the deep Italian subalpine lakes (Table 1) and oligotrophic by nature, underwent an intense process of cultural eutrophication during the 1960s, which continued at a lower rate in the course of the following decade. Phosphorus from sewage was the main cause of the process, the same as with many lakes world-wide at those times. On the basis of nearly tripled values of annual primary production, of algal biovolume and growth of blue-greens, the lake was considered to be close to eutrophy in the early 1970s (Ruggiu & Saraceni, 1977). It was still at the border between mesotrophy and eutrophy a decade later, after further P enrichment (Mosello & Ruggiu, 1985). The peak P loads and concentrations were reached in 1978–79: the maximum in-lake total phosphorus (TP) concentration of about 30 $\mu g\ l^{-1}$ was recorded in 1978 at winter mixing. Since then, P loads have been gradually reduced by various means, among which the adoption of treatment plants and the reduction of phosphorus in detergents were the most important. Parallel with this, in-lake P concentration kept falling so much that in the last three years (1993–95) the values of TP at mixing time were below 10 $\mu g\ l^{-1}$ (Calderoni et al., 1996).

This slow and steady reversal of the trophic state of Lake Maggiore is documented in a series of annual reports covering the period 1978–1992 (e.g. Calderoni et al., 1993), and by many papers. From a biological point of view, strong emphasis was put in the 1980s on the apparent resilience of the plankton communities against falling phosphorus (de Bernardi et al., 1988). However, starting from 1987–88, major biological changes were at last manifest, especially in the phytoplankton (Manca et al., 1992; Ruggiu, 1993).

At the present time, after clear changes have been occurring in the plankton for at least eight years following a comparable or longer period of stability, we felt

Table 1. Basic characteristics of Lake Maggiore. Chemical, biological, pH and Secchi disc data are from 1994

Location......Lat 45°57′22″ N; Long 8°38′40″ W
Area (A)......212.511 km^2
Volume (V)......37.502 km^3
Mean depth (V/A)......176.47 m
Maximum depth......370 m
Residence time (τ)......$\sim$ 4 yr Mixing frequency: oligomictic.
Mixed to > 10 m more than 9 months a year; to > 2.5 m practically the whole year
Ice cover......absent
pH range......7.45–8.37
Secchi disc range......4–12 m
Total P range......5–11 $\mu g\ l^{-1}$
SRP range......2–6 $\mu g\ l^{-1}$
Alkalinity range......0.68–0.80 eq l^{-1}
Total N range......770–940 $\mu g\ l^{-1}$
DIN range......616–845 $\mu g\ l^{-1}$
Annual mean phytoplankton biomass......1 cm m^{-3}
Peak:mean biomass ratio......2.42

that a long-term analysis encompassing the time span of lake recovery was necessary. In fact, published data on biological events accompanying lake oligotrophication are scarce, and agreed predictions are consequently difficult. This paper presents a long-term analysis of some basic phytoplankton characteristics such as numbers, biovolume, cell size, chlorophyll *a*, species richness and composition, as well as of recent primary production compared with early observations. The data will contribute to the factual knowledge in an actively-debated field, not least because a detailed account of variations of algal characteristics in the course of a lake oligotrophication has apparently not been published up to now.

Materials and methods

A regular monitoring of nutrient loading and in-lake concentration, and of plankton standing stock, has been under way in Lake Maggiore from 1978, covering the oligotrophication phase from its start. All the samples considered in this paper were taken at a station corresponding to the lake's maximum depth, approximately halfway along the lake's major axis. Two more stations were concurrently sampled in the first ten years of monitoring, one located in the Swiss basin in front of Locarno and the other in the southern part of the lake. However, after 1987 they were discarded and only the central station was maintained, as the differences in chemistry and in the plankton between the three stations were found, year after year, to be insignificant.

Most methods were unchanged throughout the study period. Water samples were taken every two meters in the upper 20 m layer and mixed together to give an integrated sample for phytoplankton and chlorophyll analysis. Discrete samples were taken for chemical analysis along the water column. The sampling frequency was variable, being on average about 20 times a year in the period 1981–1995. As a rule, samples were taken once a month in January and December, and more frequently (2–3 times a month) during the spring (April-May) and the summer growth (July-September). One and the same person performed the phytoplankton identification and biovolume assessment, a fact very important in our view for a long-term analysis. Phytoplankton was counted with the Utermöhl's method. The counting was made along two perpendicular diameters of the sedimentation chamber, corresponding to 100 visual fields. Unidentifiable, spherical cells of about 2–4 μm diameter (ultraplankton, not specifically picoplankton) were also included. Especially large individuals (free cells or colonies) were counted on half the chamber bottom. In this way, 300–400 individuals for each important species were usually counted. Average specific cell sizes (volume, μm^3) were calculated from microscopic length measurements. The sizes considered in this paper are those of cells, even in case of colonial species. As frequent checks were made, a data-base was gradually built, whereby changes of cell size of some species with time could be ascertained. For average community cell size the ratio (BV/N) between total biovolume and total numbers was calculated and the result expressed as μm^3. For recent (1994–95) primary production measurements the Danish Standard Association method (1982) was followed. The samples, gathered every fortnight, were incubated in 150 ml light and dark bottles (each one labelled with 5 μCi of ^{14}C) *in situ* at five depths in the euphotic zone for 4 hours symmetrically around noon. In the laboratory, 30 ml subsamples were filtered through 0.2 μm pore size polycarbonate filters that were then acidified and dissolved in the scintillation cocktail. The radioactivity was measured in a Beckman LS 6000 scintillation counter; counting efficiency was determined automatically by the use of a ^{14}C standard. The total inorganic carbon was calculated from pH and alkalinity (APHA, 1992). In the 1970s and early 1980s the field methods were similar, but in the laboratory a Geiger-Müller counter was used, and

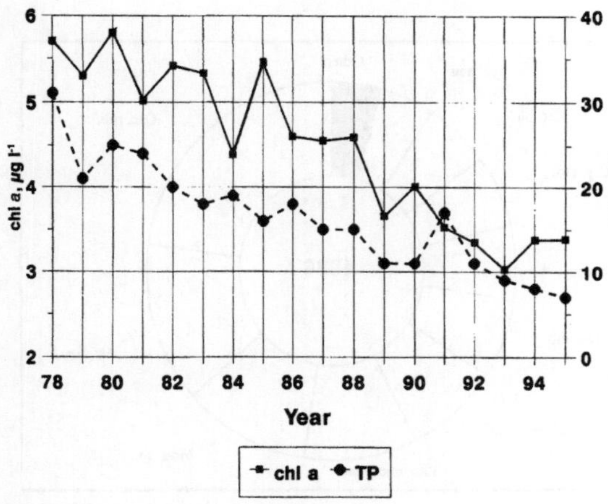

Figure 1. Time courses (1978–1995) of average annual chlorophyll *a* concentration (chl *a*: left scale) and of total phosphorus concentration at winter mixing time (TP: right scale). The values are averages over the 0–20 m water layer.

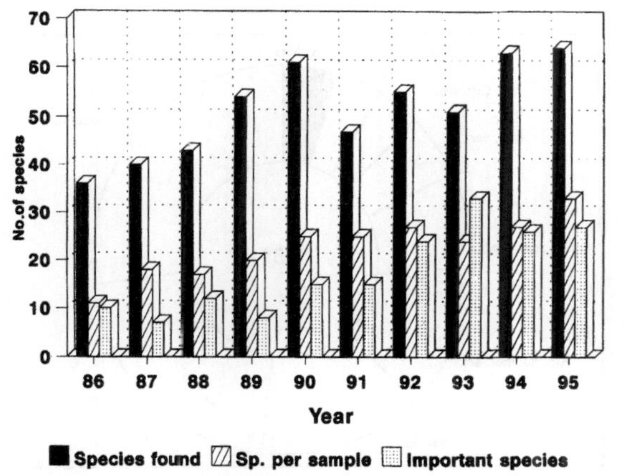

Figure 2. Numbers of differently categorised phytoplankton species from 1986 to 1995. Species found = total number of identified species; Sp. per sample = annual average number of species per sample; Important species = number of species which, at least in one occasion, participate to the group of dominant species, defined as those that make up at least 80% of the total biovolume in a series of ranked species.

sample acidification was omitted (Ruggiu & Saraceni, 1977). The calculations used to estimate daily production from the 4 hour incubations were also somewhat different in the two periods: a model proposed by Platt (1971) was adopted, because it was found to fit more exactly the experimental data than the model of Gächter (1972) used in the pre-oligotrophication period.

The methods for nutrients and chlorophyll *a* determination are described in Mosello & Ruggiu (1985). Data treatments were made by means of the software StatSoft-Statistica.

Results

Phosphorus and chlorophyll

Basic evidence of the oligotrophication process of Lake Maggiore is given in Figure 1, where total phosphorus (TP) concentrations at the winter mixing and average annual chlorophyll *a* concentrations are reported. In the long term, both curves show a clear decreasing trend. However, the sharp chlorophyll decline in 1984 was judged to be an effect of anomalous meteorological conditions (Ruggiu, 1989), since in that year the average epilimnetic water temperature was the coldest in a series of 23 years. This view was fully supported by the events of the following, exceptionally warm 1985, when the chlorophyll concentration rose again to exactly the average value of the period 1978–1983. Thus, until 1985, the chlorophyll fluctuations shown responded to meteorological vagaries rather than to any change in the phosphorus supply which, in the meantime, was reduced by a half. Only when TP values reached 15 $\mu g\ l^{-1}$ or less, did the chlorophyll show a definite drop in its values. The latest available data (1995) show 7 $\mu g\ l^{-1}$ TP at mixing time and an annual average of 3.4 $\mu g\ l^{-1}$ chl *a*, thus typifying an oligotrophic situation.

Species assemblages

The species assemblages were interannually rather stable from 1978 to 1987, as repeatedly pointed out in the annual reports (e.g. Ruggiu & Panzani, 1988). The recent changes can be illustrated in the simplest way by considering variously classified numbers of species on an annual basis (Figure 2). The increasing numbers of the recorded species, and possibly more important, of the species per sample along with a more even species distribution, demonstrate that a gradual and remarkable diversification has been occurring in the last ten years of oligotrophication. Starting from 1986, the species list passed from 36 to more than 60, the species per sample from 11 to more than 32, and the species which were dominant at one time or another, from 9 to 26–27. The phenomenon can be again illustrated at the time scale of one year as exemplified in Figure 3, where the rings represent the average annual biovolumes of the

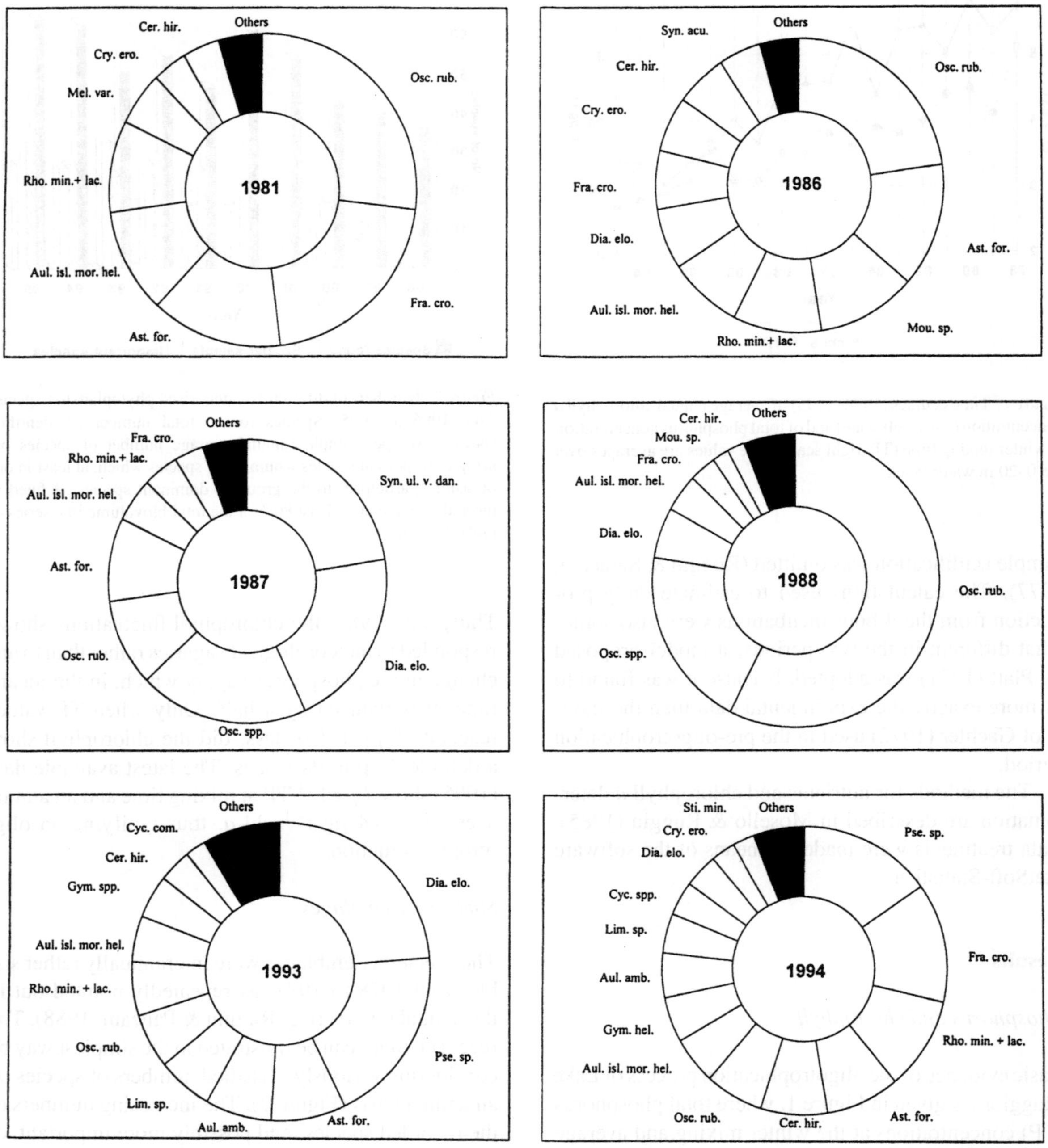

Figure 3. Phytoplankton assemblages in selected years. The sectors of each ring represent the mean annual percentage contribution of the corresponding species to the total community biomass. The average total biomasses on which the specific contributions were calculated are respectively ($cm^3 m^{-3}$): 1.88 (1981), 1.72 (1986), 2.01 (1987), 2.47 (1988), 1.46 (1993), 1.00 (1994). For the full names of species see the Appendix.

total phytoplankton, and the sectors represent the percent contribution of the single, most important species. In the Figure, one year from the stable phase (1981) is contrasted with the rapid transition phase of 1986–88 and with two most recent years where diversity attains its maximum.

Table 2. Correlations between phytoplankton variables. Marked correlations are significant at $p <$ .001. Time period: 1981–1995, n = 180, corresponding to the number of months. BV = total biovolume; N = abundance; SIZE = average cell size; CHL = chlorophyll *a* concentration; CHL/BV = percent chlorophyll per unit biovolume; CHL/N = chlorophyll amount per cell; TR = Secchi disk transparency

Variable	MONTH	BV	N	SIZE	CHL	CHL/BV	CHL/N	TR
MONTH	1.00							
BV	–0.22	1.00						
N	0.06	0.49*	1.00					
SIZE	–0.24	0.56*	–0.29*	1.00				
CHL	–0.29*	0.75*	0.33*	0.47*	1.00			
CHL/BV	–0.07	–0.50*	–0.33*	–0.26*	–0.02	1.00		
CHL/N	–0.28*	0.19	–0.47*	0.75*	0.50*	0.34*	1.00	
TR	–0.04	–0.66*	–0.49*	–0.28*	–0.63*	0.31*	0.08	1.00

Time courses and relations among variables

For a long-term analysis of some fundamental phytoplankton characteristics, i.e. numbers, biovolume, average cell sizes, variously expressed chlorophyll and water transparency, we constructed a basic set of data where each variable was represented by a value per month, obtained through interpolations and averaging. The result was 180 monthly values per variable, corresponding to the 15-year period 1981–1995. For a first exploration of the data, the linear correlation matrix reported in Table 2 was produced, where the highly significant coefficients ($p < 0.001$) are marked. All the coefficients are significant at least at $p < 0.01$ if their value is higher than 0.20. A full analysis of the form of the regressions is under way and will be reported elsewhere; however, some additional statistical considerations will be found in the following paragraphs.

Some correlations were expected: for example, total chlorophyll is obviously positively related to total biovolume or abundance. On the other hand, its being also related to average cell size depends on the fact that the latter is positively related with biovolume. Similarly, it is not unexpected that the amount of chlorophyll per cell is related to cell size, or that transparency is inversely related to chlorophyll concentration or to biovolume. However, other relations are not so obvious. CHL/BV, the chlorophyll percentage per unit biovolume, has a mean value over the long-term of 0.30%, with a range 0.08–1.18% and a standard deviation of 0.15. It is inversely related to cell size, which could not be taken for granted, nor could any effect of covariation be so either. The same can be said for the negative correlation between cell size and abundance, or for other reported relations.

Cell size

Of the time series, that of the average cell size seemed particularly interesting, because the topic of cell or organism size has been raising much interest in recent years, and we had at hand a long-term example with a very clear course. Consequently, we decided to analyse this case with some detail.

A monthly plot of the series is shown in Figure 4. Notwithstanding a great seasonal and interannual variability, it appears that, after a trend of relatively high and increasing cell sizes culminating in 1987, there came a sudden drop and a gradual overall decrease in size down to the present time. This comes about, at a visual inspection, both from the shortening of the peaks and the sinking of the troughs after 1987.

It had first to be ascertained whether the reduction of cell size after 1987 was significant, and this was done by a simple comparison of means. To this purpose, the entire set of data was divided into two subsets, the first comprising the first 75 months ending with the peak value of 340 μm^3 in March 1987, and the second comprising the remaining 105 months until the end of 1995. A Box & Whisker plot with the respective means and their 95% confidence limits is reported in Figure 5 (upper left), and shows how far apart are the limits of the respective means. A Wilcoxon-test was highly significant ($p < 0.0001$). The data distribution is lognormal; however, the sample being large, a log-transformation did not change substantially the outcome of either the plot or the Wilcoxon-test.

Thus a highly significant reduction in average algal cell size is indeed taking place in Lake Maggiore since 1987. Basic statistical data, and a look at Figure 4, also tell us that the data dispersion is lower, and shrinking

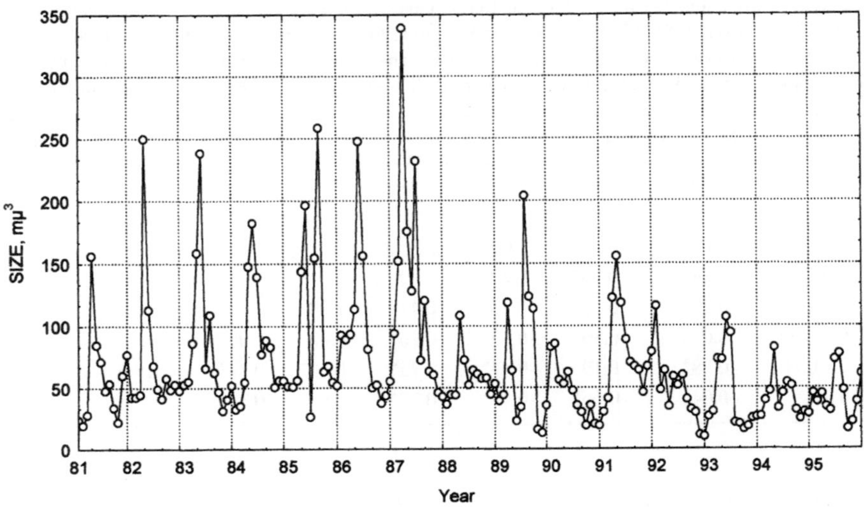

Figure 4. Time course (1981–1995) of phytoplankton average cell size.

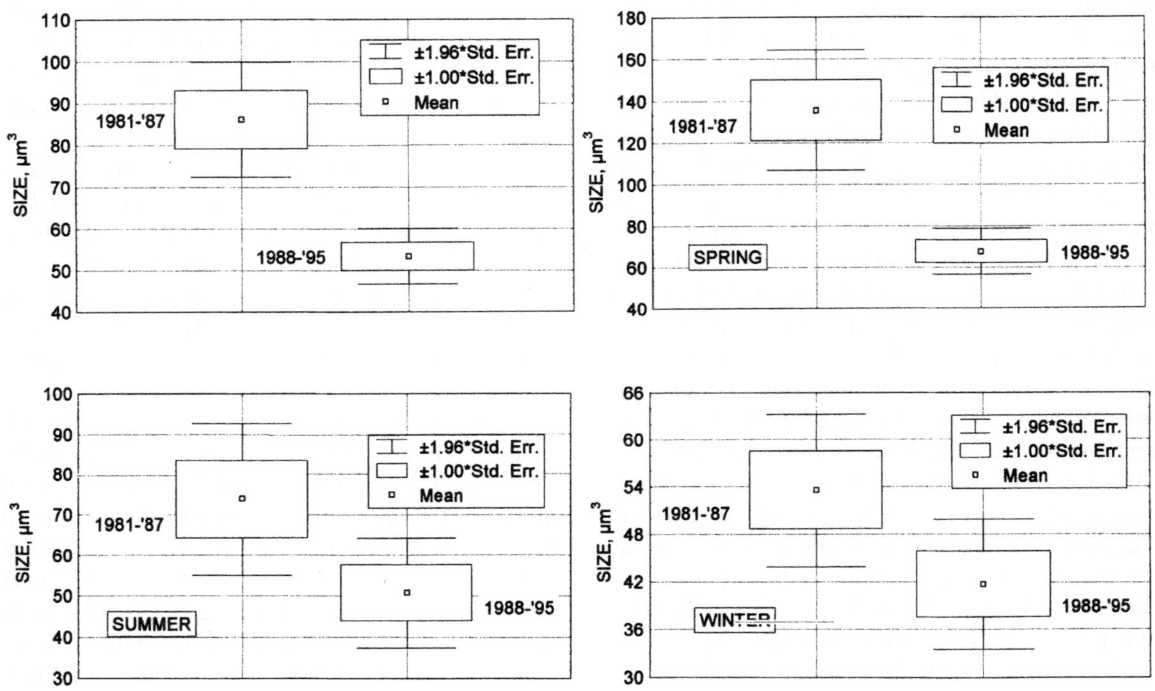

Figure 5. Box & Whisker plots of phytoplankton cell size. Upper left: all monthly data relative to the time periods 1981–87 and 1988–95 respectively. Upper right and bottom: the same time periods with a seasonal decomposition.

with time, in the more recent period. On average, the algal cells are becoming smaller and less variable in size, and this has been the trend since about the time when chlorophyll concentration started to decrease and TP winter concentration fell to 15 $\mu g\ l^{-1}$ (Figure 1).

The next question was how the size reduction was generated. It could be result of an (improbable) generalised cell size reduction, or the outcome of a reorganisation of species structure, not excluding the possibility that some species could experience a reduction in size.

Some data decomposition was necessary. We knew from experience that in Lake Maggiore, three main periods of phytoplankton growth may be recognised during a calendar year: a late-winter and spring growth (March through June), a summer and early autumn growth (July through October) and a late-autumn and winter growth (November through February). Although refinements are possible, these periods are sufficient to permit a rough but realistic classification of phytoplankton seasons in Lake Maggiore. The first period may be called *spring*, and is characterised by the most intense diatom growth, and usually also by the highest total biovolume supported during the year; the assigned time span is justified by the fact that within it the early diatom development and sudden decline takes place, although usually this phase is shorter than four months in a particular year. The second period (here called *summer*) starts in July, when the spring growth is over, and ends with October, a month into which, given the insubrian climate, the summer growth, or even blooms, can well extend. The blue-greens are usually most important in this season, but also other groups may be prominent at particular times; the period typically includes a second diatom peak (*Fragilaria crotonensis*). The third considered season (*winter*) is a period of low production and biovolume, with various groups possibly dominant, again including the blue-greens. A partial vertical water turnover (no complete mixing occurred since 1981) characterises this period.

The Box & Whisker plots of Figure 5 relative to the single seasons show that the overall, abrupt decrease in size after 1987 was largely due to the decrease in the spring communities, while the differences in summer and winter were much less significant. To ascertain how the decrease in size came about, attention was consequently directed to the changes in the spring communities.

To this end, specific spring biovolumes were averaged through the season and then ranked in decreasing order. The species which accounted for 1% or more of the community BV were considered: these were usually ten or so, about twenty at most. The average cell size of this assemblage was usually quite similar to that calculated for the entire community: this means that most of the species in a complete list were irrelevant in determining the average cell size of the community, even though they could be important in other respects.

Table 3 illustrates the resulting analysis for spring 1987, when the average cell size was maximum over the entire period of study. The entire spring community was composed by 27 identified species. This can be compared with the spring situation of the following 1988, when the first definite reduction in average cell size occurred (Table 4). The community was composed by 29 identified species.

Finally, Table 5 shows the data for spring 1994; 48 identifiable species composed the whole spring community.

The above tables illustrate an important fact: the initial, abrupt reduction of average cell size in 1987 was the result of a rearrangement of existing species, whereby some small-sized species were favoured. However, with time, compositional changes took place, and new, small species appeared and outgrew larger ones. In fact, the lists of species of both spring 1987 (Table 3) and 1988 (Table 4) are quite similar, and *Oscillatoria* sp., which made a large difference in 1988, had been constantly found in the lake since 1982. However, the 1994 list reported in Table 5 is rather different from the former two, and for instance the two taxa *Limnothrix* sp. and *Stichococcus minutissimus*, which were essential in bringing down the average cell size because of their smallness and high numbers, were recent appearances, having been respectively listed for the first time in Lake Maggiore in 1991 and in 1994.

The cells were largest in spring both before and after the size reduction, and were followed by those present in summer and, finally, by the winter species.

The other possible mechanism for average cell size decrease, i.e. a reduction of the cell size of extant species, was of small importance. Among the 150 taxa or so identified during the period of study, only six species exhibited a definite reduction in cell size in the course of time, i.e. (μm^3 range in brackets; second value more recent):
Microcystis aeruginosa (33–6.4),
Melosira varians (4800–2700),
Synedra acus v. *radians* (1400–100),
Dinobryon divergens (800–400),
Ceratium hirundinella (76500–72300),
Sphaerocystis schröteri (270–160).

Three species apparently increased their cell size, i.e.:
Cyclotella comensis (170–240),
Gymnodinium helveticum (11100–15500),
Monoraphidium contortum (100–250).

None of these species was of much importance in determining the value of the community cell size. However, starting from 1988, the unidentified ultraplankton

Table 3. Phytoplankton structure, spring 1987. Size = specific cell size, μm^3; BV_{sp} = specific biovolume, $mm^3\ m^{-3}$; % BV_{sp} = percentage of specific biovolume on total biovolume; $N_{sp} = 10^6$ cells l^{-1}. The complete species names are given in the appendix

	Size	BV_{sp}	% BV_{sp}	N_{sp}
Asterionella formosa	984	1124	40.5	1.142
Diatoma elongatum	1000	455	16.4	0.455
Synedra ulna var. *danica*	19262	433	15.6	0.023
Fragilaria crotonensis	440	173	6.2	0.393
Ultraplankton	14.14	129	4.7	9.137
Aulacoseira islandica var. *helvetica*	1237	96	3.5	0.078
Rhodomonas minuta	139	79	2.8	0.566
Synedra acus	3678	76	2.7	0.021
Oscillatoria rubescens	69	48	1.7	0.699
Rhodomonas lacustris	203	31	1.1	0.154
Cryptomonas erosa	1803	27	1.0	0.015
Totals		2672	96.3	12.683
Average size	211			
Avg. size from total community	219			

Table 4. Phytoplankton structure, spring 1988. Symbols and units as in Table 3

	Size	BV_{sp}	% BV_{sp}	N_{sp}
Oscillatoria rubescens	69	1163	32.0	16.853
Oscillatoria sp.	75	952	26.2	12.693
Diatoma elongatum	1000	372	10.2	0.372
Aulacoseira islandica var. *helvetica*	1237	297	8.2	0.240
Ultraplankton	14.14	276	7.6	19.542
Asterionella formosa	984	111	3.0	0.112
Synedra acus	3678	68	1.9	0.018
Synedra ulna v. *danica*	19262	51	1.4	0.003
Rhodomonas minuta	139	49	1.4	0.354
Mougeotia sp.	877	42	1.2	0.048
Cryptomonas erosa	1803	39	1.1	0.022
Totals		3420	94.1	50.259
Average size	68			
Avg. size from total community	69			

appeared to be composed by species of diminishing cell size. Average size had been estimated previously to be 14.14 μm^3 (see Tables 3 and 4) but, after 1988, we had to assume a gradual diminution to 3.2 μm^3 in 1994 (Table 5). The ultraplankton is undoubtedly significant in determining the final community cell size value. However, it must be remembered that in 1988, when the first, great reduction in average cell size occurred, the ultraplankton size was still unchanged and the significant fact was the predominance of the *Oscillatoria* species over the large diatoms prevailing in 1987.

Cell numbers and total biovolume

The consideration and data treatment on the time courses of the variables biovolume (BV) and abundance (N) were analogous to those reported for cell size, and will not repeated in detail. Their relations with the time variable (Table 2) are respectively significant at $p < 0.01$ (BV) or not significant (N). The time courses resemble those of cell size in that the relations are not linear, but with cell size the variations were so strong after a certain time that a highly significant linear correlation formally existed. However, just as in the case of cell

Table 5. Phytoplankton structure, spring 1994. Symbols and units as in Table 3

	Size	BV_{sp}	$\% BV_{sp}$	N_{sp}
Pseudanabaena sp.	24.5	348	25.3	14.193
Asterionella formosa	984	107	7.8	0.109
Limnothrix sp.	19	88	6.4	4.647
Aulacoseira islandica var. *helvetica*	1237	86	6.3	0.070
Fragilaria crotonensis	440	80	5.8	0.181
Ceratium hirundinella	72282	75	5.5	0.001
Oscillatoria rubescens	75	61	4.4	0.813
Stichococcus minutissimus	9.8	60	4.4	6.135
Cyclotella hakanssoniae	236	58	4.2	0.244
Rhodomonas minuta	139	52	3.8	0.377
Diatoma elongatum	1000	51	3.7	0.051
Cyclotella comensis	238	50	3.6	0.209
Aulacoseira ambigua	1237	49	3.6	0.040
Rhodomonas lacustris	203	33	2.4	0.162
Gymnodinium helveticum	15494	25	1.8	0.002
Cryptomonas erosa	1803	25	1.8	0.014
Ultraplankton	3.2	21	1.5	6.447
Uroglena americana	60	20	1.5	0.333
Mallomonas elongata	4059	14	1.0	0.003
Totals		1303	94.8	34.031
Average size	38			
Avg. size from total community	52			

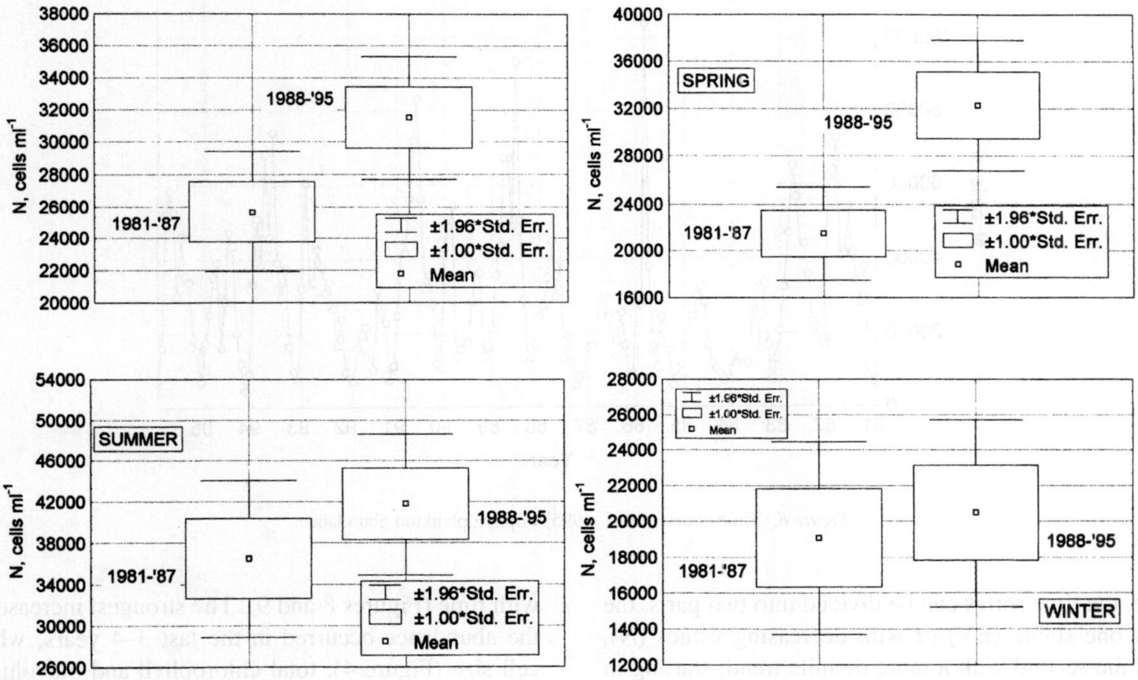

Figure 6. Box & Whisker plots of phytoplankton abundance. Specifications as in Figure 5.

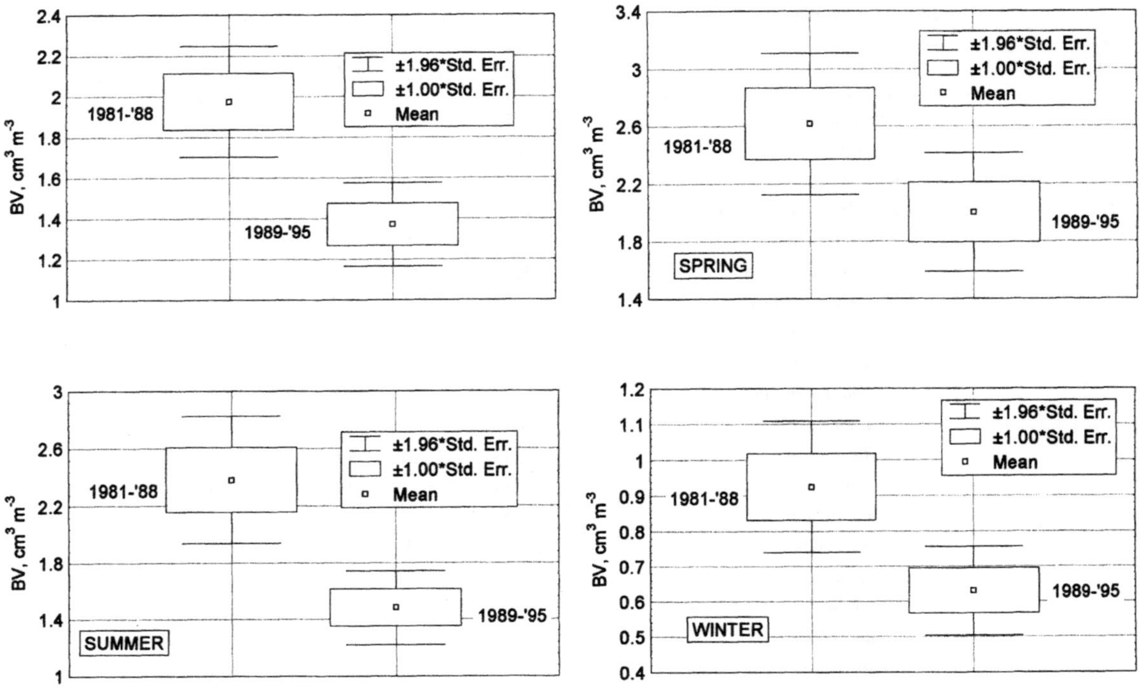

Figure 7. Box & Whisker plots of total phytoplankton biovolume. Specifications as in Figure 5 except that the compared time periods are 1981–88 and 1989–95 respectively.

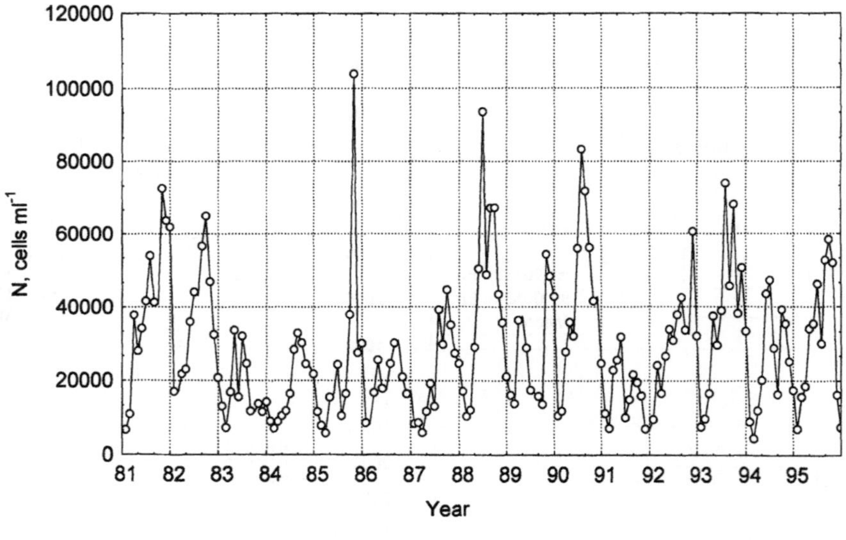

Figure 8. Time course (1981–1995) of phytoplankton abundance.

size, the time series can be divided into two parts, the first one stable (BV) or with decreasing values (N), and the second with a more definite trend, starting in 1987 for N and 1988 for BV. The respective trends are opposite, as N tends to increase and BV to diminish with time (Figures 8 and 9). The strongest increase in the abundance occurred in the last 3–4 years, when cell size (Figure 4), total chlorophyll and biovolume were at their lowest; seasonally, the most significant increase was in spring (Figure 6). The difference in the

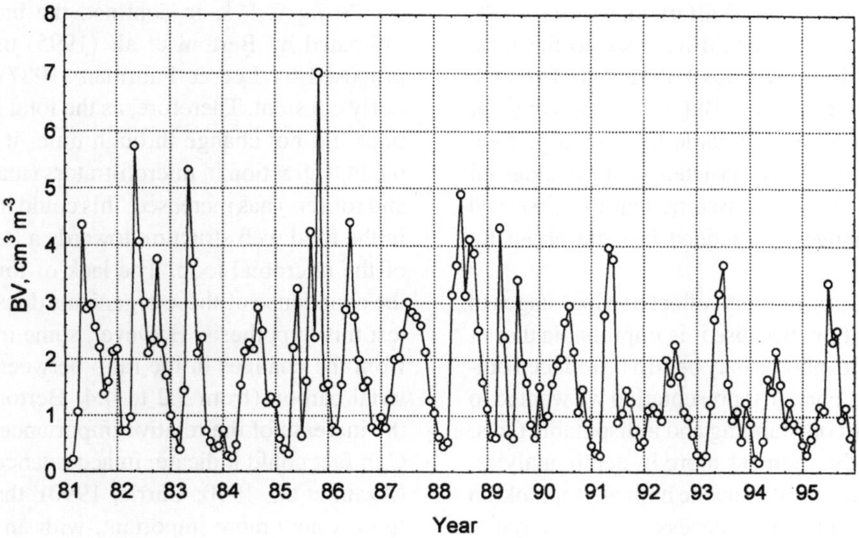

Figure 9. Time course (1981–1995) of total phytoplankton biovolume.

two time periods of BV was highly significant (Figure 7), just as in the case of cell size. On average, BV was highest in spring and lowest in winter; however, the most significant decrease occurred in summer.

Primary production

During the 1970s the annual phytoplankton production was estimated at around 300 g C m^{-2} y^{-1} (Ruggiu & Saraceni, 1977); the last measurement, made in 1981, gave 280 g C m^{-2} y^{-1} (Ruggiu et al., 1982). After less than 15 years the yearly production almost halved: it was 148 and 136 g C m^{-2} y^{-1} respectively in 1994 and in 1995. In the same period the mean chlorophyll *a* concentration passed from 5 mg m^{-3} to 3.5 mg m^{-3} and the yearly mean specific photosynthetic rate per unit of chlorophyll *a* increased from 8.4 to 11.8 (1994) and to 9.6 (1995) mg C (mg chl)$^{-1}$ d^{-1}.

Notwithstanding the large differences in the carbon photoassimilation rates, both the POC, which is mainly of autochthonous origin in this lake (Bertoni & Callieri, 1993), and the net carbon production calculated from the changes of in-lake POC concentrations, were fairly constant through the years (Bertoni & Callieri, 1993). This suggests that during the oligotrophication the decrease of photosynthetic rate has been accompanied by a reduction of carbon losses. Also the dissolved organic carbon (DOC) concentration decreased by more than 30% in these years (Bertoni & Callieri, 1993), a fact that was attributed to a reduction of the allochthonous DOC input that occurred since 1983 after the construction of treatment plants in the watershed. However, a concomitant cause could be a decrease of the DOC lost by the algae, as exudates or during cellular decay. This is suggested by the correlation between biomass per unit volume and production, not significant in 1981 but significant in the years 1994–95 (r = 0.86; $p < 0.05$), indicating that the coupling between primary production and biomass has become stricter in recent years.

Discussion

Although the literature on lake trophic state and change is very large, clear-cut case histories of biological responses to oligotrophication, i.e. to decreasing nutrient loads, are comparatively few. Whereas most lakes underwent eutrophication in the urbanised areas of the developed countries in the second half of this century, only a few of them have been subject to restoration measures, and the outcome is often uncertain (Sas, 1989). We can now see that this uncertainty often reflects the fact that restoration of lakes, and specifically the appropriate reduction of nutrient availability, generally takes a long time. One of the consequences is that short-term fluctuations may be rightly evaluated only in the context of a long-term series. It can be seen from Figure 1 that in Lake Maggiore the first clear decrease of average chlorophyll concentration took place in 1986–88. A shorter time series, ending for instance in 1984, could have led to the false conclusion that chlorophyll was falling right from the start of phosphorus reduction. But today, knowing that TP is the only monitored variable to show a parallel trend to chlorophyll in the last ten years (Calderoni et al., 1993), we can safely assume that the observed phytoplankton changes are indeed brought about by oligotrophication.

These changes presumably reflect the final stages of lake oligotrophication, because it is improbable that in Lake Maggiore phosphorus may be still be reduced other than marginally. For a first presentation we wanted to emphasise the most outstanding and indisputable facts, leaving aside for the moment more in-depth analysis. To do so, we had to compensate the high phytoplankton variability due to autogenic succession or to variable meteorological conditions by choosing a time scale corresponding to the year or to very broad seasons. An elementary data treatment was also adequate.

We can summarise the emerging pattern as follows. Since the time when TP fell to around 15 μg l^{-1} at mixing time (maximum values in the year), a series of changes took place in the phytoplankton, formerly characterised by a remarkable stability. Species diversity increased consistently, as demonstrated by Figure 2. The trend is so clear that the application of diversity indices is unnecessary. The composition of the dominant species (in biovolume) also changed (Figure 3; Tables 3–5). The most important algal groups are still the Cyanobacteria and the Bacillariophyceae, but by 1992 the first were represented by 10 species and the second by 19 species, against 3 and 8 respectively in 1981 (Ruggiu, 1993). Total biovolume (BV) is decreasing, whereas abundance (N) tends to increase. Seasonally, N increases most in spring (Figure 6), while the significant BV decrease is in summer (Figure 7. Average cell size definitely diminishes during oligotrophication, most strikingly in spring (Figure 5; Tables 3–5), and is caused first by a rearrangement of existing species, and later by the appearance of new, small-sized species. Cell size is negatively correlated both with percent chlorophyll per unit biovolume and with abundance (Table 2).

Primary production decreased during oligotrophication, while specific rates of photoassimilation increased and the POC concentrations did not show significant variations. However, looking at the different POC fractions, that ascribed to the algae, estimated as 13% of the phytoplankton biomass as fresh weight (Ruggiu et al., 1981), has decreased from 190 to 120 μg C l^{-1}; in contrast, the bacterial fraction, estimated by Bertoni et al. (1995) using the factors proposed by Lee & Fuhrman (1987), has remained fairly constant. Therefore, as the total POC concentrations did not change through time, it is possible that the POC fraction in microfiltrators (mainly protozoans and rotifers) has increased: this could indicate a change in the food web structure towards a larger importance of the microbial loop. The lack of long-term data on the abundance of these organisms does not allow us to test this hypothesis. However, some indications come from the changes in the ratio between algal and bacterial carbon (from 5.2 to 4.4; Bertoni et al., 1995): the increase of the relative importance of the bacterial C in fact could indicate, in accordance to some views (Azam et al., 1990; Currie, 1990), that the microbial loop is now more important, with an organisation of the food web more typical of oligotrophic waters. An observed increase of the coccoid cyanobacteria during 1994 and 1995 (Callieri, unpublished data) may also be significant in this respect.

Among the consequences of the increased importance of the microbial loop there could be the reduction of the metabolic losses, through a better coupling between carbon photosynthetic fixation and utilisation (Harris, 1994).The final result seems therefore a better efficiency of the whole system.

None of the reported findings, and especially those on community structure and cell characteristics, are really unexpected, as one or the other has been anticipated on theoretical or experimental grounds, or recognised as characterising trophic *states* (e.g. Harris 1986, 1994, and included references; Reynolds, 1988). Much less do they come from case histories of oligotrophication *processes* in nature.

Among the few such cases, one of the earliest and better known is that of Lake Washington (U.S.A.), with its spectacular recovery involving, among many studied effects, the decline of *Oscillatoria*, improvements in transparency, changes in the food web (Edmondson, 1991). In Europe, the investigations on reversed eutrophication of large Swedish lakes must be mentioned (Willén, 1987, 1992). This is not only because in those lakes too the major emphasis was on phytoplankton characteristics but also because some long-term results (increased species richness and evenness, decreasing cell size, changes in species composition) were similar to the events in Lake Maggiore, which is hardly comparable to the Swedish lakes as to morphology and geographic position.

One of the most intensely studied lakes in Europe is Lake Constance, a large and deep pre-alpine lake, and therefore much more similar to Lake Maggiore than the Swedish lakes. It has been undergoing oligotrophication for more than ten years, and assessments of the phytoplankton response have recently appeared (Tilzer et al. 1991; Gaedke & Schweizer, 1993; Sommer et al., 1993). A comparison with the long-term events in Lake Maggiore shows similarities and discrepancies. Among the first we can select the following. Contrary to some theoretical expectations (Sas, 1989), in Lake Constance changes in phytoplankton species composition preceded the changes in biovolume; in Lake Maggiore the two events were probably contemporary, but it can be said that in any case biovolume changes did not precede species changes. In the two lakes, biovolume decreased most significantly in summer. The peak biovolume is now shifted to spring in Lake Constance, the season where it consistently occurred in Lake Maggiore. In both lakes, there is evidence of an increased general efficiency of the system. The major discrepancies are to be found in the average cell size, which did not change in Lake Constance, in the apparent stability of species diversity (no comments are found in this respect in the above cited papers), and in the time course of primary productivity, which in Lake Constance did not follow the reduction of phosphorus load. However, oligotrophication admittedly has not reached its end in the northern lake. Notwithstanding a large reduction with time, soluble phosphorus concentration in winter is still higher than it was in Lake Maggiore at its peak in the late 1970s, so that a larger convergence between the two lakes will probably occur in future.

Lake Maggiore, having run the full circle from pristine oligotrophy through a process of cultural eutrophication and again to a complete restoration, will remain a useful reference in the subject matter of events characterising trophic changes in lakes. However, a decision on the degree of generality or specificity of observed phenomena, which would give us predictive power, is at present difficult. What we need is not so much new speculation as new case histories and/or additional elaboration of existing data, among which those from routine monitoring, often confined to technical reports, may prove invaluable.

References

American Public Health Association A.P.H.A., 1992. Standard methods for the examination of water and wastewater. 18th Edition. APHA, Washington D.C..

Azam, F., B. C. Cho, D. C. Smith & M. Simon, 1990. Bacterial cycling of matter in the pelagic zone of aquatic ecosystems. In M. M. Tilzer & C. Serruya (eds), Large lakes: ecological structure and function. Springer Verlag, New York 24: 477–488.

Bertoni, R. & C. Callieri, 1993. Organic carbon trend during the oligotrophication of Lago Maggiore. In R. de Bernardi, R. Pagnotta & A. Pugnetti (eds), Strategies for lake ecosystem beyond 2000. Mem. Ist. ital. Idrobiol. 52: 191–205.

Bertoni, R., C. Callieri, G. Morabito, M. L. Pinolini & A. Pugnetti, in press. Quali-quantitative changes in organic carbon production during the oligotrophication of Lake Maggiore, Italy. Verh. int. Ver. Limnol. 26: 300–304.

Calderoni, A., R. de Bernardi & D. Ruggiu, 1993. Osservazioni conclusive. Chimismo delle acque ed evoluzione trofica. In C.N.R. Istituto Italiano di Idrobiologia, Ricerche sull'evoluzione del Lago Maggiore. Aspetti limnologici. Campagna 1992 e Rapporto quinquennale 1988–1992 (ed.), Commissione Internazionale per la protezione delle acque italo-svizzere: 111–114.

Calderoni, A., R. Mosello, A. Pranzo & G. Tartari, in press. Chimica lacustre. In C.N.R. Istituto Italiano di Idrobiologia, Ricerche sull'evoluzione del Lago Maggiore. Aspetti limnologici. Campagna 1995 (con sintesi degli anni 1993–1994) (ed.), Commissione Internazionale per la protezione delle acque italo-svizzere.

Currie, D. J., 1990. Large scale variability and interactions among phytoplankton, bacterioplankton and phosphorus. Limnol. Oceanogr. 35: 1437–1455.

Danish Standards Association (Dansk Standardiseringsråd), 1982. Water Quality – Incubator method for determining the carbon assimilation by plankton algae using ^{14}C technique. Dansk Standard, DS/R 293.

de Bernardi, R., G. Giussani, M. Manca & D. Ruggiu, 1988. Long-term dynamics of plankton communities of Lago Maggiore (Italy). Verh. int. Ver. Limnol. 23: 729–733.

Edmondson, W. T., 1991. The Uses of Ecology. Lake Washington and Beyond. Univ. Washington Press, Seattle, 329 pp.

Harris, G. P., 1986. Phytoplankton Ecology. Chapman and Hall, London, 384 pp.

Harris, G. P., 1994. Pattern, process and prediction in aquatic ecology. A limnological view of some general ecological problems. Freshwat. Biol. 32: 143–160.

Gächter, R., 1972. Die Bestimmung der Tagesraten der planktischen Primärproduktion – Modelle und in-situ Messungen. Schweiz. Z. Hydrol. 34: 211–244.

Gaedke, U. & A. Schweizer, 1993. The first decade of oligotrophication in Lake Constance. I. The response of phytoplankton biomass and cell size. Oecologia 93: 268–275.

Lee, S. & J. A. Fuhrman, 1987. Relationships between biovolume and biomass of naturally derived marine bacterioplankton. Appl. Envir. Microbiol. 53: 1298–1303.

Manca, M., A. Calderoni & R. Mosello, 1992. Limnological research in Lago Maggiore: studies on hydrochemistry and plankton. Mem. Ist. ital. Idrobiol. 50: 171–200.

Mosello, R. & D. Ruggiu, 1985. Nutrient load, trophic conditions and restoration prospects of Lake Maggiore. Int. Revue ges. Hydrobiol. 70: 63–75.

Platt, T., 1971. The annual production of phytoplankton in St. Margaret's Bay, Nova Scotia. J. Cons. Explor. Mer 33: 324–333.

Reynolds, C. S., 1988. Functional morphology and the adaptive strategies of freshwater phytoplankton. In C. D. Sandgren (ed.), Growth and Reproductive Strategies of Freshwater Phytoplankton. Cambridge University Press, Cambridge: 388–433.

Ruggiu, D., 1989. Indagini sul fitoplancton. In Relazione conclusiva delle indagini limnologiche relative al Lago Maggiore. Commissione internazionale per la protezione delle acque italo-svizzere. Rapporto quinquennale 1983–1987: 65–77.

Ruggiu, D., 1993. Osservazioni conclusive. Considerazioni generali sull'evoluzione a lungo termine dei popolamenti planctonici. Fitoplancton. In C.N.R. Istituto Italiano di Idrobiologia, Ricerche sull'evoluzione del Lago Maggiore. Aspetti limnologici. Campagna 1992 e Rapporto quinquennale 1988–1992 (ed), Commissione Internazionale per la protezione delle acque italo-svizzere: 115–116.

Ruggiu, D. & C. Saraceni, 1977. Fitoplancton, clorofilla e produzione primaria nel Lago Maggiore durante gli anni 1972–1973. Mem. Ist. ital. Idrobiol. 34: 57–78.

Ruggiu, D. & P. Panzani, 1988. Indagini sul fitoplancton. In: C.N.R. Istituto Italiano di Idrobiologia, Ricerche sull'evoluzione del Lago Maggiore. Aspetti limnologici. Programma quinquennale 1983–1987. (Campagna 1986) (ed.), Commissione Internazionale per la protezione delle acque italo-svizzere: 65–71.

Ruggiu, D., P. Panzani & A. Candido, 1982. Indagini sul fitoplancton. In C.N.R. Istituto Italiano di Idrobiologia, Ricerche su studi e ricerche condotti nel bacino de Lago Maggiore. (Campagna 1981) (ed.), Commissione Internazionale per la protezione delle acque italo-svizzere: 189–217.

Sas, H. (ed.), 1989. Lake restoration by reduction of nutrient loading: Expectations, Experiences, Extrapolations. Academia Verlag Richarz, St. Augustin, 497 pp.

Sommer, U., U. Gaedke & A. Schweizer, 1993. The first decade of oligotrophication in Lake Constance. II. The response of phytoplankton taxonomic composition. Oecologia 93: 276–284.

Tilzer, M. M., U. Gaedke, A. Schweizer, B. Beese & T. Wieser, 1991. Interannual variability of phytoplankton productivity and related parameters in Lake Constance: no response to decreased phosphorus loading? J. Plankton Res. 13: 755–777.

Willén, E., 1987. Phytoplankton and Reversed Eutrophication in Lake Mälaren, Central Sweden, 1965–1983. Br. phycol. J. 22: 193–208.

Willén, E., 1992. Long-term changes in the phytoplankton of large lakes in response to changes in nutrient loading. Nord. J. Bot. 12: 575–587.

Appendix

Full names and abbreviations of species or genera cited in the paper.

Name	Abbrev.
Ultraplankton	
Microcystis aeruginosa Kützing	
Oscillatoria rubescens D.C.	Osc. rub.
Oscillatoria spp.	Osc. spp.
Pseudanabaena sp.	Pse. sp.
Limnothrix sp.	Lim. sp.
Melosira varians Agardh	
Aulacoseira islandica var. *helvetica* (O. Müller) Simonsen	Aul. isl. mor. hel.
Aulacoseira ambigua (Grunow) Simonsen	Aul. amb.
Cyclotella comensis Grun.	Cyc. com.
Cyclotella hakanssoniae Wendker	
Cyclotella spp.	Cyc. spp.
Diatoma elongatum (Lyngb.) Ag.	Dia. elo.
Fragilaria crotonensis Kitton	Fra. cro.
Synedra acus Kg.	Syn. acu.
Synedra acus var. *radians* (Kg.)	
Synedra ulna var. *danica* (Kg.)	Syn. ul. v. dan.
Asterionella formosa Hassall	Ast. for.
Mallomonas elongata Reverdin	
Uroglena americana Calkins	
Dinobryon divergens Imhof	
Rhodomonas minuta Skuja	Rho. min.
Rhodomonas lacustris Pasch. et Ruttn.	Rho. lac.
Cryptomonas erosa Ehrenb.	Cry. ero.
Gymnodinium helveticum Penard	Gym. hel.
Gymnodinium spp.	Gym. spp.
Ceratium hirundinella (Müller) Schrank	Cer. hir.
Sphaerocystis schröteri Chodat	
Monoraphidium contortum (Thur.) Kom.-Legn.	
Stichococcus minutissimus Skuja	Sti. min.
Mougeotia sp.	Mou. sp.

Hydrobiologia **369/370:** 259–267, 1998.
M. Alvarez-Cobelas, C. S. Reynolds, P. Sanchez-Castillo & J. Kristiansen (eds), Phytoplankton and Trophic Gradients.

Development of the phytoplankton of the shallow Srebarna Lake (north-eastern Bulgaria) across a trophic gradient

Maya Petrova Stoyneva
Department of Botany, Faculty of Biology, Sofia University 'St Kliment Ohridski', 8 bld. Dragan Zankov, 1421, Sofia, Bulgaria

Key words: phytoplankton, shallow lake, trophic gradient, eutrophication, restoration, biosphere reserve, Danube

Abstract

Results of an analysis of phytoplankton samples, collected between 1982 and 1995 from Srebarna Lake, a biosphere reserve, in north-eastern Bulgaria, are presented. The lake is polymictic, and strongly eutrophied. It has undergone an anthropogenically-forced succession. In 1994, a restoration of the lake was commenced when a canal connecting it to the Danube River was built.

Special attention is paid to the shifts in the qualitative composition of the phytoplankton (240 species), in the number of dominant species and assemblage structure, as well as to the changes in structural parameters and phytoplankton abundance during lake enrichment. The generalised changes involved a shift from a chlorococcal dominated-phytoplankton, rich in phytoplankton groups, to a chroococcal and oscillatorialean one, poor in algal groups. With recovery, from hypertrophy to eutrophy, more algal groups contributed to the phytoplankton, and dominance of chlorococcal genera resumed. An increase in total phytoplankton abundance with advancing eutrophication and a decrease during early oligotrophication were detected. The values of structural parameters were related to nutrient input and changed after restoration. However, the stages of phytoplankton development along the trophic gradient and back were not exactly reversed.

Introduction

The condition of water bodies across trophic gradients is one of the most interesting current problems in limnology. The importance placed on phytoplankton composition is governed by the fact that indices of productivity are simultaneously defining lake trophic status (e.g. Carlson, 1977; Rott 1984). The oligotrophic → eutrophic spectrum is widely accepted as a base for the typology of lakes in the temperate zone: the trophic state of a given lake also represents a stage in its evolution (Rodhe, 1969). At the same time, there are many examples of successful and not-so-successful biomanipulation and management of lakes where trophic evolution has been reversed (Bjõrk, 1994). Comparative studies of phytoplankton in various water-bodies of different trophic state, as well as long-term studies of the phytoplankton of the same limnetic system over many years, have uncovered trends in responses to trophic change of species composition, productivity and seasonal succession (Tilman et al., 1982; Moss, 1983; Rott, 1984; Trifonova, 1989, 1990; Bailey-Watts et al., 1990; Bales et al., 1993; Vörös & Göde, 1993; Dokulil & Padisák, 1994). Although much information has been collected and analysed, and the main tendencies of the changes seemed to be clear (Reynolds, 1984; Sommer et al., 1986; Alvarez Cobelas & Jacobsen, 1992; Olrik, 1994), there remain many unpredictable events in the development of the phytoplankton, especially concerning the changes at the species level (Padisák, 1993a; Reynolds, 1996). Of major interest is the fact that many of the main phylogenetic divisions are represented across the entire trophic gradient. An attempt to devise a trophic spectrum based on characteristic algal genera and species is given elsewhere in this volume (Reynolds, 1997). Since the final form of the OECD model on nutrients load and its relationship with eutrophication (Vollenweider & Kerekes,

1982) was published, classification of lake trophic state has been mainly based on this model, ignoring such site-specific characteristics as the influence of physical and biotic components in regulating algal dynamics (Reynolds & Bellinger, 1992). Shallow lakes have been always considered as special cases in attempts to apply the nutrient/trophic concept model (Clasen & Bernhardt, 1980; Hamza et al., 1995). The problem of the dynamics of the phytoplankton community across the trophic gradients is yet more complicated when a river and a lake are interlinked and influence each other. The comparison of the phytoplankton of turbid systems, especially rivers and shallow lakes, showed their close selective and compositional affinities (Reynolds et al., 1994).

In this paper, some findings from long-term studies of the phytoplankton of the shallow Srebarna Lake, which is situated on the right bank of the Danube River and is intermittently in connection with the river, are presented. Special attention has been paid to the efforts to reverse the anthropogenically-driven eutrophication of the lake through the input of nutrient-rich Danube water. Since mass loadings are sometimes difficult to assess, and nutrient concentrations in the water are misleading in view of internal cell quotas (Alvarez Cobelas & Jacobsen, 1992), the composition of the phytoplankton is potentially a better indicator of trophic state.

Description of site studied

The Lake of Srebarna, situated in North-Eastern Bulgaria on the right bank of the Danube River (44° 05′ N and 27° 07′ E – Figure 1), is located in a karstic depression surrounded by three hills. The climate in the region is continental with a temperature varying between −2.5 °C in January and 39 °C in July and with an ice cover period of 2–3 months. Annual average precipitation is 159 mm.

Over the last two centuries, a regular annual penetration of the lake by floodwater from the Danube has occurred. In 1948–49, a dike was built to isolate the lake from the river, but in 1983, this dike was partially removed. Until 1994, a strip of arable land was maintained between Srebarna and the Danube. River water has continued to enter the lake erratically, the last occasion being in the spring of 1988. The isolation from the riverine input, together with the increased abstraction of local ground water led to a sharp decline in depth (from 3.5 to 0.2 m) and surface area (from 5 to 2.4 km^2), while the invasion of reed further decreased the area of open water to 0.6 km^2. The lake volume was estimated to be 0.37 Hm^3 (Radev et al., 1993). In 1994, a canal was built: Danube waters entered the lake in May 1994 and in March–April 1995, when depth increased to a maximum of 1.4 m. Today, Srebarna remains a shallow polymictic basin. The karstic lake sediments are covered by a 7-m layer of sapropel (Radev et al., 1993; Kovachev et al., 1996).

The increase in nutrient input (more than 15-fold increase of phosphorus and more than 5-fold increase of nitrogen loading – Radev et al., 1993) since 1983 is attributed to agricultural activities in the watershed (e.g. active soil fertilization up to 1990, extirpation of vineyards surrounding the lake in 1989), combined with several strong storms and rainfall events. In addition, the overwintering of thousands of birds that justify Biosphere, Heritage and Ramsar designations contribute significantly to the nutrient budget (Stoyneva & Michev, in press).

The ranges of the main hydrochemical constituents were derived from data of Michev et al. (1993), Radev et al. (1993) and Kovachev et al. (1996). During the relevant period, PO_4^{3-} ranged between 0.2 and 1.14 mg l^{-1}; NH_4^+, between 0.07 and 3.8 mg l^{-1}; NO_2^{2-}, between 0.2 and 2.53 mg l^{-1}; NO_3^-, between 0.02 mg l^{-1} and 1.2 mg l^{-1}; Cl^-, between 32 and 150.5 mg l^{-1}; SO_4^{3-}, between 38.1 and 487.3 mg l^{-1} and HCO_3^-, between 207 and 321.9 mg l^{-1}. The average ionic concentration is high (1135 mg l^{-1}), with a sulphate character; pH ranges between 6.5 and 8.5 (Kovachev et al., 1996).

The zooplankton of the lake has been little studied. In the 1960s, it had a cladoceran character (Naidenov, 1965) but, in 1993, an almost total impoverishment in favour of ciliate dominance was noted (Michev et al., 1993; Radev et al., 1993). After the restoration activities in 1994, a slight recovery of zooplankton community, with dominance of rotifers, was detected (Kovachev et al., 1996).

Materials and methods

Phytoplankton samples were collected at the central part of the lake in glass bottles (1 l) and were fixed with 2–4% formalin. Sampling was carried out at approximately seasonal intervals from 1987 until 1995. Additional samples were also taken in 1982 and 1984.

Determination of algae and all measurements and estimations were carried out using an Amplival microscope, offering up to 1200 times magnification. Count-

Figure 1. Map of Bulgaria with location of Srebarna Lake and map of Srebarna Lake.

ing was done using a Thoma chamber, 16 big squares in 4–8 reiterations. For the phytoplankton, the units counted were cells. Biomass was estimated by the method of stereometrical approximations based on original measurements of cell sizes. Some cell enumerations of colonial algae were based on methods considered by Reynolds & Jaworski (1978). The taxonomy and nomenclature adopted followed those of previous work (see Stoyneva & Draganov, 1994). The trophic state was estimated according to OECD criteria based on phytoplankton biomass and on cell numbers according to the system of Felföldy (1974 – cited after G. Tóth & Padisák, 1982). The results obtained were compared with the OECD open-boundary system (Vollenweider, 1989). The structural parameters used are the diversity index of Shannon (H), the evenness index of Pielou (E), and the dominance index of Margalef (C).

Results and discussion

Species composition, number of species and dominance

The Lake of Srebarna is characterised by its extremely rich algal flora in which more than 410 taxa except diatoms have been found (Stoyneva & Draganov, 1994). In the samples used for this study, 240 species were determined.

The biggest number of taxa at the species and infraspecies level was found in 1987–1988 (up to 97 taxa per sample). A substantial decrease of species numbers was detected in 1989–1991 (down to 2 species per sample). This coincided with major changes in the entire lake (described above) and, especially, with declining water level, increasing nutrient input and blue-green algal blooms (see below). The decrease in the numbers of chlorococcal algae was particularly remarkable because it is atypical for water bodies along the Bulgarian Danube shore and for the Danube River (Stoyneva, 1991). The number of diatom species, both centrics and pennates, also decreased. A disappearance of desmids and other zygnemophytes characteristic of waters of lower trophic state coincided with the other changes.

An increase in total number of species and, especially of chlorococcal species, was detected in 1992 (up to 56 and 33 species per sample, respectively). It could have been coincident with the slight increase of the water column height after the strong rainfall in that period.

This self-organisation of the community seemed to be the last one before the extremely dry 1993, when there was increased abstraction of underground waters. Then the depth of the lake reached its most critical point of 20 cm and the number of species sharply decreased again (down to 7 species per sample), heterocystous blue-green algae were excluded from the phytoplankton assemblages, but the number of chroococcal cyanophytes and pyrrhophytes increased.

A renewed increase in overall species numbers (up to 43 species per sample) was noted after reconnection with the Danube River, with introduction of phytoplankton from the river. In that period, the numbers of chlorococcal algae and diatoms increased, the presence of nostocalean blue-green algae and of cryptophytes were comparable to the situation before 1989, and silica-scaled chrysophytes appeared in the phytoplankton. This first significant influx of Danube waters in May 1994 was recognised to be a major event in the development of the lake which, possibly, led to its oligotrophication.

During the investigation, the phytoplankton was dominated by 35 different species from 28 genera. Even these dominant forms had exhibited marked interannual and within-year variation in the timing of their maxima and of their numerical densities. Moreover, there are differences in the dominant species estimated by biomass instead of cell numbers. The fact that some dominants so estimated also belong to different phylogenetic groups is noteworthy. This is especially true for the most critical periods in the development of the lake after 1989, when there was no more normal influx of Danube waters and high amounts of fertilizers had been carried by strong rainfall to the lake (e.g. in September 1993 *Plectonema* was dominant according to cell numbers, whereas *Peridinium* dominated according to biomass). In several cases, differences appeared at order level (e.g. *Aphanocapsa* and *Oscillatoria*) or only at generic level (e.g. *Crucigenia* and *Dictyosphaerium*). Regarding the dominant structure of the phytoplankton according to the percentage contribution of species (how many species contribute to phytoplankton quantity less than 0.5% or more than 0.5%, 5% and 25%), it could be stated that in spite of the differences at species level, the general structure was similar, whether it was estimated on biomass or on cell numbers (Figure 2).

Even if eutrophication is ending with nuisance blooms in lakes (Sommer, 1993) we cannot neglect the fact that the establishment of blooms depends on the continuity of appropriate physical conditions (Reynolds, 1987, 1993, 1994) and on sediment enrichment by nutrients (e.g. Kawabata & Kagawa, 1986). Thus, it is difficult to equate the succession of dominant plankters only with the trophic conditions in the lake.

Generally, during the eutrophication process dominant green chlorococcal algae were replaced by chroococcal and oscillatorial blue-greens. Diatom dominance (*Aulacoseira sp.*) was detected only once, at the beginning of the investigated period. Blue-green algae remained dominant under hypertrophic conditions and were replaced only once, by pyrrhophytes (*Peridinium*). With the recovery towards eutrophic conditions, blue-green algae were partially replaced by chlorococcaleans.

During the development of eutrophication, the role of genera like *Scenedesmus*, *Tetrastrum*, *Crucigenia*, *Monoraphidium* and *Schroederia* decreased. They appeared again in the sub-dominant complexes when the hypertrophic conditions tend to recede. It is interesting to note that with the ongoing eutrophication *Scenedesmus* was partially replaced by *Golenkinia* and *Pediastrum* and *vice versa*.

In hypertrophic conditions, cyanophytes of the generic complexes *Microcystis* and *Gomphosphaeria* were replaced by smaller representatives like *Woronichinia*, *Lemmermaniella*, *Cyanodictyon*, *Chroococcus*, *Aphanocapsa* and *Merismopedia*, and filamentous *Leptolyngbya* and *Pseudanabaena* were replaced by smaller representatives of *Lyngbya*, *Planktolyngbya* and unidentified filaments of *Plectonema*-type less than 1 μm wide. Nostocalean genera, previously represented by *Anabaena* and *Aphanizomenon*, were replaced by *Cylindrospermopsis* and *Anabaenopsis* though these disappeared in 1993. After April 1994 *Anabaena* and *Aphanizomenon* re-appeared.

As the trophic state fluctuated, the phytoplankton changed at all taxonomic levels and altered the proportional representation.

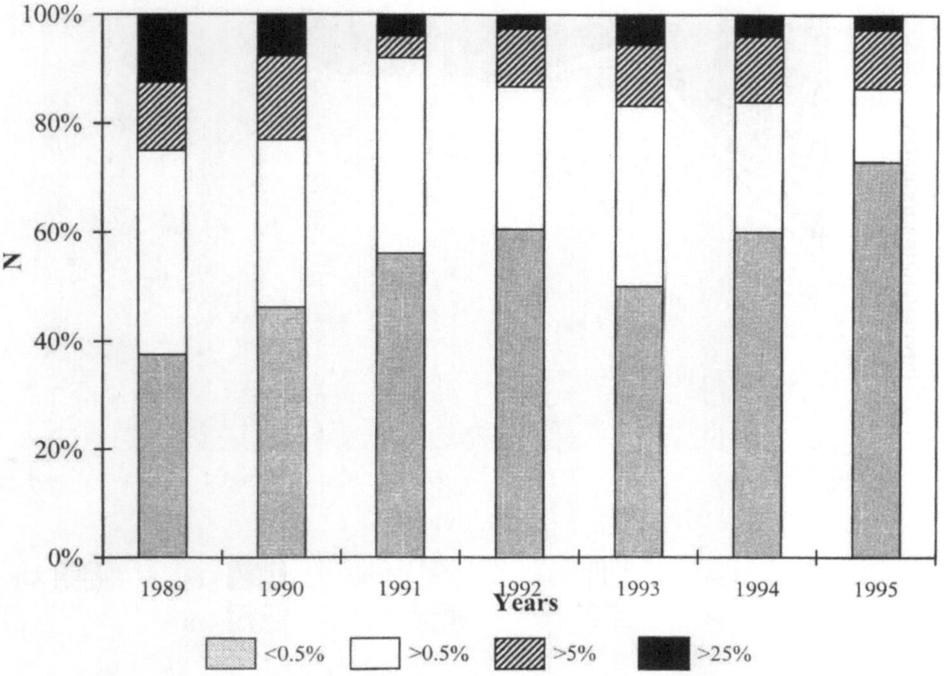

Figure 2. Phytoplankton structure according to the contribution of algae to phytoplankton biomass in Srebarna Lake during the period 1989–1995 (N – number of species which contributed <0.5%, >0.5%, >5% and >25% to the total biomass).

Changes in phytoplankton quantitie

As annual averages, the total numbers of phytoplankton cells in Srebarna varied from 1.9×10^8 cells l^{-1} to 18.5×10^9 cells l^{-1}. The algal quantities rose from 1982 to 1993, but the situation before and after the last dilution of lake water by flow from the Danube (1988) was different. For the 1987–1988 period, the average numbers were 2.6×10^8 cells l^{-1}, and for the 1989–1993 period they were 4.8×10^9 cells l^{-1}. The lack of this dilution together with the increased pumping of underground water in the vicinity of Srebarna became critical at the background of the special agricultural activities in the region that led to increased nutrient input. After the Danube waters entered again the lake in May 1994 and in April–March 1995, the algal numbers tended slowly to decrease to 3.9×10^9 cells l^{-1} on average. The peak/mean ratio generally tended to decrease in the period 1987–1993 (values between 2.1 and 1.3) and clearly increased in 1994 and 1995 (up to value of 4.0 in 1995).

Thus, according to the system of Felföldy (1974-cited after G. Tóth & Padisák, 1982), the lake is characterised as progressively eu-, eu-politrophic and hypertrophic up to 1993 and began to reverse to eu-politrophic conditions after 1994.

The changes concerned not only total cell numbers but also the proportions among different algal groups and their appearance or disappearance (Figure 3). Changes were well-expressed at order level, especially when they are combined with some functional differences and properties (e.g. general trend of disappearance of heterocystous nostocalean algae from phytoplankton assemblages with the progress of the eutrophication and their replacement by chroococcal ones). The analysis of the percentage contribution to total phytoplankton by algal group (Figure 3) shows a shift from a chlorococcal-dominated community which includes a high number of groups to one dominated by Chroococcales and an impoverished community during increased eutrophication, followed by a subsequent tendency to revert to a chlorococcal plankton thereafter.

The phytoplankton biomass, as annual averages, varied between 15.3 and 110.7 mg l^{-1}. These values clearly show that during the investigated period the boundary between eutrophic and hypertrophic conditions was crossed. It has to be mentioned that in several cases (before 1994–1995), biomass values varied between 170 and 200 mg l^{-1}.

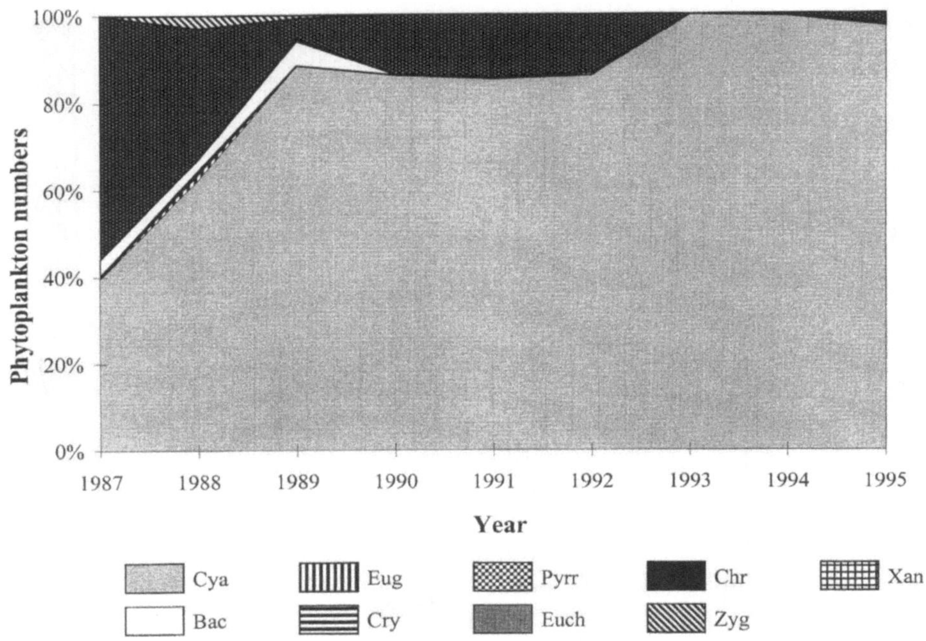

Figure 3. Percentage contribution of different algal groups to total phytoplankton numbers in Srebarna Lake during the period 1987–1995 (Cya – cyanophytes, Eug – euglenophytes, Pyr – pyrrhophytes, Chr – chrysophytes, Xan – xanthophytes, Bac – bacillariophytes, Cry – cryptophytes, Euch – chlorophytes, Zyg – zygnemophytes).

As in the case of phytoplankton numbers, the average biomass value tended to rise from 1982 to 1993 and to decrease in 1994–1995. The average values for selected periods were as follows: 38.5 mg l^{-1} in 1987–1988, 67.9 mg l^{-1}in 1990–1993 and 40.8 mg l^{-1} in 1994–1995. The peak/mean ratio generally tended to decrease during the first period and increased during the last one, reaching value of 3.2.

In spite of this common trend of forced eutrophication up to 1993 and of slight oligotrophication in 1994–1995, some differences appear with the changes in total phytoplankton biomass. When the contributions of the various algal groups (Figure 4) are compared with the situation estimated from cell numbers, a different impression is given. There is a detailed difference in the fluctuations of total biomass and groups like pyrrhophytes, diatoms, desmids and nostocalean blue-green algae, whose role is more important in contribution to biomass than as cell numbers. All these differences are explainable by the variations in cell sizes and volumes and they confirm the need to use both criteria in such studies. According to biomass changes, the succession through eutrophication leads from chlorococcal- through cyanophyte- to pyrrhophyte-dominated assemblages. With the shift to oligotrophication, the restoration of both the chlorococcal dominated-assemblages and of the nostocalean algae was detected. As the oligotrophication proceeds, the role of diatoms, both pennate and centric, clearly increases (Figure 4).

Structural parameters

The value of Shannon's diversity index (H) as annual averages varied from 3.13 to 0.86 (Figure 5). The diversity – disturbance relations have been studied in nutrient-rich rivers (e.g. Descy, 1993) and shallow lakes (e.g. Padisák, 1993b), and special attention has been paid to these relations in hypertrophic habitats (e.g. Olrik & Nauwerck, 1993; Rojo & Alvarez-Cobelas, 1993). In the case of Srebarna, both high-frequency events, such as input of nutrients from the catchment area, and infrequent events, like the entrance of Danube waters, affected phytoplankton structure and dynamics. The values of H tended to decrease with increased nutrient input, combined with decreased water level, and slowly tended to rise again after the dilution by Danube waters. Annual averages of the phytoplankton biomass and values of H generally were inverse relations.

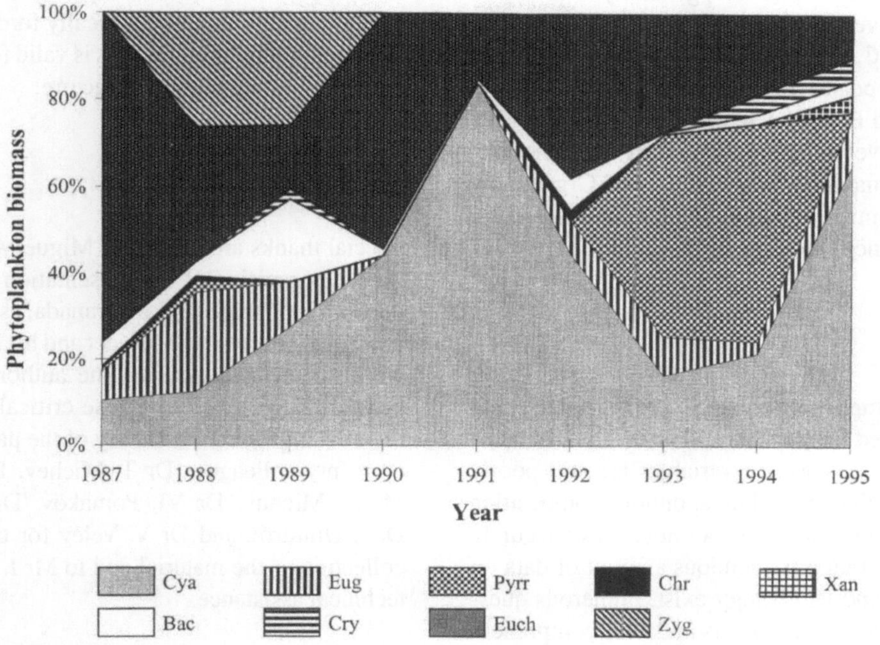

Figure 4. Percentage contribution of different algal groups to total phytoplankton biomass in Srebarna Lake during the period 1987–1995 (Cya – cyanophytes, Eug – euglenophytes, Pyr – pyrrhophytes, Chr – chrysophytes, Xan – xanthophytes, Bac – bacillariophytes, Cry – cryptophytes, Euch – chlorophytes, Zyg – zygnemophytes).

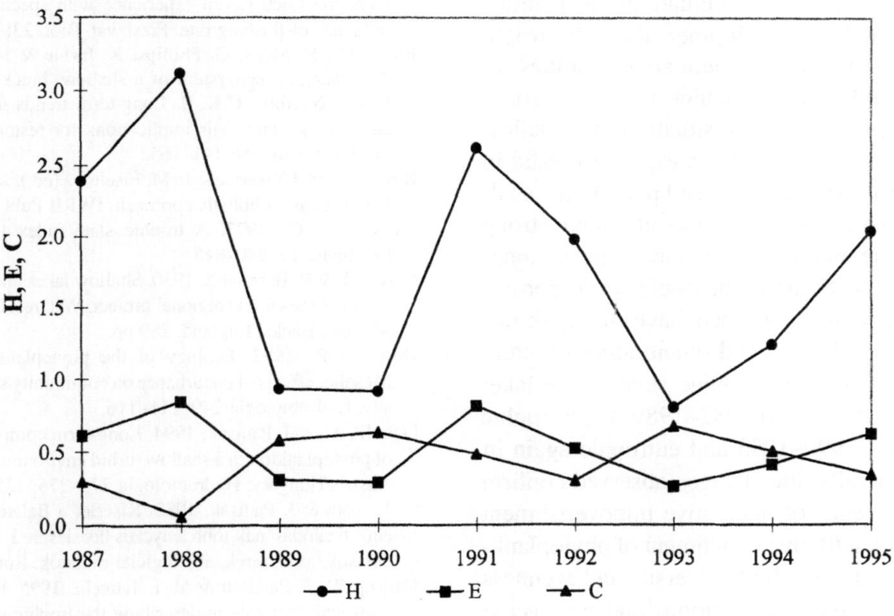

Figure 5. Structural parameters of Srebarna Lake phytoplankton during the period 1987–1995 (H – Shannon's index of diversity, E – Pielou's index of evenness, C – Margalef's index of dominance).

As might be expected, the values of evenness of Pielou (E; between 0.85 and 0.2) generally followed the dynamics of diversity, whereas the Margalef's index of dominance (C, with values between 0.64 and 0.05) followed an opposite trend (Figure 5). The minimal values of H and E were detected during the period of lowest water level in 1993, whereas at the same time C reached its maximal value. Values of C generally followed the annual averages of algal biomass but a strong dependence could not be demonstrated.

Conclusions

That species composition changes across trophic gradients can be traced back to Pearsall (1932) but the operational dynamics of driving variables are still poorly understood and the reversal in taxonomic composition along with oligotrophication does not always occur. In spite of the fact that a tremendous amount of data on phytoplankton and its ecology exist, numerous questions remain open. There is no updated, comprehensive account of phytoplankton ecology across trophic gradients since Kalff & Knoechel's (1978) preliminary review. Long-term studies on phytoplankton are regarded by many authors (e.g. Smith, 1990) among the possible tools for solving this problem. Following this opinion, the development of a phytoplankton community with significant fluctuations in quantity, could be considered as a development through a trophic gradient. It is obvious that there are difficulties for explaining phytoplankton dynamics across the trophic gradients in the complicated situation of a shallow lake (Hamza et al., 1995). This is especially valid in the case of anthropogenically forced polymictic, shallow, turbid Srebarna Lake, with the additional strong disturbance of the entrance of Danube waters. Long-term studies of the Srebarna phytoplankton, permitting the main trends to be outlined, have clearly shown that changes in qualitative and quantitative structure are linked to the changing trophic state of the lake: eutrophic during the period 1982–1989, hypertrophic during the period 1990–1993 and eutrophic again in 1994–1995. Generally, the changes observed confirm the well known trends of qualitative impoverishment combined with quantitative enrichment of phytoplankton and decrease of community diversity and evenness with the ongoing hypertrophication, and *vice versa* – enrichment by species and groups combined with decreasing of phytoplankton abundance and increased community diversity and evenness during oligotrophication. However, the steps of phytoplankton development to the extreme of the trophic gradient and back in the case of this anthropogenically forced shallow lake were not exactly reverse. This is valid for its qualitative as well as its quantitative structure.

Acknowledgements

Special thanks are due to Dr Miguel Alvarez Cobelas for the organising of the presentation of this study at the 10th IAP Workshop in Granada, as well as for providing literature on the subject and his helpful editorial work on the manuscript. The author wish to thank both unknown referees whose critical comments and remarks improved the quality of the paper. I am obliged to my colleagues Dr T. Michev, Dr M. Marinov, Mr N. Michov, Dr Vl. Pomakov, Dr St. Kovachev, Dr I. Dimitrov and Dr V. Velev for their help in the collecting of the material and to Mr I. Stoynev for his technical assistance.

References

Alvarez-Cobelas, M. & B. A. Jacobsen, 1992. Hypertrophic plankton: an overview. Freshwater Forum 2: 184–192.

Bailey-Watts, A. E., A. Kirika, L. May & D. H. Jones, 1990. Changes im phytoplankton over various time-scales in a shallow, eutrophic loch: the Loch Leven experience with special reference to the influence of flushing rate. Freshwat. Biol. 23: 85–112.

Bales, M., B. Moss, G. Phillips, K. Irvine & J. Stansfield, 1993. The changing ecosystem of a shallow, brackish lake, Hickling Broad, Norfolk, U.K. II. Long-term trends in water chemistry and ecology and their implications for restoration of the lake. Freshwat. Biol. 29: 141–165.

Bjõrk, S., 1994. Overview. In M. Eiseltová (ed.), Restoration of Lake Ecosystems. A holistic approach. IWRB Publ. 32: 1–5.

Carlson, R. E., 1977. A trophic state index for lakes. Limnol. Oceanogr. 22: 840–845.

Clasen, J. & H. Bernhardt, 1980. Shallow lakes and Reservoirs. Final report of the OECD regional project. Wat. res. Cent. Medmenham Marlow, Bucks, England, 289 pp.

Descy, J.-P., 1993. Ecology of the phytoplankton of the River Moselle: effects of disturbance on community structure and diversity. Hydrobiologia 249: 111–116.

Dokulil, M. & J. Padisák, 1994. Long-term compositional response of phytoplankton in a shallow turbid environment, Neusiedlersee, Austria/Hungary. Hydrobiologia 275–276: 125–137.

G. L. Tóth & J. Padisák, 1982. Kísérlet a Balaton Tihanyi térsége eutrofizálódásának több tényezös becslésére I. Bevesetö meggondolások, módszerek, algológiciai mutatók. Bot. Közl. 69: 71–84.

Hamza, W., P. Pandolfi & M. I. Taticchi, 1995. Planktonic interactions and their role in describing the trophic status of a shallow lake in Central Italy (Lago Trasimeno). Mem. Ist. ital. Idrobiol. 53: 125–140.

Kalff, J. & R. Knoechel, 1978. Phytoplankton and their dynamics in oligotrophic and eutrophic lakes. Ann. Rev. Ecol. Syst. 9: 475–495.

Kawabata, Z. & H. Kagawa, 1986. Succession of phytoplankton species in indoor reservoir models varying in sediment composition and water inlet site. Hydrobiologia 133: 31–38.

Kovachev, St., T. Michev, M. Stoyneva, V. Velev, G. Baeva & R. Bardova., 1996. Preliminary report on Srebarna monitoring. – Ministry of Environment, 56 pp (manuscript, in Bulgarian).

Michev, T., M. Stoyneva, V. Velev & N. Michov, 1993. Preliminary report on ecological study of Srebarna Lake, Ministry of Education, 62 pp (manuscript, in Bulgarian).

Moss, B., 1983. The Norfolk Broadland: Experiments in the restoration of a complex wetland. Biol. Rev. 58: 521–561.

Naidenov, V., 1965. Beitrag zur Kenntnis der Copepoden und Brachiopodenfauna der bulgarischen Donauueberschwemmungsgebiete. Bull. zool. Inst. (Sofia), XIX: 5–54 (in Bulgarian, German summ.).

Olrik, K., 1994. Phytoplankton – Ecology. Ministry of the Environment, Denmark, Danish Environmental Protection Agency, 183 pp.

Olrik, K. & A. Nauwerck, 1993. Stress and disturbance in the phytoplankton community of a shallow, hypertrophic lake. Hydrobiologia 249: 15–24.

Padisák, J., 1993a. Species composition, spatial distribution, and the seasonal and interannual dynamics of phytoplankton in brownwater lakes enclosed with reed-belts (Neusiedlersee/Ferto; Austria/Hungary). BFB – Bericht 79: 13–29.

Padisák, J., 1993b. The influence of different disturbance frequences on the species richness, diversity and equitability of phytoplankton in shallow lakes. Hydrobiologia 249: 135–156.

Pearsall, W. H., 1932. Phytoplankton in the English Lakes. II. The composition of the phytoplankton in relation to dissolved substances. J. Ecol. 20: 241–262.

Radev, V. (ed.), 1993. Project 'Recovery of the biosphere reserve Srebarna'. Waterengineering Ltd., Sofia, October 1993 (in Bulgarian).

Reynolds, C. S., 1984. The ecology of freshwater phytoplankton. Cambridge Univ. Press, Cambridge, 384 pp.

Reynolds, C. S., 1987. The response of phytoplankton communities to changing lake environments and the response of phytoplankton. Freshwat. Biol. 23: 25–53.

Reynolds, C. S., 1993. Scales of disturbance and their role in plankton ecology. Hydrobiologia. 249: 157–171.

Reynolds, C. S., 1994. The long, the short and the stalled: on the attributes of phytoplankton selected by physical mixing in lakes and rivers. In J.-P. Descy, C. S. Reynolds & J. Padisák (eds), Phytoplankton in Turbid Environments: Rivers and Shallow lakes. Hydrobiologia 289: 9–22.

Reynolds, C. S., 1996. Further remarks on phytoplankton ecology and trophic degree (manuscript), 7 pp.

Reynolds, C. S. & E. G. Bellinger, 1992. Pattern of abundance and dominance of the phytoplankton of Rostherne Mere, England: evidence from an 18 year data set. Aquat. Sci. 54: 11–36.

Reynolds, C. S., J.-P. Descy & J. Padisák, 1994. Are phytoplankton dynamics in rivers so different from those in shallow lakes? In J.-P. Descy, C. S. Reynolds & J. Padisák (eds), Phytoplankton in Turbid Environments: Rivers and Shallow lakes. Hydrobiologia 289: 1–8.

Reynolds, C. S. & G. H. Jaworski, 1978. Enumeration of natural *Microcystis* population. Br. Phycol. J. 13: 269–277.

Rodhe, W., 1969. Crystallisation of eutrophication concepts in northern Europe. In G. A. Rochlich (ed.), Eutrophication: causes, consequences, correctives. Washington: 50–64.

Rojo, C. & M. Alvarez-Cobelas, 1993. Hypertrophic phytoplanktona and the Intermediate Disturbance Hypothesis. Hydrobiologia 249: 43–57.

Rott, E., 1984. Phytoplankton as biological parameter for the trophic characterization of lakes. Verh. int. Ver. Limnol. 22: 1078–1085.

Smith, V. H., 1990. Phytoplankton responses to eutrophication in inland waters. In I. Akatsuka (ed.), Introduction to Applied Phycology, SPB Acad. Publ., The Hague, The Netherlands: 231–249.

Sommer, U., 1993. The scientific bases of eutrophication management: reconcycling basic physiology and empirical biomass models. In R. de Bernardi, R. Pagnotta, A. Pugnetti (eds), Strategies for lake ecosystems beyond 2000. Mem. Ist. ital. Idrobiol. 52: 89–111.

Sommer, U., Z. M. Gliwicz, W. Lampert & A. Duncan, 1986. The PEG-model of seasonal succession of planctonic events in freshwaters. Arch. Hydrobiol. 106: 433–471.

Stoyneva, M., 1991. Algal flora of the River Danube (Bulgarian sector) and its adjacent water basins. Ph. D. Thesis, Sofia University, Sofia, 372 pp. (in Bulgarian, English abstr.).

Stoyneva, M. & St. Draganov, 1994. Algal flora of the River Danube (Bulgarian sector) and its adjoined water basins. IV. Contribution to the algal flora of some adjoined water basins. Ann. Univ. Sof. 84: 15–24.

Stoyneva, M. & T. Michev. Srebarna case: habitat changes as reflected by waterfowl. In J. Kerekes & S. Farago (eds), Proceed. SIL Workshop on Aquatic birds, Sopron, Hungary, 16–23 November, 1994 (in press).

Tilman, D., S. S. Kilham & P. Kilham, 1982. Phytoplankton community ecology: the role of limiting nutrients. Ann. Rev. Ecol. Syst. 13: 343–372.

Trifonova, I. S., 1989. Changes in community structure and productivity of phytoplankton as indicators of lake and reservoir eutrophication. Arch. Hydrobiol. Beih. Ergebn. Limnol. 33: 363–371.

Trifonova, I. S., 1990. Ecology and succession of lake phytoplankton. Nauka, Leningrad, 184 (in Russian).

Vollenweider, R. A., 1989. Global problems of eutrophication and its control. In J. Salanki & S. Herodek (eds), Conservation and management of lakes, Akad. Kiadó, Budapest: 19–42.

Vollenweider, R. & J. Kerekes, 1982. Eutrophication of water: Monitoring, assessment and control. OECD, Paris, 154 pp.

Vörös, L. & P. N. Göde, 1993. Long term changes of phytoplankton in Lake Balaton (Hungary). Verh. int. Ver. Limnol. 25: 682–686.

Hydrobiologia **369/370:** 269–275, 1998.
M. Alvarez-Cobelas, C. S. Reynolds, P. Sanchez-Castillo & J. Kristiansen (eds), Phytcplankton and Trophic Gradients.

Life-cycle studies and culture experiments for identification of doubtful volvocacean organisms – observations on *Pandorina* and *Eudorina*

David G. Angeler
Institute of Plant Physiology, Department of Hydrobotany, Althanstraße 14, A-1090 Vienna, Austria

Key words: Pandorina, Eudorina, identification problems, life cycles, culture experiments

Abstract

Traditionally, the volvocacean genera *Pandorina* and *Eudorina* are distinguished by vegetative morphology: *Eudorina* has hollow coenobia with loosely arranged spherical to ovoid cells, whereas *Pandorina* coenobia are compact with pear- to keystoneshaped cells contiguous in the centre. However, sometimes intermediate forms are found which cannot be identified to genus by traditional morphological characters. Especially the degree of compactness of *Pandorina* coenobia may be highly variable. Such abberant forms must be taken into culture in order to establish their identity, by studying (1) their life cycles, both sexual and asexual processes, and (2) the influence of varied culture conditions (temperature, daylenght, nutrients) on their morphology. In these ways, the taxonomic position may be clarified.

Introduction

The Volvocaceae include a series of coenobial organisms, show a range of vegetative morphologies and reproductive cycles, as well as an increasing complexity of somatic differentiation (cf. Nozaki & Itoh, 1994). The differences between one genus and the next are often quite gradual, thus showing a continuum of features, from basal to quite complex characteristics. From the systematic point of view, this generates problems of correct identification of certain species, as variations in vegetative morphology lead to overlaps in the characters of individual taxa. For example, strains of *Pandorina morum* Bory growing under artificial culture conditions and even freshly collected material brought into laboratory conditions will deviate far from their typical vegetative morphology and lose the compact coenobial arrangement of their cells (see Figure 3). However, even in the genus of *Pandorina* the compact arrangement of the coenobia is not unique since *Pandorina charkowiensis* Korshikov (1923) and *Pandorina minodii* Chodat (1931) were described as having hollow coenobia with loosely-arranged ovoid to spherical cells. The appearance of such coenobia resembles more the abnormal hollow coenobia of *Eudorina* than spheroids of *Pandorina*. Consequently, many phycologists are left to distinguish *Pandorina* from *Eudorina* by the presence of 'sperm packets' produced in the latter during sexual reproduction (Thompson, 1954; Rayburn & Starr, 1974).

In this study, light microscopical examinations of stages in the life-cycles were undertaken to distinguish questionable strains of *Pandorina morum* Bory, *Eudorina elegans* Ehrenberg and *E. illinoisensis* (Kofoid) Pascher as circumscribed by Nozaki (1981) and Nozaki & Kuroiwa (1992). Observing both asexual and sexual reproduction, it proved possible to distinguish species of *Pandorina*, even when showing a *Eudorina*-like habit.

Culture experiments were also undertaken to study the influence of various culture conditions on the phenotypes of these algae. Culture experiments may serve as a systematic tool since the altered appearances of certain features can be observed in modified culture conditions. In this study, the morphological reactions of volvocacean algae to modified abiotic factors are also reported. Although the results of the culture experiments are preliminary, the approach may be extended to future investigations. As little work has been carried out on this topic, further autecological observations should provide valuable information.

Materials and methods

Culture conditions

The strains listed in Table 1 were obtained from the culture collection of algae from the Institute of Plant Physiology at the University of Vienna (Kusel-Fetzmann & Schagerl, 1992) and grown in clonal but not axenic cultures. The maintenance of these strains was carried out in the media listed in Table 2 at 20 ± 2 °C under a light regime of 25 μmol photons m^{-2} s^{-1} provided by Phillips fluorescent tubes (36W/82) on a diurnal 16L:8D hours light-dark cycle.

Life cycle studies

Vegetative reproduction occurred under non-limiting conditions. The induction of sexual reproduction and zygote germination was carried out under nitrogen limitation as described in Nozaki (1983).

Culture experiments

To test the influence of varied abiotical factors, modifications in daylength and temperature as well as the chemical composition were imposed, as follows: (1) Modifications to daylength and temperature were achieved by decreasing the light period and the temperature. In intervals of two weeks the day-night cycles and temperatures were consecutively altered from 16L:8D hr + 20 °C (weeks 1; 2), over 14L:10D hr + 15 °C (weeks 3; 4) and 10L:14D hr + 10 °C (weeks 5; 6) to 8L:16D hr + 5 °C (weeks 7; 8).

(2) Effects of altered chemical composition on the appearance of the algae were observed on clones grown in the media listed in Table 2.

Results

Life cycles

Vegetative reproduction: During the vegetative-growth phase, a continuous increase in cell- and coenobium-diameter was observable. The generation time (time within which young coenobia reaching their maturity resulting in autocolony formation) was approximately 8 to 11 days. Ripe coenobia of *P. morum* (see Figure 1) and *E. elegans* (see Figure 2) reached cell dimensions of about 20 μm in diameter. Before the onset of the first divisions, the appearance of the cell content became granular. Autocolony formation in *Pandorina*- *E. elegans*- and *E. illinoisensis* coenobia started at approximately 55 μm, 95 μm and 115 μm in diameter respectively. The number of pyrenoids increased continuously in *E. elegans* and *E. illinoisensis* but remained single in *P. morum*. Adult cells of *Eudorina* contained more than five pyrenoids. However, observation by light microscopy of the behaviour of the pyrenoids during the division processes was unsatisfactory. Gottlieb & Goldstein (1977) have shown pyrenoids are present during asexual reproduction in *E. elegans*, leading to the conclusions that pyrenoids do not dissappear during the division processes. Thus, the single pyrenoids in cells of the daughters are not formed *de novo*.

After reaching the maximum cell dimensions, asexual reproduction began. After 4–6 longitudinal sections of each cell, a 16-celled curved plate (plakea) is formed in *Pandorina* and, normally, a 32-celled plakea in *Eudorina*. By a subsequent inversion process, the peripheral cells of the plakea become attached to one another forming a new coenobium. These remain embedded in the parental gelatinous envelope for a period of time, but are eventually released to form a new generation of free swimming coenobia (for further details on the asexual reproduction the reader is referred to Nozaki, 1980 and 1981; Gottlieb & Goldstein, 1977; Marchant, 1977 and Angeler, 1995a).

A feature, crucial for delimiting *Pandorina* from *Eudorina*, is the extracellular matrix of the gelatinous parental envelope. During the division process of *Eudorina*, coenobium formation takes place in a 'transparent vesicle'. Nozaki & Kuroiwa (1992) have shown that this transparent vesicle (cellular envelope) (see Figure 5) has a dense fibrillar zone, visible in electron micrographs, to surround each cell tightly. This structure is common to the genera *Eudorina* and *Pleodorina* but is lacking in *Pandorina* (Fulton, 1978; Nozaki & Kuroiwa, 1991). In the latter, cell division takes place in a 'keystone-shaped space' in the parental gelatinous matrix only. This is revealed by staining with methylene blue (see Figure 7); no such structure is seen in *Eudorina* (see Figure 8).

Sexual reproduction: Pandorina has an iso- to anisogamous reproduction cycle (Coleman, 1959; Nozaki & Kazaki, 1979). However, the sexual process

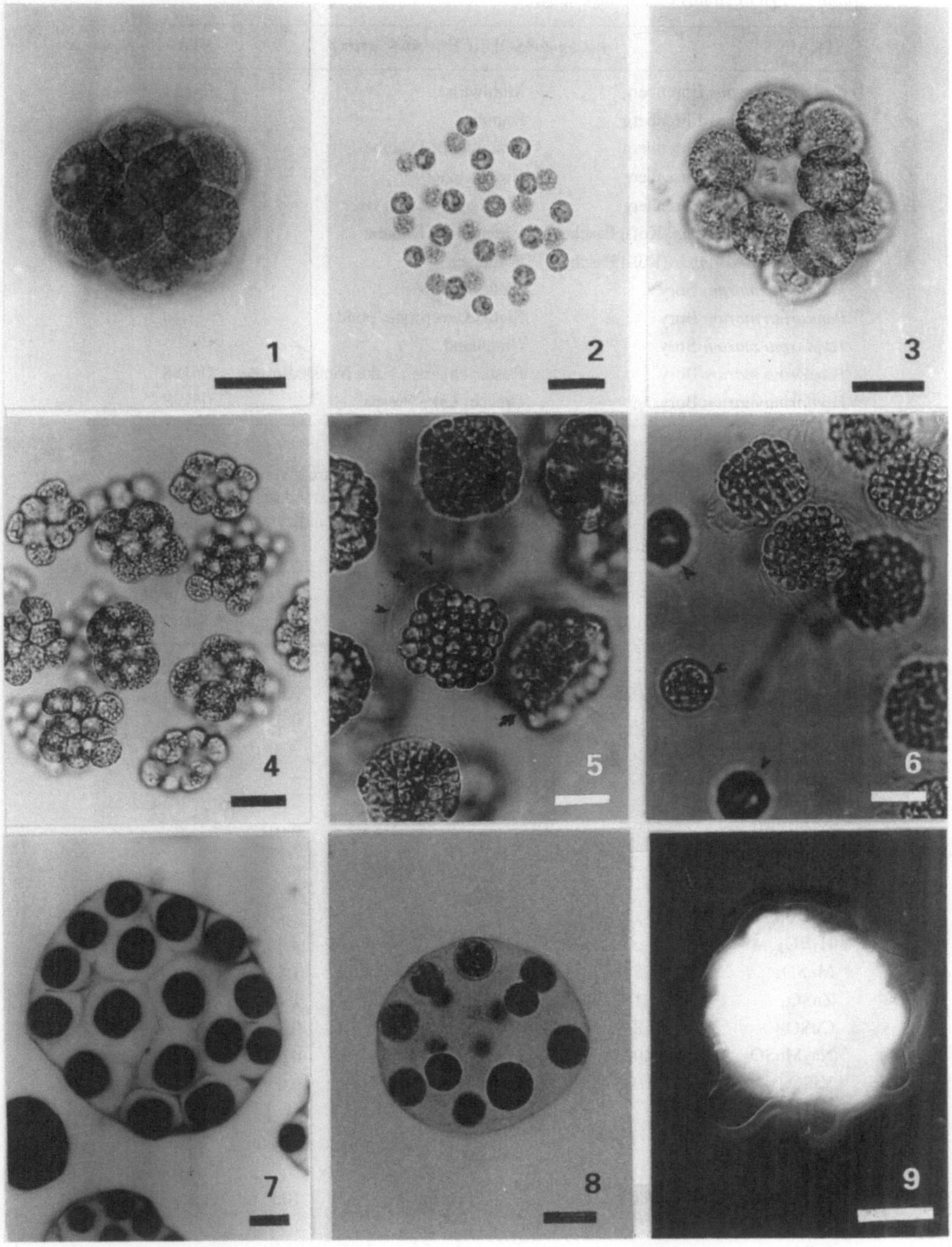

Figures 1–9. 1. Mature coenobium of *Pandorina morum*. The coenobium is a compact spheroid whose cells are keystone shaped and contiguous in the center. Scale bar = 20 μm; 2. Coenobium of *Eudorina elegans*. Note the hollow coenobium with spherical cells situated at the periphery of the gelatinous matrix. Scale bar = 30 μm; 3. Abnormal coenobium of *Pandorina morum*. The coenobium has lost its compact cell arrangement. The cells are spherical and situated at the periphery of the hollow coenobium (*Eudorina*-habitus). Scale bar = 20 μm; 4. Different division stages during autocolony formation in *Pandorina morum*. The plates are not surrounded by a cellular envelope. Scale bar = 10 μm; 5. 32-cell stages in *Eudorina illinoisensis* during autocolony formation. Each plakea is surrounded by a cellular envelope (arrowheads). Note the bowl like shape of a late inversion stage (arrow). Scale bar = 10 μm; 6. *Eudorina illinoisensis*: Male coenobium exhibiting antheridia with differentiated 'sperm packets'. Each plate consists of 32 to 62 spindle-shaped, biflagellated gametes and is surrounded by a cellular envelope. Note: somatic cells situated at the anterior pole of the coenobium (arrowheads). Scale bar = 10 μm; 7. *Pandorina morum* prior to liberation of daughter coenobia. Each daughter coenobium is located within a keystone-shaped space; contrasted with methylene-blue. Note: no cellular envelope but each cell is surrounded by a keystoneshaped space in the matrix. Scale bar = 10 μm; 8. *Eudorina illinoisensis*: contrasted with methylene blue. Note: keystoneshaped rooms are absent. Scale bar = 20μm; 9. Optical section through *Pandorina morum* exhibiting mound like protuberances of the gelatinous matrix. Photographed with interference contrast optics. Scale bar = 20 μm.

Table 1. List of strains used in this study.

STRAIN	SOURCE	STRAIN-NO.
Eudorina elegans Ehrenberg	Mühlwasser	05139
Eudorina elegans Ehrenberg	Danube	05141
Eudorina elegans Ehrenberg	Altenwörth; Danube	05142
Eudorina elegans Ehrenberg	Haslau; pool	05107
Eudorina elegans Ehrenberg	Marchegg; March-river	05106
Eudorina illinoisensis (Kof.) Pascher	Altenwörth; Danube	05143
Eudorina illinoisensis (Kof.) Pascher	Wallersee	05144
Pandorina morum Bory	Melk; Danube	05118
Pandorina morum Bory	Finland; eutrophic pond	–
Pandorina morum Bory	Greenland	–
Pandorina morum Bory	Ruster-channel; Lake Neusiedlersee	05188
Pandorina morum Bory	Greece; Lake Prespa	05128

Table 2. List of culture media used in this study. The ingredients are given in g l^{-1}. The values for the trace element mix are the amounts for preparing 1 liter of standard solution. One ml of the standard solution was used for one liter of culture medium. + indicates the presence in the medium; – indicates the absence from the medium.

Compounds	Mineralic medium	Med. CHU 10 mod	Med. M3 mod. after Rayburn & Starr (1974)
KNO_3	0,2	–	0,1
K_2HPO_4	0,02	0,01	0,087
$MgSO_4*7H_2O$	0,02	0,025	0,05
$Ca(NO_3)_2$	–	0,04	–
$Na_2CO_3*10H_2O$	–	0,054	–
Na_2SiO_3	–	0,025	–
$CaCl_2*2H_2O$	–	–	0,01
Trace element mix	+	–	+
H_3BO_3	3	–	3
$MnSO_4$	1,5	–	1,5
$ZnSO_4$	1,1	–	1,1
$CuSO_4$	0,8	–	0,8
Na_2MoSO_4	0,025	–	0,025
$VaSO_4$	0,035	–	0,035
Vit.B_{12}	1 ml	–	1 ml
Aqua dest.	980 ml	1000 ml	1000 ml
Soilextract	10 ml	–	–

in *P. morum* was not observable in this study, as the strains used failed to mate with one another. Possible reasons for the failure of the sexual reproduction are discussed later.

In *Eudorina*, the mode of sexual reproduction is more complex than in *Pandorina*. The strains of *Eudorina* used in this study were heterothallic. In their asexual phase, the male and female clones appeared morphologically identical. During sexual development the male clones were clearly distinguished by the formation of 'sperm packets'. These were produced by a series of divisions of a cell resulting in a 32- or 64- celled plate. However, inversion was incomplete and the individual cells of the plakea differentiated into spindle-shaped, biflagellated sperms but remained grouped (see Figure 6). The colour of the cells changed from green to yellowish. Female coenobia showed no morphological differentiation during sexual development but individual cells acted as egg cells. The 'sperm packets' formed in the male antheridia were released from the parental gelatinous boundary to swim actively by the synchronous movement of the flagella. On becoming attached to a female coenobium, the 'sperm packets' disintegrated and each individualized gamete penetrated the gelatinous matrix of the female organism. Subsequently, each male fused with a female gamete resulting in the formation of a quadriflagellated planozygote. Next, the flagella gradually disappeared and within one week the green colour changed to a reddish brown, and a hypnozygote surrounded by a thick wall was formed. Zygote germination seemed to follow the same division- and inversion patterns as of a single cell, such as was observed during the asexual reproduction-cycle. Detailed reports are given in Waters (1960), Goldstein (1964), Nozaki (1983) and Angeler (1995b).

Culture experiments

The experiment with decreasing temperature and daylength provided the opportunity to examine the different patterns of adaptation of the strains. *E. illinoisensis* failed to live at 5 °C and 8L:16D hours. The strains of *E. elegans* retained the *Eudorina*-like habit, although the cells were sometimes slightly narrowed to the center of the coenobia. Under these conditions, the strain of *P. morum* from Greenland showed a loosening of the compact arrangement of the cells in the coenobia, thus resembling *Eudorina*-like organisms, although it seemed to be in a healthy physiological condition. The coenobial gelatinous matrix exhibited mound-like protuberances (see Figure 9). Tiffany (1934) referred spheroids of *Pandorina* with such an appearance of the gelatinous border to *Pandorina protuberans*. Although Ettl (1983) suggested that the doubtful *P. protuberans* may be identical with *P. morum*, many other authors continue treating it as a good species. The results of my preliminary investigation show that the protuberances of the gelatinous matrix of *P. morum* are formed in response to modified environmental conditions and do not persist throughout all conditions. However the current state of investigation does not allow a firm conclusion about the validity of *P. protuberans*. The low-temperature test led to the conclusion that the strains of *P. morum* represent different races, as their reaction to these conditions were so different. The clones from northern regions (Finland and Greenland) showed higher division rates as well as intensive locomotion of the coenobia, whereas the clones from central Europe (Ruster channel; Austria) formed resting stages. When exposed to normal growth conditions (16L:8D hours + 20 °C), the culture recovered, resulting in the formation of normal coenobia of *Pandorina*. Strains reacted differently to the variously altered chemical environments. In life, these algae usually occur in waters loaded with higher amounts of nutrients, especially phosphate and nitrate (Pringsheim & Pringsheim, 1959). Hence, the best growth of all clones was obtained in the mineral medium with the highest amounts of nutrients. The medium CHU_{10} mod., with its lower nutrient composition, was also suitable for the culture of these algae, although degeneration of the biomass appeared earlier than in the mineral medium. The modified M3-medium did not support growth. Even seven weeks after inoculation, the colour of the suspension remained colourless to very light green. All strains of *Eudorina* showed extreme degeneration. The misshapen coenobia, which under normal conditions consisted of 16 or 32 cells, now had only 2 to 8 cells. Furthermore, the pyrenoid formation was significantly disturbed. Even in adult cells, the strains showed not more than 3 pyrenoids. In contrast, the strains of *Pandorina* exhibited a better adaptation to these conditions by keeping their regular shape and number of cells per coenobium.

Discussion

The observations on the life cycles reported distinguish clearly among genera. Although the sexual process was incomplete in *P. morum*, features observable during asexual reproduction were adequate to place it firmly within the genus *Pandorina*. The presence of a 'cellular envelope' in *Eudorina*, which can be clearly seen during asexual reproduction, and the mode of the sexual processes offer clear grounds for generic differentiation.

As already stated, the clones of *P. morum* used in this study did not show mating reactions when mixed with one another. Coleman (1977), Coleman & Zollner (1977) and Coleman et al. (1994) pointed out that *P. morum* is presumably a species swarm, consisting of several breeding groups or syngens (*sensu* Sonneborn, 1957), which are more or less limited to certain geographic regions. The members of a single syngen are capable of mating with one another whereas members of different syngens cannot be crossed, owing to genetic barriers (e.g. different chromosome numbers etc).

The studies of Coleman and her coworkers demonstrate the presence of local races primarily based on the results of molecular and caryological analyses. This study also suggests the presence of geographically isolated tribes of *P. morum*, as shown by the different reactions to environmental conditions and the failure of an intercrossing of the strains used. It seems possible that the strains of *P. morum* used in this study may belong to different syngens because of their geographical source (Greenland, Iceland, central and south Europe). Until caryological and molecular analyses confirm this hypothesis remains speculative.

Little information is available on the autecology of *Pandorina* and *Eudorina*. The most common autecological reports are nutritional observations (Pringsheim & Pringsheim, 1954; Palmer & Starr, 1971). In the course of this investigation it was found that adaptations of strains representing a single species may vary enormously. The different adaptation of the clones of *P. morum* to low temperature indicate the existence of local tribes, although their vegetative morphology was essentially similar. The first impression of this comparative study leads to the conclusion that *P. morum* is broadly more tolerant of extreme situations than *E. elegans* and *E. illinoisensis*. In conclusion, it should be pointed out that ecological and ecophysiological investigations of a series of strains isolated throughout the world should be the subject of further studies with adequate methods to determine the role of environmental parameters on the biology of these and other volvocacean species.

Acknowledgements

The author wishes to thank Prof. Dr Elsa Kusel-Fetzmann and Mag. Dr Michael Schagerl for critical reading of the manuscript and Mr Warwick Bottomley for the correction of the English version of the manuscript. This work was supported by a grant from the Siegfried Ludwig Fonds für universitäre Einrichtungen in Niederösterreich.

References

Angeler, D., 1995a. Asexuelle Reproduktion bei *Eudorina illinoisensis* (KOFOID) PASCHER (Volvocales, Chlorophyta). Carinthia II (Klagenfurth, Austria) 53: 1–2.

Angeler, D. G., 1995b. Untersuchungen zur sexuellen Vermehrung von *Eudorina illinoisensis* (KOFOID) PASCHER (Volvocaceae, Chlorophyta). Lauterbornia (Dinkelscherben, Germany) 21: 151–158.

Chodat, R., 1931. Sur quelques algues nouvelles du plancton du lac geneve. In Recueil de Travaux Cryptogamiques. Volume dédié a Louis Mangin. Laboratoire de Cryptogamie, Muséum National d'histoire Naturelle, Paris: 1–11.

Coleman, A. W., 1959. Sexual isolation in *Pandorina morum*. J. Protozool. 6: 249–264.

Coleman, A. W., 1977. Sexual and genetic isolation in the cosmopolitan algal species *Pandorina morum*. Am. J. Bot. 64: 361–368.

Coleman, A. W. & J. Zollner, 1977. Cytogenetic polymorphism within the species *Pandorina morum* Bory de Saint Vicent (VOLVOCACEAE). Arch. Protistenk. 119: 224–232.

Coleman, A. W., Suarez, A. & L. J. Goff, 1994. Molecular delineation of species and syngenes in volvocacean green algae. J. Phycol. 30: 80–90.

Ettl, H., 1983. Chlorophyta I. Phytomonadina. In H. Ettl, J. Gerloff, H. Heynig & D. Mollenhauer (eds), Süßwasserflora von Mitteleuropa, 9, Gustav Fischer Verlag, Stuttgart, 807 pp.

Fulton, A. B., 1978. Colonial development in *Pandorina morum*. I. Structure and composition of the extracellular matrix. Dev. Biol. 64: 224–235.

Goldstein, M., 1964. Speciation and mating behaviour in *Eudorina*. J. Protozool. 11: 317–344.

Gottlieb, B. & M. E. Goldstein, 1977. Colony development in *Eudorina elegans*. J. Phycol. 13: 358–364.

Korshikov, A. A., 1923. Über zwei neue Organismen aus der Gruppe der Volvocales. Arch. russ. Protistol. 2: 170–178.

Kusel-Fetzmann, E. L. & M. Schagerl, 1992. Verzeichnis der Sammlung von Algen-Kulturen an der Abteilung für Hydrobotanik am Institut für Pflanzenphysiologie der Universität Wien. Phyton 32: 209–234.

Marchant, H. J., 1977. Colony formation and inversion in the green flagellate *Eudorina elegans*. Protoplasma 93: 325–339.

Nozaki, H., 1980. The asexual reproduction of Japanese *Pandorina morum* BORY (Volvocales, Chlorophyta). Jap. J. Phycol. 28: 157–158.

Nozaki, H., 1981. The life history of Japanese *Pandorina unicocca* (Chlorophyta, Volvocales). J. Jap. Bot. 56: 65–74.

Nozaki, H., 1983. Sexual reproduction in *Eudorina elegans*. Bot. Mag. Tokyo 96: 103–110.

Nozaki, H. & M. Itoh, 1994. Phylogenetic relationships within the colonial Volvocales inferred from cladistic analyses based on morphological data. J. Phycol. 30: 353–365.

Nozaki, H. & H. Kazaki, 1979. The sexual process of Japanese *Pandorina morum* BORY (Chlorophyta). J. Jap. Bot. 54: 363–371.

Nozaki, H. & T. Kuroiwa, 1991. *Pandorina colemaniae* sp. nov. from Japan. Phycologia 30: 381–393.

Nozaki, H. & T. Kuroiwa, 1992. Ultrastructure of the cellular matrix and taxonomy of *Eudorina*, *Pleodorina* and *Yamagishiella* gen. nov. Phycologia 31: 529–541.

Palmer, E. G. & R. C. Starr, 1971. Nutrition of *Pandorina morum*. J. Phycol. 7: 85–89.

Pringsheim, E. G. & O. Pringsheim, 1959. Die Ernährung koloniebildender Volvocales. Biol. Zentralblatt 78: 937–971.

Rayburn, W. R. & R. C. Starr, 1974. Morphology and nutrition of *Pandorina unicocca* sp. nov. J. Phycol. 10: 42–49.
Sonnerborn, T. W., 1957. Breeding systems, reproductive methods, and species problems in Protozoa. In E. Mayr (ed.): The species problem. American Association for the Advancement of Science, Washington D.C.: 125–324.
Thompson, R. H., 1954. Studies on the Volvocales. I. Sexual reproduction of *Pandorina charkowiensis* and observations on *Volvulina steinii*. Am. J. Bot. 41: 142–145.
Tiffany, L. H., 1934. The plankton algae of the west end of Lake Erie. The Ohio State University. Contributions to the Franz Theodore Stone Laboratory 6: 1–112.
Waters, A. J., 1960. Studies on *Eudorina*. Thesis Univ. of California, Berkeley, 25 pp.

Hydrobiologia **369/370**: 277–285, 1998.
M. Alvarez-Cobelas, C. S. Reynolds, P. Sanchez-Castillo & J. Kristiansen (eds), Phytoplankton and Trophic Gradients.

Morphological changes of Euglenophyta in response to organic enrichment

Visitación Conforti
Depto. de Ciencias Biológicas, Facultad de Ciencias Exactas y Naturales, Ciudad Universitaria, Pab. II, 1428, Buenos Aires, Argentina

Key words: Euglenophyta, bioassay, organic matter, paramylon bodies

Abstract

Euglena acus, Euglena spirogyra, Phacus curvicaudus, Phacus pyrum and *Phacus tortus* isolated from Matanza River, Buenos Aires, Argentina, were grown in soil-water medium. When transferred to soil-water medium enriched with different beef-extract concentrations, a marked cell deformation could be observed in every instance, owing either to an extraordinary accumulation of paramylon grains or to a great development of some of them. In this paper, these morphological changes are described and their ecological implications are discussed.

Introduction

Morphological plasticity of the Euglenophyte cell and its sensitivity to environmental conditions has been recognised for many years. Lefèvre (1931), analysing the principal specific characters in some Euglenophytes, drew attention to the morphological variability of the organisms. Conrad (1943) reported the influence of the development of paramylon grains on the shape of the cell. Kiss et al. (1986) described morphological changes in *Euglena gracilis,* detected by light microscopy, and associated ultrastructural changes as well, after cells had been transferred from an incomplete to a complete medium.

In previous works (Conforti, 1991; Conforti et al., 1995), I have presented certain aspects of the responses of phytoplankton to the environmental conditions in the Matanza River, Buenos Aires, Argentina. This river suffers a high degree of organic contamination, derived from waste waters of local meat industries. Among the Euglenophytes, organisms could be observed to contain a great number of paramylon bodies or to contain rather fewer but more conspicuous ones. A marked morphological divergence among these specimens was noted, and which were supposed to be related to the content of organic matter in the river.

In the present paper, light-microscope observations of the morphological variability in *Euglena acus, Euglena spirogyra, Phacus curvicaudus, Phacus pyrum* and *Phacus tortus,* cultured in soil water medium enriched with different levels of organic matter, are presented.

Materials and methods

The algae investigated in this study were sampled at various sites on the Matanza River. Water was sieved with a 20-μm mesh net. The organisms were observed under a Leitz inverted microscope, identified and isolated, using micropipettes. The selected organisms were washed with culture medium and maintained in soil- water medium (Pringsheim, 1946), stabilized with CO_3Ca. Cultures were non-axenic, each tube (10 ml) containing many Euglenophyte species. Beef extract was used to enrich the organic content of the culture medium, because the main source of pollution in the river is the untreated discharge of wastewaters from meat industries. Thirty tubes containing 5 ml of soil water – CO_3Ca medium were prepared. Each was inoculated with 10 organisms in a sterile chamber, 2 each of *Euglena acus, Euglena spirogyra, Phacus curvicaudus, Phacus pyrum* and *Phacus tortus*. A week later, when algal growth was verified, five tubes were separated as controls; the remainder were enriched with a sterile beef extract solution (BES) of 50 g in 1500 ml

double-distilled water, according to the following protocol:
5 tubes with medium + 0.5 ml of BES.
5 tubes with medium + 1.0 ml of BES.
5 tubes with medium + 1.5 ml of BES.
5 tubes with medium + 2.0 ml of BES.
5 tubes with medium + 2.5 ml of BES.

Tubes were incubated at room temperature (ca. 20 °C), with a 12-hour photoperiod. After two days, and before fungal and bacterial proliferation, the liquid phase of each tube was removed to another flask and fixed with a solution of 3% formaldehyde. Three 0.1 ml aliquots of each culture tube were observed, photographed and drawn with the aid of a Zeiss optical microscope. Normal and altered specimens were counted and measured. Table 1 shows the percentage of deformed organisms, the standard deviation and the number of replica considered. From these data, sectorial diagrams and dose-response curves were derived.

Results

Extensive morphological alterations of specimens were observed. These changes involved modifications to the paramylon (ß, 1:3 glucans) bodies. This material is normally accumulated in the Euglenophyta, usually as disks, rings, rods or links, just outside the chloroplasts. In all the species studied, dramatic cell changes were closely related to the increase in the number and size of paramylon grains accumulated. Table 1 shows the percentage of deformed organisms, the standard deviation and the number of replicates considered. The data were used to construct sectorial diagrams and dose-response curves.

Euglena acus Ehrenb. normally has fusiform cells, very slender, ribbon-like, 65–300 μm long, 7–28 μm wide; length/width ratio, 7–20. Two or more rod-shaped paramylon bodies are usually located before and behind the nucleus (Figures 1–7, 10–12). When organisms were cultured in enriched medium, significant shortening and widening of the cells was observed, attributable to a marked increase in the number of paramylon grains. The length/width ratio fell to < 5 (extremes, 1.66, 2.06, 3.95, 4.66). This widening follows a transversal axis, maintaining the flattened shape in the apical and lateral views. In addition, a change in the pellicle striae was detected: normally almost longitudinal (Figures 1, 2), they appeared clearly helicoidal, as a result of deformation (Figures 4–7, 11–12).

Similar modifications were observed in *Euglena spirogyra* Ehrenb. (Figures 8–9, 13–14). Its dimensions in enriched medium were 56–76 μm long, 16–27 μm wide, compared to literature values (Gojdics, 1953; Huber-Pestalozzi, 1955; Tell & Conforti, 1986; Starmach, 1983) of 70–95 μm long and 8–13 μm wide. Cell widening was caused by the greater size of the paramylon grains rather than by an increase in their number. The direction of the pellicle striations were also altered (Figure 13).

No shortening or widening was observed among the *Phacus* spp. Either the increase in paramylon bodies or their exaggerated development gave these flattened organisms, altered lateral and apical morphology.

Phacus tortus (Lemm.) Skv. cells are commonly orbicular, broadly ovate to broadly elliptic, 54–66 μm long, 31–37 μm wide, thin, moderately to strongly twisted, with a long slender caudus. Two central, concentric discs, or a single ring or disc of paramylon can be seen. When cultured in enriched medium, in many cases, we observe a large, central bobbin–like grain, surrounded by many smaller bodies develops, increasing the thickness of the cell (Figures 17b, 18a, 19b). Some cells produced large bodies of alternative shapes (Figures 16a,b).

Phacus pyrum (Ehrenb.) Stein cells are normally clavate to fusiform, flattened to ovate in apical view, measuring 20–31 μm long, 12–16 μm wide. Pellicle ribs spiral to the right. This type of pellicle stiffens the cell. Despite this, cells showed a remarkable increase in thickness, when they were cultured in enriched medium, from 5–9 μm to 11–14 μm. Two paramylon bodies, large lateral rods or thick plates are normally present (Figures 20a–b, 21). In altered medium, cells gained a big central grain and some smaller (Figures 22–25) or numerous grains of assorted sizes (Figure 26).

Phacus curvicaudus Swir. cells are commonly orbicular to broadly ovate, 28–30 μm long, 24–30 μm wide, asymmetrical, inflated and comprising two overlapping halves. Two spheres or rings of paramylon appeared, one lying in each cell half. When the alga was cultured in enriched medium, these grains increased in volume and numerous smaller ones appeared, forcing a dorsi –ventral widening of the cell. The cell length / width relationship was unchanged, only the thickness was increased, from 10–12 μm to 11–23 μm (Figures 28–34).

In some taxa, like *E. spirogyra* and *P. curvicaudus*, the modifications were evident, even at the lowest organic enrichment (0.5 ml); in other species, the mod-

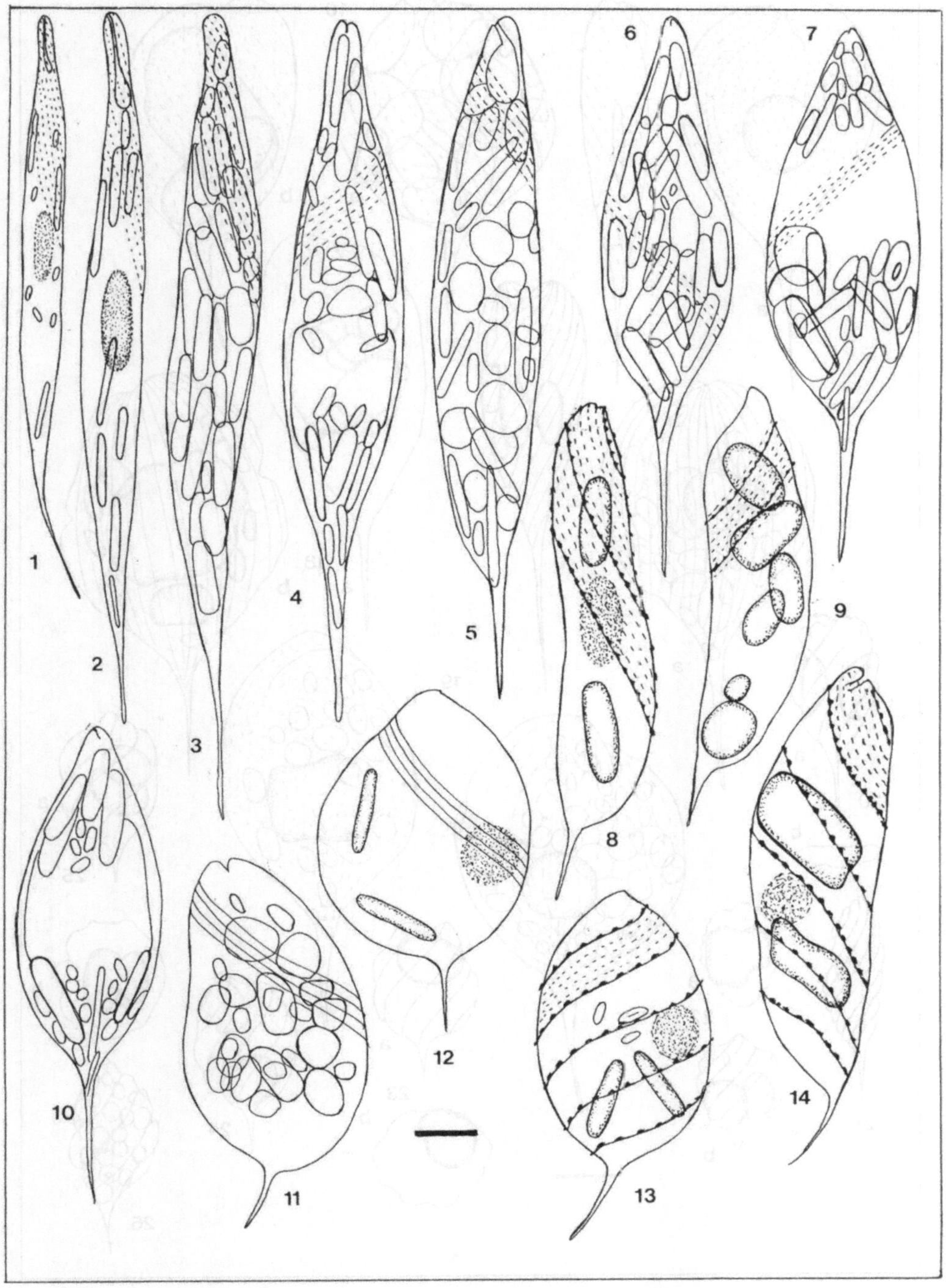

Figures 1–7, 10–12. E. acus, 1–2. normal morphology; 3–7, 10–12. different altered organisms; 8–9, 13–14. *E. spirogyra*, 8. normal morphology; 9,13–14, several altered specimens. All scale bars = 10 μm.

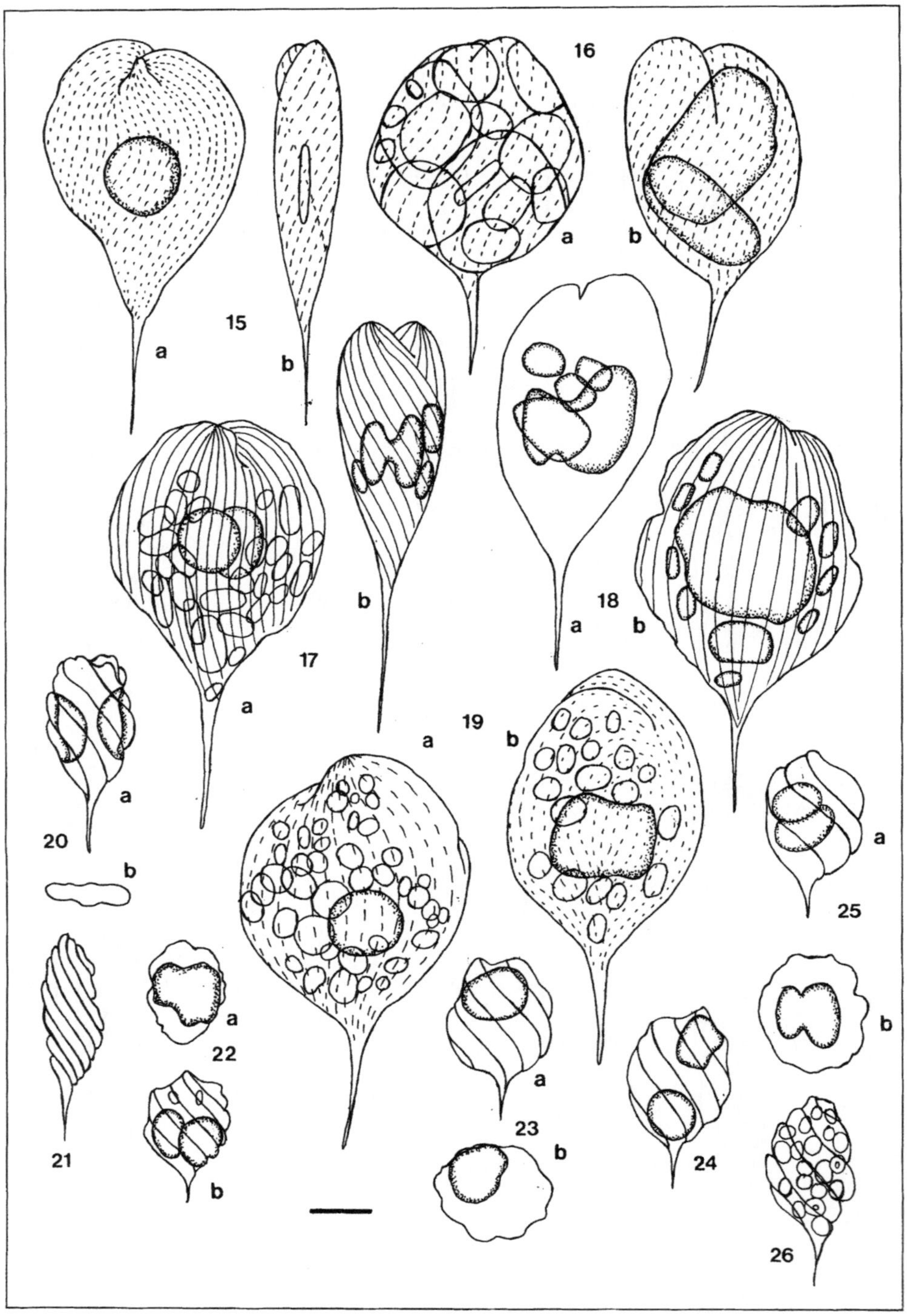

Figures 15–19. P. tortus; 15a–b. normal morphology, a- frontal view, b- lateral view; 16a–b, 17a–b, 18a–b, 19a–b. frontal and lateral views of altered specimens; 20–26. *P. pyrum*; 20a–b. normal morphology, a- frontal view, b- apical view; 21. normal specimen, lateral view; 22a–b, 23a–b, 25a–b. apical and frontal views of altered specimens, 24. altered specimen; 26. specimen showing numerous paramylon grains, lateral view. All scale bars = 10 μm.

Table 1. Statistical parameters of the response to different concentrations added (expressed as percentage of deformed organisms)

Solution (ml)	$g\,l^{-1}$	Mean	Standard deviation	*n*	Coefficient of variation (%)
0.5	2.2	35	26.9	*15*	76.8
	4.4	69	15.7	*15*	22.7
1.5	6.6	62	24.7	*15*	39.8
	8.8	71	25.9	*15*	36.4
2.5	11.1	81	12.4	*15*	15.3

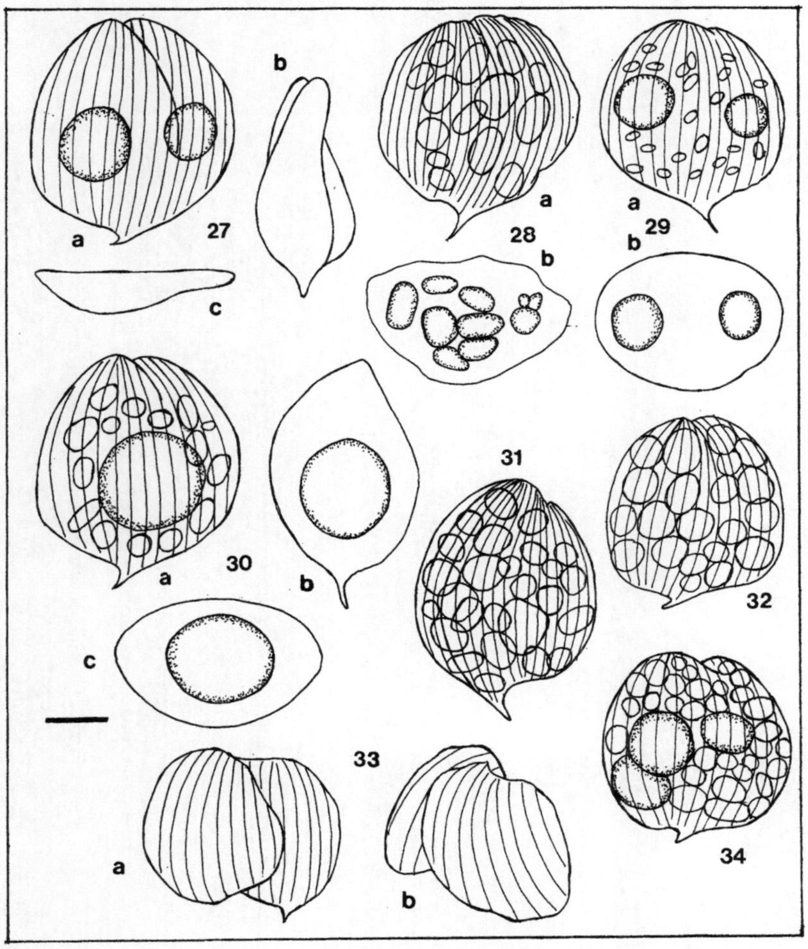

Figures 27–34. P. curvicaudus; 27a–c. normal morphology, a- frontal view, b- lateral view, c- apical view; 28- 34. altered specimens, 28. a- frontal view, b- apical view; 29. a- frontal view, b- apical view; 30. a- frontal view, b- lateral view, c- apical view; 31–32 & 34. general views; 33 a- frontal view, b- apical view. All scale bars = 10 μm.

ifications affected cells only in the higher concentrations (> 1 ml added).

Regardless of the organic concentration added, morphological alteration was consistent. As may be judged from the dose-response curve (Figure 51), the percentage of malformed organisms increased linearly with the greater additions of beef extract.

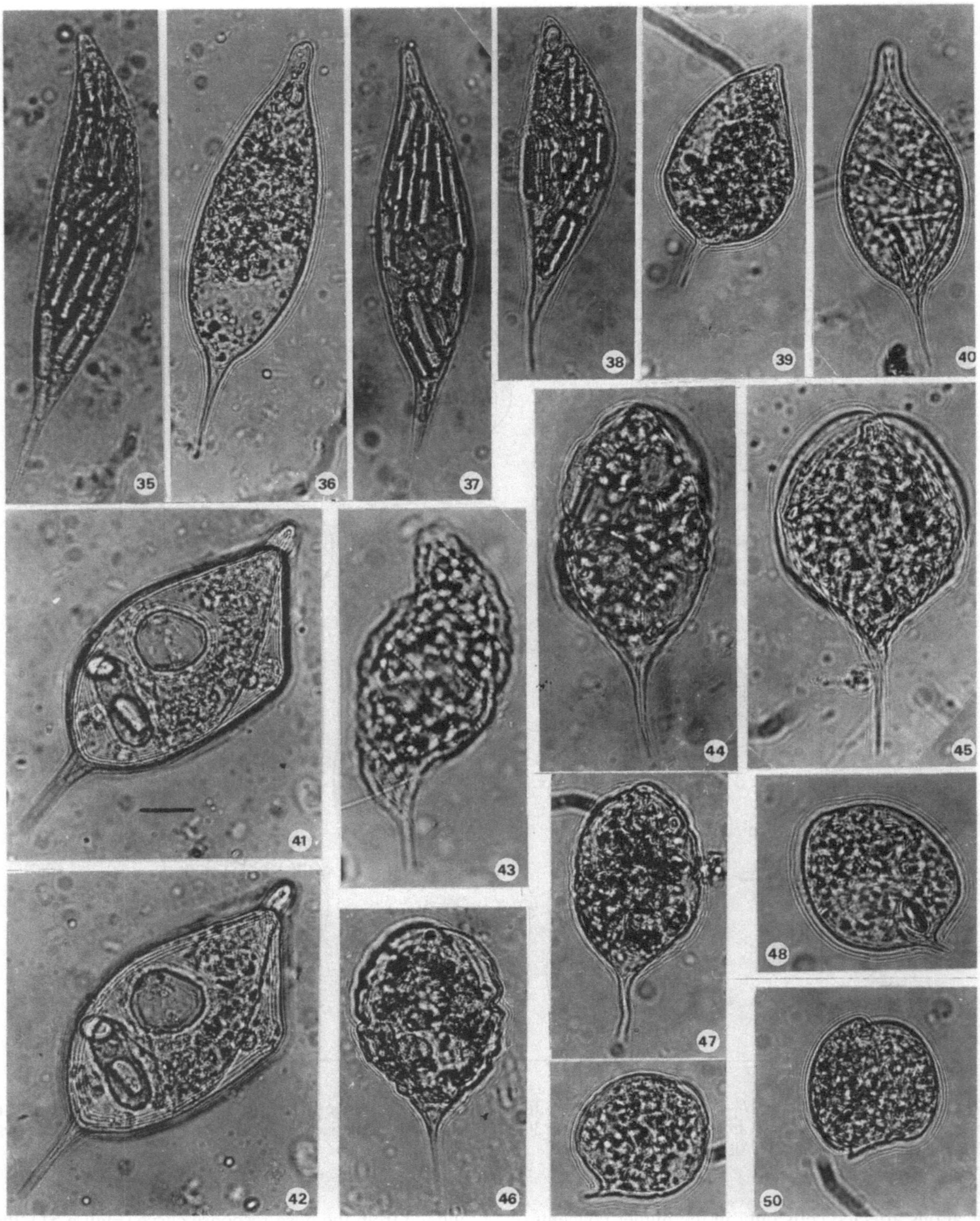

Figures 35–50. Optic micrographs. 35–42. *E. acus*, several altered specimens; 43. *E. spirogyra*, altered specimen; 44–47. *P. tortus*, 44–45. Altered specimens, 44 & 47. Lateral views, 45 & 46. frontal views; 48–50. *P. curvicaudus*, altered specimens showed an extraordinary number of paramylon grains. All scale bars = 10 μm.

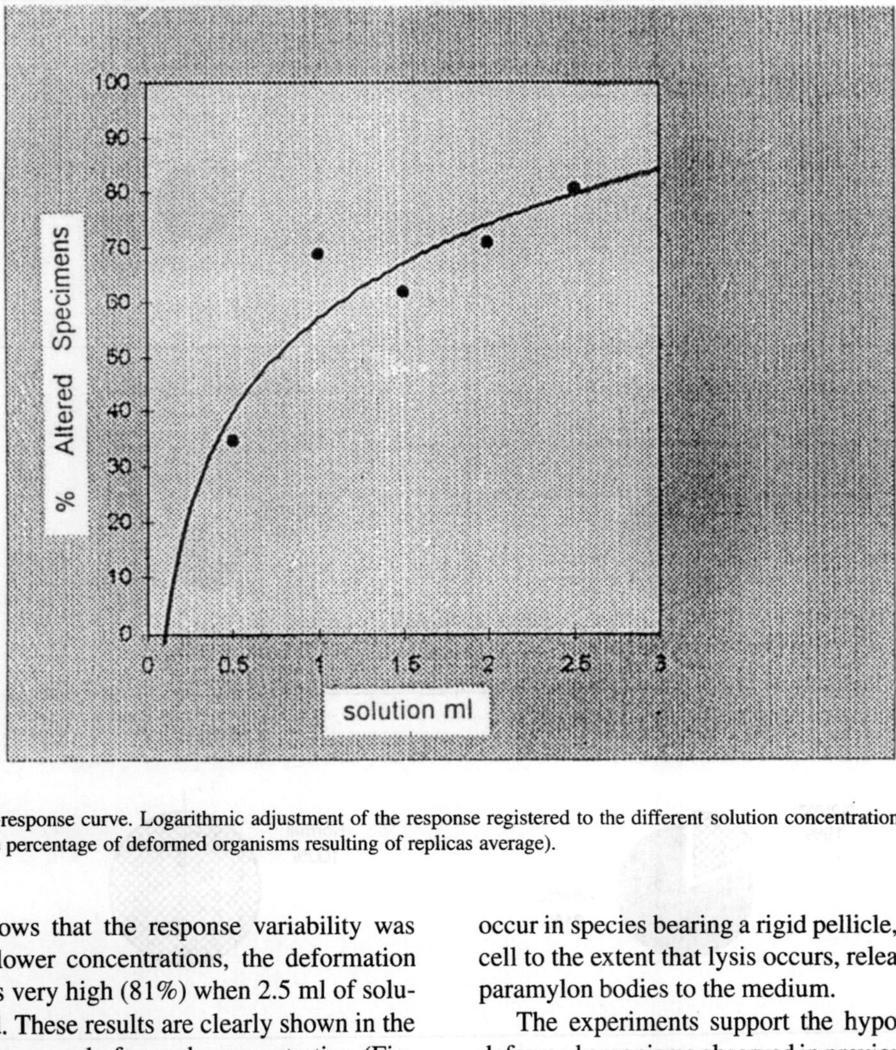

Figure 51. Doses-response curve. Logarithmic adjustment of the response registered to the different solution concentration added (expressed the response as the percentage of deformed organisms resulting of replicas average).

Table 1 shows that the response variability was higher in the lower concentrations, the deformation percentage was very high (81%) when 2.5 ml of solution was added. These results are clearly shown in the sectorial diagrams made for each concentration (Figure 52).

Conclusions

All the Euglenophyte species studied in our bioassays have shown morphological changes in response to additions of beef extract. These changes resulted from an increase in the number or size of paramylon bodies, some of which were extremely large. As a consequence, the direction of the pellicle striae in the species of *Euglena* was modified; moreover, the cell length/width relationship diminished without affecting cell thickness. Among the species of the genus *Phacus*, a remarkable thickening along the cellular dorsi – ventral axis was observed, thus strongly altering the apical and lateral views of the organisms. The modifications occur in species bearing a rigid pellicle, deforming the cell to the extent that lysis occurs, releasing numerous paramylon bodies to the medium.

The experiments support the hypothesis that the deformed organisms observed in previous studies in the Matanza River (Conforti, 1991; Conforti et al., 1995) could, in part, have been attributed to excess organic matter in the water. In *Phacus tortus*, the malformations detected in the enriched cultures were closely similar to those observed in the river.

On the basis of these observations, it is possible to deduce that paramylon grains have little diagnostic value as species-specific characters in the Euglenophyta, as many authors once supposed (Van Oye, 1924; Lefèbre, 1931; Conrad, 1943): their numbers and shapes are too sensitive to variation, in response to environmental conditions.

In asexual organisms, such as the ones studied here, where taxonomic determination is based mainly on morphological characteristics, these variations are particularly worrying. Several species, previously considered as different, could turn out to be ecomorphs

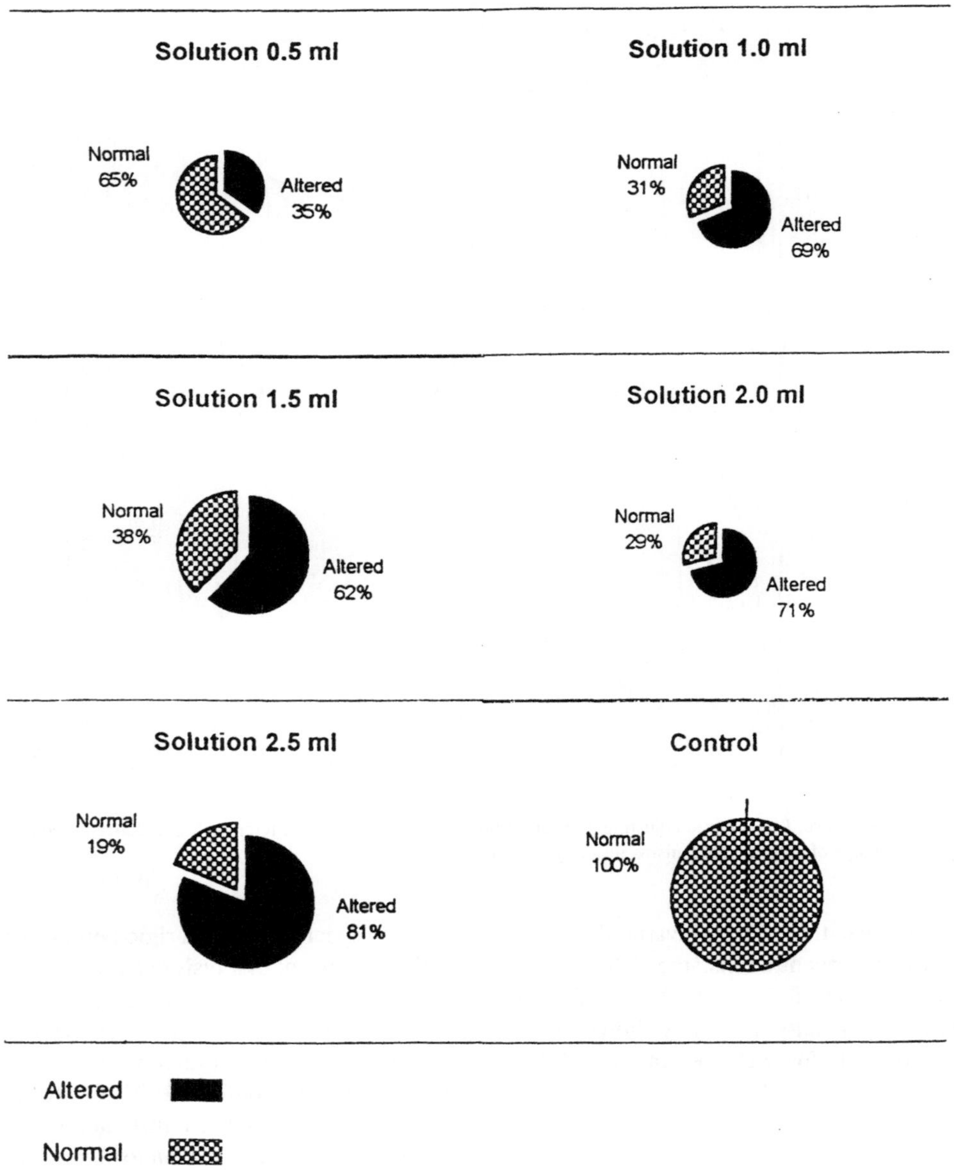

Figure 52. Sectorial diagrams showed the deformed specimens percentage as related to each concentration used.

of the same taxon. A thorough re-examination of the systematics of the group is becoming essential.

On the other hand, the changes observed were generally similar, being independent of the doses of organic matter added. Only the numbers of individuals affected varied, being greater in the case of the higher concentrations of organic substrate. All cells showed deformation with the addition of 2.5 ml of beef extract solution. These changes were fast, being observed within two days of exposure to the enhanced levels. In a natural environment, any action leading to a significant increase in the organic matter present will rapidly affect the local Euglenophytes. These results suggest that the presence of enormous intracellular paramylon bodies, or the deformation of cells in natural samples, could be used as a good environmental bioindicator.

Acknowledgements

This investigation was supported by U.B.A. grants, Ex 035, 91/93. The author is indebted to Lic. Ruben Lombardo for this technical assistance and wishes to extend her gratitude to Lic. Alejandra Nudelman and Prof. Ana María Tosellio for linguistic help with English translation.

References

Conforti, V., 1991. Taxonomic study of the Euglenophyta of a highly polluted river of Argentina. Nov. Hedwigia 53: 73–98.

Conforti, V., J. Alberghina & E. Gonzalez Urda, 1995. Structural changes and dynamics of the phytoplankton along a highly polluted lowland river of Argentina. J. Aquat. Ecosyst. Health 4: 59–75.

Conrad, W., 1943. Remarques sur le genre *Phacus* Duj., 1841. Bull. Mus. r. Hist. nat. Belg. 19: 1–16.

Gojdics, M., 1953. The genus *Euglena*. The Univ. of Wisconsin Press. Madison. 268 pp.

Huber-Pestalozzi, G., 1955. Die Binnengewässer. Das Phytoplankton des Süsswassers 16(4). Stuttgart, 1135 pp.

Kiss, J., A. Vasconcelos & R. Triemer, 1986. Paramylon synthesis and chloroplast structure associated with nutrient levels in *Euglena* (Euglenophyceae). J. Phycol. 22: 327–333.

Lefèvre, M., 1931. De la valeur des caracteres specifiques chez quelques Eugleniens. Rec. Trav. Cryptog. Dediés a L. Manguin: 343–354.

Pringsheim, E., 1946. The biphasic or soil-water cultured method for growing algae and flagellates. J. Ecol. 33: 193–204.

Starmach, K., 1983. Euglenophyta-Eugleniny. Flora slodkowodna Polski. 3. Polska Akademia Nauk. Warszawa-Krakow. 594 pp.

Tell, G. & V. Conforti, 1986. Euglenophyta pigmentadas de la Argentina. Bibliotheca Phycologica, Cramer (ed.), B75, 301 pp.

Van Oye, P., 1924. Note sur l'*Euglena acus* Ehrenberg. Bull. Soc. r. Bot. Belg. 56: 124–132.

Hydrobiologia **369/370:** 287–295, 1998.
M. Alvarez-Cobelas, C. S. Reynolds, P. Sanchez-Castillo & J. Kristiansen (eds), Phytoplankton and Trophic Gradients.

Specific composition and ecology of chrysophyte flagellates in Lake Sanabria (NW Spain)

Caridad De Hoyos[1], Juan José Aldasoro[2], Manuel Toro[3] & Francisco A. Comín[4]
[1] *Department of Ecology, Faculty of Biology, University of Salamanca, 37071 Salamanca, Spain*
[2] *Real Jardín Botánico de Madrid (CSIC), Plaza Murillo, 2, 28014 Madrid, Spain*
[3] *Department of Ecology, Faculty of Biology, Universidad Autónoma de Madrid, 28049 Madrid, Spain*
[4] *Department of Ecology, Faculty of Biology, University of Barcelona, C/ Diagonal, 645, 08017 Barcelona, Spain*

Key words: chrysophyte flagellates, ecology, cysts, spatial-temporal distribution, Lake Sanabria

Abstract

Lake Sanabria is a glacial lake located at 1000 m a.s.l. in NW Spain. Its water is characterised by a low content of mineral salts, rather low pH, and oligotrophic status, which are all currently viewed as being favorable for Chrysophyte growth and proliferation. The study of the phytoplankton at different depths of the water column for three years showed that Chrysophytes were one of the groups of algae less well-represented in the phytoplankton (1% and 8% respectively, for numbers of cells and biovolume).

Ochromonas, Pseudopedinella, Mallomonas and *Dinobryon* were the more representative chrysophyte genera. *D. cylindricum, D. divergens and D. crenulatum* were some of the *Dinobryon* species observed. *M. akrokomos, M. crassisquama* and *M. costata* were among the most frequent *Mallomonas* species. Observations on the Chrysophyte cysts from samples of the lake sediment suggest that variability of representation is higher than in planktonic samples.

The maximum values of Chrysophyte biovolume were found in the nutrient-exhausted epilimnion of summer and autumn which could be related to the mixotrophic capabilities of several species.

Introduction

Chrysophyte flagellates are abundant, both in number of species and in biomass, in the plankton of most oligotrophic lakes located at high latitudes (Nygaard, 1978; Moore, 1978; Duthie & Hart, 1987; Ilmavirta, 1980; Rosén, 1981; Eloranta, 1989, 1995). They also dominate the phytoplankton of many high mountain oligotrophic lakes located in temperate latitudes (Capblancq, 1972; Vilaseca, 1978). This geographical distribution is related to their capacity to develop in waters of very low phosphorus concentrations and to their mixotrophic capabilities (Sandgren, 1988).

Some chrysophyte species, particularly silica-bearing species, have been identified as environmental indicators (Kristiansen, 1986; Siver, 1995) and some of them have been closely studied in relation to the trophic status or the acidity of waters (Rosén, 1981; Smol et al., 1984a; Kristiansen, 1986; Eloranta, 1989; Siver, 1989; Siver et al., 1990).

Many chrysophyte species are identified by the microscopic structure of their siliceous components. In addition, chrysophyte microfossils (siliceous scales and cysts) are often well preserved in lake sediments, which means they are highly valuable as paleolimnological indicators (Sandgren & Carney, 1983; Sandgren, 1988), particularly in the study of lake acidification (Dixit & Dixit, 1989; Smol et al., 1984b). The main problem here, though, is related to the difficulties of cyst identification (Kristiansen, 1986). There are many cysts described in the literature (Sandgren & Carney, 1983; Cronberg, 1986). Most of them are described as morphotypes and are not related to known species (Cronberg & Sandgren, 1986; Duff et al., 1995).

This paper reports on the specific composition and ecological characteristics of chrysophytes in Lake Sanabria, the largest Spanish lake, and contributes to the definition of the environmental conditions under which chrysophyte populations develop in warm monomictic lakes, a type for which scarce information about chrysophyte flagellates is available (Sandgren, 1988). Further limnological characteristics of Lake Sanabria and its phytoplankton have been described previously (Vega et al., 1992; Aldasoro et al., 1992; De Hoyos, 1996.

Materials and methods

Phytoplankton sampling was carried out with a Van Dorn bottle from a boat anchored in the deepest part of the lake. Water samples were collected monthly, from May 1987 to December 1989, at 0, 2, 5, 10, 15, 20, 25, 35 and 45 m depth. Subsamples preserved with Lugol were studied under an inverted microscope at 600× and 1 000× after sedimentation, and the abundance of chrysophytes was calculated from the cell countings of at least 20 fields. The dimensions of thirty individuals from each species were measured and their biovolume was estimated, according to standard geometrical figures. The average biovolume of these individuals was used to calculate that of the whole population. Samples containing a high number of *Mallomonas* were observed in a scanning electron microscope in order to identify distinctive characteristics. The same was done for cyst identification with surface sediment samples collected in August 1993.

Physical and chemical characteristics of the water were determined with electronic equipment or following standard analytical methods (APHA, 1989). Pigment determinations were carried out following Jeffrey & Humphrey (1975).

Lake characteristics

Lake Sanabria (42°07′ N, 6°43′ W) is a glacial lake located at 1 000 m.a.s.l. in the valley of the River Tera (River Duero watershed, NW Spain). The weather of the zone is characterized by low temperatures in winter (freezing temperatures may occur during November-May) and high temperatures during summer (absolute maximum 36 °C). Rainfall occurs with a high interannual variability. For a period of 50 years (1940–1990), the average annual rainfall was 1 422 mm, ranging between 544–2 223 mm (De Hoyos, 1996). Most of the precipitation comes from the Atlantic Ocean and falls from September to February. The catchment area of the lake is 127.3 km^2. The geology is unyielding, comprising gneiss (66.6%), granodiorite (21.9%), slate (6.5%), together with residual deposits of glacial, fluvial and lacustrine origin (5%). The lake water is slightly acid (annual average pH = 6.2) and has low mineral content (annual average conductivity = 14.9 μS cm^{-1}). The relative abundance of anions and cations indicate that the lake experiences an oceanic influence (De Hoyos, 1996).

The lake volume is 93.3 Hm^3 and the area is 3.46 km^2. The maximum depth is 51 m and the average depth 27 m. The average annual water residence time for the period 1940–1990 was 0.76 years. During the period of this study, the annual water residence time was lower than the average indicated. Water residence time during September-February 1987–88 and 1989–90 was very low (0.3 years) and during September 1988–February 1989 was higher (0.96 years) than average (De Hoyos, 1996). Lake Sanabria is a warm (4.5 °C–24.8 °C) monomictic lake. The thermocline begins to establish in March-April and breaks in November-December. The photic zone is 13m deep on average, nearly always extending beyond the depth of the summer thermocline.

The lake is oligotrophic. The maximum chlorophyll *a* concentration is about 5–6 μg l^{-1} in the metalimnion in summer. Major nutrients are present at very low concentrations (spatial-temporal averages: 51 μg N–NO_3^- l^{-1}, 6 μg soluble reactive phosphorus l^{-1}). The atomic ratio N–NO_3/SRP is low (it ranges from 0 to 15 in the epilimnion during the period of stratification). The water flowing into the lake is low in inorganic nitrogen relative to phosphorus when compared to the relative proportion of these elements required by phytoplankton (Harris, 1986). The proportion of nitrogen relative to phosphorus increases inside the lake, which is related to the mineralization of organic matter (De Hoyos, 1996). Hypolimnetic dissolved oxygen is never lacking during the annual stratification period (De Hoyos, 1996). A large part of the total nitrogen is dissolved organic nitrogen (71%) and most of the dissolved inorganic fraction consists of nitrates. These data and some others (e.g., colour of the water: 21 mg Pt l^{-1} in summer and 41 mg Pt l^{-1} in winter, extinction of red light at depths greater than other wavelengths during periods of high water input) also indicate that the lake is slightly brown-watered (De Hoyos, 1996).

In this lake, chlorophytes, especially the order Chlorococcales is the group of algae with most species, followed by diatoms. As regards the cell number, the most abundant algae are cyanobacteria (chroococcales), followed by chlorophytes chlorococcaleans. In terms of biovolume, chlorophytes and cryptophytes are the most important algae. The average biovolume of all the phytoplankton cells (years 1987–1989) in the lake was 184 mm^3 m^{-3} but ranged between 56 and 1 020 mm^3 m^{-3} depending on the time of the year and on the depth (De Hoyos, 1996).

Results and discussion

Chrysophyte specific composition and taxonomy

Fifteen chrysophyte species were observed in the water column of Lake Sanabria during the period of this study: *Dinobryon cylindricum* Imhof, *Dinobryon divergens* Imhof, *Dinobryon crenulatum* W. & G. S. West, *Dinobryon* cf. *borgei* Lemmermann, *Dinobryon* cf. *marssonii* (Lemmermann) Lemmermann, *Dinobryon* cf. *eurystoma* (Stoke) Lemmermann, *Mallomonas akrokomos* Ruttner, *Mallomonas crassisquama* (Asmund) Fott, *Mallomonas costata* Dürrschmidt, *Pseudopedinella* cf. *erkensis* Skuja, *Ochromonas globosa* Skuja, *Ochromonas* sp(1), *Ochromonas* sp(2), *Bitrichia ochridana* (Fott) Bourrelly and *Kephyrion rubri-claustri* Conrad.

All the *Mallomonas* species observed in Lake Sanabria during this study are common species with wide distributions (Asmund & Kristiansen, 1986). *M. akrokomos* and *M. crassisquama* are widespread species. *M. costata* is less common, although it has been found in various zones of N. Europe and N. America (Dürrschmidt, 1984; Asmund & Kristiansen, 1986; Eloranta, 1989; Nicholls, 1990). These three species have also been found in N. Portugal (Calado & Craveiro, 1995). *M. crassisquama* can live in waters with a wide range of environmental characteristics (pH, nutrients, conductivity, humic substances) (Siver et al., 1990; Siver, 1991) although it grows better in slightly acidic waters and is not present at a pH lower than 5.5 (Siver, 1989). *M. akrokomos* has been reported in oligotrophic and eutrophic waters, mostly in winter (Hutchinson, 1967; Dürrschmidt, 1984), although it can be very common also in summer (Eloranta, 1989). Ilmavirta (1980) describes this species as being typical of oligotrophic and humic waters.

The relatively high number of *Dinobryon* species observed in Lake Sanabria is a common feature of oligotrophic water bodies. Frequently, 3 to 6 species of *Dinobryon* are present at the same time in the plankton of lakes in Finland, typically in summer (Eloranta, 1989), which was also the case in this study. Some of the species observed in Sanabria (*D. cylindricum, D. divergens, D. crenulatum, D. borgei*) are among the most frequent and abundant species in Nordic lakes.

Many authors note the presence of *Dinobryon* species in oligotrophic waters: *D. cylindricum, D. divergens* (Järnefelt, 1952), *D. crenulatum* (Rosén, 1981), *D. borgei* (Eloranta, 1989) and oligo-mesotrophic: *D. borgei* (Rosén, 1981). However, some species (e.g., *D. divergens*) have also been found in waters with high concentrations of nitrogen and phosphorus (Rosén, 1981; Eloranta, 1989).

Pseudopedinella and *Ochromonas* are genera described from oligotrophic lakes (Rosén, 1981). Abundant populations of several species of *Ochromonas* have been reported as being present in Pyrenean lakes in summer (Capblancq, 1972; Vilaseca, 1978).

Twenty seven different cysts were observed in the sediment samples. They have been distributed according to the collar and ornamentation of the cyst (Duff et al., 1995) in the following groups. In the group 'sphaerical stomatocyst lacking ornamentation but with a collar', three cysts were included. Four different types of cysts were ascribed to the group 'stomatocysts ornamented with scabrae or verrucae'. In the group 'stomatocysts ornamented with spinae less than 1 μm in length', four types of cysts were observed and in the group 'cysts ornamented with spinae more than 1 μm in length', five types. One of the cysts belonging to the last group has been identified as *Mallomonas crassisquama* (Duff & Smol, 1994); another cyst (Figure 1b) is quite similar to that described for *Mallomonas pseudocoronata*, although its dimensions (14 × 11 μm) are smaller than the species description (16–18 × 14–15 μm: Smol, 1984). *M. psedocoronata* is one of the most common chrysophyte taxa in North America (Kling & Kristiansen, 1983) but, in Europe, it has been found only in the sediments of an Austrian meromictic lake (Smol, 1988).

In the group 'stomatocysts ornamented with both spinae and depressions', 4 types of cysts were observed; in the group 'stomatocysts ornamented with ridges', four types and in group 'stomatocysts ornamented with a reticulum', three types of cyst were

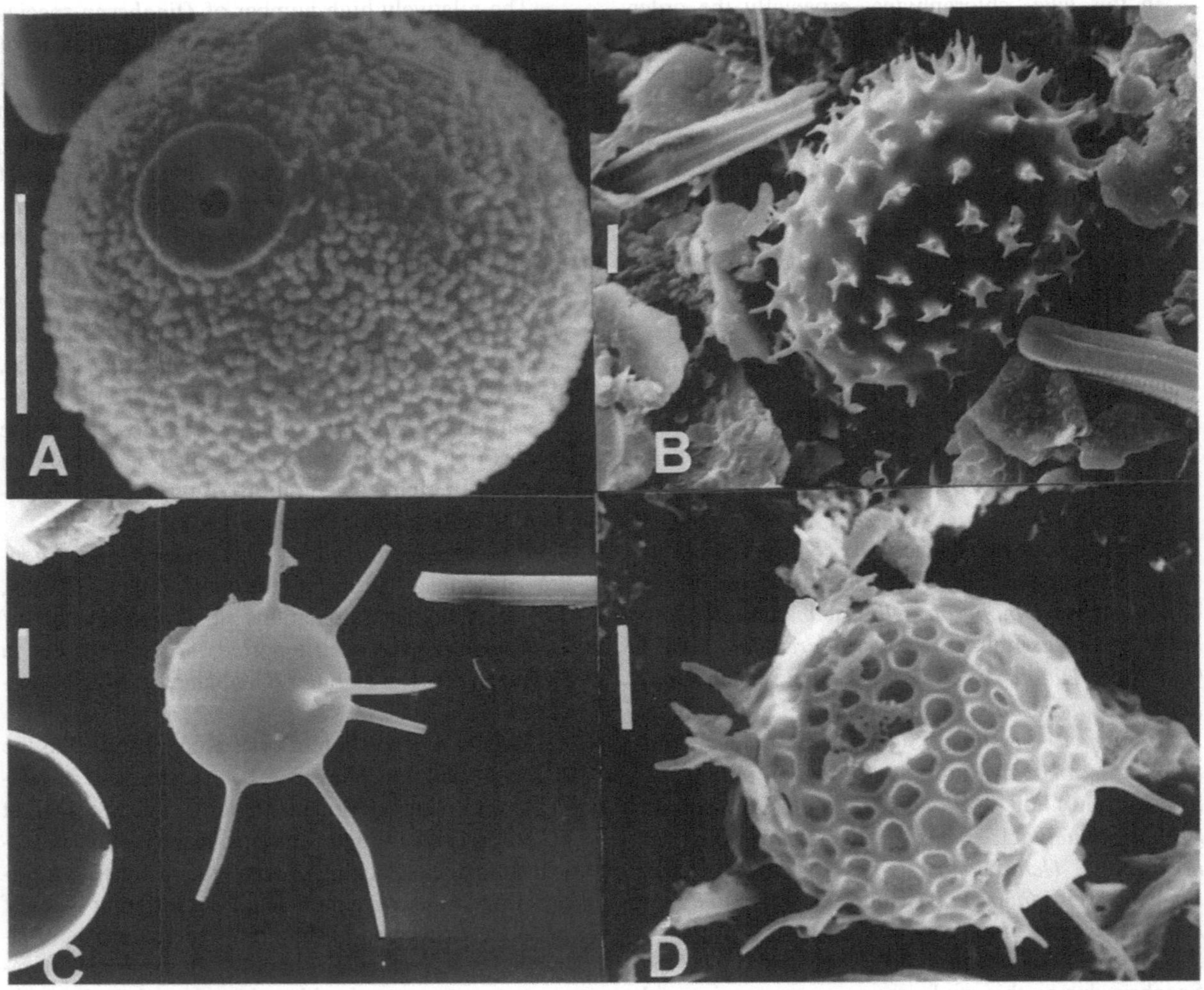

Figure 1. Some chrysophyte cysts found in the surface sediments of Lake Sanabria. Following groups of Duff et al (1993): a) 'stomatocysts ornamented with scabrae or verrucae'. b and c) 'stomatocysts ornamented with spinae more than 1 μm in length'. d) 'stomatocysts ornamented with a reticulum'.

observed. Some of the cysts more scarcely mentioned in literature are shown in Figure 1.

Chrysophyte temporal and spatial distribution

It is emphasised that Chrysophytes are generally scarce in Lake Sanabria, both in absolute abundance (Figure 2) and relative to that of other groups of algae (Figure 3). The average biovolume of chrysophytes in Lake Sanabria was 15 mm^3 m^{-3} during the period of the study.

The low absolute and relative abundances of chrysophytes in Lake Sanabria contrasts with their dominance in some other slightly acid oligotrophic lakes (Rosén, 1981; Sandgren, 1988; Siver, 1995). Chrysophytes are superior to most other phytoplankton species with regard to both their storage and growth potential under conditions of low phosphorus availabity (Rodhe, 1948; Sandgren, 1988) In Lake Sanabria the nutrient concentrations are very low, particularly the dissolved inorganic nitrogen, and so is also the N:P ratio. The latter can be a key factor to explain the scarcity of chrysophytes in Lake Sanabria. Pearsall (1932) only observed high percentages of Dinobryon at high N:P ratios, and Reynolds (1984) pointed out that chrysophyte populations are said to require a high N:P ratio for their development.

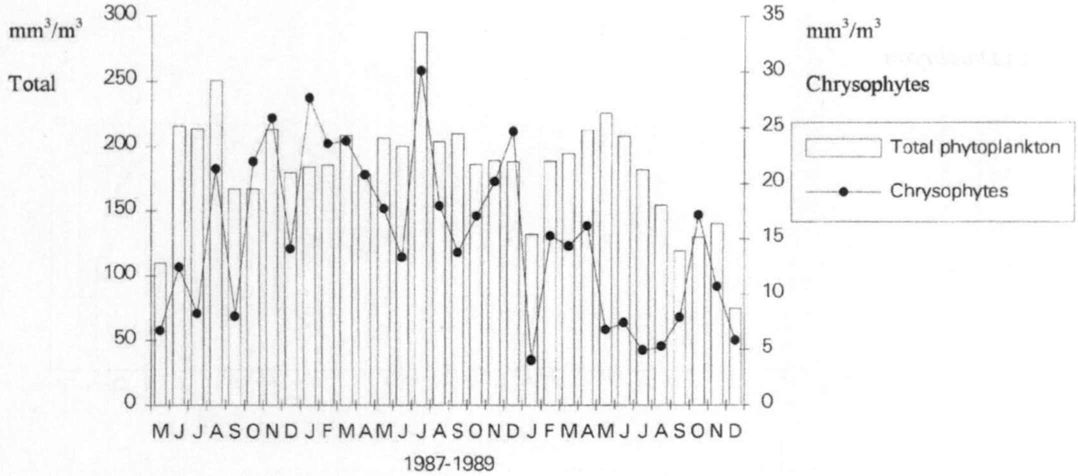

Figure 2. Variation in chrysophyte and total phytoplankton biovolume (integrated mean of the water column) in Lake Sanabria.

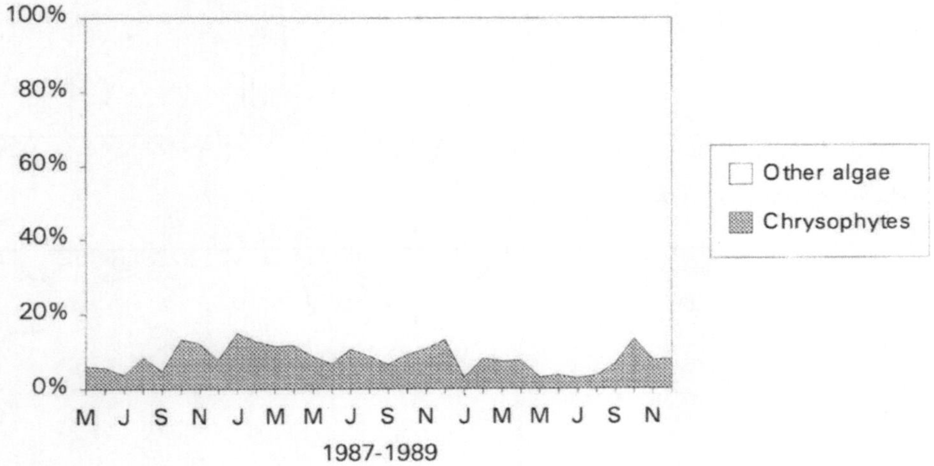

Figure 3. Relative chrysophyte abundance as percentage of the total phytoplankton biovolume in Lake Sanabria.

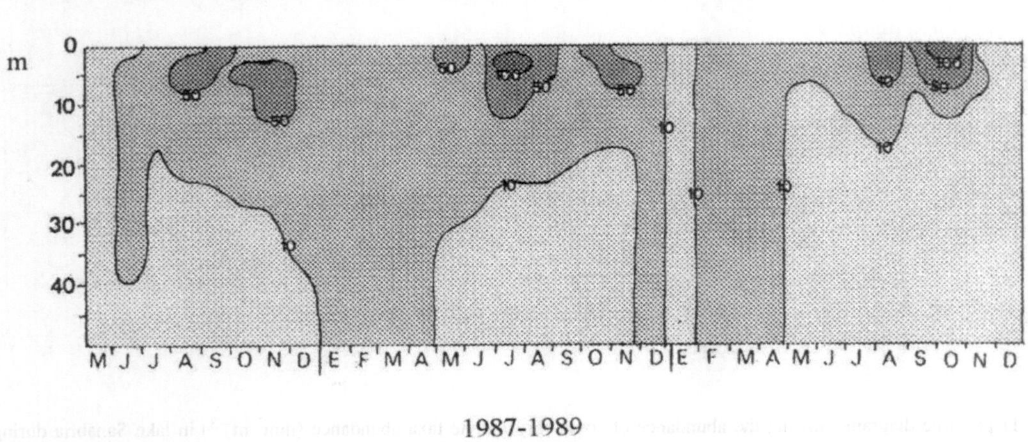

Figure 4. Depth-time diagram of chrysophyte abundance (mm^3 m^{-3}) in Lake Sanabria during 1987–89.

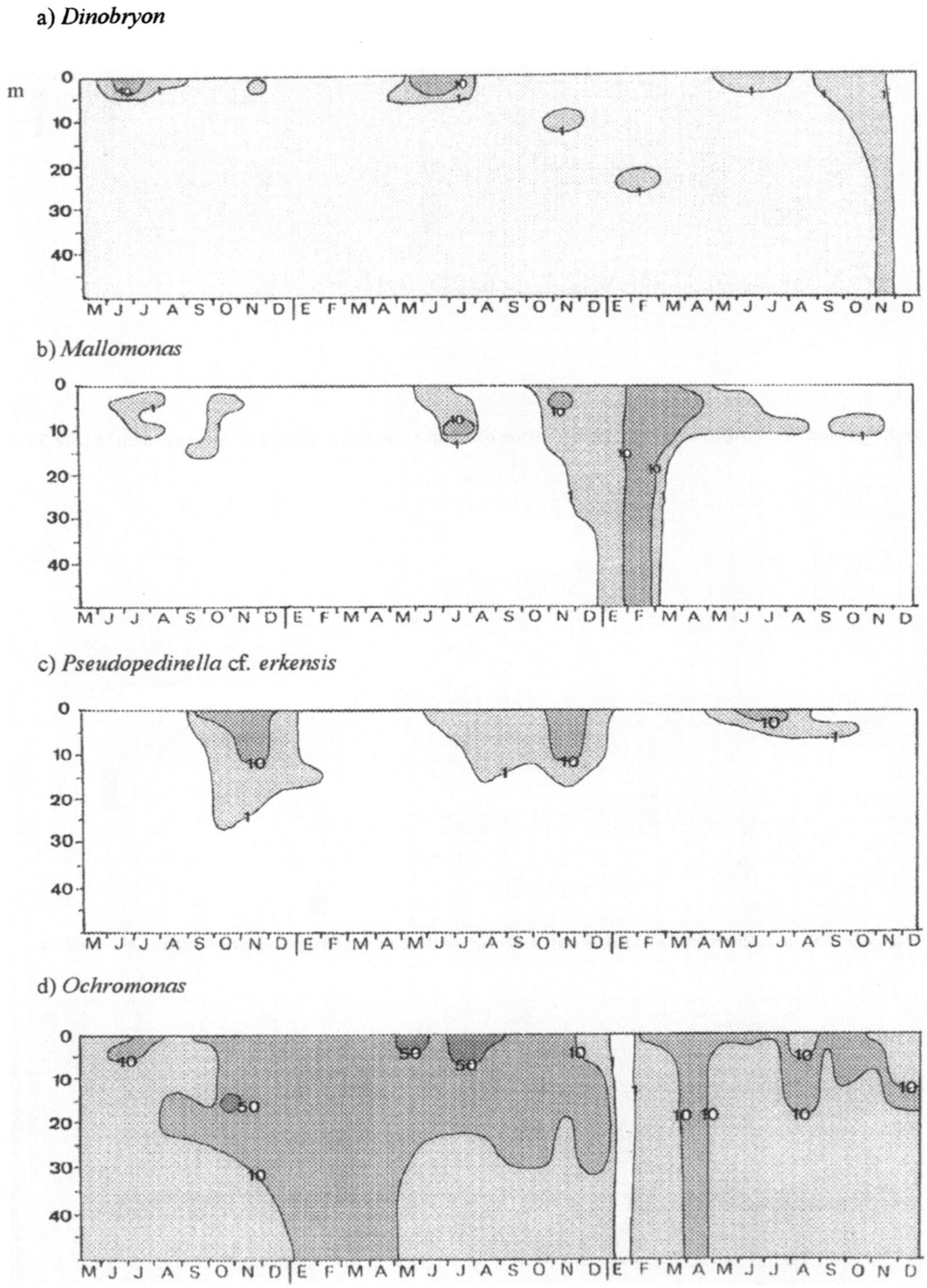

Figure 5. Depth-time diagram showing the abundance of some chrysophyte taxa abundance ($mm^3\ m^{-3}$) in lake Sanabria during 1987–89. (*Dinobryon, Mallomonas* and *Ochromonas* includes all the species of the genera).

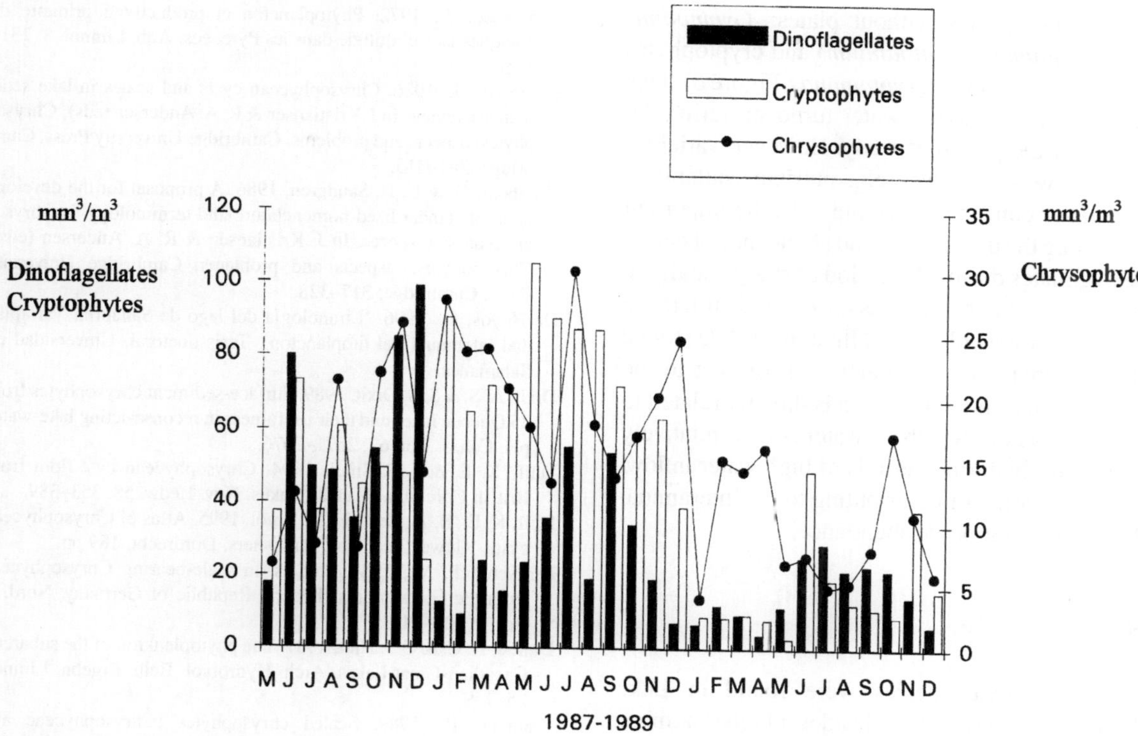

Figure 6. Seasonal changes in biovolume of dinoflagellates, cryptophytes and chrysophytes (integrated mean of the water column) in Lake Sanabria.

As a group, chrysophytes reached biovolume peaks between 50 and 100 mm^3 m^{-3} in summer (1987 and 1988), and autumn (Figure 4). These peaks were observed in the upper 10 m of the water column. In the rest of the water column, abundances were higher in winter than in summer (Figure 4). This pattern of seasonal distribution has been also observed in lakes of Lapponia, Alps and the Pyrenees (Nauwerck, 1966; Capblancq, 1972). The preferential distribution of chrysophytes in layers of the water column of Lake Sanabria with low nutrient availability and high organic matter content (De Hoyos, unpublished data) may be related to mixotrophic capabilities of several species (Sandgren, 1988).

Pseudopedinella cf. *erkensis*, the species with the most restricted spatial-temporal distribution in Lake Sanabria (Figure 5), was limited to the upper 20 m of the water column in summer 1989 and in the autumns of 1987 and 1988. *Ochromonas* (Figure 5) had the widest distribution both in space and time. It was always present along the whole water column abounding during the summer. *Ochromonas* genus accounted for most of the chrysophyte abundance during 1988, a year with a higher water turnover rate than the other two years of the study. In small peatland lakes located in N. Canada, two biomass peaks, both of them in summer, have been observed: one was caused by *Ochromonas* sp., the other by *P. erkensis* (Sheath & Hellebust, 1978).

Dinobryon and *Mallomonas* presented a similar spatial-temporal distribution (Figure 5, each figure includes all the species of the genera), characterized by two epilimnetic peaks, one in summer, the other in autumn. *Dinobryon* was preferently located in the upper parts of the epilimnion and *Mallomonas* occupied the deepest layers of it.

Other chrysophyte species (*Bitrichia ochridana* and *Kephyrion rubri-claustri*) are scarce and found only sporadically; their total biovolume never exceeded 1 mm^3 m^{-3}.

De Hoyos (1996) and De Hoyos & Comín (submitted) describe the seasonal interannual variability of physical and chemical characteristics of Lake Sanabria and demonstrate that the water turnover is a key factor regulating the seasonal changes of phytoplankton. Chrysophytes were more abundant during 1987–88 than during 1988–1989, and the same difference was observed in the temporal distribution of dinoflagellates (especially the species without plates: *Gymnodinium, Amphidinium* and *Katodinium*) and cryptophytes (especially the species of *Cryptomonas*, Figure 6). This may be related to a higher water turnover during the earlier part of the period of study. However, variables related to the water turnover (e.g., nutrient availability, organic matter concentration), may also be important in determining the distribution and abundance of chrysophyte flagellates during the period of study. Sandgren (1988) found that chrysophytes have a wide tolerance with regard to water colour, and Ilmavirta (1983) found that the percentage of flagellates with respect to the total abundance of phytoplankton is directly related to water colour. In Lake Sanabria, water color is relatively high, particularly during periods of high water inflow, which is another factor contributing to the interannual variability of chrysophyte abundance.

Conclusions

Fifteen chrysophyte species were found in Lake Sanabria during 1987–1989. Besides a higher number of cist forms (27) were found in the surface sediment. These species were reported in other oligotrophic lakes located in mountains or at high northern latitudes. Lake Sanabria provides good general environmental conditions for chrysophyte development in the epilimnion during summer stratification: slightly acid waters, low phosphorus concentrations and relatively high organic matter content. However, chrysophyte abundance, both absolute and relative to total phytoplankton abundance, is very low compared to those observed in other lakes. This may be related to the scarcity of nitrogen in Lake Sanabria compared to other oligotrophic lakes where chrysophytes are abundant and where phosphorus is the major nutrient controlling phytoplankton growth.

References

Aldasoro, J. J., C. de Hoyos & J. C. Vega, 1992. El lago de Sanabria. Estudio limnológico. Monografias de la Red de Espacios Naturales de Castilla y León. Junta de Castilla y León. Consejería de Medio Ambiente y Ordenación del Territorio, 135 pp.

APHA, 1989. Standard methods for the examination of Water, sewage and Wastewater. 17th ed. American Public Health Association, Washington, D.C., 1550 pp.

Asmund, B. & J. Kristiansen, 1986. The genus *Mallomonas* (Chrysophyceae). Opera Bot. 85: 5–128.

Calado, A. J. & S. C. Craveiro, 1995. Notes on the ecology of Synurophycean algae found in Portugal. Nord. J. Bot. 15: 641–654.

Capblancq, J., 1972. Phytoplancton et productivité primaire de quelques lacs d'altitude dans les Pyrénées. Ann. Limnol. 8: 231–321.

Cronberg, G., 1986. Chrysophycean cysts and scales in lake sediments: a review. In J. Kristiansen & R. A. Andersen (eds), Chrysophytes: aspects and problems. Cambridge University Press, Cambridge: 281–315.

Cronberg, G. & C. D. Sandgren, 1986. A proposal for the development of standardized nomenclature and terminology for chrysophycean statospores. In J. Kristiansen & R. A. Andersen (eds), Chrysophytes: aspects and problems. Cambridge University Press, Cambridge: 317–328.

De Hoyos, C., 1996. 'Limnología del lago de Sanabria. Variabilidad interanual del fitoplancton'. Tesis doctoral, Universidad de Salamanca.

Dixit, A. S. & S. S. Dixit, 1989. Surface-sediment chrysophytes from 35 Quebec lakes and their usefulness in reconstructing lake-water pH. Can. J. Bot. 67: 2071–2076.

Duff, K. E. & J. P. Smol, 1994. Chrysophycean cyst flora from British Columbia (Canada) lakes. Nov. Hedw. 58: 353–389.

Duff, K. E., B. A. Zeeb & J. P. Smol, 1995. Atlas of Chrysophycean cysts. Kluwer Academic Publishers, Dordrecht, 189 pp.

Dürrschmidt, M., 1984. Studies on scale-bearing Chrysophyceae from the Giessen area, Federal Republic of Germany. Nord. J. Bot. 4: 123–143.

Duthie, H. C. & C. J. Hart, 1987. The phytoplankton of the subarctic Canadian Great Lakes. Arch. Hydrobiol. Beih. Ergebn. Limnol. 25: 1–9.

Eloranta, P., 1989. Scaled chrysophytes (Chrysophyceae and Synurophyceae) from national park lakes in souther and central Finland. Nord. J. Bot. 8: 671–681.

Eloranta, P., 1995. Phytoplankton of the national park lakes in central and southern Finland. Ann. Bot. Fenn. 32: 193–209.

Harris, G. P., 1986. Phytoplankton ecology. Chapman and Hall, London, 384 pp.

Hutchinson, G. E., 1967. *A treatise on Limnology*. Volumen II. Introduction to lake biology and the limnoplankton (ed.), John Wiley & Sons, 1114 pp.

Ilmavirta, V., 1980. Phytoplankton in 35 finnish brown-water lakes of different trophic status. Dev. Hydrobiol. 3: 121–130.

Ilmavirta, V., 1983. The role of flagellated phytoplankton in chains of small brown-water lakes in southern Finland. Ann. Bot. Fenn. 20: 187–195.

Järnefelt, H., 1952. Plankton als Indikator der Trophiegruppen der Seen. Ann. Acad. Sci. Fenn. (A IV)18: 1–29.

Jeffrey, S. W. & G. F. Humphrey, 1975. New spectrophotometric equations for determining chlorophylls a, b, c1 and c2 in higher plants, algae and natural phytoplankton. Biochem. Physiol. Pflanzen. 167: 191–194.

Kling, H. & J. Kristiansen, 1983. Scale-bearing Chrysophyceae (Mallomonadaceae) from Central and Northern Canada. Nord. J. Bot. 3: 269–290.

Kristiansen, J., 1986. Silica-scale bearing Chrysophytes as environmental indicators. Br. phycol. J. 21: 425–436.

Moore, J. M., 1978. Distribution and abundance of phytoplankton in 153 lakes, rivers and pools in the Northwest Territories. Can. J. Bot. 56: 1765–1773.

Nauwerck, A., 1966. Beobachtungen über das Phytoplankton klarer Hochgebirgsseen. Schweiz. Z. Hydrol. 28: 4–28.

Nicholls, K. H., 1990. The taxonomy of problematic *Mallomonas* species (Synurophyceae). II Mallomonas costata and M. galeiformis. Nord. J. Bot. 10: 243–247.

Nygaard, G., 1978. Freshwater phytoplankton from the Narssaq Area, South Greenland. Bot. Tidsskr. 73: 191–238.

Pearsall, W. H., 1932. Phytoplankton in the English lakes. II. The composition of the phytoplankton in relation to dissolved substances. J. Ecol. 20: 241–262.

Reynolds, C. S., 1984. The ecology of freshwater phytoplankton. Cambridge University Press, 384 pp.

Rodhe, W., 1948. Environmental requirements of fresh-water plankton algae. Symbolae Botanicae Upsalienses 10, 149 pp.

Rosén, G., 1981. Phytoplankton indicators and their relations to certain chemical and physical factors. Limnologica 13: 263–290.

Sandgren, C. D., 1988. The ecology of chrysophyte flagellates: their growth and perennation strategies as freshwater phytoplankton. In: C. D. Sandgren (ed.), Growth and reproductive strategies of freshwater phytoplankton. Cambridge University Press: 9–104.

Sandgren, C. D. & H. J. Carney, 1983. A flora of fossil chrysophycean cysts from the recent sediments of Frains lake, Michigan, USA. Nova Hedwigia 38: 129–165.

Sheath, R. G. & J. A. Hellebust, 1978. Comparison of algae in the euplankton, tychoplankton and periphyton of a tundra pond. Can. J. Bot. 56: 1472–1483.

Siver, P. A., 1989. The distribution of scaled chrysophytes along a PH gradient. Can J. Bot. 67: 2120–2130.

Siver, P. A., 1991. The Biology of *Mallomonas*: Morphology, Taxonomy and Ecology. Kluwer Academic Publishers, Dordrecht, 230 pp.

Siver, P. A., 1995. The distribution of chrysophytes along environmental gradients: their use as biological indicators. In: C. D. Sandgren, J. P. Smol & J. Kristiansen (eds), Chrysophyte Algae. Ecology, Phylogeny and Development. Cambridge University Press: 232–268.

Siver, P. A., J. S. Hamer & H. Kling, 1990. Separation of *Mallomonas duerrschmidtiae* sp. nov. from *M. crassisquama* and *M. pseudocoronata*: implications for paleolimnological research. J. Phycol. 26: 728–740.

Smol, J. P., 1984. The statospore of *Mallomonas pseudocoronata* (Mallomonadaceae, Chrysophyceae). Nord. J. Bot. 4: 827–831.

Smol, J. P., 1988. The North-American 'Endemic' *Mallomonas pseudocoronata* (Mallomonadaceae, Chrysophyta) in an Austrian Lake. Phycologia 27: 427–429.

Smol, J. P., D. F. Charles & D. R. Whitehead, 1984a. Mallomonadacean microfossils provide evidence of recent lake acidification. Nature 307: 628–630.

Smol, J. P., D. F. Charles & D. R. Whitehead, 1984b. Mallomonadacean (Chrysophyceae) assemblages and their relationships with limnological characteristics in 38 Adirondack (New York) lakes. Can. J. Bot. 62: 911–923.

Vega, J. C., C. de Hoyos & J. J. Aldasoro, 1992. The Sanabria lake. The largest natural freshwater lake in Spain. Limnetica 8: 49–57.

Vilaseca, J. M., 1978. Fitoplancton de los lagos pirenaicos. Unpublished. Tesina de Licenciatura. Departamento de Ecología, Universidad Central de Barcelona, 102 pp.

Hydrobiologia **369/370:** 297–313, 1998.
M. Alvarez-Cobelas, C. S. Reynolds, P. Sanchez-Castillo & J. Kristiansen (eds), Phytoplankton and Trophic Gradients.

Taxonomic remarks on planktonic phytoflagellates in a hypertrophic tropical lagoon (Brazil)

Patrícia Domingos[1] & Mariângela Menezes[2]
[1] *Laboratório de Cultivo e Fisiologia de Microalgas, NPPN, CCS, Bloco H, Universidade Federal do Rio de Janeiro, Ilha do Fundão, 21941-590, Rio de Janeiro, Brazil*
[2] *Laboratório de Ficologia, Departamento de Botânica, Museu Nacional, Universidade Federal do Rio de Janeiro, Quinta da Boa Vista, São Cristovão, 20940-040, Rio de Janeiro, Brazil*

Key words: planktonic phytoflagellates, taxonomy, seasonality, hypertrophic brackish water, tropical coastal lagoon, Brazil

Abstract

The phototrophic flagellate flora of the Barra Lagoon (22°57′ S and 42°47′ W), Rio de Janeiro State, southeastern Brazil, is described. The lagoon is a shallow, oligohaline, hypertrophic ecosystem in which Cyanophyceae, Chlorophyceae and phytoflagellates are dominant. Between October 1990 and August 1993, weekly integrated samples were collected from a bay, using a plexiglass tube, 8 cm diameter and 1m length. All identifications, descriptions and measurements were carried out on live or in Lugol-fixed material. A total of 30 taxa, belonging to Euglenophyceae (5), Cryptophyceae (7), Raphidophyceae (1), Dinophyceae (4), Chrysophyceae (3), Prymnesiophyceae (2), Prasinophyceae (7) and Chlorophyceae (1) were identified. All the phytoflagellate species identified have been recorded previously from temperate sites. Nine of them were registered in a tropical water for the first time, suggesting that they have much broader distributions. *Gymnodinium sanguineum*, *Pavlova lutheri* and *Pyramimonas grossii* were the most frequently observed phytoflagellates in this study. The seasonal occurrence and some environmental requirements of selected phytoflagellates species are discussed.

Introduction

In brackish waters, phytoflagellates are of great interest since their presence is characteristic. However, the complexity and the restricted representation in the algal communities of brackish environments, relative to those of other ecosystems, increase the difficulties of defining the biological group (Caljon, 1983). Although there are few descriptive accounts on phytoplankton in Brazilian coastal lagoon systems, a significant diversity and biomass of phytoflagellates has been documented from brackish waters in the coastal region of Rio de Janeiro State (Huszar & Silva, 1992; Menezes & Domingos, 1994). These ecosystems show symptoms of accelerated eutrophication, mainly anthropogenic in origin (Carmouze et al., 1995).

In the course of an ecological investigation of Barra Lagoon, a tropical coastal system in the southeastern Brazil, we carried out a taxonomic survey of the planktonic algal flora. So far, only those species recorded for the first time in Brazil have been described (Menezes & Domingos, 1994).

This paper offers a more comprehensive account of the phototrophic flagellates of the Barra Lagoon. Moreover we discuss the abundance of the more common phytoflagellate species in relation to some of the environmental variables.

Study site

The Barra Lagoon is a coastal system situated in the northern part of Rio de Janeiro State, southeastern Brazil (22°57′ S, 42°47′ W). It is a shallow, oligohaline lagoon, strongly impacted by the discharge of domestic sewage, with nitrogen and phosphorous concentrations

Table 1. Some limnological data of Barra Lagoon. * Average values for 1990–1993

Surface area	6.2 km^2
Mean depth	1 m
Retention time	45 m
Mixing frequency	Continuous to daily
TN	1527 $\mu g\,l^{-1}$ *
pH	8.2*
Salinity	5.4‰ *
DIN	101.5 $\mu g\,l^{-1}$ *
TP	136.5 $\mu g\,l^{-1}$ *
DOM	11.5 $\mu g\,l^{-1}$ *

increasing in consequence. According to the OECD classification (1982), it is a hypertrophic system; like many others, it experiences high pH values associated with episodes of high primary production (Domingos, unpublished data). The Barra Lagoon is indirectly connected with the sea. A small salinity fluctuation was detected during the period of study is indicative of a weak tidal influence. The phytoplankton is dominated by the Cyanophyceae, *Synechocystis aquatilis* Sauvageau f. *salina* (Wislouch) Komárek and *Synechococcus elongatus* Nägeli, which may produce toxins resulting in summer fish kills, as observed in 1991 (Azevedo et al., 1994). From September to December 1992, the lagoon was subject to a large development of macroalgae, mainly of *Cladophora vagabunda* (Linnaeus) Hoek and *Enteromorpha* sp.

The limnological characters of Barra Lagoon are summarised in Table 1.

Materials and methods

Integrated water samples were collected weekly for three years, from October 1990 to August 1993 at one bay of lagoon, using a plexiglass tube, 8 cm in diameter and 1m in length. For the qualitative analysis, the samples were carried to the laboratory and kept unpreserved under refrigeration (±4 °C). For the quantitative analyses, subsamples were fixed immediately with Lugol's solution. All identifications, descriptions, measurements and drawings were made on live and fixed material under light microscope with phase-contrast equipment. The techniques for phytoflagellate identification followed those described in Menezes (1994).

Phytoplankton samples were counted according to Utermöhl (1958), and the biovolumes were calculated according to Edler (1979). Correlation analysis between fourteen selected species and some environmental variables (nitrate, DOC, DON, DOP, TN:TP ratio, SRP, TP, TN, and Secchi depth) used Pearson-'s statistic (Sokal & Rolf, 1981) with N = 115 and $p < 0.05$.

The classification system used was taken essentially from Throndsen (1993). In the classification of Dinophyceae we adopted the criteria of Steidinger & Tangen (1996). The salinity tolerances follow the criteria of Caljon (1983).

The identification and geographical distribution of taxa have been based on the general and regional works for freshwater, brackish and marine waters and on morphological characters observed under light microscopy.

Results

Floristic composition

In the plankton of Barra Lagoon, phototrophic flagellates were represented by 30 taxa accounting 37% of total phytoplankton (Figure 1). When considering the main taxonomic phytoflagellates classes, Prasinophyceae, Cryptophyceae and Euglenophyceae were the most species-rich, followed by Dinophyceae, Chrysophyceae and Prymnesiophyceae; Raphidophyceae and Chlorophyceae were the least represented (Figure 2).

With regard to the salinity tolerance of the phytoflagellates, limnobiont species dominated in number, followed by brackish and marine species; only one taxon (*Durinskia baltica* (Levander) Carty & Cox) was classified as holoeurihaline (Figure 3). Most freshwater taxa recorded occurred sporadically or rarely in the lagoon, especially species of Chrysophyceae, Euglenophyceae and the single Chlorophyte, *Selenochloris stigmatophora* (Korsikov) Ettl, with two individuals found in one sample in September, 1992. However, some taxa which usually live in freshwater were common throughout the period investigated, such as *Euglena proxima* Dangeard, *Cryptomonas erosa* Ehrenberg, *Ochromonas mutabilis* Klebs, *Kephyrion ovale* (Lackey) Huber-Pestalozzi and *Pedinomonas minutissima* Skuja. Prasinophyceae, Cryptophyceae, Dinophyceae and Prymnesiophyceae were the most common groups represented; most species were typical from brackish (*Pyramimonas orientalis* Butcher, *Hemiselmis simplex* Butcher, *Gymnodinium sanguineum* Hirasaka, *Pavlova lutheri* (Droop) Green) or marine environments (*Resultor mikron* Throndsen,

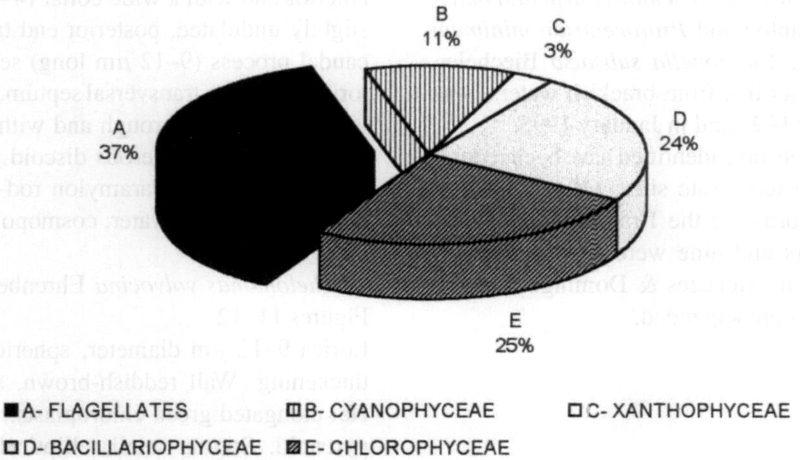

Figure 1. Phytoplankton species richness (relative proportion of taxa) in Barra Lagoon, 1990–1993.

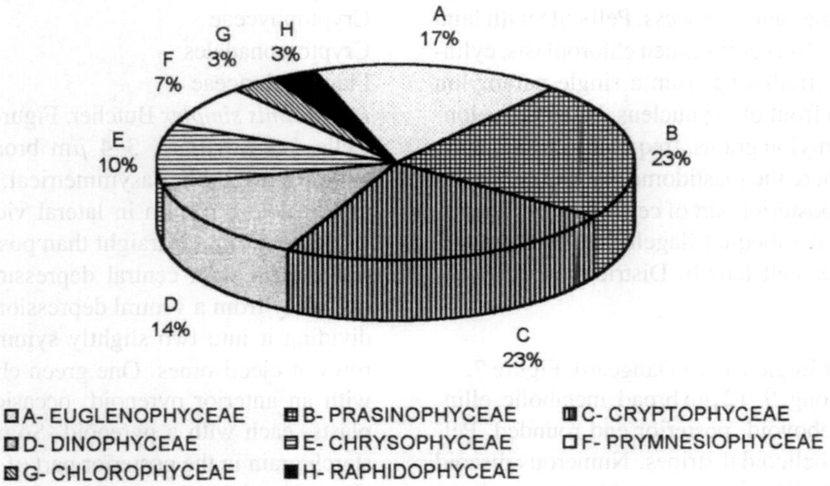

Figure 2. Phytoflagellate species richness (relative proportion of taxa) in Barra Lagoon, 1990–1993.

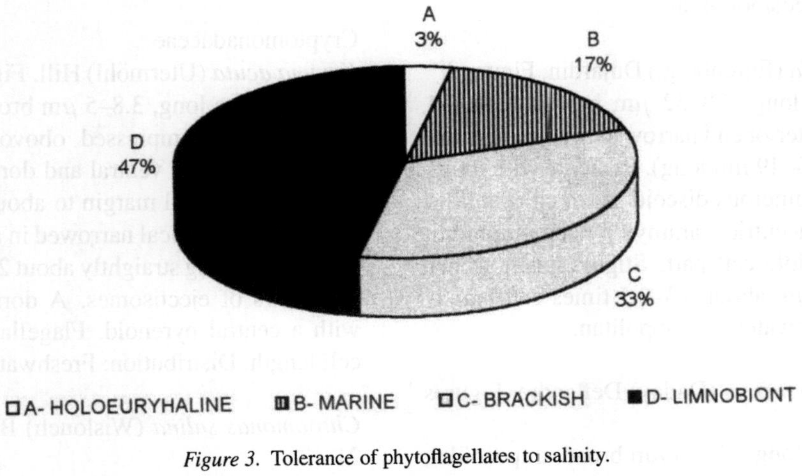

Figure 3. Tolerance of phytoflagellates to salinity.

Pyramimonas grossii Parke, *Pseudoscourfieldia marina* (Throndsen) Manton and *Prorocentrum minimum* (Pavillard) Schiller. *Chattonella subsalsa* Biecheler, which had been described from brackish waters, was found in December 1992 and in January 1993.

All phytoflagellate taxa identified have been recorded previously from temperate sites. Of those, eighteen were new records for the Brazilian continental and brackish waters and nine were first records for any tropical location (Menezes & Domingos, 1994). Species descriptions are appended.

Euglenophyceae
Euglenales
Eutreptiaceae
Eutreptia viridis Perty. Figures 4–6.
Cells 40–53 μm long, 4–9 μm broad, strongly metaboly, fusiform to elliptical, posterior end narrowed in an acute, hyaline caudal process. Pellicule with faint helicoidal stripes. Numerous green chloroplasts, cylindrical to oblong irradiating from a single paramylon center situated in front of the nucleus. Numerous elongated ovoid paramylon grains, frequently closed in the central region where the plastidome converges. Nucleus oblong in the posterior part of cell; stigma elliptical, conspicuous. Two subequal flagella, heterodynamic, about 1–1,5 times cell length. Distribution: Coastal; cosmopolitan.

Euglena proxima Euglenaceae Dangeard. Figure 7.
Cells 53–62 μm long, 9–12 μm broad, metabolic, elliptical to slightly obovoid, posterior end rounded. Pellicule with faint helicoidal stripes. Numerous discoid green chloroplasts. Elongated paramylon grains scant. Nucleus elliptical in the cell middle; stigma small. Flagellum about 0.5–0.7 times cell length. Distribution: Freshwater; cosmopolitan.

Phacus longicauda (Ehrenberg.) Dujardin. Figure 8.
Cells 92–96 μm long 28–32 μm broad, flattened, heart–shaped, posterior end narrowed in a long hyaline caudal process (45–49 μm long). Pellicule with longitudinal stripes. Numerous discoid green chloroplasts. Two rounded concentric paramylon grains approximately in the middle cell part. Stigma has not been observed. Flagellum about 0.3–0.5 times cell length. Distribution: Freshwater; cosmopolitan.

Strombomonas verrucosa (Daday) Deflandre. Figures 9–10.
Lorica 35–39 μm long, 17–23 μm broad, trapezoidal, anterior end with a wide collar (4–6 μm broad), sides slightly undulated, posterior end tapered to a conical caudal process (9–12 μm long) separeted from main lorica body by a transversal septum. Wall thick, yellowish to dark brown, rough and with adhering particles on its surface. Numerous discoid green chloroplasts, without pyrenoid. Paramylon rod-shaped or discoid. Distribution: Freshwater; cosmopolitan.

Trachelomonas volvocina Ehrenberg. var. *volvocina*. Figures 11–12.
Lorica 9–12 μm diameter, spherical, pore with ring thickening. Wall reddish-brown, smooth. Two parietal elongated green chloroplasts, each with an inner pyrenoid. Stigma small, elliptical. Flagellum 1.5–2 times diameter of the lorica. Distribution: Freshwater; cosmopolitan.

Cryptophyceae
Cryptomonadales
Planonephraceae
Hemiselmis simplex Butcher. Figures 13–17.
Cells 4–9 μm long, 3–4 μm broad, 3–4 μm thick, strongly metabolic, asymmetrical, flattened laterally, elliptical to reniform in lateral view, rounded poles, anterior end more straight than posterior, convex ventral margin with central depression. Gullet extends obliquely from a ventral depression to the central cell dividing it into two slightly symmetrical parts. Two rows of ejectisomes. One green chloroplast, parietal, with an anterior pyrenoid, occasionally two chloroplasts, each with a pyrenoid. Sometimes a spherical starch grain in the posterior part of cell. Flagella laterally inserted which extend forwards when the cell is in rest, 0.5–1.5 times cell length. Distribution: Coastal; Europe, Japan, Brazil.

Cryptomonadaceae
Komma acuta (Utermöhl) Hill. Figures 18–19.
Cells 6–8 μm long, 3.8–5 μm broad, 3.5–4 μm thick, dorsiventrally compressed, obovoid in lateral, ventral and dorsal views, ventral and dorsal margins convex, the convex ventral margin to about 2/3 of cell length, posterior end conical narrowed in a curved hyaline tip. Gullet extending straightly about 2/3 of the cell length. Two rows of ejectisomes. A dorsal blue chloroplast with a central pyrenoid. Flagella 0.6 times to about cell length. Distribution: Freshwater; cosmopolitan.

Chroomonas salina (Wislouch) Butcher. Figures 20–21.

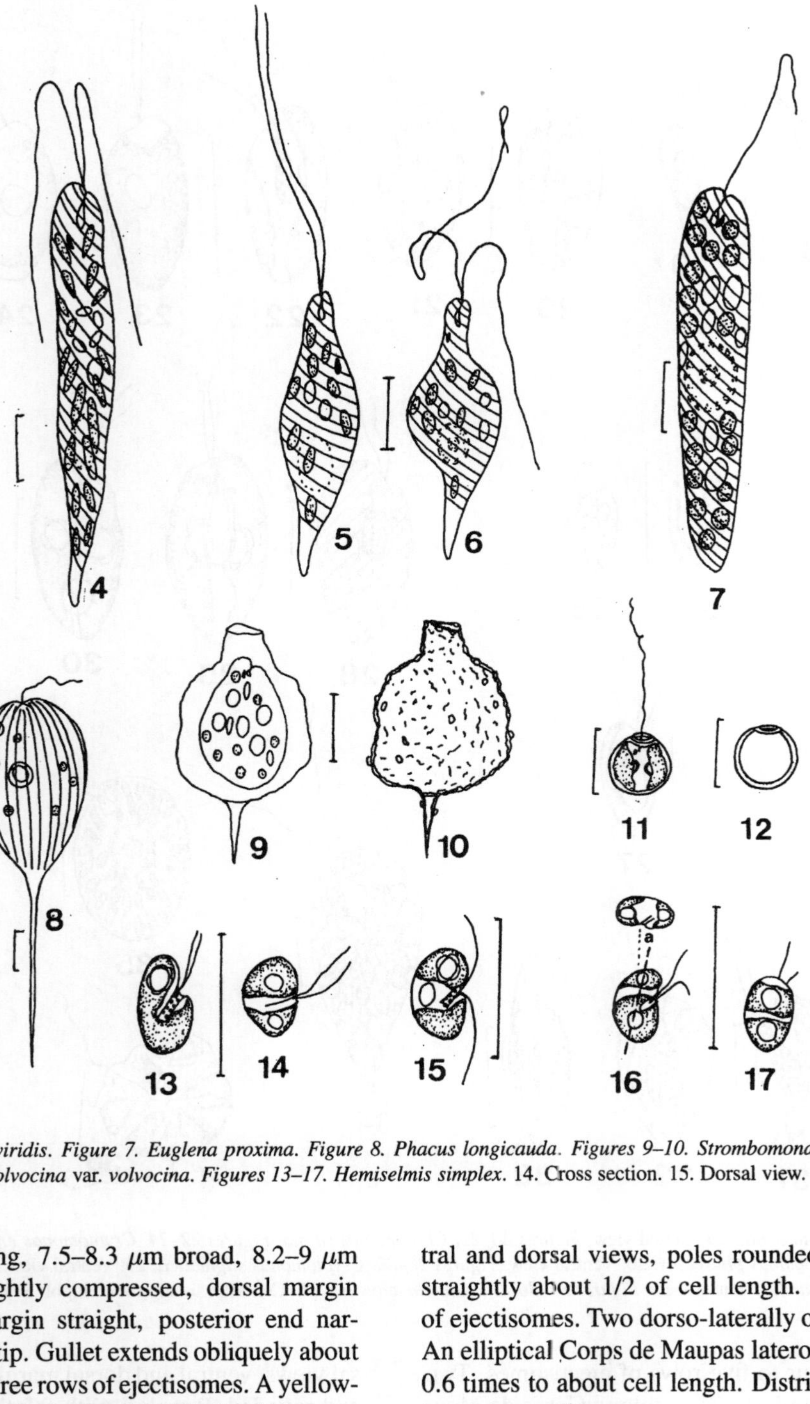

Figures 4–6. Eutreptia viridis. Figure 7. Euglena proxima. Figure 8. Phacus longicauda. Figures 9–10. Strombomonas verrucosa. Figures 11–12. Trachelomonas volvocina var. *volvocina. Figures 13–17. Hemiselmis simplex.* 14. Cross section. 15. Dorsal view. Scales = 10 μm.

Cells 15–18 μm long, 7.5–8.3 μm broad, 8.2–9 μm thick, obovoid, slightly compressed, dorsal margin convex, ventral margin straight, posterior end narrowed to a rounded tip. Gullet extends obliquely about 1/2 of cell. Two or three rows of ejectisomes. A yellow-olive parietal chloroplast, margins lobed with a central pyrenoid. Flagella about cell length. Distribution: Coastal; cosmopolitan.

Cryptomonas erosa Ehrenberg. Figures 22–24.
Cells 10–15 μm long, 4.5–7 μm broad, 4–6 μm thick, laterally flattened, obovoid to elliptical in lateral, ventral and dorsal views, poles rounded. Gullet extends straightly about 1/2 of cell length. Five to six rows of ejectisomes. Two dorso-laterally olive chloroplasts. An elliptical Corps de Maupas laterodorsally. Flagella 0.6 times to about cell length. Distribution: Freshwater; cosmopolitan.

Cryptomonas phaseolus Skuja. Figures 25–27.
Cells 18–20 μm long, 8–10 μm broad, 8–10 μm thick, dorsiventrally flattened, obovoid in lateral view, elliptical in ventral and dorsal views, convex margins, posterior end rounded. Gullet extends obliquely about 1/2

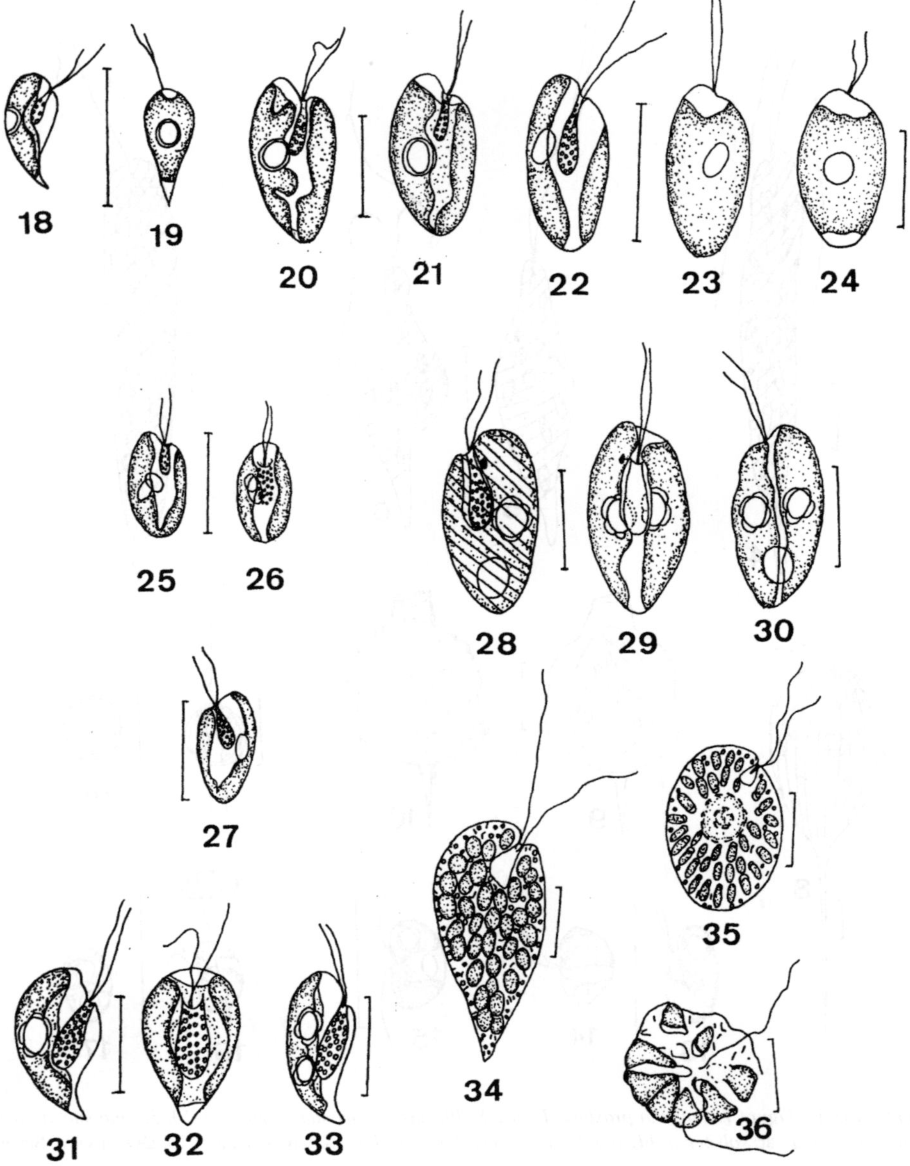

Figures 18–19. Komma acuta. 19. Dorsal view. *Figures 20–21. Chroomonas salina. Figures 22–24. Cryptomonas erosa.* 23–24. Dorsal view. *Figures 25–27. Cryptomonas phaseolus.* 26. Ventral view. *Figures 28–30. Cryptomonas stigmatica.* 29. Ventral view. 30. Dorsal view. *Figures 31–33. Teleaulax acuta.* 32. Ventral view. *Figures 34–36. Chattonella subsalsa.* 36. Cell after fixation with lugol. Scales = 10μm.

of cell length. Four to five rows of ejectisomes. Two lateral olive chloroplasts. Two oblong Corps de Maupas. Flagella about 0.5 times cell length. Distribution: Freshwater; cosmopolitan.

Cryptomonas stigmatica Wislouch. Figures 28–30. Cells 18–21 μm long, 9–10 μm broad, 8–9 μm thick, laterally flattened, obovoid in lateral, ventral and dorsal views, ventral and dorsal margins convex, posterior end rounded. Periplast with faintly helicoidal stripes. Gullet extends straightly about 1/2 cell length. Five to six rows of ejectisomes. Two lateral olive chloroplasts, each with a pyrenoid surrounded by a sheath of 3–4 starch grains. A basal starch grain. Stigma elliptical, conspicuous. Flagella about cell length. Distribution: Coastal; cosmopolitan.

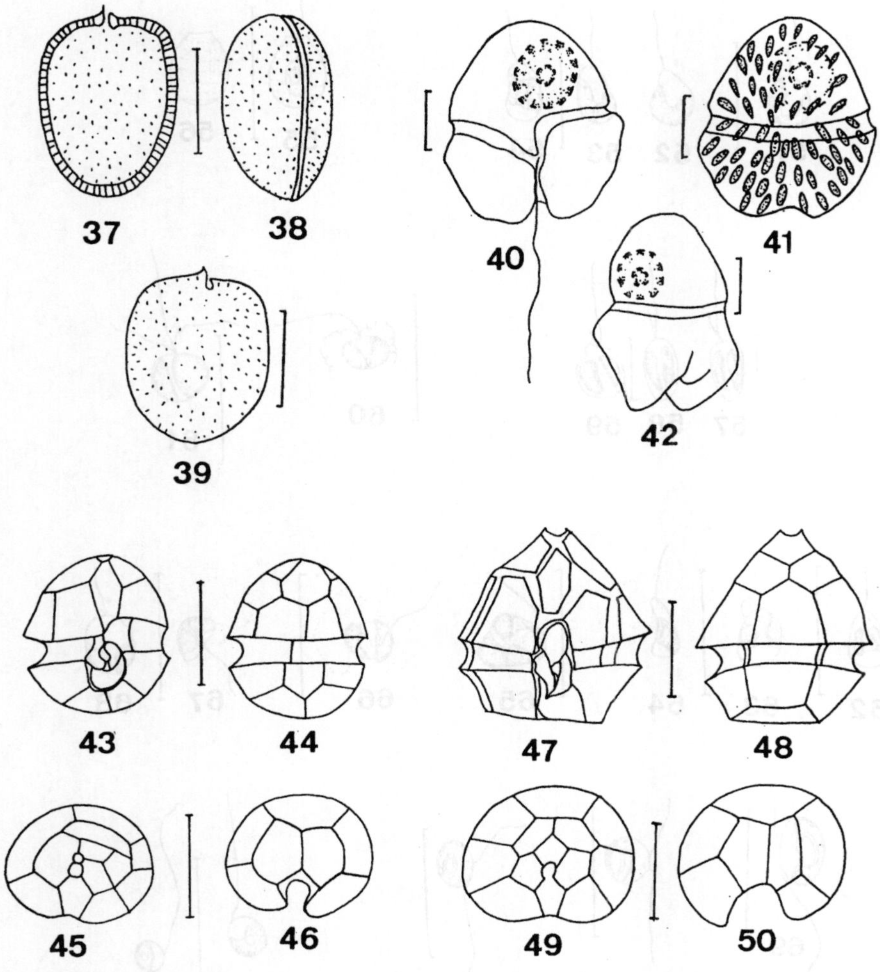

Figures 37–39. Prorocentrum minimum. Figures 40–42. Gymnodinium sanguineum. 40. Ventral view. 41 Dorsal view. 42. Laterodorsally view. *Figures 43–46. Durinskia baltica.* 43. Ventral view. 44. Dorsal view. 45. Apical view of epitheca. 46. Apical view of hypotheca. *Figures 47–50. Peridinium umbonatum.* 47. Ventral view. 48. Dorsal view. 49. Apical view of epitheca. 50. Apical view of hypotheca. Scales = 10μm.

Teleaulax acuta (Butcher) Hill. Figures 31–33.
Cells 16–18 μm long, 8–10 μm broad, 6–9 μm thick, flattened laterally, obovoid in lateral and ventral views, elliptical in dorsal view, convex ventral margin to about 2/3 of cell length, posterior end narrowed to an acute blunt short hyaline curved tail. Gullet extends obliquely, longer than 1/2 of cell length. Four to five rows of ejectisomes. A brown dorsal chloroplast with a central pyrenoid surrounded by a sheath of 3–4 cup shaped starch grains, sometimes two chloroplasts, each with a pyrenoid. Flagella about cell length. Distribution: Coastal, off-shore; England, Norway, Denmark, Japan, Brazil.

Raphidophyceae
Chattonellales
Chattonellaceae
Chattonella subsalsa Biecheler. Figures 34–36.
Cells 28–38 μm long, 14–18 μm broad, very fragile, laterally flattened, obovoid to obpyriform, margins convex or rounded, posterior end frequently narrowed in an acute hyaline tip. Subapical ventral notch extended by an oblique groove. Numerous lenticular yellow-green chloroplasts in the peripheral cytoplasm, each with a pyrenoid. Small refringents lipid globules peripherically situated. A large central nucleus. Numerous elongated mucocysts in the peripheral cytoplasm. Flagella heterodynamic, about 1.5 times cell

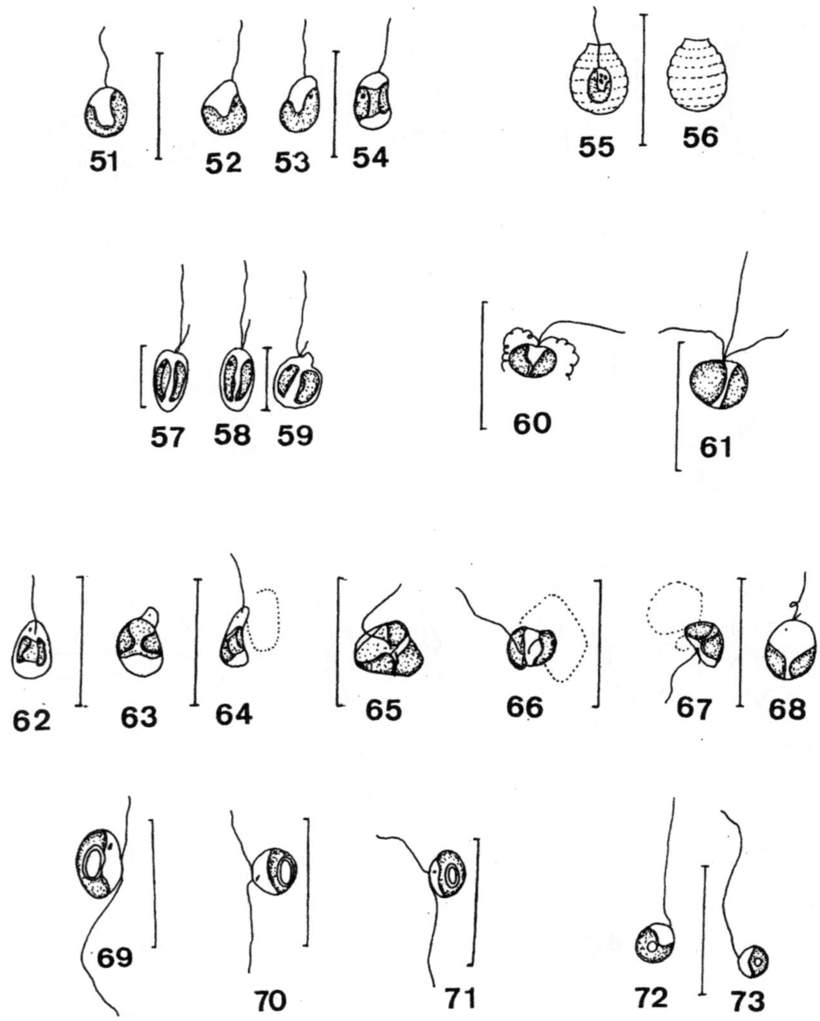

Figures 51–54. Chromulina pygmea. Figures 55–56. Kephyrion ovale. Figures 57–59. Ochromonas mutabilis. Figures 60–61. Chrysochromulina parva. Figures 62–68. Pavlova lutheri. 64, 66, 67. Cell with periplast separated, spotted line. *Figures 69–71. Nephroselmis minuta. Figures 72–73. Pedinomonas minutissima.* Scales = 10μm.

length, the anterior flagellum undulated, the recurrent flagellum lying in the ventral groove extension. Distribution: Coastal; France, Spain, South India, Brazil.

Dinophyceae
Prorocentrales
Prorocentraceae
Prorocentrum minimum (Pavillard) Schiller. Figures 37–39.
Cells 15–17.5 μm long, 11–15 μm broad, heart-shape to ovoid, anterior end truncated with a weak short spine, posterior end rounded. Theca ornamented with short spines. Distribution: Coastal, oceanic; cosmopolitan.

Peridiniales
Gymnodiniaceae
Gymnodinium sanguineum Hirasaka. Figures 40–42.
Cells 33–36 μm long, 25–33 μm broad, laterally flattened, ovoid. Epicone conical, rounded, hypocone trapezoidal, bilobed. Cingulum large, helicoidal, displaced about 0.2 times cell length. Sulcus does not reach the epicone, deeply notches hypocone. Numerous elongated yellowish green chloroplasts disposed in

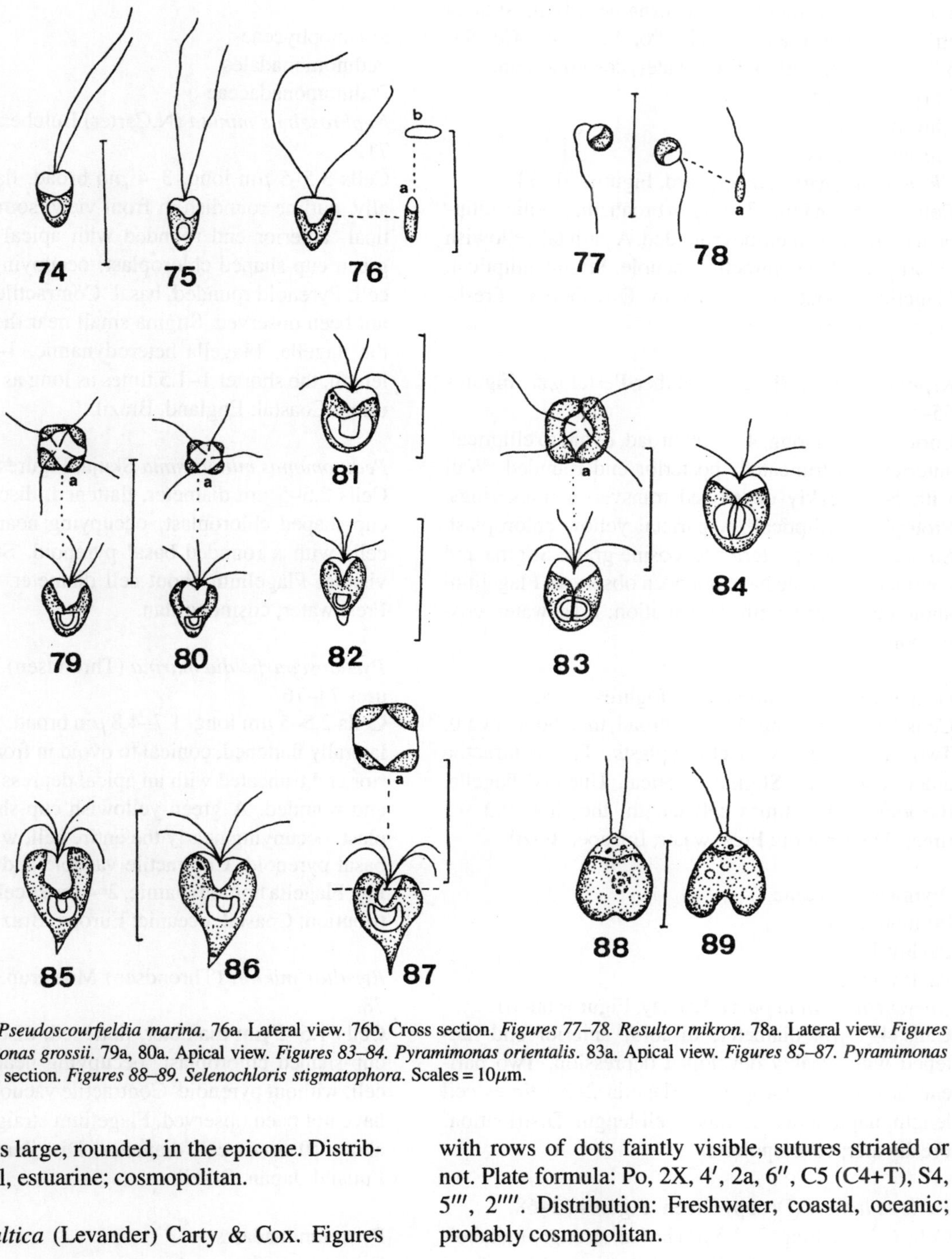

Figures 74–76. Pseudoscourfieldia marina. 76a. Lateral view. 76b. Cross section. *Figures 77–78. Resultor mikron.* 78a. Lateral view. *Figures 79–82. Pyramimonas grossii.* 79a, 80a. Apical view. *Figures 83–84. Pyramimonas orientalis.* 83a. Apical view. *Figures 85–87. Pyramimonas torta.* 87a. Cross section. *Figures 88–89. Selenochloris stigmatophora.* Scales = 10μm.

rows. Nucleus large, rounded, in the epicone. Distribution: Coastal, estuarine; cosmopolitan.

Durinskia baltica (Levander) Carty & Cox. Figures 43–46.
Cells 11–17 μm long, 11–14 μm broad, ovoid or spherical, slightly flattened dorsiventrally. Epiteca slightly larger than hypotheca. Cingulum wide, helicoidal. Sulcus large extending only onto hypotheca and does not reach the antapex, anterior sulcal plate (Sa) absent. Pattern plate asymmetrical, plate 3′ triangular. Plate with rows of dots faintly visible, sutures striated or not. Plate formula: Po, 2X, 4′, 2a, 6″, C5 (C4+T), S4, 5‴, 2⁗. Distribution: Freshwater, coastal, oceanic; probably cosmopolitan.

Peridinium umbonatum Stein. Figures 47–50.
Cells 19–24 μm long, 16–19.5 μm broad, egg-shaped, flattened dorsiventrally. Epitheca conical, hypotheca roundish truncate, the former usually larger than the later. Cingulum wide, helicoidal. Sulcus large extending onto both theca and reach the antapex. Pattern plate

symmetrical without visible ornamentation, sutures striated or not. Plate formula: Po, 4′, 2a, 7″, C6, S5, 5‴, 2⁗. Distribution: Freshwater; cosmopolitan.

Chrysophyceae
Chromulinales
Chromulinaceae

Chromulina pygmaea Nygaard. Figures 51–54.
Cells 4–4.5 μm long, 3–3.5 μm broad, metabolic, elliptical to ovoid, sometimes rounded. A parietal yellowish chloroplast. No contractile vacuole. Stigma elliptical. Flagellum about the cell length. Distribution: Freshwater; Denmark, Brazil.

Kephyrion ovale (Lackey) Huber-Pestalozzi. Figures 55–56.
Lorica 5–7 μm long, 4–5 μm broad, ovoid to elliptical, anterior end truncated, posterior end rounded. Wall with 5–6 weakly developed transversal thick rings. Protoplasm elliptical. A parietal yellow chloroplast. Sometimes one posterior leucosine grain. Stigma and contractile vacuole have not been observed. Flagellum about the lorica length. Distribution: Freshwater; cosmopolitan.

Ochromonas mutabilis Klebs. Figures 57–59.
Cells 10–13 μm long, 5–7 μm broad, metabolic, ovoid. Two lateral yellowish chloroplasts. Two contractile anterior vacuole. Stigma elliptical. Unequal flagella, the longest 1–2 times cell length, the short 0.3–0.5 times. Distribution: Freshwater; Europe, Brazil.

Prymnesiophyceae
Prymnesiales
Pavlovales
Pavlovaceae

Chrysochromulina parva Lackey. Figures 60–61.
Cells 4–7 μm diameter, discoid, anterior end flattened with weakly developed depression. Two parietal brownish chloroplasts. Flagella 2–2.5 times cell length, haptonema 1.5 times cell length. Distribution: Freshwater; cosmopolitan.

Pavlova lutheri (Droop) Green. Figures 62–68.
Cells 6–7 μm long, 3–3.5 μm broad, strongly metabolic, laterally flattened, elliptical to triangular in front view, sometimes square. Two parietal yellowish to olive chloroplasts. Stigma small, near insertion of flagella. Two unequal flagella, the longer 1.5 times cell length, the shorter 0.1–0.2 times cell length, haptonema very short and rarely observed. Distribution: Coastal; Europe, Brazil.

Prasinophyceae
Pedinomonadales
Pedinomonadaceae

Nephroselmis minuta (N.Carter) Butcher. Figures 69–71.
Cells 3.5–5 μm long, 3–4 μm broad, flattened laterally, outline roundish in front view, sometimes elliptical, anterior end rounded with apical swelling. A green cup-shaped chloroplast, occupying half of the cell. Pyrenoid rounded, basal. Contractile vacuole has not been observed. Stigma small near the insertion of the flagella. Flagella heterodynamic, 1–2 times cell length, the shorter 1–1.5 times as long as cell. Distribution: Coastal; England, Brazil.

Pedinomonas minutissima Skuja. Figures 72–73.
Cells 2.5–5 μm diameter, flattened, discoid. A green cup-shaped chloroplast, occupying nearly the entire cell, with a rounded basal pyrenoid. Stigma faintly visible. Flagellum about cell diameter. Distribution: Freshwater; cosmopolitan.

Pseudoscourfieldia marina (Throndsen) Manton. Figures 74–76.
Cells 2.5–5 μm long, 1.7–4.8 μm broad, 1.5 μm thick, laterally flattened, conical to ovoid in front view, anterior end truncated with an apical depression, posterior end rounded. A green-yellowish cup-shaped chloroplast, occupying nearly the entire cell, with a rounded basal pyrenoid. Contractile vacuole and stigma lacking. Flagella homodynamic, 2–4 times cell length. Distribution: Coastal, oceanic; Europe, Brazil.

Resultor mikron (Throndsen) Moestrup. Figures 77–78.
Cells 1.5–2 μm diameter, flattened, discoid. A green cup-shaped chloroplast, occupying nearly the entire cell, without pyrenoid. Contractile vacuole and stigma have not been observed. Flagellum straight or curvate about 10 times cell diameter. Distribution: Coastal; Finland, Japan, Brazil.

Pyramimonadales
Pyramimonadaceae

Pyramimonas grossii Parke. Figures 79–82.
Cells 4.5–10 μm long, 3–6 μm broad, 3–5 μm thick, obovoid or pyramidal, quadrangular in apical view, anterior end truncated with 4 equal lobes weakly developed surrounding a central depression, posterior end rounded sometimes narrowed to a rounded tip. A green

urn-shaped chloroplast, 4-lobed, corresponding to the 4 periplast lobes. Pyrenoid cup-shaped, basal surrounded by a starch sheath. One or two contractile vacuoles near the insertion of flagella. Stigma small in the anterior cell part. Flagella homodynamic, 0.6–1 times cell length arising from the base of the apical depression. Distribution: Coastal, oceanic; probably cosmopolitan.

Pyramimonas orientalis Butcher. Figures 83–84.
Cells 6–7 μm long, 4.5–6 μm broad, obovoid, anterior end truncated with 4 equal lobes, weakly developed, quadrangular with rounded angles in apical view, posterior end bluntly rounded. A green urn-shaped chloroplast with 2 lobes. Pyrenoid rounded, basal, surrounded by a sheath of two cup-shaped starch grains. Contractile vacuole has not been observed. Stigma absent. Flagella as long as the cell. Distribution: Coastal; word-wide.

Pyramimonas torta Conrad & Kufferath. Figures 85–87.
Cells 5.5–9 μm long, 4.7 μm broad, 4.5–5.5 μm thick, heart-shaped, anterior end truncated with 4 lobes weakly developed, surrounding the central depression, quadrangular with rounded angles in apical view, posterior end narrowed to a hyaline acute tip. A green urn-shaped chloroplast with 4 lobes. Pyrenoid cup-shaped surrounded by a starch sheath. Contractile vacuole has not been observed. Stigma elliptical, in the anterior cell part. Flagella as long as the cell. Distribution: Coastal; Belgium, Brazil.

Chlorophyceae
Chlamydomonadales
Chlamydomonadaceae
Selenochloris stigmatophora (Korsikov) Ettl. Figures 88–89.
Cells 10–12 μm long, 7–9 μm broad, lunate with short anterior papilla. A yellowish green faintly urn-shaped chloroplast. Pyrenoid lacking. Numerous contractile vacuole dispersed in the cytoplasm. Stigma small, in the anterior cell part. Flagella about the cell length. Distribution: Freshwater; cosmopolitan.

Seasonality of selected phytoflagellate species

Over the three years as a whole, the phytoplankton abundance showed a relatively weak seasonality (Domingos, unpublished data). The phytoplankton biovolume (Figure 90) was characterized by a greater contribution of Cyanophyceae (average 39%) and Dinophyceae (average 28%) followed by Bacillariophyceae (average 10%), Chlorococcales (average 7%) and Prasinophyceae (average 6%). The other algal classes did not contribute significantly to the total biovolume. However, if the phytoflagellates are expressed as a distinct biological group, they were quantitatively the most important in the phytoplankton biovolume (average 44%) (Figure 91). Among the phytoflagellates (Figure 92), the dominating class was the Dinophyceae. Raphidophyceae, exclusively represented by *Chattonella subsalsa*, and Euglenophyceae with *Eutreptia viridis* and *Euglena proxima* were the classes which contributed less. All other remaining classes contributed between 4% and 14% to the total phytoflagellate biovolume. Species of Prasinophyceae (*Pyramimonas grossii*, *P. orientalis* and *Pedinomonas minutissima*) and Cryptophyceae (*Hemiselmis simplex*) were the most abundant ones.

The event of a mass fish-kill (*Brevoortia tyrannus*-filter) in summer (February 1991) due to the production of toxins by *Synechocystis aquatilis* f. *salina* (Azevedo et al. 1994) resulted in high ammonia input to the water column as a consequence of the fish decomposition. The increase in ammonia concentration resulted in drastic decline of Cyanophyceae biovolume, whereas the proportion of Chlorococcales largely increased. Phytoflagellates and the other classes did not contribute significantly (Domingos et al., 1994).

During the development of the macroalgal stand (September-December 1992) phytoflagellates dominated the total phytoplankton biovolume, especially *Gymnodinium sanguineum* (Domingos, unpublished data)

Table 2 shows some of the abiotic variables values from Barra Lagoon, at the times of peak phytoflagellate biomass in the three years. In general, there is no strong seasonal pattern to which any conforms (Figure 93), although *Pavlova lutheri* showed some slight recurrence each summer. *Pyramimonas grossii* and *P. orientalis* often co-occurred with peaks at the beginning of the study period.

Numerous small flagellates exhibited their highest peaks at the beginning of the period, before the fish-kill, when the transparency of water was low (0.3m). The larger species, such as *Gymnodinium sanguineum*, *Chattonella subsalsa*, *Durinskia baltica* and *Eutreptia viridis* had the maximum peak towards the end of the period, during the development of macroalgal phase when transparency was greatest and light could penetrate through the entire water column. With the

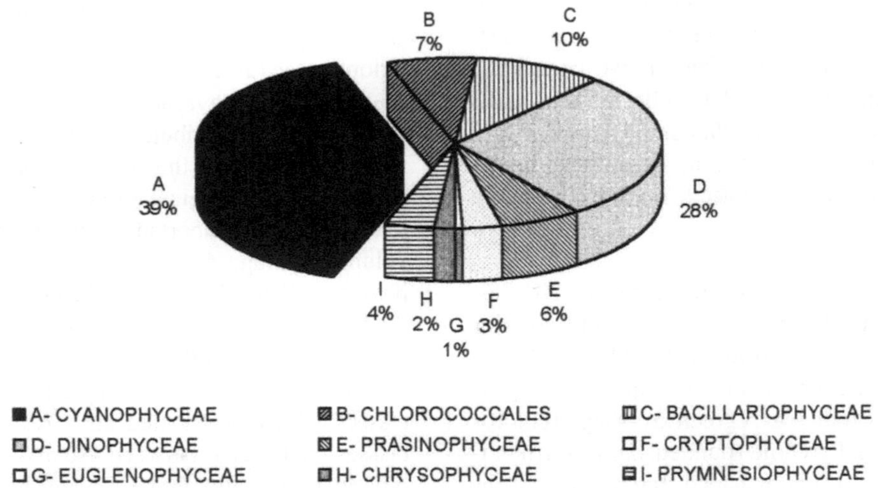

Figure 90. Percentage contribution of different taxonomic groups to total biovolume (average of three years).

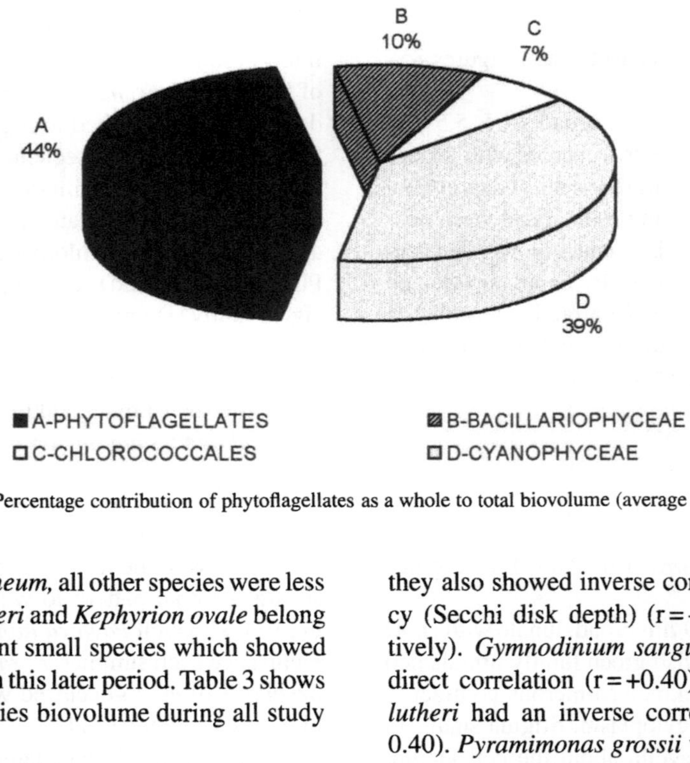

Figure 91. Percentage contribution of phytoflagellates as a whole to total biovolume (average of three years).

exception of *G. sanguineum,* all other species were less abundant. *Pavlova lutheri* and *Kephyrion ovale* belong to the group of abundant small species which showed maximum biovolume in this later period. Table 3 shows the mean selected species biovolume during all study periods.

No statistically significant correlation between phytoflagellate biovolume and either ammonia, DOM or dissolved inorganic nutrients was found. A direct correlation was observed between DOP and *Euglena proxima* (r = +0.37) and *Cryptomonas erosa* (r = +0.43). Salinity and *Pyramimonas orientalis* (r = –0.43) and *P. grossii* (r = –0.46) were inversely related. These species were correlated with each other and they also showed inverse correlation with transparency (Secchi disk depth) (r = –0.42 and -0.37, respectively). *Gymnodinium sanguineum* presented closely direct correlation (r = +0.40) with turbidity. *Pavlova lutheri* had an inverse correlation with depth (r = –0.40). *Pyramimonas grossii* was the only phytoflagellate species which showed significant inverse correlation (r = –0.37) with TN:TP ratio.

Discussion

The floristic composition of the phytoplankton of Barra Lagoon was characterized by a small number (81)

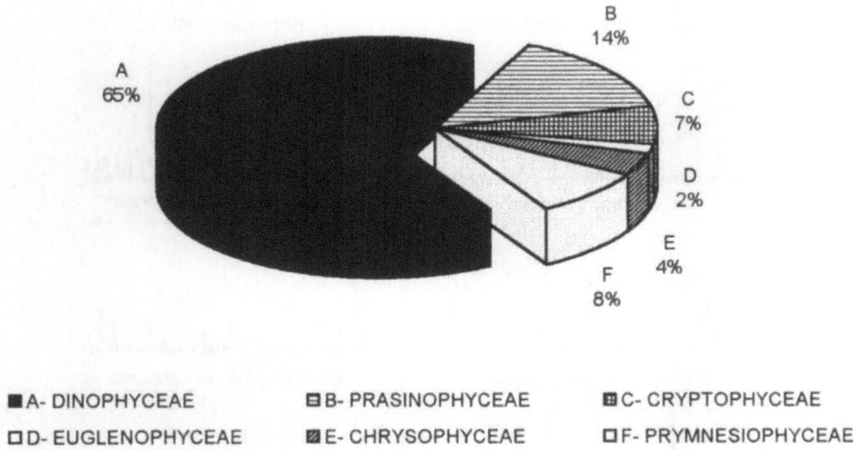

Figure 92. Percentage contribution of different biovolume phytoflagellate classes (average of three years).

Table 2. Some abiotic variables measured at the peak growth of fourteen selected phytoflagellate species

	Date	S‰	pH	NH_4 $\mu g\,l^{-1}$	TN $mg\,l^{-1}$	TP $\mu g\,l^{-1}$	DOM $mg\,l^{-1}$
Eutreptia viridis	15/06/93	9.8	8.7	22.7	1.1	48.8	6.6
Euglena proxima	24/02/92	8.1	8.9	14.0	3.0	515.2	15.4
Hemiselmis simplex	26/03/91	6.5	8.4	30.8	1.9	184.7	14.2
Cryptomonas erosa	24/02/92	8.1	8.9	14.0	3.0	515.2	15.4
Chattonella subsalsa	12/01/93	7.1	8.4	26.6	1.9	129.3	10.7
Gymnodinium sanguineum	01/12/92	6.9	9.3	30.9	1.5	72.1	9.1
Durinskia baltica	09/03/93	9.7	8.5	26.9	1.3	95.5	5.5
Kephyrion ovale	28/10/92	9.5	8.5	26.4	1.6	57.0	12.5
Ochromonas mutabilis	29/04/91	3.8	8.5	18.2	1.2	196.7	18.8
Pavlova lutheri	16/03/93	10.1	8.5	68.0	1.8	152.5	10.0
Pedinomonas minutissima	15/11/90	3.1	9.3	7.0	2.4	137.5	16.2
Pseudoscourfieldia marina	06/12/90	3.9	9.3	15.4	2.5	168.5	19.7
Pyramimonas grossii	29/04/91	3.8	8.5	18.2	1.2	196.7	18.8
Pyramimonas orientalis	09/04/91	3.2	9.1	14.0	1.2	172.0	7.6

of taxa, when compared to marine and freshwater environments, and by species drawn from freshwater, brackish, marine and holoeuryhaline primary habitats (Menezes & Domingos, 1994). The low number of taxa is common in continental brackish lakes (Caljon, 1983). In the case of Barra Lagoon, the indirect connection to the sea is possibly critical.

The phytoflagellate assemblage and the other planktonic non-flagellates observed in Barra Lagoon clearly reflect the influence of salinity on the specific composition. The low salinity with weak fluctuation (3–8‰ and mean value 5.4‰) over the three years of study favoured a higher number of limnobionts, although the majority of these taxa are classified as weakly euryhaline limnobionts (Caljon, 1983). This is compounded by their sporadic or rare occurrence and their quantitatively insignificant contribution. Although a relatively large species number (eleven) of limnobiontic phytoflagellates have been recorded, the quantitative contribution represented by salinity-tolerant organisms which may tolerate low salinity, was slight, some do occur in abundance in dilute brackish waters: *Euglena proxima*, *Cryptomonas erosa*, *Ochromonas mutabilis*, *Kephyrion ovale* and *Pedinomonas minutissima*. Coastal species were the richest taxonomic groups and they also showed a higher frequency and abundance; some are typical of brackish waters (*Pyramimonas grossii*, *P.*

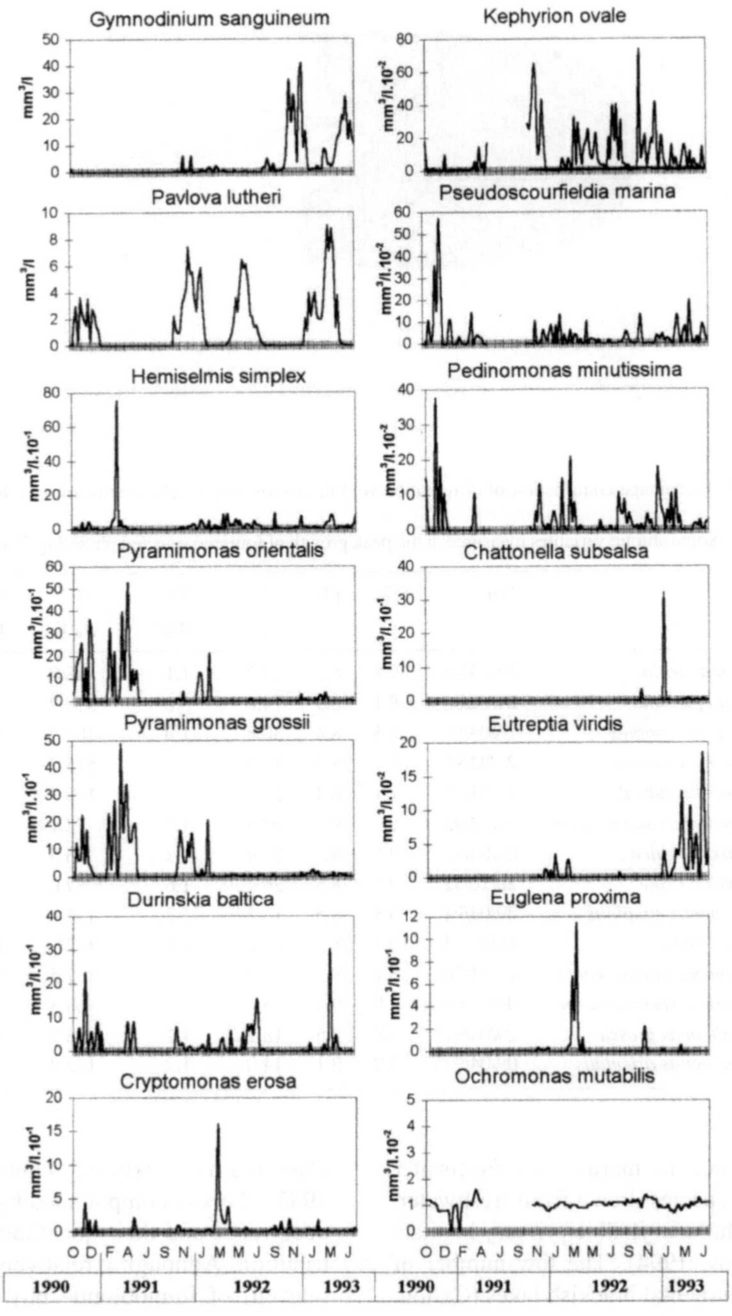

Figure 93. Biovolume seasonality of the fourteen selected phytoflagellate species in Barra Lagoon, 1990–1993.

orientalis, *Hemiselmis simplex*, *Gymnodinium sanguineum*), others of marine waters (*Pavlova lutheri*, *Pseudoscourfieldia marina*). This latter group occurs in marine littoral areas with a somewhat reduced salinity but with good development in brackish waters. Besides *Synechocystis aquatilis* f. *salina*, other typically brackish non-flagellate species were frequent in Barra Lagoon, among them: *Oocystis submarina*

Table 3. Biovolume averages of fourteen selected phytoflagellate species, 1990–1993

	Length × width (μm)	Biovolume (μm^3)	Standard deviation	*n*
Eutreptia viridis	30.0 × 9.5	1483.0	2.2	20
Euglena proxima	49.0 × 11.0	2925.0	1.4	20
Hemiselmis simplex	8.9 × 4.2	53.5	0.6	35
Cryptomonas erosa	14.0 × 7.2	477.5	0.5	30
Chattonella subsalsa	29.2 × 15.0	653.0	4.7	10
Gymnodinium sanguineum	35.0 × 26.0	6800.0	3.0	27
Durinskia baltica	12.1 × 10.8	647.8	1.5	25
Kephyrion ovale	5.4 × 4.8	18.0	2.1	25
Ochromonas mutabilis	11.2 × 5.5	88.5	2.1	20
Pavlova lutheri	4.5 × 2.7	23.5	0.6	29
Pedinomonas minutissima	4.0 × 3.5	19.5	0.3	30
Pseudoscourfieldia marina	4.0 × 2.8	20.5	2.3	25
Pyramimonas grossii	5.0 × 4.5	99.0	1.3	35
Pyramimonas orientalis	6.7 × 5.1	165.0	0.6	40

Lagerheim (Chlorococcales), *Actinocyclus normanii* (Gregory) Hustedt f. *subsalsa* (Juhl-Dannf) Hustedt, *Chaetoceros muellerii* Lemmermann and *Nitzschia closterium* (Ehrenberg) W. Smith (Bacillariophyceae) (Menezes & Domingos, 1994).

From a biogeographic point of view, the presence and abundance of phytoflagellate taxa previously recorded only from temperate areas (*Hemiselmis simplex, Teleaulax acuta, Ochromonas mutabilis, Chromulina pygmea, Pavlova lutheri, Nephroselmis minuta, Resultor mikron, Pseudoscourfieldia marina, Pyramimonas torta* is interesting. But for *Closteriopsis acicularis* (G.M. Smith) Belcher & Swale var. *africana* Hindák (Chlorococcales), none of the remaining non-flagellate taxa from Barra Lagoon, is restricted to the tropics (Menezes & Domingos, 1994), suggesting that the flora is cosmopolitan.

Comparing the taxonomic composition and biomass results with available data from literature, the more frequent and/or abundant phytoflagellates identified here have been recorded from other hypertrophic brackish environments (Domingos, 1991; Romo & Miracle, 1994) and from coastal polluted areas (Throndsen, 1969).

Although there are few studies on the abundance of phytoflagellates under richer trophic conditions, features such as motility and varied trophic behaviour have been suggested to confer competitive advantages to these algae in organic matter-rich environments (Porter, 1990; Rojo & Alvarez-Cobelas, 1993). Particularly in relation to nutritional requirements, Alvarez-Cobelas & Jacobsen (1992) proposed a diagram based upon the dominant phytoplankton biomass, which related phytoflagellate increase in hypertrophic environments to the ammonia- and DOM-content, including brackish coastal lagoons.

The lack of correlation between phytoflagellates biovolume and ammonia, DOM and dissolved inorganic nutrients may demonstrate that these resource concentrations were not the main factors for the phytoflagellate composition and abundance in Barra Lagoon. The high temperatures in tropical environments and the usually shallow depth of coastal lagoons contribute to high and fast organic matter recycling. In the Barra Lagoon, up to 70% of the primary production is sustained by recycling of organic matter in the water column (Carmouze et al. 1994a); such recycling leaves a low content of dissolved inorganic nutrients.

Taking into account that pigmented flagellates contributed 44% of the total phytoplankton biovolume over the three years of study, mixotrophic behavior is a likely explanation for the maintenance of the phytoflagellate populations in Barra Lagoon. Many phytoflagellate species identified appear to be mixotrophic (Porter, 1988; Jones, 1997) and, although there is a high dissolved organic matter content in Barra Lagoon (Carmouze et al., 1994b), the lack of correlation between DOM and phytoflagellates could support the idea that the mixotrophic algae are poor competitors for dissolved nutrients, as pointed out by Holen & Boraas (1995).

The maximal biovolume of pigmented flagellates with predominance of larger cells during the development of benthic macroalgae might be attributable to the alternative source of nutrients made possible through phagotrophy (bacterivory?). Although shallow (1 m at the collecting station), the benthic macroalgae stand affects the vertical heterogeneity of nutrient availability, while resources in the water column are shared with growing macroalgae. This is supported by the low planktonic biomass verified on this occasion (Domingos, unpublished data).

The dominance of *Gymnodinium sanguineum*, a large mixotrophic species, might be favoured by swimming velocity and displacement amplitude towards the bottom for nutrient uptake, clearly demonstrated by its closely positive relation with water transparency, which reached maximal values during the development of macroalgae stand. Motility is an advantage of acquisition of particles by phagotrophic planktonic algae (Fenchel, 1982; Raven, 1995) and because swimming velocity increases with algal size (Sommer, 1988), larger cells offer a larger boundary surrounding them facilitating particle ingestion (Raven, 1995).

The abundance of larger cells during the development of macroalgae stand at the latter period of study contrasts with that observed at the beginning of study. In the latter there was a dominance of generally smaller species which, in addition to their mixotrophic behaviour, were probably favoured by low water transparency, as demonstrated by inverse correlation between *Pyramimonas grossii* and *P. orientalis* and Secchi depth. The contrast between the peak of larger flagellates during the highest transparency *versus* the peak of smaller cells during the lowest water transparency is in agreement with the literature. Sommer (1988) and Jones & Ilmavirta (1988) pointed out that there is selective advantage for larger flagellates in heterogeneous environments while the smaller cells have ability to exploit homogeneous conditions.

The lack of seasonal dynamics for both phytoflagellates and other planktonic algae was expected (Domingos, unpublished data). The irregular pattern of phytoplankton seasonality is common in shallow lakes (Reynolds et al., 1994), particularly in tropical environments, where the succession is frequently affected by abrupt events (Melack, 1979).

It seems that, in Barra Lagoon, the main factors driving abundance and occurrence of planktonic phytoflagellates are water transparency, the distribution of nutrients in the water column and nutritional sources. However, there is little information available on the environmental requirements of phytoflagellates in brackish tropical ecosystems under hypertrophic conditions. More accurate investigations are required to evaluate potencial nutritional resources of the phototrophic flagellates in Barra Lagoon.

Our results support different physiological or behavioral phytoflagellate responses to environmental factors (Jones & Ilmavirta, 1988) and the importance of taxonomy for a better understanding of the occurrence and the dynamics of this biological group, especially in tropical brackish ecosystems with high trophic status.

On the other hand, major information about the interactions between natural populations of organisms and their environment will contribute, undoubtedly, to a better understanding in the taxonomic studies. The similar ecological requirements and dynamics observed between *Pyramimonas grossii* and *P. orientalis* and their morphological similarity under light microscopy suggest that the species are closely related, although an electron microscopy examination of the material will be necessary in order to make a more reliable identification.

Acknowledgements

We thank Dr Jean-Pierre Carmouze for his essential support during the field and laboratory work and his suggestions to the manuscript. We also want to thank two anonymous referees for critically reading the manuscript and for revising the English language. This study was supported by ORSTOM, CNPq and FUJB-UFRJ.

References

Alvarez-Cobelas, M. & B. A. Jacobsen, 1992. Hypertrophic phytoplankton: an overview. Freshwat. Forum 2: 184–199.

Azevedo, S. M. F. O. & J-P. Carmouze, 1994. Une mortalité de poissons dans une lagune tropicale (Brésil) durant une période de dominance de Cyanophycées. Coincidence ou conséquence?. Rev. Hydrobiol. trop. 27: 265–272.

Caljon, A. G., 1983. Brackish-water phytoplankton of the Flemish lowland. Dev. Hydrobiol. 48: 1–272.

Carmouze, J-P., B. Farias & P. Domingos, 1994a. Evolution du metabolism d'une lagune tropicale (Brèsil) au cours d'une période marquée par une mortalité de poisson. Rev. Hydrobiol. trop. 27: 199–215.

Carmouze, J-P., C. D'Elia Sampaio & P. Domingos, 1994b. Evolution des pools de matière organique et des nutriments dans une lagune tropicale (Brésil) aux cours d'une période marquée par une mortalité de poissons. Rev. Hydrobiol. trop. 27: 217–234.

Carmouze, J-P., M. Bernardes & P. Domingos, 1995. Asphyxie des lagunes cotières de l'état de Rio de Janeiro. ORSTOM Actualités 46: 23–26.

Domingos, P., 1991. Estrutura da comunidade fitoplanctônica e produção primária na lagoa de Saquarema, Rio de Janeiro. Dissertação de Mestrado em Geoquímica, Universidade Federal Fluminense, Niterói, 155 pp.

Domingos, P., V. L. M. Huszar & J. P. Carmouze, 1994. Composition et biomasse du phytoplankton d'une lagune tropicale (Brésil) au cours d'une période marquée par une mortalité de poisson. Rev. Hydrobiol. trop. 27: 235–250.

Edler, L. (ed.), 1979. Recommendations for marine biological studies in the Baltic Sea. Phytoplankton and chlorophyll. 38p. (UNESCO), Working Group 11, Baltic Marine Biologists.

Fenchel, T., 1982. Ecology of heterotrophic microflagellates. I. Some important forms and their functional morphology. Mar. Ecol. Prog. Ser. 8: 211–223.

Holen, D. A. & M. E. Boraas, 1995. Mixotrophy in chrysophytes. In C. D. Sandgren, J. P. Smol & J. Kristiansen (eds), Chrysophyte algae: ecology, phylogeny and development. Cambridge University Press: 119–140.

Huszar, V. L. M. & L. H. S. Silva, 1992. Estrutura das comunidades de quatro lagoas do litoral norte do Estado do Rio de Janeiro. Acta limnol. bras. 4: 291–314.

Jones, H. L. J., 1997. A classification of mixotrophic protists based on their behavior. Freshwat. Biol. 37: 35–43.

Jones, R. I. & V. Ilmavirta, 1988. Flagellates in freshswater ecosystems- Concluding remarks. Hydrobiologia 161: 271–274.

Melack, J. M., 1979. Temporal variability of phytoplankton in tropical lakes. Oecologia 44: 1–7.

Menezes, M., 1994. Fitoflagelados pigmentados de quatro corpos d'água da região sul do Município do Rio de Janeiro, Estado do Rio de Janeiro, Brasil. Tese de Doutorado, Universidade de São Paulo, São Paulo. 2v, 701 pp.

Menezes, M. & P. Domingos, 1994. La flore planctonique d'une lagune tropicale (Brésil). Rev. Hydrobiol. trop. 27: 273–297.

OECD, 1992. Eutrophication of waters. Monitoring, assessment and control. OECD, Paris.

Porter, K. G., 1988. Phagotrophic phytoflagellates in microbial food webs. Hydrobiologia 159: 89–97.

Porter, K. G., 1990. Heterotrophic, autotrophic, and mixotrophic nanoflagellates: seasonal abundances and bactevory in a eutrophic lake. Limnol. Oceanogr. 35: 1821–1832.

Raven, J. A., 1995. Comparative aspects of chrysophyte nutrition with emphasis on carbon, phosphorous and nitrogen. In C. D. Sandgren, J. P. Smol & J. Kristiansen (eds), Chrysophyte algae: ecology, phylogeny and development. Cambridge University Press: 95–118.

Reynolds, C. S., J. Padisák & I. Kóbor, 1994. A localized bloom of *Dinobryon sociale* in Lake Balaton: Some implications for the perception of patchiness and the maintenance of species richness. Abstracta Bot. 17: 251–260.

Rojo, C. & M. Alvarez-Cobelas, 1993. Taxonomy and ecology of phytoplankton in a hypertrophic gravel-pit lake II. Cryptophyceae, Euglenophyceae, Dinophyceae. Nova Hedwigia 57: 47–63.

Romo, S. & R. Miracle, 1994. Population dynamics and ecology of subdominant phytoplankton species in a shallow hypertrophic lake (Albufera of Valencia, Spain). Hydrobiologia 273: 37–56.

Sokal, R. R. & F. J. Rohlf, 1981. Biometry. The principles and practice of statistics in Biological research. W.H. Freeman and Company, New York, 859 pp.

Sommer, U., 1988. Some size relationships in phytoflagellate motility. Hydrobiologia 161: 125–131.

Steidinger, K. A. & K. Tangen, 1996. Dinoflagellates. In C. R. Tomas (ed.), Identifying Marine Diatoms and Dinoflagellates. Academic Press, California: 387–598.

Throndsen, J., 1969. Flagellates of Norwegian coastal waters. Nytt. Mag. Bot. 16: 161–216.

Throndsen, J., 1993. The planktonic marine flagellates. In C. R. Tomas (ed.), Marine Phytoplankton. A guide to naked flagellates and coccolithophorids. Academic Press, California: 1–145.

Utermöhl, H., 1958. Zur Vervollkommnung der quantitativen Phytoplankton Methodik. Mitt. int. Ver. Limnol. 9: 1–38.

Hydrobiologia **369/370:** 315–327, 1998.
M. Alvarez-Cobelas, C. S. Reynolds, P. Sanchez-Castillo & J. Kristiansen (eds), Phytoplankton and Trophic Gradients.

Chrysophytes and chlamydomonads: pioneer colonists in extremely acidic mining lakes (pH <3) in Lusatia (Germany)

Brigitte Nixdorf, Ute Mischke & Dieter Leßmann
Brandenburg Technical University of Cottbus, Faculty of Environmental Sciences and Process Engineering, Chair of Conservation, Seestr. 45, D-15526 Bad Saarow, Germany

Key words: acidic mining lakes, phytoplankton, *Chlamydomonas, Ochromonas,* water chemistry, limiting factors

Abstract

Twenty-three extremely acidic (pH between 2.5 and 3.5) mining lakes in Lusatia (Germany) were analysed in order to classify their hydrochemistries and to assist the understanding of phytoplankton colonization of these extreme environments. Neither morphometric nor physical parameters influence phytoplankton composition but determine the extent to which the nutrient supply supports the mass development of Chrysophyceae and Chlorophyceae in certain layers of the water (hypo- or epilimnetic chlorophyll maxima and short mass developments). Conventional trophic classification is not readily applicable to these lakes but a chemical classification on the basis of hydrogen, total iron and acidity is proposed. Species of *Ochromonas* and *Chlamydomonas* dominate the phytoplankton in fourteen of the most acid lakes; dinoflagellates occurre additionally in four; a more diverse algal assemblage with diatoms and cryptophytes is found in lakes with moderately acidic (pH 5.7–7.0) or alkaline conditions (pH 7.0–9.4). The lake chemistry is the main determinant for the planktonic composition of the water bodies whereas the trophic state mainly determines the level of algal biomass.

Introduction

Two main types of lakes characterize the landscape of Lusatia (Brandenburg, Saxony, Germany): more or less shallow, eutrophic waters (more than 2000 > 10 ha) of glacial origin and artificial lakes created by mining activities (more than 400, with a total area of several thousand ha). As a consequence of the geogenic potential for acidification by pyrite oxidation most of these lakes are extremely acidic (pH < 3.5), rich in iron, sulphate and, in many cases, also heavy metals. The extreme acidity of certain of the lakes (pH between 2 and 3, base capacity $K_{B4.3}$ up to 15–30 mmol l^{-1}) is problematic. Whereas phosphorus concentrations (mostly < 10 μg l^{-1}) and CO_2 are at oligotrophic to mesotrophic levels, inorganic nitrogen (mostly > 1 mg l^{-1}) is usually available in excess, generally as ammonia. Under these severe conditions, protozoans and metazoans are not abundant and grazing pressure on phytoplankton is expected to be low (Eriksson et al., 1980; Klapper & Schultze, 1995; Nixdorf et al., 1998).

It is known that species richness, diversity and biomass of pelagic primary producers in very acidic lakes are low. Blouin (1989) found that pH tends to affect plankton diversity more than abundance or standing crop. Most published observations are confined to pH-ranges between 4 and 5.5, not to extremes (pH < 3) as often measured in the geogenically acidified lakes of Lusatia. Since 1993, we have been investigating the biotic and abiotic conditions in a number of mining lakes to determine the basic mechanisms of primary succession in relation to the load of groundwater and surface water and nutrients, as well as to morphometric, topographic and hydrographic conditions. The reasons for the dominance of particular algal groups in these extremely acidic systems are presumed to lie in their adaptive capabilities, their ecophysiological specialisation and the ecological advantages that may be offered. At present, the database does not provide much resolution on long-term planktonic succession. The aim of this paper is therefore to show the general character and potential of algal development across the gradient of acidities furnished by these mining lakes.

Study sites

The region of Lusatia comprises the south-eastern part of Brandenburg and the north-eastern part of Saxony in eastern Germany (Figure 1). Many artificial waters were created here as a result of the open-cast lignite mining activities that began in the late 19th century. As a consequence of, first, the intensive scale of exploitation and, since 1990, the dramatic fall in production of brown coal, water now fills a large number of abandoned workings. Some of these are popular recreational lakes or are expected to become so in the near future. The restoration and rehabilitation of such waters are under investigation in several regions (Schlabendorfer Felder (B), around Drebkau near Cottbus (E), Muskauer Faltenbogen (F), Grünewalde/Plessa (C), around Senftenberg (D) and Hoyerswerda (G): see Table 1 and Figure 1 for location) in addition to Lusatia (A).

Methods

Water samples for chemical and biological analyses (planktonic composition, chlorophyll, oxygen demand) were taken monthly (Plessa 107, Plessa 111, Plessa 117, Lichtenauer See, and Schlabendorf RL B) or irregularly (elsewhere). Samples were taken with a 2 l – LIMNOS-sampler and integrated vertically, from surface to bottom for fully-mixed columns, or as epilimnetic and hypolimnetic samples when lakes were stratified. The sampling stations were generally at the deepest point of each lake (Nixdorf et al., 1995a; Schultze et al., 1994), where the Secchi-disk transparency was also measured. Depth profiles of temperature, oxygen (absolute and relative as saturation), pH, redox potential, conductivity were determined. Since 1996, chlorophyll fluorescence has also been measured routinely by means of a fluorescence probe (HAARDT), linked to a field computer (HUSKY Hunter).

The concentration of anions (among them the nutrients, as phosphate, nitrate) silicate and cations, the base capacity up to pH 4.3 ($K_{B4.3}$) or the acid capacity down to pH 4.3 ($K_{S4.3}$), total inorganic carbon (TIC), and metal concentrations (Al, Fe, Mn, Cu, Cd, Ni, Zn) were determined by standard methods (DEV 1986–1993) and following the recommendations of Schultze et al. (1994).

Chlorophyll *a* (including seston) was measured according to standard method, but with some modifications related to the peculiarities of acidic water samples (Nixdorf et al., 1995a). Lugol-fixed phytoplankton was counted by means of Utermöhl technique (1958). To avoid heavy precipitation of iron-complexes in the acidic waters, samples were diluted 1:1 with distilled water before settling in the counting-chambers. The phytoplankton composition of five lakes (Waldsee, Felixsee, Plessa 111, Lichtenauer See, Schlabendorf RL B) was studied monthly for more than one year. For most of the lakes, however, phytoplankton was analysed only for the period October 1995 to May 1996. Since the determination of *Ochromonas* to species level is difficult, cell counts were made at the genus level. To give the typical phytoplankton of each lake, the data sets were sorted by highest lake or epilimnetic biovolume. The algal composition at the sampling date of highest biovolume is supposed to be typical for the lake within the period of investigation. Where relevant, significant departures from this 'norm' in samples on other occasions are noted explicitly. Special examinations on autotrophic picoplankton, bacteria and phytoflagellates morphology were carried out on DAPI-stained membrane-filtrates, following the procedure of Porter & Feig (1980).

Results

Morphometry and mixis of mining lakes

As can be seen from Table 1, there is considerable variation in lake morphometry. After flooding, Lusatia will have many large lakes to contrast with the natural and more shallow lakes characteristic of Brandenburg. Although the numerous older and smaller mining lakes have maximum depths of less than 20 m, many are dimictic or even meromictic, with major chemical differentiation within individual water bodies (e.g. Waldsee, Lugteich). In general, the developing lakes belong to a type which does not occur naturally in this region. The intended maximum depth of the new, larger lakes will rarely be less than 20 m and about a half of them will have more than 40 m. Today, the lakes differ greatly in size and shape and therefore some are well stratified while others are polymictic (Mischke et al., 1994; Nixdorf et al., 1998). After filling, all these lakes will be dimictic. Prediction of trophic state from

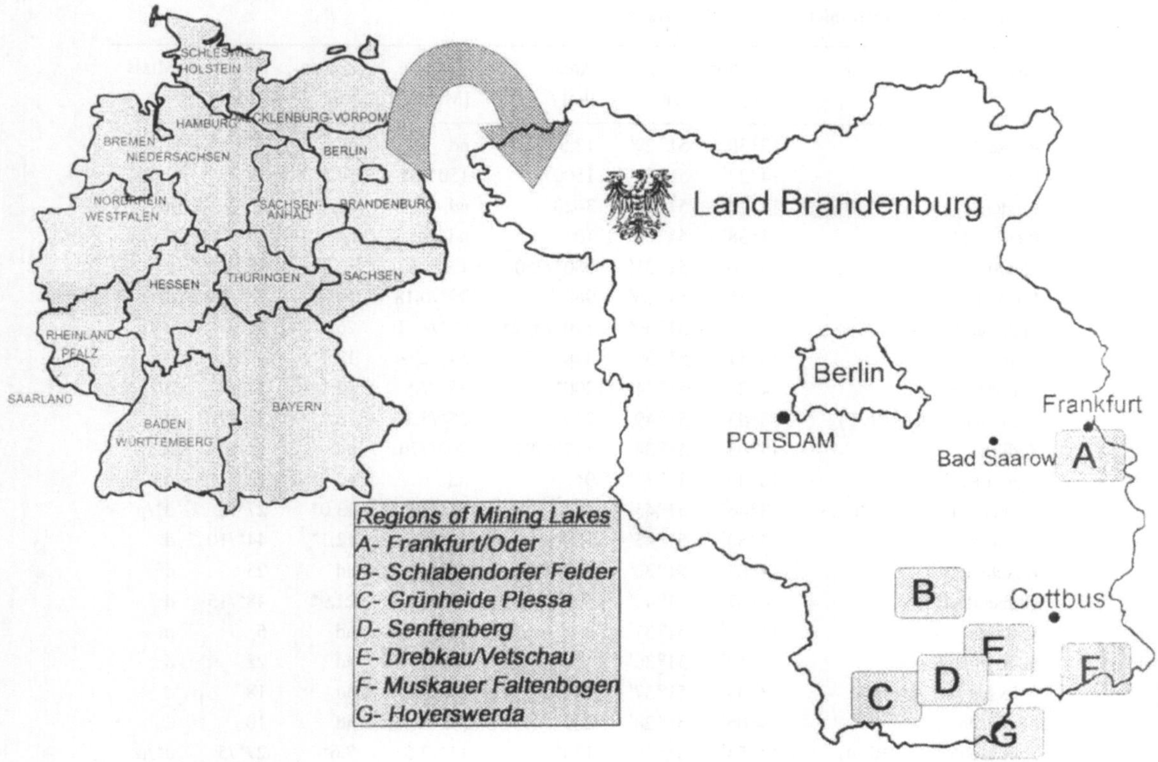

Figure 1. Map of regions in Brandenburg and Saxonia where mining lakes are investigated.

lake morphometry according to the approach of Klapper (TGL 27885/01, 1982) using the ratio between epi- and hypolimnion volume works well for the large, deep lakes (e.g. Gräbendorf, Bärwalde, Scheibe, Burghammer). On the basis of morphometry, these lakes will become potentially oligotrophic or mesotrophic.

Chemical environment

The dominant characteristic of mining lakes' chemistry is the extremely high proton concentration which influences all other chemical parameters. Due to oxidation of pyrite and marcasite under aerobic conditions by bacteria (*Thiobacillus*), waters become extremely acidic (pH between 3.5 and 4.2); many of the lakes are, indeed, unusually acidic (pH $<$ 2.6–3.5). This situation is stabilized by the permanent inflow of acidic groundwater and the buffering capacity of the dissolved Fe (III).

On the basis of the concentration of hydrogen ions, acidity and the concentration of total-iron, two main groups of lakes may be distinguished: extremely and very acidic lakes and circumneutral lakes (Table 2). The aluminium buffering of Felixsee sets this lake apart from the others and we put this into a third category of mining lake. A gradient of pH permits a further subdivision of the lakes in the acidic category (Ia to Ic). The full classification is as shown below.

Group I: Lakes I0 to Ic of Table 2 belong to this group. With pH-values < 4.0 they are all iron-buffered. Within this group Plessa 107 must be considered as an outsider (I0) because of its extremely high acidity ($K_{B4.3}$ > 20 mmol l^{-1}) and iron concentrations > 400 mg l^{-1}. While the majority of lakes (*group 1b*) shows an acidity between 1 and 5 mmol l^{-1}, there are some lakes with high $K_{B4.3}$-values (> 5 mmol l^{-1}) associated with high iron concentrations (*group 1a*). Another subgroup comprises the acidic lakes which show a pH-minimum > 3.0 and an acidity < 1 mmol l^{-1} (*group 1c*). In this range of pH there is an obvious reduction in dissolved iron due to precipitation of Fe (III) oxihydroxide. Lakes buffered by aluminium are not encountered in the range pH 4.0–5.6.

Table 1. Topography and morphometry of mining lakes in Lusatia (* = intended). Mixis: d = dimictic, p = polymictic, m = meromictic. nd = no data available

Lakes	Location Long	Lat	Area [ha]	Volume [Mio m^3]	Zmean [m]	Zmax [m]	Mixis
Plessa 107	13°30′	51°29′	12.5	nd	nd	4	p
Bärwalde	14°33′	51°23′	1500*	150*/65	14.5*	59*/5	d*/p
Lugteich	14°10′	51°28′	3–20	nd	nd	5	m
Plessa 111	13°38′	51°30′	10	nd	nd	10	d
Skado	14°10′	51°31′	980*/750	130*/60	13.2*	46*/20	d
Lohsa II	14°25′	51°26′	960*	99*/0.18	10.3*	56*/13	d
Spreetal-NO	14°20′	51°30′	310*/< 25	97*/< 1	26.9*	58*/5	d*/p
Bluno	14°11′	51°30′	446*/270	54*/29.4	12.1*	43*/10	d
Dreiweibern	14°26′	51°24′	290*	35*/7.5	12.1*	38*/6	d*
Lichtenauer See (RL F)	13°53′	51°49′	230*	25*/8.3	10.7*	32*/7	d
Sedlitz	14°05′	51°34′	1170*/850	202*/70	nd	12	d
Plessa RL 117	13°40′	51°31′	94	nd	nd	13	d
Bergen-Niederhof (RL 13)	13°46′	51°45′	48*	5*/1.68	13.0*	27*/5	d*/p
Drehnasee (RL 12)	13°50′	51°45′	180*	17*/5.5	12.1*	44*/10	d
Koschen	14°07′	51°30′	620*/590	97*/40	nd	25	d
Gräbendorf	14°07′	51°42′	500*	90*/9	21.9*	48*/15	d
Waldsee	14°34′	51°37′	0.1	nd	nd	5	m
Senftenberger See	13°59′	51°30′	1300	160	nd	22	d
Felixsee	13°33′	51°37′	11.2	nd	nd	18	d
Laubusch	14°06′	51°28′	200	3.3	nd	10	d
Schönfelder See (RL 4)	13°54′	51°49′	138*	11*/2.5	7.6*	27*/5	d*/p
Burghammer	14°21′	51°28′	390*	65*/37	17.6*	49*/15	d
Schlabendorf (RL B)	13°51′	51°51′	6	nd	nd	11	d

Group II: Up to now we have found only one lake (Felixsee) in the moderate acidic state (pH around 4 in 1994) and likely in the Al-buffer system. The lake water varied in pH between 3.6 and 4.3 from 1993 to 1996.

Group III: The lakes of group III are all in the carbonate buffering system. The pH is always > 5.6, but with the exception of Stöbritzer See (Schlabendorf RL B), alkalinities with $K_{S4.3}$-values normally < 1 mmol l^{-1} must be considered as low.

Nearly all lakes have high conductivities, owing to the high concentrations of calcium and sulphate and, in the case of most acid lakes, iron. Only in the two basins of lake Senftenberg and in Felixsee conductivities of < 1000 μS cm^{-1} were measured. Following the removal of dissolved iron through precipitation of Fe (III) oxyhydroxide, the water escapes the iron buffering system which holds pH generally < 4.

High concentrations of hydrogen and dissolved Fe (III) are responsible for the extremely high values for base capacity ($K_{B4.3}$). High inputs of bases are needed to achieve the pH range of the carbonate buffering system. The acidity of the mining lakes varies considerably (Table 2, $K_{B4.3}$ and Figure 2) whereas the pH-values are generally within a narrow range.

Trophic conditions, temporary developments and seasonal layering of chlorophyll

All the lakes are oligotrophic or mesotrophic. In the acidic lakes, concentrations of total phosphorus (TP) are typically around 5 μg l^{-1}. Only temporary increases in TP may occur (e.g. Lichtenauer See in 1994, Nixdorf et al., 1998). Exceptions are the hypertrophic Lugteich lakes, which were loaded with sewage from the town of Hoyerswerda in the 1970s and 1980s.

The portion of ammonia in the total dissolved inorganic nitrogen is very high in lakes belonging to group I (Table 2). In contrast to natural hard water lakes, inorganic carbon in mining lakes is often available only at very low concentrations for primary production in the euphotic zone but may be high locally, through

Table 2. Physical and chemical parameters of mining lakes in Lusatia (TP – total phosphorus, TN – total nitrogen, SD – Secchi depth, italic values = KS4.3)

	Lakes	pH	Conductivity	$K_{B4.3}/K_{S4.3}$	Ca^{+}	SO_4^{2-}	Fe_{tot}	Al	TIC	TP	TN	SD
			[μS cm^{-1}]	[mmol l^{-1}]	[mg l^{-1}]	[mg l^{-1}]	[mg l^{-1}]	[mg l^{-1}]	[mg l^{-1}]	[μg l^{-1}]	[mg l^{-1}]	[m]
I 0	Plessa 107	2.3–2.8	2600–5300	20.7–34.6	260–530	1800–3700	400 –840	44 –60	<0.5– 2.0	<5– 7	0.2– 9.1	1.4– 2.8
I a	Bärwalde	2.6–3.0	2400–4200	11.2–14.9	100–250	1600–2100	60 –230	10 –33	<0.5	<5– 7	3.5– 9.5	shore
I a	Lugteich (Epi)	2.6–2.8	2900	12.8–15.0	290–350	1900	260	4.3–8.5	3.8	63–230	12 –19	0.5
I a	Plessa 111	2.6–3.0	2300–3100	9.7–12.0	200–300	1100–1800	130 –200	26 –44	<0.5– 5.3	<5– 6	3.1–11.3	2.4– 4.2
I a	Skado	2.5–3.3	2300–3100	7.3– 8.1	50–500	1100–1400	110 –130	7.0–11	<0.5– 6.1	<5– 7	1.3– 3.3	1.4– 3.5
I a	Lohsa II	2.5–2.9	2300–3400	4.3–12.0	70–250	1500–1800	20 –170	9.4–30	<0.5– 6.2	<5– 7	1.2– 6.2	0.5– 3.0
I a	Spreetal-NO	2.8–3.4	2300–3100	1.4– 9.0	50–250	1100–1500	30 –250	3.0–11	<0.5–21.1	<5– 7	1.3– 1.7	shore
I b	Bluno	2.7–2.9	2100–2600	1.7– 5.0	60–250	900–1100	15 – 80	1.3– 5.1	<0.5– 2.9	<5–13	1.4– 2.4	shore
I b	Dreiweibern	2.4–2.5	2400–2900	3.3– 4.2	100–300	1100–1400	10 – 40	2.3–11	<0.5– 0.6	<5– 9	3.8– 3.9	shore
I b	Lichtenauer See	2.6–3.0	2800–3300	3.4– 3.9	150–400	900–1600	6 – 25	3.3–17	<0.5– 1.7	<5– 7	1.1– 5.0	2.4– 4.0
I b	Sedlitz	2.7–3.1	2400–2700	2.7– 2.9	80–300	900–1000	7 – 30	2.8–11	<0.5– 1.9	<5– 7	2.3– 2.6	2.6– 4.1
I b	Plessa 117	2.8–3.0	1000–1200	2.2– 2.5	63–104	360– 390	15 – 23	0.6– 1.6	<0.5– 5.1	10–28	1.6– 1.9	4.5– 8.4
I b	Bergen-Niederhof	2.8–3.0	1100–1300	2.0– 2.1	30–90	350– 400	6 – 16	1.7– 2.7	<0.5– 1.0	<5–10	0.3– 2.0	4.3– 4.5
I b	Drehnasee (RL 12)	3.0–3.3	1600–1900	1.1– 1.8	60–240	500– 900	2 – 25	2.4– 7.1	<0.5– 4.2	<5–17	1.1– 5.1	0.8– 2.6
I b	Koschen	2.9–3.5	1700–1900	1.3– 1.8	70–150	450– 700	5 – 60	0.6– 0.9	<0.5– 1.7	<5–10	1.3– 2.6	2.5– 5.0
I c	Gräbendorf	3.1–3.7	1300–1600	0.5– 0.9	100–250	400– 700	0.7– 50	1.0– 3.2	<0.5– 7.3	<5–13	0.2– 2.5	1.8– 3.5
I c	Waldsee (Epi)	3.0–3.4	550–1000	0.7– 3.8	80–100	250– 350	6 – 12	<0.2	<0.5– 8.2	<5– 7	1.0– 2.0	1.3– 2.3
I c	Senftenberg(Südf.)	3.0–3.4	750– 970	0.6– 0.7	40–80	200– 300	0.3– 9	<0.2– 0.6	<0.5– 1.8	<5– 6	1.3– 2.1	2.6–10.4
II	Felixsee	3.6–3.9	520– 600	0.1– 0.8	40–90	150– 250	0.4– 5	1.8– 3.8	<0.5– 2.1	<5– 7	1.1– 5.0	4.0– 8.8
III	Senftenberg (Elfd)	5.7–7.2	630– 800	*0.1– 0.2*	50–140	200– 300	0.2– 4.5	<0.1– 0.4	<0.5– 3.3	8–10	0.3– 2.3	2.3– 5.0
III	Laubusch	5.7–7.4	1100–2000	*0.1– 0.9*	30–170	500– 700	0.1– 0.8	0.1– 0.3	1.3–11.5	12–38	1.4– 4.0	0.6– 3.0
III	Schönfeld (RL 4)	6.1–7.9	1500–2900	*0.04–1.5*	140–250	550– 650	0.4– 7	0.2	32.9–64.5	16–20	0.1– 1.4	0.2– 1.0
III	Burghammer	7.9–9.4	1600–2400	*0.7– 0.8*	120–140	800–1000	0.2– 0.8	0.1– 0.3	7.6–12.6	6–20	1.4– 1.9	1.3– 2.2
III	RL B	7.3–7.7	1100–1200	*1.8– 2.3*	140–200	150– 500	0.2– 0.5	<0.1	18.9–32.4	<5–10	<1 – 5.0	3.0– 4.8
nc	Waldsee (Monim.)	5.8–6.3	900–1550	*3.5–10.6*	90–120	200– 350	130 –180	<0.2	2.0–160	8–42	3.0– 4.5	–
nc	Lugteich (Monim.)	6.0–6.2	3500–4000	*12.5*	340–450	2200–2900	750 –1000	0.4– 1.9	240–390	33–90	32 – 53	–

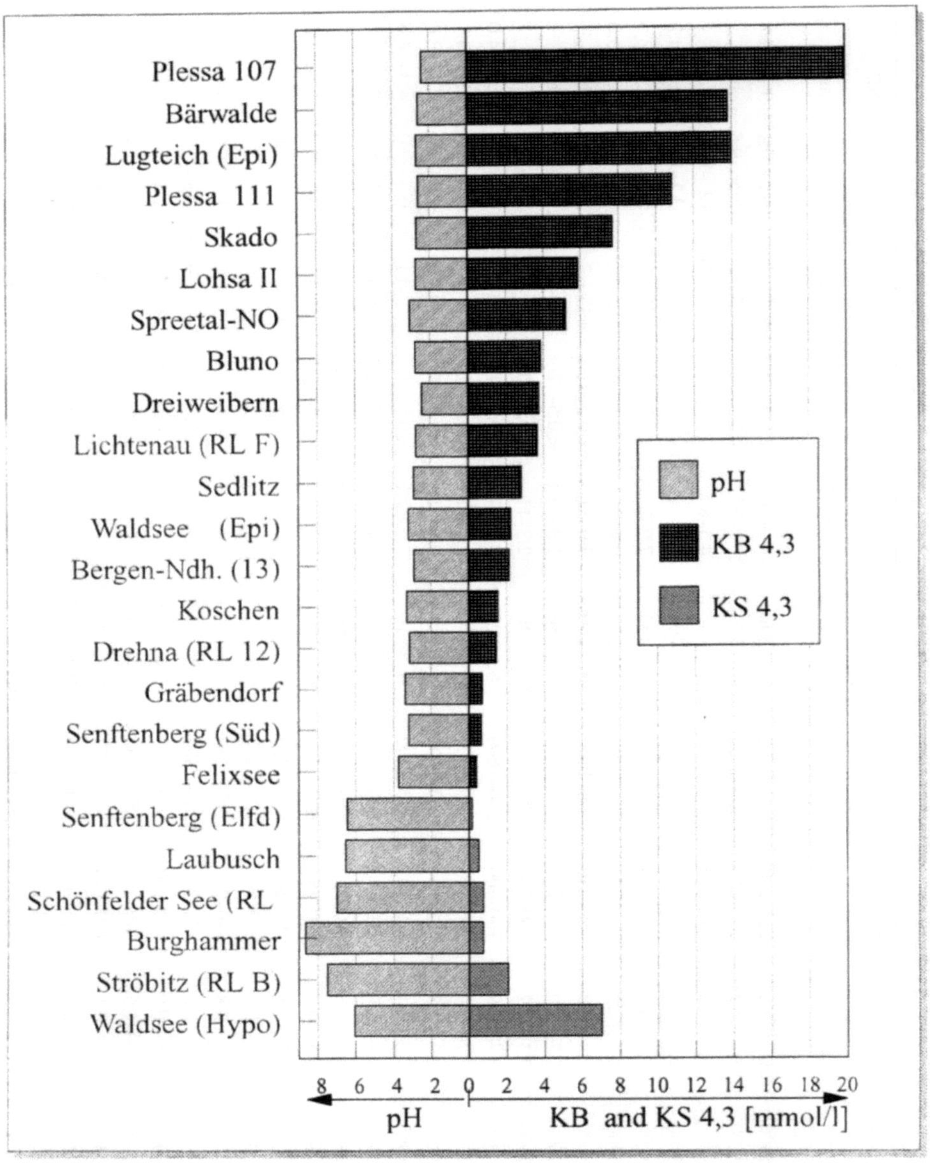

Figure 2. pH and acidity (KB = base capacity) or alkalinity (KS = acid capacity), respectively, in mining lakes.

respiration, decomposition at the sediment surface, or allochthonous carbon input from groundwater inflow or under ice cover (Table 2).

The temporal development of chlorophyll concentration in normally plankton-poor lakes and the permanent or seasonal development of chlorophyll-rich layers are represented in Figure 3. We can distinguish among lakes with episodic chlorophyll maxima (Lichtenau, Felix and the epilimnion of Waldsee) and lakes with seasonal deep chlorophyll maxima (Plessa 111) or permanent maxima in monimolimnia (Waldsee). Vertical distribution of *Ochromonas* was analysed in Plessa 111 on three separate dates. In the epilimnion of this lake, peaks of *Ochromonas* were rare, whereas in deeper layers (6 m and 7 m water depth) cells of *Ochromonas* could be observed (Plessa 111). On two sampling dates the proportion of chlamydomonads increased considerably with water depth, with maxi-

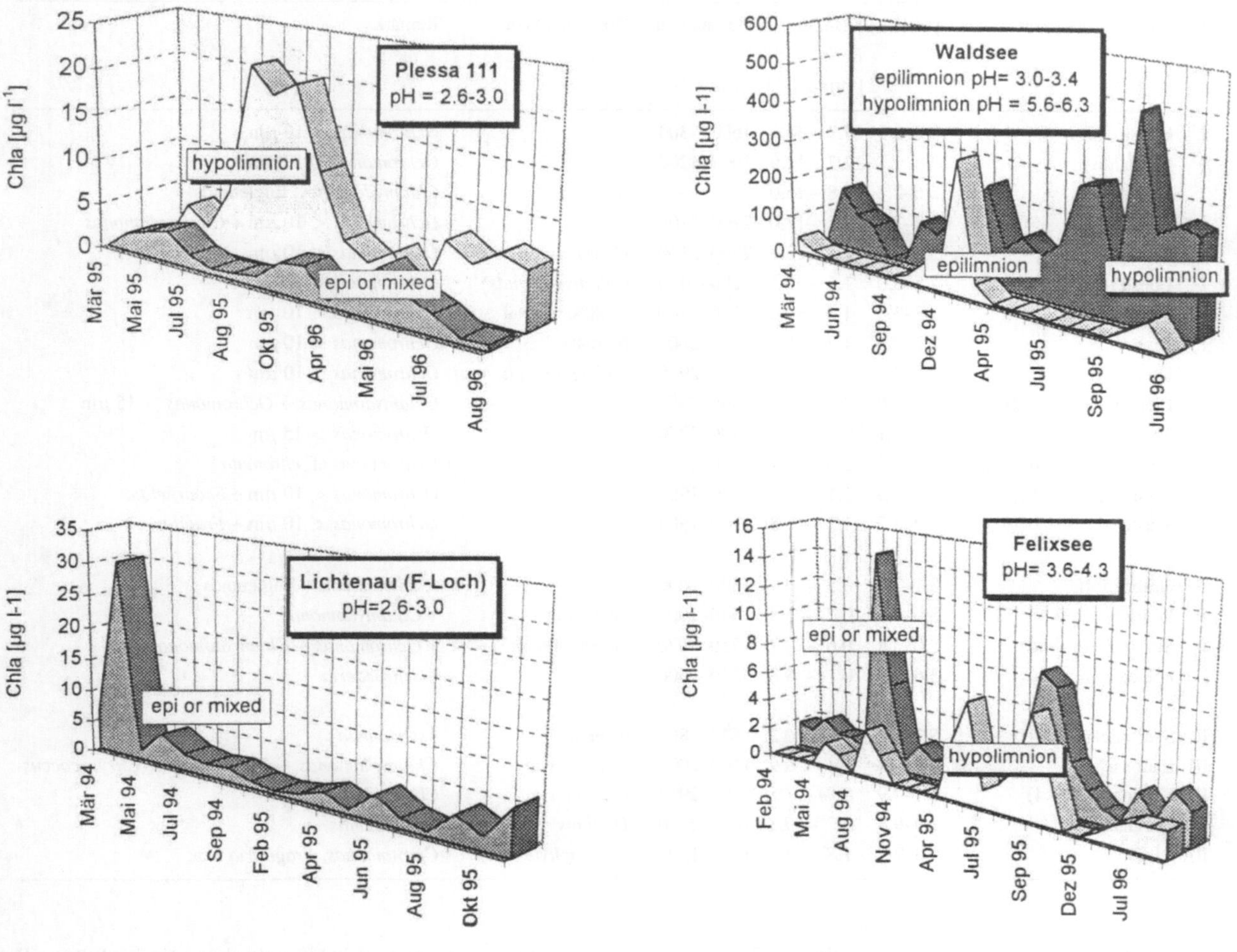

Figure 3. Temporary (Lake Lichtenau and Felixsee), deep or summer chlorophyll maxima in meromictic Waldsee and in Plessa 111.

ma just above the sediment surface (80% of the total biovolume from surface to bottom).

A further feature was observed in the acidic Lugteich lakes which are meromictic. These lakes showed extreme vertical gradients of chemistry and vertical phytoplankton distribution. We measured chlorophyll concentrations in the upper part (20–50 cm) of the acidic epilimnion between 104 and 523 μg l^{-1} Chl a in summer 1996. To be sure that these unexpected large quantities of algae were living and photosynthetically active, we tested the chlorophyll fluorescence and measured the primary production with the 14-carbon method. Both tests were positive: primary production rate was between 1.52 and 14.4 mg C (mg Chl $a * $h)$^{-1}$ indicating a high production potential even in this very acidic waters (pH between 2.7 and 2.8).

Phytoplankton composition and diversity

During the investigation period, from October 1995 to May 1996 all lakes, even the extremely acidic mining lakes (pH < 2.7), were populated by algae. The algal composition at the date with highest biovolume during the investigation period is shown in Figure 4. The phytoplankton composition of all lakes can be pooled into three groups, distinguished by the algal assemblages they supported (Table 3).

Group I: *Ochromonas/Chlamydomonas*-group. In these fourteen lakes, total biovolumes of phytoplankton ranged between 0.04 and 17.9 cm^3 m^{-3}. Only in lake Lugteich did phytoplankton biovolume exceed 2 cm^3 m^{-3}. Lakes belonging to this group

Table 3. Phytoplankton species and taxa arranged according to their dominant occurrence in acidic lakes in Lusatia in 1995 and 1996

	Lakes	pH	KB4,3/ *KS4,3* [mmol l^{-1}]	Conductivity [$\mu S\ cm^{-1}$]	Dominant taxa	Remarks
I	Plessa 107	2.3–2.5	27.3 –34.6	4600–5300		*Ochromonas* < 10 μm
I	Bärwalde	2.6	12.7 –14.9	3000–4200		*Ochromonas* < 10 μm
I	Lugteich	2.6–2.8	12.8 –15.0			*Chlamydomonas*, Euglenoids
I	Plessa 111	2.5–2.7	9.7 –12.0	2700–3100		*Ochromonas* < 10 μm + *Chlamydomonas*
I	Skado	2.5–2.9	7.3 – 8.1	2300–3100	*Ochromonas/*	*Ochromonas* < 10 μm
I	Lohsa II	2.5–3.0	1.5 – 9.2	1200–3400	*Chlamydomonas**	*Ochromonas* < 10 μm
I	Spreetal-NO	2.8–3.4	1.4 – 9.0	2300–3100	> 80% of total	*Ochromonas* < 10 μm
I	Bluno	2.7–2.9	1.7 – 5.0	2100–2600	biomass	*Ochromonas* < 10 μm
I	Dreiweibern	2.4–2.5	3.3 – 4.2	2400–2900	*(*C. cf. minutissima*)*	*Ochromonas* < 10 μm
I	Lichtenauer See (RL F)	2.6–3.0	3.4 – 3.9	2800–3300		*Chlamydomonas* + *Ochromonas* > 15 μm
I	Sedlitz	2.7–3.1	2.7 – 2.9	2400–2700		*Ochromonas* > 15 μm
I	Bergen-Niederhof (RL13)	2.8–3.0	2.0 – 2.1	1100–1300		*Ochromonas* cf. *chromata*
I	Drehnasee (RL 12)	3.0–3.3	1.1 – 1.8	1600–1900		*Ochromonas* < 10 μm + *Scourfieldia*
I	Koschen	2.9–3.5	1.3 – 1.8	1700–1900		*Ochromonas* < 10 μm + *Fragilaria*
II	Gräbendorf	3.1–3.7	0.5 – 0.9	1300–1600		*P. umbonatum* + Euglenoids
II	Waldsee (Epi)	3.0–3.4	0.7 – 3.8	550–1000	*Peridinium*	+ *Chlamydomonas*
II	Senftenberg (Süd)	3.0–3.4	0.6 – 0.7	750– 970	*Gymnodinium*	+ *Ochromonas* + *Chlamydomonas*
II	Felixsee	3.6–3.9	0.1 – 0.8	520– 600		+ *Schroederia*
III	Senftenberg (Elst)	5.7–7.2	*0.1 – 0.2*	630– 800	diverse	*Asterionella*
III	Laubusch	5.7–7.4	*0.1 – 0.9*	1100–2000	*Cryptophyta*	*Chlamydomonas* + *Ochromonas* + *Chrysococcus*
III	Schönfeld (RL 4)	6.1–7.9	*0.04– 1.5*	1500–2900	*Chlorophyta*	*Chrysochromulina*
III	Burghammer	7.9–9.4	*0.7 – 0.8*	1600–2400	*Diatoms*	*Rhodomonas*
III	RL B	7.3–7.7	*1.8 – 2.3*	1100–1200	*(Haptophyta)*	*Cryptomonas, Fragilaria* u. a.

were inhabited by small phytoflagellates of the genus *Ochromonas* and *Chlamydomonas* (Figure 4). The phytoplankton biovolumes of 12 lakes were dominated by *Ochromonas*, though absolute cell numbers were very low and ranged from to 80–7,010 cells ml^{-1}. At least three different morphotypes of *Ochromonas* could be found but they were not distinguished during the routine work (see Methods). Usually, *Ochromonas* cells were smaller than 8 μm in diameter but, in the lakes Lichtenau and Sedlitz, *Ochromonas* were greater than 15 μm in diameter. In some lakes, food vacuoles of *Ochromonas* contained high numbers of bacteria, identified in DAPI-stained filter samples.

Species of *Chlamydomonas* were not as frequent as *Ochromonas* in the acidic lakes, but in the lakes Plessa 107, Lichtenau (not shown), Lugteich and Laubusch, *Chlamydomonas* occasionally dominated the phytoplankton. One very small species of *Chlamydomonas* was able to build up deep chlorophyll maxima and to be common in lake Plessa 111. A bloom of another *Chlamydomonas* species resulted in the highest cell number ever observed in an acidic lake in Lusatia (Lugteich: > 100 000 cells ml^{-1}). Additionally, some lakes of the *Ochromonas/Chlamydomonas* type were colonized by *Chlorella*-like nannoalgae having cells of about 3 μm in diameter (Lichtenau), by the chlorophyte *Scourfieldia cordiformis* Takeda and by the euglenoid *Lepocinclis* teres (Schmitz) Francé. No cryptophytes or cyanobacteria could be detected in these lakes. In contrast to this Steinberg et al. (1998) found that Lake Lichtenau was inhabited by filamentous cyanobacteria, resembling *Oscillatoria/Limnothrix* and *Spirulina* spp.. The occurrence of dinoflagellates was restricted to few sampling dates with very low abundance in spring. The contribution of diatoms was negligible (less than 4% of total biovolume).

Group II: Dinoflagellates, chrysophytes, chlorophytes and euglenoids. In four lakes, dinoflagellates were found at the times of the biovolume maxima (Fig-

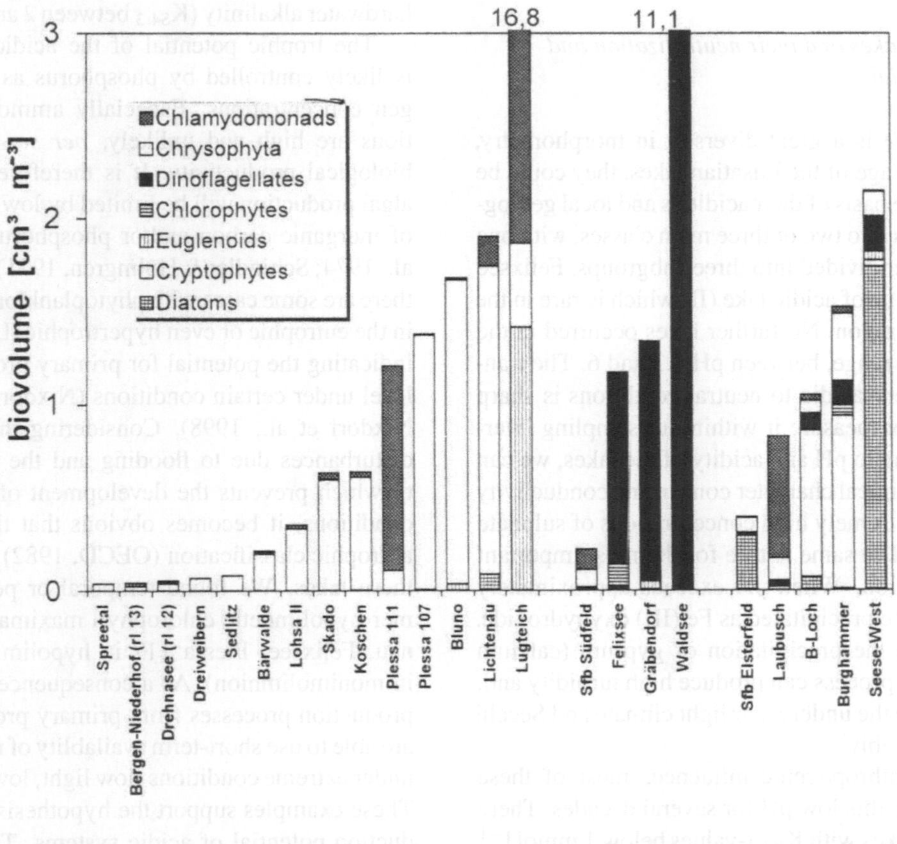

Figure 4. Phytoplankton biovolume (maximum biomass) of different taxa and species during the investigation period 1995/96.

ure 4). In lake Senftenberg Südfeld, the algal composition was similar to that in group-I lakes, except for the occurrence of *Peridinium*. In Felixsee, *Peridinium umbonatum* Stein was the most common alga throughout the investigation period. It was accompanied by filaments of Ulotrichs and occasionally by a significant biomass of the green alga, *Schroederia* (Figure 3). Phytoplankton of the lake Gräbendorf was dominated by *Gymnodinium* in October 1995, but otherwise was similar to that of group I (*Chlamydomonas/Ochromonas*) on all other sampling dates. Epilimnetic water of lake Waldsee supported blooms of *Gymnodinium sp.* in the winter months with maxima under the ice cover in January. Species biovolume exceeded more than 50 cm^3 m^{-3} in integrated epilimnion samples during the winters of 1994/95 and 1995/96. Diatoms, cryptophytes and cyanobacteria never accounted for more than 5% of total biovolume in group-II lakes and the number of common species (> 1% of total biovolume) was always fewer than 6.

Group III: Cryptophytes/chlorophytes/diatoms. Lakes of this group were far more diverse in their plankton composition (Figure 4), dominated by diatoms (*Asterionella*, *Cyclotella*) or by cryptophytes (*Cryptomonas*, *Rhodomonas*). These algal classes are also common in phytoplankton of non-acidic mining lakes in winter and early spring. Lake Laubusch is also assigned to group III, although the main contribution to phytoplankton biovolume came from chrysophytes, chlamydomonads and dinoflagellates, as in group II. As the number of relevant species was significantly higher (8–10 species) than in group II and diatoms and cryptophytes are occasionally abundant, lake Laubusch shows more similarities to group III.

Discussion

Chemistry of lakes and their neutralization and trophic potential

Although there is a great diversity in morphometry, chemistry and age of the Lusatian lakes, they could be grouped on the basis of their acidities and local geological conditions into two or three main classes, with one of these further divided into three subgroups. Felixsee represents a type of acidic lake (II) which is rare in the Brandenburg region. No further lakes occurred in the weakly-acidic range, between pH 4.3 and 6. The transition from very acidic to neutral conditions is sharp and we did not measure it within our sampling intervals. According to pH and acidity of the lakes, we can extend the chemical character concerning conductivity and high or extremely high concentrations of sulphate and calcium. The same is true for the most important heavy metal iron. When pH exceeds approximately 3.2, iron will be precipitated as Fe (III) oxyhydroxide. Together with the precipitation of gypsum (calcium sulphate), this process can produce high turbidity and, thus, influence the underwater light climate and Secchi depth considerably.

Without anthropogenic influence, most of these lakes will keep this low pH for several decades. There are only few lakes with $K_{B4.3}$-values below 1 mmol l^{-1} for which an increase in pH is easy to initiate. We found low $K_{B4.3}$-values in Gräbendorf, Senftenberg-Südfeld and Felixsee. Lake Senftenberg has two parts: one is influenced by eutrophic river water running through Elsterfeld and is neutral, the other part (Südfeld) is dominated by groundwater with a low pH (pH 3.0–3.4). This lake is an example of a chemical instability, with sensitivity to environmental changes, as was shown in an obvious way in the spring of 1995 when the flooding with neutralizing eutrophic surface water was stopped. After a short period the pH in Elsterfeld started to decrease until the river was pumped again through this part of the lake and supported rapid neutralization as a consequence of autocatalytic processes (Benndorf, 1994; Klapper et al., 1998).

By long-term maturation, avoiding anthropogenic influence, flushing with alkaline ashes (Burghammer) or permanent flooding with eutrophic river water (Lake Senftenberger, Laubusch), some of the lakes have moved to neutrality. These lakes (pH > 6) exhibit a wide range of alkalinities. Whereas the original acidic lakes (Senftenberg-Elsterfeld, Laubusch and Burghammer) have $K_{S4.3}$-values below 1 mmol l^{-1}, Schlabendorf Restloch B is characterized by typical hardwater alkalinity ($K_{S4.3}$ between 2 and 3 mmol l^{-1}).

The trophic potential of the acidic mining lakes is likely controlled by phosphorus as well as nitrogen concentrations. Especially ammonia concentrations are high and unlikely, *per se*, to be limiting biological productivity. It is therefore expected that algal production will be limited by low concentrations of inorganic carbon and/or phosphorus (Goldman et al., 1974; Schindler & Holmgren, 1971). Nevertheless, there are some cases with phytoplankton developments in the eutrophic or even hypertrophic (Lugteiche) state indicating the potential for primary production at this level under certain conditions (Nixdorf et al., 1995b; Nixdorf et al., 1998). Considering the considerable disturbances due to flooding and the extreme acidity which prevents the development of stable trophic conditions, it becomes obvious that the conventional trophic classification (OECD, 1982) is unhelpful in these lakes. We found temporal or permanent summer hypolimnetic chlorophyll maxima (Lake Lichtenau, Felixsee, Plessa 117 in hypolimnion, Lugteich in monimolimnion). As a consequence of decoupling production processes from primary production, algae are able to use short-term availablity of resources, even under extreme conditions (low light, low temperature). These examples support the hypothesis of a high production potential of acidic systems. That means that there must be sometimes (or permanently) enough phosphorus and carbon available for production. We assume an occasional increase of one or more of the production limiting factors due to inflow of CO_2-rich groundwater or higher rates of respiration. Whereas the concentration of total inorganic carbon in the very acidic waters is < 0.5 mg l^{-1}, we could detect inorganic carbon at more than 1 mg l^{-1} C in other layers of the water. One reason for this high value may be an allochthonous or autochthonous organic matter input. In one case (Felixsee), the outwash of the shore material by heavy rain was sufficient to improve the production conditions. On the other hand these phenomena indicate that pH influences plankton diversity more than abundance or biomass.

Chemistry determines algal assemblages in acidic lakes

Correspondingly, three different assemblages of phytoplankton were characterized which fitted closely to the division of the lakes. A more general classification considering the origin and the buffering mechanisms of acidic lakes is given in Geller et al. (1998). Whereas group I of the phytoplankton corresponds to subgroups I0 to Ic of the abiotic classification, group II and group III coincide with group II and III on abiotic factors.

Although the biomass of phytoplankton is generally low, as reported from similar systems elsewhere (Almer et al., 1974; Kwiatkowski & Roff, 1976; Yan, 1979), it is obvious that the potential for phytoplankton abundance and standing crop is not restricted by the acidity, as is evident from a comparison of the maxima of total phytoplankton biomass (Figure 3) and that in all groups (Figure 4). If conditions are favourable, production of phytoplankton can produce a biomass in extremely acidic waters which is comparable with that possible in hypertrophic systems (Waldsee, Lugteiche). On the other hand, the algal composition and diversity within group I seems to be strongly restricted by the acidity (range of pH 2.5–3.0), since lake morphometry and nutrient availability differ in a wide range in the corresponding lakes whilst the diversity of algae is very low. *Ochromonas* and *Chlamydomonas* represent the algal community in the extremely acidic lakes in Lusatia; they are sharply distinguished from algae in waters influenced by volcanic soils. These habitats frequently have pH values above 2, commonly accompanied by extremely high temperatures. Even in volcanic lakes without geothermal acitivity, a completely different algal community from that of Lusatia is found. In lakes with pH < 2, the main phytoplankton contributors are *Chlamydomonas acidophila* Negoro, *Dunaliella acidophila* Kalina et Masjuk, *Ochromonas vulcanica* Gromov and *Spermatozopsis acidophila* Kalina (Satake & Saijo, 1978; Pinto & Taddei, 1978; Albertano et al., 1981). These taxa were either not found or were negligible in the phytoplankton of our investigation. These observations confirm the hypothesis of Blouin (1989) that the pH tends to affect plankton diversity more than algal biomass.

Problems in the taxonomy of Ochromonas and Chlamydomonas

The taxonomy of the *Ochromonas* and *Chlamydomonas* species found in the investigated lakes in Lusatia, is unclear. The most common *Ochromonas* morphotype shares some characters with the *Ochromonas* species isolated from surface soils and described by Albertano et al. (1994). This tolerates a pH of 1.8, but grows better in the range between pH 2.0–6.0. Since no stigma could be found in cells of our species and phagotrophy can occur, we may have found a new acid-tolerant species. It also differs from another very acid- and heat- tolerant species *O. vulcania* Gromov (Gromov et al., 1991) by the length of the flagella.

Special abilities of the common algae in the extreme environmental conditions or: what enables phytoplankton to survive and to grow in extreme acidic environments?

Ochromonas, *Chlamydomonas* and *Euglena* are already known from waters polluted by acid mine waste (Lackey, 1939; Sheath et al., 1982). Normally, in waters of pH below 4 the concentration of inorganic carbon is primarily regulated by Henry's law and would not exceed about 0.31 mg l^{-1} C (Satake & Saijo, 1974). One metabolic strategy to overcome limiting carbon concentration for photosynthesis is a capacity for mixotrophy. Various chrysophytes and many *Ochromonas* species are known to be mixotrophic (Andersson et al., 1989; Jones, 1994), which can provide an important carbon source in inorganic carbon-limited waters, which seem common among acidic mining lakes. The role of bacterivorous algae in these lakes has not been established. Furthermore, many chrysophytes are low-phosphorus and low-light strategists, which enables them to grow in the Lusatian lakes. In highly acidic streams (pH < 3), *Euglena mutabilis* Schmitz and various diatoms are frequently found, whilst reports of *Ochromonas* and *Chlamydomonas* are rare (Whitton & Diaz, 1981).

Chlamydomonas acidophila Negoro is commonly found in extreme acidic environments (Rhodes, 1981; Satake & Saijo, 1978; Sheath et al., 1982; Twiss, 1990). Neither type of *Chlamydomonas* which occurs regularly in our lakes matches with the description of *C. acidophila* Negoro. Still the widespread distribution of *Chlamydomonas* spp. in acidic ponds is demonstrated by Lackey (1939). Mixotrophy (Laliberté & Noüe, 1993), and the ability of migration in deep layers to build chlorophyll maxima (Arvola et al., 1992) is known for some *Chlamydomonas* species.

A further metabolic advantage is the ability of the organisms to produce phosphatases which is an apparent adaption to the poor nutrient situation in these lakes. The induction of phosphatase activity in acidic waters by aluminium was described by Jannson (1981). *C. acidophila* Negoro responds to phosphate deficiency not only by enhancing the phosphatase activities but also by producing another phosphatase with an optimum at pH 2 (Boavida & Heath, 1986).

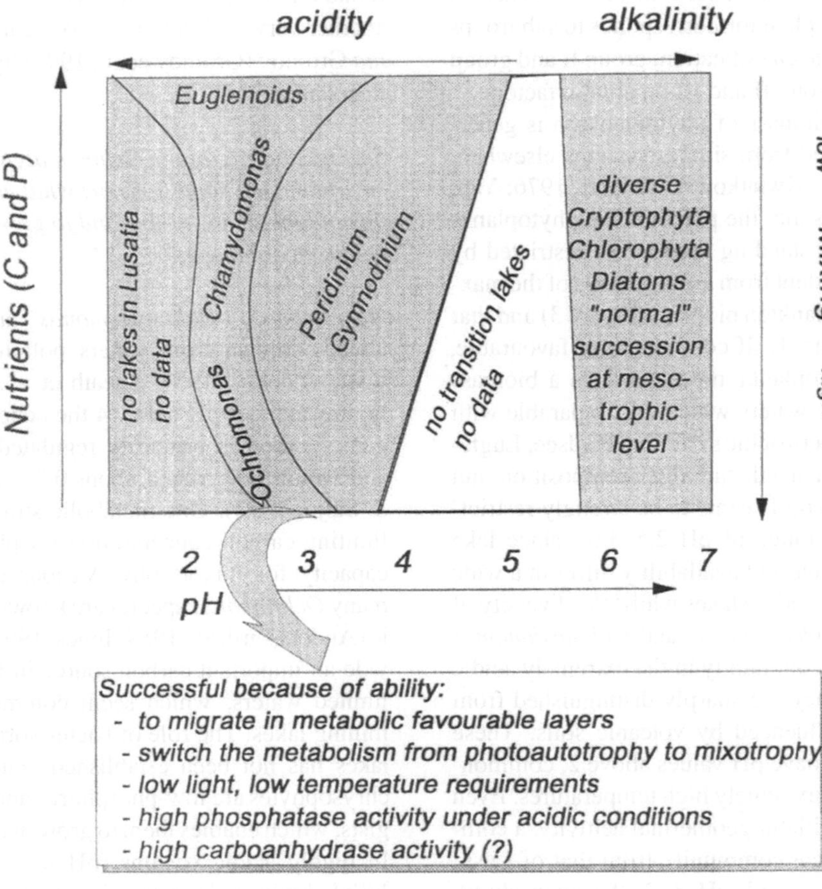

Figure 5. Strategies of phytoplankton to oppose acidity in mining lakes.

Including the chlorophyte *Scourfieldia cordiformis* Takeda and the euglenid *Lepocinclis* teres (Schmitz) Francé which are common in group I and dinoflagellates of group II, phytoflagellates are the most successful members under the given conditions: Their motility enables them to migrate to layers of the lakes where carbon and nutrient sources in the form of dissolved substrates or particles (e.g. bacteria) are available.

In Figure 5, a pictorial summary is presented which shows the chemical background and (up to now, more hypothetical) the strategies of adaptation of the common phytoplankters in very acid environments.

Acknowledgements

We thank our colleagues from IDUS-Lab, Ottendorf-Okrilla who determined and counted phytoplankton from October 1995 to May 1996, Mike Kühne, Jörg Koebcke and Ingo Henschke for their careful work during sampling; we also thank Simone Petersohn and Roswitha Pech for technical assistance in sampling preparation, and last but not least the whole staff of the research station at Bad Saarow.

This work was partially supported by BMBF (FKZ 0339648), LMBV and Volkswagen-Stiftung (Nr. II/69 854).

References

Albertano, P., G. Pinto, S. Santisi & R. Taddei, 1981. *Spermatozopsis acidophila* Kalina (Chlorophyta, Volvocales), a little known algae from highly acidic environments. Giorn. Bot. Ital. 115: 65–76.

Albertano, P., G. Pinto & A. Pollio, 1994. Ecophysiology and ultrastructure of an acidophilic species of *Ochromonas* (Chrysophyceae, Ochromonadales). Arch. Protistenk. 144: 75–82.

Almer, B., W. Dickson, C. Ekström, E. Hörnström & U. Miller, 1974. Effects of acidification on Swedish lakes. Ambio 3: 30–36.

Andersson, A., S. Falk, G. Samuelsson & A. Hagström, 1989. Nutritional characteristics of a mixotrophic nanoflagellate, *Ochromonas* sp. Microb. Ecol. 17: 251–262.

Arvola, L., K. Salonen, P. Kankaala & A. Lehtovaara, 1992. Vertical distributions of bacteria and algae in a steeply stratified humic lake under high grazing pressure from *Daphnia longispina*. Hydrobiologia 229: 253–269.

Benndorf, J., 1994. Sanierungsmaßnahmen in Binnengewässern: Auswirkungen auf die trophische Struktur. Limnologica 24: 121–135.

Blouin, A. C., 1989. Patterns of plankton species, pH and associated water chemistry in Nova Scotia lakes. Wat. Air Soil Pollut. 46: 343–358.

Boavida, M. J. & R. T. Heath, 1986. Phosphatase activity of *Chlamydomonas acidophila* Negoro (Volvocales, Chlorophyceae). Phycologia 25: 400–404.

DEV, 1986–1993. Deutsche Einheitsverfahren zur Wasser-, Abwasser- und Schlammuntersuchung. Verlag Chemie, Weinheim.

Eriksson, M. O. G., L. Henrikson, B. I. Nilsson, G. Nyman, H. G. Oscarson & A. E. Stenson, 1980. Predator-prey relations, important for biotic changes in acidified lakes. Ambio 9: 248–249.

Geller, W., H. Klapper & M. Schultze, 1998. Natural and anthropogenic sulfuric acidification of lakes. In W. Geller, H. Klapper & W. Salomons (eds), Acidic mining lakes. Springer-Verlag: 3–14.

Goldman, J. C., W. J. Oswald & D. Jenkins, 1974. The kinetics of inorganic carbon limited algal growth. J. WPCF 46: 554–574.

Gromov, B., V. N. Nikitina, & K. A. Mamkayeva, 1991. *Ochromonas vulcania* sp. nov. (Chrysophyceae) from the acidic spring on the Kunashir Island (Kurile Islands). Algologia 1: 76–79.

Jannson, M., 1981. Induction of high phosphatase activity by aluminium in acid lakes. Arch. Hydrobiol. 93: 32–44.

Jones, R. I., 1994. Mixotrophy in planctonic protists as a spectrum of nutritional strategies. Mar. Microb. Food Webs 8: 87–96.

Klapper, H. & M. Schultze, 1995. Geogenically acidified mining lakes – living conditions and possibilities of restoration. Int. Revue ges. Hydrobiol. 80: 639–653.

Klapper, H., K. Friese, B. Scharf, M. Schimmele & M. Schultze, 1998. Ways of controlling acid by ecotechnology. In W. Geller, H. Klapper & W. Salomons (eds), Acidic mining lakes. Springer-Verlag: 401–416.

Kwiatkowski, R. E. & J. C. Roff, 1976. Effects of acidity on the phytoplankton and primary productivity of selected northern Ontario lakes. Can. J. Bot. 54: 2546–2561.

Lackey, J. B., 1939. Aquatic life in waters polluted by acid mine waste. Meet. Limnol. Soc. Am. 5: 740–746.

Laliberté, G. & J. de la Noüe, 1993. Auto-, hetero- and mixotrophic growth of *Chlamydomonas humicola* (Chlorophyceae) on acetate. J. Phycol. 29: 612–620.

Mischke, U., J. Rücker, M. Kapfer & B. Nixdorf, 1994. Besiedlungsstruktur und Interaktionen im Plankton geogen versauerter Tagebaurestseen der Lausitz. Deutsche Gesellschaft für Limnologie: Erweiterte Zusammenfassungen der Jahrestagung 1994 in Hamburg: 700–704.

Nixdorf, B., J. Rücker, R. Deneke & P. Zippel, 1995a. Limnologische Zustandsanalyse von Standgewässern im Scharmützelseegebiet – Teil 1. BTU Cottbus-UW aktuelle Reihe 1/95.

Nixdorf, B., J. Rücker, B. Köcher & R. Deneke, 1995b. Erste Ergebnisse zur Limnologie von Tagebaurestseen in Brandenburg unter besonderer Berücksichtigung der Besiedlung im Pelagial. Limnologie Aktuell No.7: 39–52.

Nixdorf, B. K., K. Wollmann & R. Deneke, 1998: Ecological potentials for planktonic development and food web interactions in extremely acidic mining lakes in Lusatia (Eastern Germany). In W. Geller, H. Klapper & W. Salomons (eds), Acidic mining lakes. Springer-Verlag: 147–168.

OECD, 1982. Eutrophication of waters – Monitoring, assessment and control. OECD, Paris.

Pinto, G. & R. Taddei, 1978. Le alghe delle acque e dei suoli acidi italiani. Delpinoa 18–19: 77–106.

Porter, K. G. & Y. S. Feig, 1980. The use of DAPI for identifying and counting aquatic microflora. Limnol. Oceanogr. 25: 943–948.

Rhodes, R. G., 1981. Heterothallism in *Chlamydomonas acidophila* Negoro isolated from acidic strip-mine ponds. Phycologia 20: 81–82.

Satake, K. & Y. Saijo, 1974. Carbon dioxide content and metabolic activity of microorganisms in some acid lakes in Japan. Limnol. Oceanogr. 19: 331–338.

Satake, K. & Y. Saijo, 1978. Mechanism of lamination in bottom sediment of the strongly acid Lake Katanuma. Arch. Hydrobiol. 83: 429–442.

Schindler, D. W. & S. K. Holmgren, 1971. Primary production and phytoplankton in the Experimental Lakes Area, northwestern Ontario, and other low carbonate waters, and a liquid scintillation method for determining C14 activity in photosynthesis. J. Fish. Res. Bd. Can. 28: 189–201.

Schultze, M., H. Klapper, B. Nixdorf, U. Mischke & U. Grünewald, 1994. Methodik zur limnologischen Untersuchung und Bewertung von Bergbaurestseen. Bund-Länder Arbeitsgruppe Wasserwirtschaftliche Planung.

Sheath, R. G., M. Havas, J. A. Hellebust & T. C. Hutchinson, 1982. Effects of long-term natural acidification on the algal communities of tundra ponds at the Smoking Hills, N.W.T., Canada. Can. J. Bot. 60: 58–72.

Steinberg, C., H. Schäfer & W. Beisler, 1998. Do acid-tolerant cyanobacteria exist? Acta hydrochim. hydrobiol. 26: 13–19.

TGL 27885/01, 1982. Fachbereichstandard-Nutzung und Schutz der Gewässer-Stehende Binnengewässer, Klassifizierung, DDR.

Twiss, M. R., 1990. Copper tolerance of *Chlamydomonas acidophila* (Chlorophyceae) isolated from acidic, copper-contaminated soils. J. Phycol. 26: 655–659.

Utermöhl, H., 1958. Zur Vervollkommnung der quantitativen Phytoplanktonmethodik. Mitt. int. Ver. Limnol. 9: 1–38.

Whitton, B. A. & B. M. Diaz, 1981. Influence of environmental factors on photosynthetic species composition in highly acidic waters. Verh. int. Ver. Limnol. 21: 1459–1465.

Yan, N. D., 1979. Phytoplankton community of an acidified, heavy metal-contaminated lake near Sudbury, Ontario: 1973–1977. Wat. Air Soil Pollut. 11: 43–55.

Hydrobiologia **369/370:** 329–338, 1998.
M. Alvarez-Cobelas, C. S. Reynolds, P. Sanchez-Castillo & J. Kristiansen (eds), Phytoplankton and Trophic Gradients.

Ecology of mixotrophic flagellates with special reference to Chrysophyceae in Danish lakes

Kirsten Olrik
Laboratory of Environmental Biology ApS, Baunebjergvej 5, DK-3050 Humlebæk, Denmark

Key words: mixotrophy, mixotrophic, Chrysophyceae

Abstract

Although intensively studied in Danish lakes, Chrysophytes constituted only a minor part of the phytoplankton in the lakes studied in the Danish Survey Programme of the Water Environment during 1989–94. However, in the lakes Holm Sø, Maglesø by Brorfelde, and Bastrup Sø, populations of naked and loricated (mixotrophic) Chrysophytes exhibited 2–3 maxima yr^{-1} and contributed 2–36% to the yearly mean phytoplankton biomass. The mixotrophic Chrysophyte biomass in these lakes increased with increasing biomass of the entire phytoplankton community up to 5 mm^3 l^{-1}. Above this phytoplankton biomass, the mixotrophic Chrysophyte biomass became irregular and scarce.

Mixotrophic Chrysophytes were mainly found at TP concentrations below 0.050 mg l^{-1} and at SRP concentrations below or at the detection limit (0.010 mg l^{-1}). There was a slight increase in the biomass when SRP climbed over the detection limit but above a concentration of 0.015 mg SRP l^{-1}, mixotrophic Chrysophytes disappeared. Mixotrophic Chrysophytes increased in the interval of 2–6 mg COD l^{-1} and in the interval of 2.5–6 mg suspended matter l^{-1}. The mixotrophic Chrysophytes are hardly dependent on bacteria uptake for C but rather for P. Under substrate limitation, bacteria have a much lower C:P ratio than the minimum C:P ratio (Redfield ratio) of 106:1 for the optimum growth of algae. Under P-limitation, the C:P ratio of algae is normally higher. The difference in C:P ratio between bacteria and algae makes it possible for mixotrophic Chrysophytes to solve their demand of P by ingestion of bacteria.

Introduction

During the last decade, Lindeman's trophic-level concept has proven inappropriate for plankton organisms and attempts to force these organisms into trophic categories such as autotrophic primary producers, herbivorous primary producers, and carnivorous secondary or tertiary consumers have been challenged by mixotrophy, omnivory, conspecific cannibalism, ontogenic changes in feeding ecology, and ontogenic reversals of predator/prey relationships (Turner & Roff, 1995).

Although intensively studied in Danish lakes (Kristiansen 1991), auto- and mixotrophic Chrysophytes constituted an average of only 0.02–0.14 mm^3 l^{-1}, equal to 0.1–1.0% of the summer phytoplankton biomass of all the lakes studied by the Danish Survey Programme of the Water Environment during 1989–94 (Jensen et al., 1995). Thus, apart from several clear exceptions, Chrysophytes are not quantitatively important in the Danish lake phytoplankton.

The aim of this study is to elucidate the conditions for life of the mixotrophic Chrysophytes in three Danish lakes where these organisms constitute a prominent part of the phytoplankton.

Mixotrophy

Mixotrophy is the combined ability to live as an autotrophic primary producer and as a secondary or tertiary consumer by ingestion and digestion of organic particles (Sanders 1991). This means of nutrition is found among several plankton flagellates, both primarily autotrophic and primarily heterotrophic ones. Recent research has demonstrated the widespread

Table 1. Mixotrophic phytoplankton organisms. (From: Eloranta & Räike, 1995; Jacobson & Andersen, 1994; Jones et al., 1995; Jones & Rees, 1995a, 1995b; McKenzie et al., 1995; Kirchhoff & Meyer, 1995; Bockstahler & Coats, 1993a, 1993b; Dolan, 1993; Jones et al., 1993; Rees & Jones, 1992; Sibbald & Albright, 1992; Eloranta & Järvinen, 1991; Fields & Rhodes, 1991; Bird & Kalff, 1989; Porter, 1988; Sanders & Porter, 1988)

CRYPTOPHYCEAE – CRYPTOPHYTES	RAPHIDOPHYCEAE
Cryptomonas	*Gonyostomum semen*
DINOPHYCEAE – DINOFLAGELLATES	TRIBOPHYCEAE – TRIBOPHYTES
Amphidinium	*Chlorochromonas*
Ceratium	
Dinophysis	DICTYOCHOPHYCEAE
Gymnodinium	*Dichtyocha*
Gyrodinium	
Massartia	PRYMNESIOPHYCEAE
Peridinium	*Chrysochromulina*
	Coccolithus
CHRYSOPHYCEAE – CHRYSOPHYTES	
Catenochrysis	EUGLENOPHYCEAE
Chromulina	*Euglena*
Chrysococcus	*Phacus*
Chrysosphaerella	
Chrysamoeba	CHLOROPHYCEAE – GREEN ALGAE
Chrysostephanosphaera	*Chlamydomonas*
Cyrtophora	*Chlorogonium*
Dinobryon	
Mallomonas	
Ochromonas	
Palatinella	
Pedinella	
Phaeaster	
Poteriochromonas	
Pseudopedinella	
Uroglena	

occurrence of mixotrophic flagellates in both freshwater plankton (e.g Veen, 1991) and marine plankton (e.g. Havskum, 1996). These organisms vary widely in their photosynthetic and ingestion capabilities and may be considered to occupy different points along an axis of nutritional strategies from absolute autotrophy to absolute heterotrophy. At least seven algal classes (Chrysophyceae, Cryptophyceae, Dictyophyceae, Dinophyceae, Raphidophyceae, Xanthophyceae, and Prymnesiophyceae) include mixotrophic genera (Table 1).

Chrysophytes

Within the Chrysophytes *sensu lato*, there is a wide spectrum of nutritional capabilities from strict phototrophy (*Synurophyceae*), through varying levels of mixotrophy (*Dinobryon, Uroglena, Ochromonas danica, Chrysamoeba*) to strict osmotrophy (*Paraphysomonas*) and strict phagotrophy (*Spumella, Ochromonas vallesiaca*). The mixotrophic Chrysophytes have a special structure to the flagellar apparatus as they possess extra microtubular roots with which is formed a 'feeding cup' or a 'cave'. When a food particle is captured between the flagella, the cell forms the so-called 'feeding cup' to engulf the particle. The strictly autotrophic Synurophyceae (e.g. *Mallomonas* and *Synura*) do not possess this structure (Andersen & Wetherbee, 1992). Furthermore, the mixotrophic Chrysophytes are all naked or loricated, whereas all of the Synurophyceae are covered by scales and some of them possess spines.

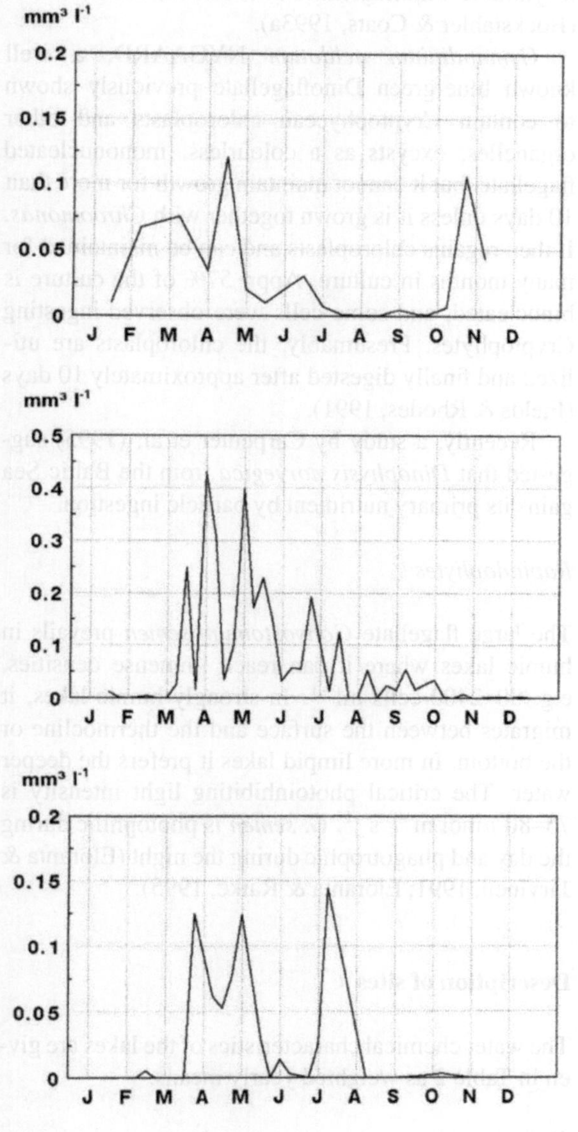

Figure 1. Annual cycles of mixotrophic Chrysophytes (mm^3 l^{-1}). Bi-monthly averages from 1989-95. Top: Holm Sø. Centre: Maglesø by Brorfelde, Bottom: Bastrup Sø.

The grazing capacity of mixotrophic Chrysophyte on bacteria was estimated by Porter (1988) and Salonen & Jokinen (1988) to represent 55% of the entire microflagellate- and zooplankton grazing on bacteria. *Ochromonas* (Figure 1) and *Chromulina* (Chrysophyceae) were capable of digesting 75–203% of their body weight in bacterial carbon per day (Salonen & Jokinen, 1988).

Although primarily autotrophic, a marine flagellate *Ochromonas* sp. can fufil its diet by selective phagotrophic ingestion of different types of bacteria under conditions of low or limiting light and nutrients. Nevertheless, the growth of *Ochromonas* decreased in periods of prolonged darkness (Keller et al., 1994). *Ochromonas* utilized C, N, P, and vitamins from phagotrophic ingestion of bacteria. In contrast to light-grown cultures, phagotrophic rates increased in the dark in N-, P-, or vitamin-limited cultures (Sibbald & Albright, 1992).

Two freshwater strains of the Chrysophyte *Dinobryon* were studied by Jones & Rees (1995a, 1995b). The cells were grown in clonal batch cultures and offered flourescent beads in suspension. The preferred particle size for ingestion was 0.49–0.92 μm. Smaller particles (0.22 μm) were ingested to a lesser extent, while larger particles (> 2.5 μm) were not ingested at all. Clearence rates were found to be strongly dependent on temperature, the optimum being at 20 °C. Clearance rate was slightly higher during the dark phase of the diurnal cycle. Nevertheless, photosynthesis was still the primary source of carbon for *Dinobryon* in these experiments.

The marine Chrysophyte *Dinobryon balticum* was capable of phagotrophy and twice as many healthy *Dinobryon* cells were ingesting fluorescent latex beads at 40 m depth than at 5 m depth, although the ingestion rates for those cells actively ingesting beads were similar at both depth (McKenzie et al., 1995). In their extensive observations on simultaneous phagotrophy and photosynthesis, Bird & Kalff (1986, 1987, 1989) found that, subject to ambient light conditions, phagotrophy might account for up to 80% of the algal community carbon assimilation. This high proportion was measured only on a heavily overcast day in a highly concentrated metalimnetic layer of *Dinobryon* and *Ochromonas*. The absolute amount of assimilated bacterial carbon was rather low, less than 10% of the algal carbon content. Veen (1991) measured only insignificant amounts of bacterial carbon to be assimilated by *Dinobryon* (1–5 nl cell^{-1} h^{-1}).

Prymnesiophytes

Mixotrophy in *Chrysochromulina* appears to be a complex mode of nutrition, as dependent upon environmental variables as on nutritional status. The feeding mechanism has been studied in *Chrysochromulina spinifera* by Kawachi & Inouye (1995): A water current generated by flagellar movement brings food particles to the cell. The particles adhere to the long spine scales that cover the cell. They are then moved to the tip of the haptonema where an aggregate is formed. The haptonema bends to transfer the aggregate to a specific area of the cell surface, where it is ingested by phagocytosis. Some species take up dissolved organic carbon as well as ingesting particulate prey. Other species are purely autotrophic. The marine species *Chrysochromulina brevifilum* was grown on living cells of a small green flagellate. The rate of ingestion of the flagellate was proportional to the concentration of the green flagellate and inversely proportional to light intensity. The species changed its cellular chlorophyll *a* content and ingestion of prey according to the photon flux density. Under a relatively low photon flux density (25 μmol photons m^{-2} s^{-1}) both the concentration of chlorophyll *a* and the ingestion of prey were greater than under higher photon flux density (100 μmol photons m^{-2} s^{-1}). Ingestion was higher by P-starved than by P-saturated *C. brevifilum* (Jones 1995). *Chrysochromulina* sp. was also capable of digesting small Diatoms (Jones, 1990).

Dinoflagellates

Phagotrophy in Dinoflagellates is particularly prominent among the colourless specimens, but is found among pigmented forms as well. The freshwater species *Ceratium hirundinella* has two different mechanisms of phagotrophy seldom observed (Hofeneder in Gaines & Elbraechter, 1987). One mechanism is extra-cellular digestion in a fine network of cytoplasmic threads emerging from the pores of the cell wall. In this network Diatoms, other algae and small flagellates are trapped and digested outside the *Ceratium* cell itself, with subsequent uptake of the dissolved products. The other form is 'true' phagotrophy, in which the cell emits a pseudopod through the ventral opening in the longitudinal furrow and engulfs a food item (blue-green alga, diatom etc.). The food item is subsequently dragged into the cell through the ventral opening in the longitudinal furrow. Dodge & Crawford (1970) made electron microscopic studies revealing remains of Bacteria, Cyanophytes and Diatoms in *Ceratium*-cells.

In Cheasapeake Bay, the large marine phototrophic red tide organism *Gymnodinium sanguineum* (Dinophyceae) removed ciliate biomass which represented 6–67% of the oligotrich standing stock. Daily consumption of ciliate biomass by *G. sanguineum* averaged 2.5% of body C and 4% of body N with maximal values of 12 and 19%, respectively. Ingestion of ciliates may give *G. sanguineum* an advantage over purely photosynthetic Dinoflagellates in N-limited environments (Bockstahler & Coats, 1993a).

Gymnodinium acidotum NYGAARD, a well known blue-green Dinoflagellate previously shown to contain Cryptophycean chloroplasts and other organelles, excysts as a colourless, mononucleated flagellate, but it cannot maintain growth for more than 10 days unless it is grown together with *Chroomonas*. It then regains chloroplasts and can be maintained for many months in culture. Appr. 57% of the culture is binucleated, and some cells were observed ingesting Cryptophytes. Presumably, the chloroplasts are utilized and finally digested after approximately 10 days (Fields & Rhodes, 1991).

Recently, a study by Carpenter et al. (1995) suggested that *Dinophysis norvegica* from the Baltic Sea gains its primary nutriment by particle ingestion.

Raphidophytes

The large flagellate *Gonyostomum semen* prevails in humic lakes where it can reach immense densities, e.g. 80–2400 cells ml^{-1}. In strongly humic lakes, it migrates between the surface and the thermocline or the bottom. In more limpid lakes it prefers the deeper water. The critical photoinhibiting light intensity is 75–80 μmol m^{-2} s^{-1}. *G. semen* is photophilic during the day and phagotrophic during the night (Eloranta & Järvinen, 1991; Eloranta & Räike, 1995).

Description of sites

The water chemical characteristics of the lakes are given in Table 2 as weighted yearly means.

Holm Sø

Holm Sø in Western Jutland, The County of Ribe, is a soft-watered and shallow *Lobelia* lake on poor, sandy soil and with a small drainage area. The lake is situated near Oxbøl close to the North Sea, surrounded by sand dunes, heath, and grassland. Holm Sø has no outlet: there is a balance between winter influx and summer evaporation. In consequence, ammonia from precipitation and from a small ditch gradually accumulates in the lake.

Maglesø by Brorfelde

Maglesø by Brorfelde in the County of West Zealand is situated in a dead-ice moraine landscape. The lake has

Table 2. Water chemical characteristics from Holm Sø, Maglesø by Brorfelde, and Bastrup Sø. pH ranges and yearly means from 1989–92 (Holm Sø and Maglesø by Brorfelde) and from 1989, 1991–94 (Bastrup Sø)

		Holm Sø 1989–92	Maglesø by Brorfelde 1989–92	Bastrup Sø 1989, 1991–94
pH (range)		3.9–5.8	7.5–8.7	7.9–9.1
Total-N (yl mean)	mg l^{-1}	1.048	1.303	1.060
NH_4/NH_3-N (yl mean)	mg l^{-1}	0.482	0.036	0.057
NO_3/NO_3-N (yl mean)	mg l^{-1}	0.140	0.407	0.115
Total-P (yl mean)	mg l^{-1}	0.019	0.027	0.066
PO_4-P (yl mean)	mg l^{-1}	0.006	0.010	0.009
Inorg N/Total P ratio		32.916	17.754	3.472
Total alkalinity (yl mean)	mmol l^{-1}	0.014	2.655	2.674
Part. COD (yl mean)	mg l^{-1}	–	3.716	–
Susp. matter (yl mean)	mg l^{-1}	–	3.045	–
Total phytoplankton biomass (yl mean)	mm^3 l^{-1}	0.425	1.108	5.713
Mixotrophic Chrysophytes (yl mean)	mm^3 l^{-1}	0.011	0.077	0.037

a very small drainage area. In spite of a maximum depth of only 6.0 m, the lake develops a distinct thermocline every summer. The lake bottom is covered by a dense vegetation of e.g. *Chara*.

Bastrup Sø

Bastrup Sø in the county of Frederiksborg, North Zealand, is situated in a tunnel valley in a moraine landscape. The lake has steep slopes close to the shore where the depth decreases to 3–4 m below water level. Mean depth is 3.3 m and maximum depth 7.0 m. The drainage area consists of 80% agricultural areas and 12% woody areas. 55 estates within the drainage area are without sewage regulation. The lake receives water from two small creeks that dry out during summer. The retention time of water in the lake is 2–5 years (Hansen & Hansen, 1995).

Materials and methods

Samples for phytoplankton analysis and water chemistry analysis used in this study were taken at a central sampling station in each of the three lakes Holm Sø (County of Ribe), Maglesø by Brorfelde (County of West Zealand), and Bastrup Sø (County of Frederiksborg). The sampling frequency was fortnightly during the vegetation period, Mar-Oct (1989–94) and monthly during the rest of the year. The samples were taken by

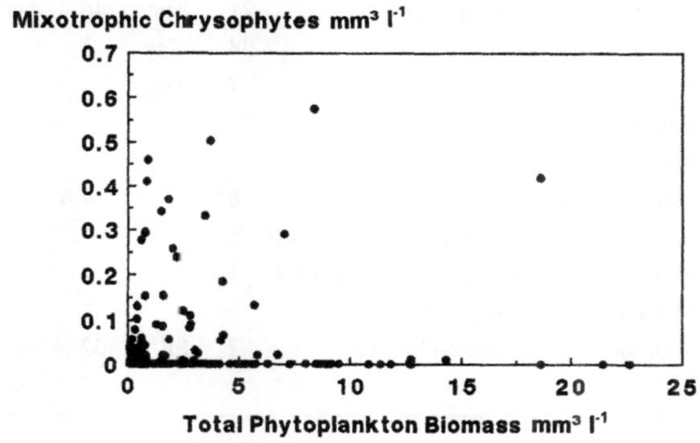

Figure 2. Distribution of mixotrophic Chrysophyte biomass (mm^3 l^{-1}) in relation to total phytoplankton biomass (mm^3 l^1).

the staff of the environmental departments in the three counties.

Phytoplankton samples were preserved with Lugol's solution and acetic acid. After sedimentation in counting chambers, the samples were enumerated in a Leitz inverted microscope equipped with phase contrast. At least 100 individuals of the more common species were counted. On each sampling date, cell dimensions and mean cell volumes of the phytoplankton species were calculated for all counted species. The calculations were based on measurements of 10–20 individuals of each species.

The chemical analyses were performed according to Danish Standards (Dansk Standardiseringsråd 1975a, b, c; 1977a, b; 1978; 1984; 1985a, b).

Table 3. Yearly mean biomass of total phytoplankton and of mixotrophic Chrysophytes exclusively (mm^3 l^{-1} and %). Mixotrophic Chrysophyte maxima (mm^3 l^{-1}), their percentage part of the total phytoplankton biomass, and the seasonal period of mixotrophic Chrysophyte maxima in Holm Sø, Maglesø by Brorfelde, and Bastrup Sø

	1989	Season of max	1990	Season of max	1991	Season of max	1992	Season of max	1993	Season max of	1994	Season max of	1995	Season of max
Holmsø														
Phytoplankton biomass mm^3 l^{-1}														
Mean Jan–Dec	0,4		0,2		0,2		0,9		0,6		0,3		0,2	
Mean mixotrophic Chrysophyte biomass														
% of mean total phytoplankton biomass	1		1		10	2	12	36	26					
Mixotrophic Chrysophyte maxima														
% of total phytoplankton biomass	6	Jul	4	Apr	88	Jan	2	Feb	56	Feb	87	Mar	51	Feb
	4	Aug	3	Nov	39	Apr	2	Jun	13	Apr	95	Apr	82	Mar
					8	Nov	4	Nov	76	Oct	81	Oct	71	Jun
Maglesø by Brorfelde														
Phytoplankton biomass mm^3 l^{-1}														
Mean Jan–Dec	1,3		1,3		1,3		1,3		3,3		1,8		1,9	
Mean mixotrophic Chrysophyte biomass														
% of mean total phytoplankton biomass	2		1		8		10		4		20		8	
Mixotrophic Chrysophyte maxima														
% of total phytoplankton biomass	21	Apr	46	Apr	51	May	84	May	56	May	77	Apr	18	May
	28	May			37	Jun	10	Jun			36	May	31	Jul
	7	Aug									8	Sep	15	Oct
Bastrup Sø														
Phytoplankton biomass mm^3 l^{-1}														
Mean Jan–Dec	8,0		4,6		–		6,6		4,1		6,3		0	
Mean mixotrophic Chrysophyte biomass														
% of mean total phytoplankton biomass	1		2		–		0,2		1		0,2		0	
Mixotrophic Chrysophyte maxima														
% of total phytoplankton biomass	2	Jul	52	Apr	–		6	Feb	14	May	2	Mar	0	
			7	Jul										

Results

Occurrence of Mixotrophic Chrysophytes in Holm Sø, Maglesø by Brorfelde, and Bastrup Sø

In the lakes Holm Sø, Maglesø by Brorfelde and Bastrup Sø, populations of mixotrophic Chrysophytes developed 2–3 maxima yr^{-1} (Figures 1–2, Table 3) and contributed 2–36% to the yearly mean phytoplankton biomass. The mixotrophic Chrysophyte biomass in these lakes increased with increasing biomass of the entire phytoplankton community up to 5 mm^3 l^{-1}. Above this phytoplankton biomass, the mixotrophic Chrysophyte biomass became irregular and scarce (Figure 2).

Holm Sø

The average phytoplankton biomass through the year in Holm Sø 1989–95 was 0.2–0.9 mm^3 l^{-1}, and mixotrophic Chrysophytes contributed 1–36% (Table 3). The mixotrophic Chrysophytes had 2–3 peaks per year, most often two maxima in spring and one in autumn (Figure 1). During their maxima, mixotrophic Chrysophytes constituted from 2% to 95% of the phytoplankton biomass. The most important mixotrophic Chrysophyte taxa were *Ochromonas* spp., *Chromulina* spp., *Dinobryon sertularia*, and *D. pediforme* (Table 4).

Maglesø by Brorfelde

The average phytoplankton biomass through the year in Maglesø by Brorfelde 1989–95 was 1.3–3.3 mm^3 l^{-1} of which mixotrophic Chrysophytes contributed 1–20% (Table 3). In this lake, mixotrophic Chrysophytes

Table 4. List of species of mixotrophic Chrysophytes from the lakes Holm Sø, Maglesø by Brorfelde and Bastrup Sø 1989–95

		Holm Sø	Maglesø by Brorfelde	Bastrup Sø
Bitrichia chodatii	(rev.) chod.		×	
Chromulina spp.		×	×	
Chrysamoeba spp.				×
Chrysococcus spp.				×
Chrysolykos planktonicus	mack		×	
Dinobryon bavaricum	imh.		×	
Dinobryon crenulatum	w. & g.s. west		×	
Dinobryon cylindricum	imh.		×	
Dinobryon divergens	imh.		×	×
Dinobryon pediforme	(lemm.) steinecke	×		
Dinobryon sertularia	ehr.	×		
Dinobryon sociale	ehr.		×	×
Kephyrion rubri-claustrii	conr.		×	
Kephyrion spp.		×	×	
Ochromonas spp.		×		×
Uroglena spp.		×		

developed 2–3 peaks per year, most often two maxima during spring and one in autumn (Figure 1). During their maxima, mixotrophic Chrysophytes constituted from 7% up to 84% of the phytoplankton biomass. Maglesø by Brorfelde is rich in mixotrophic Chrysophyte species. The quantitatively most important taxa were *Dinobryon sociale*, *D. divergens*, *D. cylindricum*, *D. bavaricum*, and *D. crenulatum* (Table 4).

Bastrup Sø

The average phytoplankton biomass in Bastrup Sø 1989–90 and 1992–95 was 4.1–8.0 $mm^3\ l^{-1}$ and mixotrophic Chrysophytes contributed 0.2–2% (Table 3). In this lake the mixotrophic Chrysophytes developed 1–2 peaks per year, most often in spring and summer (Figure 1). During their maxima, the mixotrophic Chrysophytes constituted from 2% to 52% of the phytoplankton biomass. The most important mixotrophic Chrysophyte taxon was *Dinobryon divergens* (Table 4).

Occurrence of mixotrophic Chrysophytes according to water chemical variables in the lakes Holm Sø, Maglesø by Brorfelde, and Bastrup Sø

pH and Alcalinity

Mixotrophic Chrysophytes prevailed both at pH 4–6, at very low alkalinities close to zero, and at pH 8–9 and alkalinities between 2 and 4 mmol l^{-1}. Thus, mixotrophic Chrysophytes had a bipolar distribution in relation to pH and alkalinity.

Nitrogen

Mixotrophic Chrysophytes were found at TN concentrations between 0.3 and 3.5 mg l^{-1}. Most of the occurrences were in the interval 0.5–1.5 mg l^{-1}. The biomass increased slightly in the interval 1–2 mg TN l^{-1}. Above this concentration, the mixotrophic Chrysophytes disappeared.

Mixotrophic Chrysophytes were found at inorg. N concentrations between zero and 1.5 mg l^{-1}. Most of the occurrences were below 0.5 mg inorg. N l^{-1}. The mixotrophic Chrysophyte biomass increased slightly with decrease of the concentration of inorg. N. Maxima occurred at inorg. N concentrations close to the detection limit. Above a concentration of 1.5 mg inorg. N l^{-1}, mixotrophic Chrysophytes did not occur at all.

Phosphorus

Mixotrophic Chrysophytes were found at TP concentrations between detection limit for P (0.010 mg l^{-1}) and 0.150 mg l^{-1} (Figure 3). Most occurrences were at TP concentrations below 0.050 mg l^{-1}. Mixotrophic Chrysophyte biomass increased in the interval 0.010–0.050 mg TP l^{-1}. Above 0.050 mg TP l^{-1} the mixotrophic Chrysophytes disappeared.

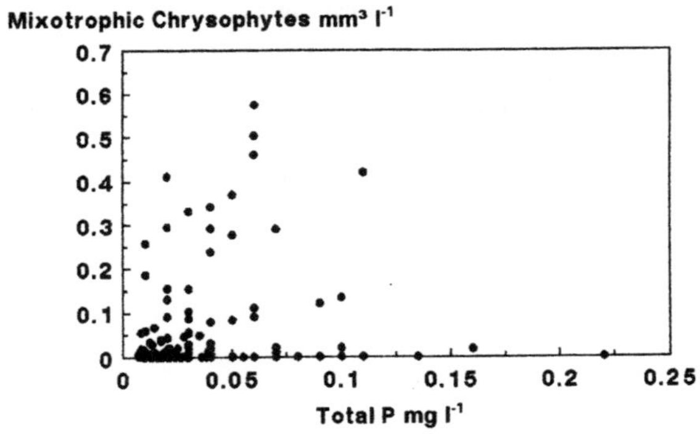

Figure 3. Distribution of mixotrophic Chrysophyte biomass (mm^3 l^{-1}) in relation to TP (mg l^{-1}).

Mixotrophic Chrysophytes were found almost exclusively at concentrations of SRP at or below the detection limit (0.010 mg l^{-1}). There was a slight increase in the biomass when SRP climbed over the detection limit but, above a concentration of 0.015 mg SRP l^{-1}, mixotrophic Chrysophytes disappeared.

COD (Chemical Oxygen Demand)
Mixotrophic Chrysophytes were found at COD concentrations between 2 and 8 mg COD l^{-1}), their biomass increased in the interval 2–6 mg COD l^{-1}.

Suspended Matter
Mixotrophic Chrysophytes were found at concentrations of suspended matter between 1 and 8 mg l^{-1}, and their biomass increased in the interval of 2.5–6 mg suspended matter l^{-1}.

Discussion

The relative contributions of autotrophy and heterotrophy to the growth of mixotrophic organisms under various environmental conditions are still poorly understood. Consequently, the ecological advantages of mixotrophy as a strategy for planktonic organisms still remain largely on speculation (Jones, 1995).

Primarily autotrophic organisms might shift to particle or bacterial uptake for competitive advantage in carbon- and energy (light)-limited surroundings or in surroundings limited in nutrients other than carbon. Primarily heterotrophic organisms might shift to autotrophic nutrition for competitive advantage in prey deficient surroundings or in competition with other (more efficient) grazers.

The mixotrophic Chrysophytes in this study declined in biomass coincidentally with increasing phosphorus levels: They have an affinity to SRP close to or below the detection limit and TP concentrations below 0.050 mg l^{-1}. Similar results were found by Smith (1990).

The mixotrophic Chrysophytes in this study obviously increase with increasing concentration of COD and suspended matter. The assimilation of bacterial carbon does not automatically imply that the gain of carbon is the ultimate reason for phagotrophy. Logically, the assimilation of bacterial carbon will only be of advantage under carbon- and light-limiting conditions. As in this investigation, *Dinobryon* and other mixotrophic Chrysophytes normally occur in rather clear water under P-deficient conditions, so light would hardly be limiting (e.g. Veen, 1991).

Several investigations during the last two decades have shown that under P-deficient conditions, less than 20% of the SRP is taken up by phytoplankton. The pool is exploited by bacteria which are much more efficient in this respect at low concentrations of SRP (Currie & Kalff, 1984a, 1984b; Veen, 1991). Bacteria would outcompete phytoplankton if they were not substrate-limited. Under substrate limitation however, bacteria have a much lower C:P ratio than the minimum C:P ratio (Redfield ratio) of 106:1 for the optimum growth of algae. Under P-limitation, the C:P ratio of algae is normally higher (Vadstein et al., 1988). This difference between bacteria and algae makes it quite possible for *Dinobryon* (and other phytoplankters) to solve their demand of P by ingestion of bacteria (Veen, 1991).

Acknowledgements

I am most obliged to the County of Ribe, the County of West Zealand, and the County of Frederiksborg for permission to use their data on water- and nutrient balance, water chemistry, and bottom vegetation. Due thanks to Dr Colin Reynolds, The Ferry House, Ambleside, Cumbria, U.K., who kindly revised the English language.

References

Andersen, R. A. & R. Wetherbee, 1992. Microtubules of the flagellar apparatus are active during prey capture in the chrysophycean alga *Epipyxis pulcra*. Protoplasma 166: 8–20.
Bird, D. F. & J. Kalff, 1986. Bacterial grazing by planktonic lake algae. Science 231: 493–495.

Bird, D. F. & J. Kalff, 1987. Algal phagotrophy: Regulating factors and importance relative to photosynthesis in *Dinobryon* (Chrysophyceae). Limnol. Oceanogr. 32: 277–284.

Bird, D. F. & J. Kalff, 1989. Phagotrophic sustainance of a metalimnic phytoplankton peak. Limnol. Oceanogr. 34: 155–162.

Bockstahler, K. R. & D. W. Coats, 1993a. Grazing of the Dinoflagellate *Gymnodinium sanguineum* on ciliate populations of Chesapeake Bay. Mar. Biol. (Berlin) 116: 477–487.

Bockstahler, K. R. & D. W. Coats, 1993b. Spatial and temporal aspects of mixotrophy in Chesapeake Bay Dinoflagellates. J. Euc. Microbiol. 40: 49–60.

Carpenter, E. J., S. Janson, R. Boje, F. Pollehne & J. Chang, 1995. The Dinoflagellate *Dinophysis norvegica*: biological and ecological observations in the Baltic Sea. Europ. J. Phycol. 30: 1–9.

Currie, D. J. & J. Kalff, 1984a. A comparison of the abilities of freshwater algae and bacteria to acquire and retain phosphorus. Limnol. Oceanogr. 29: 298–310.

Currie, D. J. & J. Kalff, 1984b. The relative importance of bacterioplankton and phytoplankton in phosphorus uptake in freshwater. Limnol. Oceanogr. 29: 311–321.

Dansk Standardiseringsråd, 1975a. Water analysis. Determination of nitrogen content by peroxodisulphate. Dansk Standard DS 221. 1. udg.

Dansk Standardiseringsråd, 1975b. Water analysis. Determination of the sum of nitrite- and nitrate-nitrogen. Dansk Standard DS 223. 1. udg.

Dansk Standardiseringsråd, 1975c. Water analysis. Determination of ammonia-nitrogen. Dansk Standard DS 224. 1. udg.

Dansk Standardiseringsråd, 1977a. Water analysis. Determination of alkalinity. Dansk Standard DS 253. 1. udg.

Dansk Standardiseringsråd, 1977b. Water analysis. Components of the carbonatesystem (Calculation method). Dansk Standard DS 256. 1. udg.

Dansk Standardiseringsråd, 1978. Water analysis. Determination of pH. Dansk Standard DS 287. 2. udg.

Dansk Standardiseringsråd, 1984. Water analysis. Chloride, potentiometric method. Dansk Standard DS 239. 1. udg.

Dansk Standardiseringsråd, 1985a. Water analysis. Phosphate. Photometric method. Dansk Standard DS 291. 2. udg.

Dansk Standardiseringsråd, 1985b. Water analysis. Total phosphor. Photometric method. Dansk Standard DS 292. 2. udg.

Dodge, J. D. & R. M. Crawford, 1970. The morphology and fine structure of *Ceratium hirundinella* (Dinophyceae). J. Phycol. 6: 137–149.

Dolan, J., 1993. Mixotrophy in ciliates: A review of *Chlorella* symbiosis and chloroplast retension. Mar. Microb. Food Webs 6: 115–132.

Eloranta, P. & M. Järvinen, 1991. Growth of *Gonyostomum semen* (EHR.) DIESING (Raphidophyceae): Results from culture experiments. Verh. int. Ver. Limnol. 24: 2657–2659.

Eloranta, P. & A. Räike, 1995. Light as a factor affecting the vertical distribution of *Gonyostomum semen* (EHR.) DIESING (Raphidophyceae) in lakes. Aqua Fenn. 25: 15–22.

Fields, S. D. & R. G. Rhodes, 1991. Ingestion and retention of *Chroomonas* spp. (Cryptophyceae) by *Gymnodinium acidotum* (Dinophyceae). J. Phycol. 27: 525–529.

Gaines, G. & M. Elbrächter 1987. Heterotrophic nutrition. In: F. J. R. Taylor (ed.), The Biology of Dinoflagellates. Bot. Monogr. 21. Blackwell Scientific Publications, 785 pp.

Hansen, L. Reersø & R. Stehn Hansen, 1995. Bastrup Sø – tilstand og udvikling 1994. Teknik & Miljø, Frederiksborg Amt. Vandmiljøovervågning nr. 20, 103 pp.

Havskum, H., 1996. Ecological importance of phagotrophic, pigmented flagellates (mixotrophs) in marine plankton. Ph.D. thesis. Univ. of Copenhagen.

Jacobson, D. M. & R. A. Andersen, 1994. The discovery of mixotrophy in photosynthetic species of *Dinophysis* (Dinophyceae): Light and electron microscopial observation of food vacuoles *Dinophysis acuminata, D. norvegica* and two heterotrophic dinophysoid dinoflagellates. Phycologia 33: 97–110.

Jensen, J. P., E. Jeppesen, M. Søndergaard, J. Windolf, T. Lauridsen & L. Sortkjær, 1995. Vandmiljøplanens Overvågningsprogram 1994. Ferske vandområder. Søer. Faglig rapport fra DMU, nr. 139, 115 pp.

Jones, H., 1990. Particle ingestion by a marine species of *Chrysochromulina* (Prymnesiophyceae). Abstract from: First International Symp. on Free-living Heterotrophic Flagellates, Aug. 1990, Helsingør, Denmark.

Jones, R. I., 1995. Mixotrophy in phytoplankton protists as a spectrum of nutritional strategies. Mar. Microb. Food Webs 8: 87–96.

Jones, R. I. & S. Rees, 1995a. Characteristics of particle uptake by the phagotrophic phytoflagellate *Dinobryon divergens*. Mar. Microb. Food Webs 8: 97–110.

Jones, R. I. & S. Rees, 1995b. Influence of temperature and light on particle ingestion by the freshwater phytoflagellate *Dinobryon*. Arch. Hydrobiol. 132: 203–211.

Jones, H. L. J., P. Durjun, B. S. C. Leadbeater & J. C. Green, 1995. The relationship between photoacclimation and phagotrophy with respect to chlorophyll *a*, carbon and nitrogen content, and cell size of *Chrysochromulina brevifilum* (Prymnesiophyceae). Phycologia 34: 128–134.

Jones, H. L. J., B. S. C. Leadbeater & J. C. Green, 1993. Mixotrophy in marine species of *Chrysochromulina* (Prymnesiophyceae): Ingestion and digestion of a small green flagellate. J. mar. biol. Ass. U.K. 73: 283–296.

Kawachi, M. & I. Inouye, 1995. Functional roles of the haptonema and spine scales in the feeding process of *Chrysochromulina spinifera* (Fournier) Pienaar et Norris, (Haptophyta = Prymnesiophyta). Phycologia 34: 193–200.

Keller, M. D., L. P. Shapiro, E. M. Haugen, T. L. Cucci, E. B. Sherr & B. F. Sherr, 1994. Phagotrophy of fluorescently labeled bacteria by an oceanic phytoplankter. Microb. Ecol. 28: 39–52.

Kirchhoff, B. & B. Meyer, 1995. A new phagotrophic species of *Katodinium* (Dinophyceae) from hypertrophic shallow lakes in North Germany. Nova Hedwigia 60: 179–185.

Kristiansen, J., 1991. A checklist of Danish freshwater Chrysophytes. Chrysophyceae – Synurophyceae – Prymnesiophyceae – Bicocoecophyceae. Third ed. Institut for Sporeplanter. Univ. Copenhagen, 54 pp.

McKenzie, C. H., D. Deibel, M. A. Paranjape & R. J. Thompson, 1995. The marine mixotroph *Dinobryon balticum* (Chrysophyceae): Phagotrophy and survival in a cold ocean. J. Phycol. 31: 19–24.

Olrik, K. & A. Nauwerck, 1993. Stress and disturbance in the phytoplankton community of a shallow, hypertrophic lake. Hydrobiologia 249: 15–24.

Porter, K. G., 1988. Phagotrophic phytoflagellates in microbial food webs. Hydrobiologia 159: 89–97.

Rees, S. & R. I. Jones, 1992. Phagotrophic nutrition in phytoflagellates. Br. Phycol. J. 27: 100.

Salonen, K. & S. Jokinen 1988. Flagellate grazing on bacteria in a small dystrophic lake. Hydrobiologia 161: 203–209.

Sanders, R. W., 1991. Trophic strategies among heterotrophic flagellates. In D. J. Patterson & J. Larsen (eds), The Biology of Free-living Heterotrophic Flagellates. The Systematics Ass., Spec. Vol. 45, Clarendon Press, Oxford: 21–38.

Sanders, R. W. & K. G. Porter, 1988. Phagotrophic phytoflagellates. Adv. Microb. Ecol. 10: 167–192.
Sibbald, M. J. & L. J. Albright, 1992. The influence of light and nutrients on the nanoflagellate, *Ochromonas* sp. Mar. Microb. Food Webs 5: 39–48.
Smith, V. H., 1990. Phytoplankton responses to eutrophication in inland waters. In I. Atatsuka (ed.), Introduction to Applied Phycology. SPB Academic Publishing bv, The Hague, The Netherlands: 231–249.
Turner, J. T. & J. C. Roff, 1995. Trophic levels and trophospecies in marine plankton: Lessons from the microbial food web. Mar. Microb. Food Webs 7: 225–248.
Vadstein, O., A. Jensen, Y. Olsen & H. Reinertsen, 1988. Growth and phosphorus status of limnetic phytoplankton and bacteria. Limnol. Oceanogr. 33: 489–503.
Veen, A., 1991. Ecophysiological studies on the phagotrophic phytoflagellate *Dinobryon divergens* Imhof. Doc. Dissert., Dept. Fundament. Appl. Ecol., Univ. Amsterdam, Kruislaan 320, 1098 SM Amsterdam, The Netherlands, 125 pp.

Hydrobiologia **369/370:** 339–351, 1998.
M. Alvarez-Cobelas, C. S. Reynolds, P. Sanchez-Castillo & J. Kristiansen (eds), Phytoplankton and Trophic Gradients.

Silica-scaled chrysophytes from permanent and temporary waters of Hortobágy, eastern Hungary

Leontin Stefan Péterfi[1], Laura Momeu[2], Judit Padisák[3] & Viktoria Varga[4]
[1] *Plant Biology Chair, Biology Department, Babeş-Bolyai University, RO-3400 Cluj-Napoca, Romania*
[2] *Institute of Biological Research, 48 Republicii Str., RO-3400 Cluj-Napoca, Romania*
[3] *Balaton Limnological Institute, Hungarian Academy of Science, H-8237 Tihany, Hungary*
[4] *Microbiological Department, Eötvös Loránd University, Múzeum krt. 4/a, H-1088 Budapest, Hungary*

Key words: Silica scales, Chrysophyceae, Synurophyceae, distribution

Abstract

The silica-scaled flagellate flora of permanent (reservoirs, fishponds) and temporary (roadside *Caricetum*) waters of the Hortobágy National Park, eastern Hungary, was studied based on materials collected in summer 1995 and spring 1996. Altogether, twenty-six taxa were identified in TEM studies: two belong to Chrysophyceae (*Chrysosphaerella brevispina, Paraphysomonas vestita*) and twenty-four to the Synurophyceae of which eighteen *Mallomonas* (*M. acaroides, M. akrokomos, M. alpina, M. annulata, M. areolata, M. calceolus, M. caudata, M. corymbosa, M. crassisquama, M. cyathellata, M. elongata, M. heterospina, M. intermedia, M. paxillata, M. portae-ferreae, M. punctifera, M. teilingii* and *M. tonsurata*) and six *Synura* (*S. curtispina, S. petersenii* var. *petersenii, S. petersenii* var. *glabra, S. spinosa, S. splendida* and *S. uvella*).

Introduction

The silica-scaled chrysophyte flora of Hungary has not been well documented. Although many data have been published during the twentieth century, few of them can be verified without appropriate electron microscopy (Kristiansen & Padisák, 1992). There are some records from summer collections from the Kis-Balaton reservoir (Kristiansen & Padisák, 1992) and some sporadic data from rivers and shallow lakes (Kiss, 1987; Ács & Kiss, 1991; Kiss & Kristiansen, 1994). The occurrences of three species of Chrysophyceae (*Chrysosphaerella annulata, Paraphysomonas vestita, Spiniferomonas trioralis*) and seven of the Synurophyceae (*Mallomonas acaroides, M. alpina, M. caudata, M. tonsurata, Synura curtispina, S. echinulata, S. petersenii*) were recorded in these papers. In this paper, we report records of twenty-six taxa of silica-scaled chrysophytes from the Hortobágy National Park (Eastern Hungary): one belongs to *Chrysosphaerella*, one to *Paraphysomonas*, eighteen to *Mallomonas* and six to *Synura*.

The Hortobágy National Park is a large lowland (88–92 m a.s.l.) in Eastern Hungary. It lies in a semiarid-semihumid climatic region with different types of saline soils, which are, periodically, deeply fissured. The differences in the distribution and thickness of these soils are related mainly to channel changes of the river Tisza and its tributaries. Beside soil factors, temperature and the seasonal distribution of precipitation are the most important factors for the plant communities of the park, mainly dry, halophilic grassland communities and forest-steppe (Szujkó-Lacza, 1982). There are several permanent aquatic habitats (River Hortobágy, fishponds), temporary ponds that dry out periodically by late summer and wetland areas that dry out very quickly after spring. The algal vegetation of the fishponds was thoroughly studied by the famous Hungarian phycologist, Tibor Hortobágyi (1954, 1957) and a check-list of algal species found in the area has been compiled by Hajdu and Zrupkó (1982).

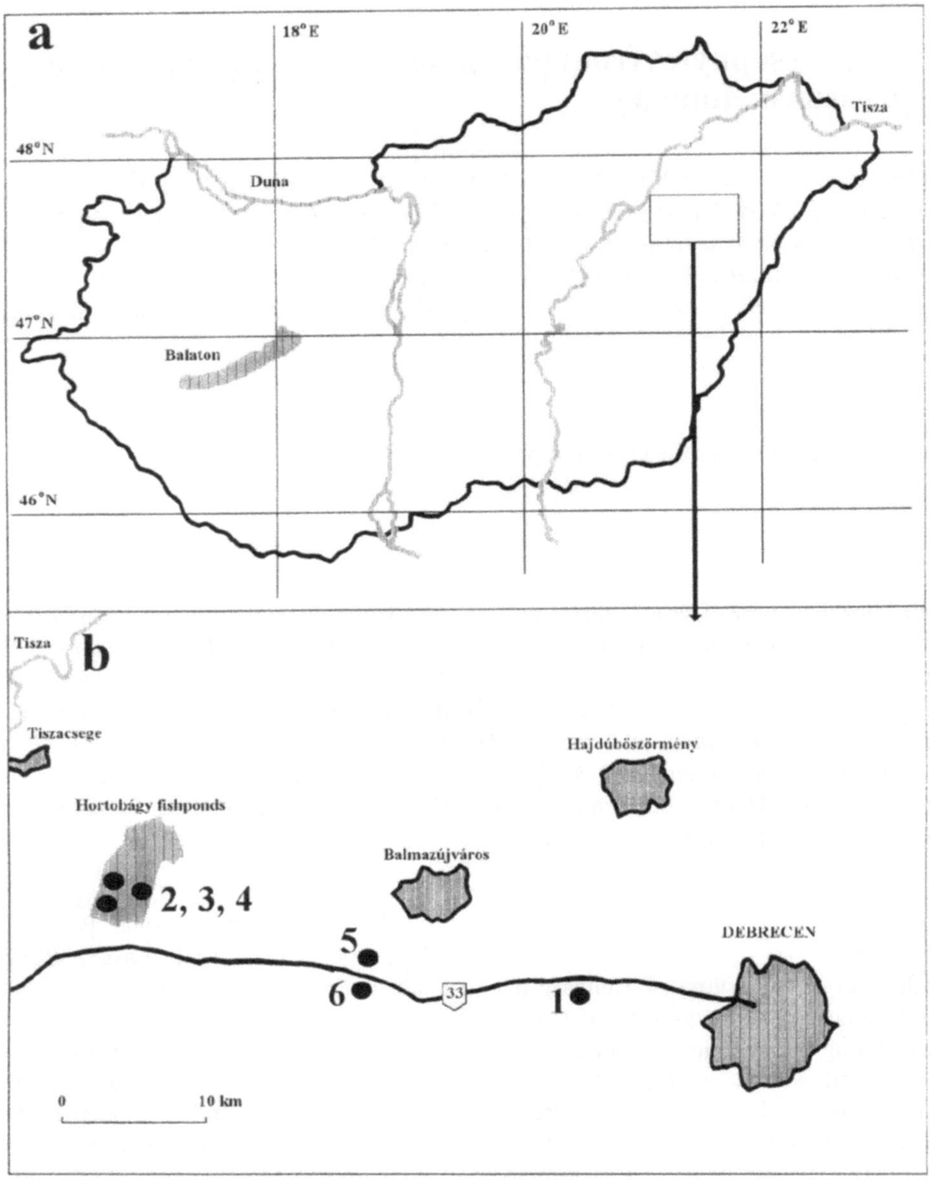

Figure 1. Location of sampling sites. a: Location of sampling area in Hungary; b: sampling sites. 1: Debrecen-Hortobágy road (no. 33), 96th km section, reservoir; 2: large fishpond; 3. small fishpond; 4. fish breeding pond; 5. roadside *Caricetum* I. (temporary water); 6. roadside *Caricetum* II. (temporary water).

Materials and methods

Samples were collected with 20–μm plankton net at sampling sites indicated in Figure 1 between 1–6 August 1995 and 7–8 April 1996. They were preserved in 4% formaldehyde solution. Drops of the material previously rinsed with distilled water were placed on formvar- or parlodion- coated grids by means of a pipette. In some cases the material was treated with conc. nitric acid and then repeatedly washed with distilled water in order to remove the acid and other soluble remnants. After drying, the grids were examined in a TESLA BS-500 transmission electron microscope.

Table 1. Water chemical data of the roadside *Caricetum*

Variable	
water temperature (°C)	10.0
pH	6.33
conductivity (μS cm^{-1})	133
acidity (mmol l^{-1})	1.3
NH_4-N (μg l^{-1})	77
NO_2-N (μg l^{-1})	19
NO_3-N (μg l^{-1})	417
PO_4-P (μg l^{-1})	1722
Ca^2 (mg l^{-1})	14.29
Mg^2 (mg l^{-1})	8.67
Na^+ (mg l^{-1})	9.00
K^+ (mg l^{-1})	12.00
Cl^- (mg l^{-1})	21.27
SO_4^2 (μg l^{-1})	19.20

Among the investigated localities the Hortobágy fishponds and a small lake are permanent waters (sites 1, 2, 3 and 4 on Figure 1). Each are shallow, eutrophic with low transparency and emergent macrophyte stands (primarily *Phragmitetum*) on the shores. Localities 5 and 6 (Figure 1) are temporary ditches. The macrovegetation indicates that the area is wet during summer. Water chemical data for locality 6, a roadside *Caricetum* (floristically the most interesting locality), are given in Table 1.

Results

Altogether tweny-six taxa of silica scaled chrysophytes were found in the samples, data of occurrences are given in Table 2.

Chrysophyceae

1. *Chrysosphaerella brevispina* Korshikov em. Harris et Bradley
This is a common member of springtime phytoplankton in most natural and artificial freshwater habitats (lakes, ponds, bogs, fishponds, dam reservoirs etc.). New record for Hungary.

2. *Paraphysomonas vestita* (Stokes) De Saedeleer
Widely distributed species occurring in eutrophic habitats. It was previously recorded in Hungary by EM examinations (Hajdu, 1975; Kristiansen & Padisák, 1992).

Synurophyceae

3. *Mallomonas acaroides* Perty em. Ivanov
(Figure 6A)
The nominate variety of 'acaroides' is widely distributed, occurring in eutrophic habitats (lakes, fishponds, channels etc.). It was previously recorded based on EM observations from the Kis-Balaton (Kristiansen & Padisák, 1992), River Danube and Lake-Balaton (Kiss & Kristiansen, 1994). Earlier LM records cannot be trusted.

4. *Mallomonas akrokomos* Ruttner in Pascher
(Figure 5D)
This species has a very wide geographic distribution, occurring mainly in winter and early spring at low water temperatures and in slightly acid pH conditions. It was previously recorded in Hungary based solely on reliable LM observations (Schmidt, 1976; Hamar, 1976). A single scale was observed in a bloom of *Synura petersenii*, together with several other synurophycean species.

5. *Mallomonas alpina* Pascher et Ruttner em.
Asmund et Kristiansen
Cosmopolitan, tolerant species, distributed in various habitat types mainly with high nutrient content. It was previously recorded in Hungary by EM observations (Kristiansen & Padisák, 1992; Kiss & Kristiansen, 1994) from Kis-Balaton water protecting reservoir, the rivers Danube and Tisza, as well as from the Eastern Main Channel. The present EM record is based on its occurrence in all investigated fishponds and larger water reservoirs of the Hortobágy.

6. *Mallomonas annulata* (Bradley) Harris (Figure 2D)
Apparently a cold water form, widely distributed in various, mostly mesotrophic, habitats. A single apical scale found in a bloom of *Synura petersenii* in roadside, temporarily flooded *Caricetum*, April 1996. New record for Hungary.

7. *Mallomonas areolata* Nygaard (Figure 2C)
Cells, scales and bristles are typical. Not very common and not so widely distributed occurring in alkaline, rather eutrophic waters, mostly during spring and early summer. It formed a scarce population in the largest

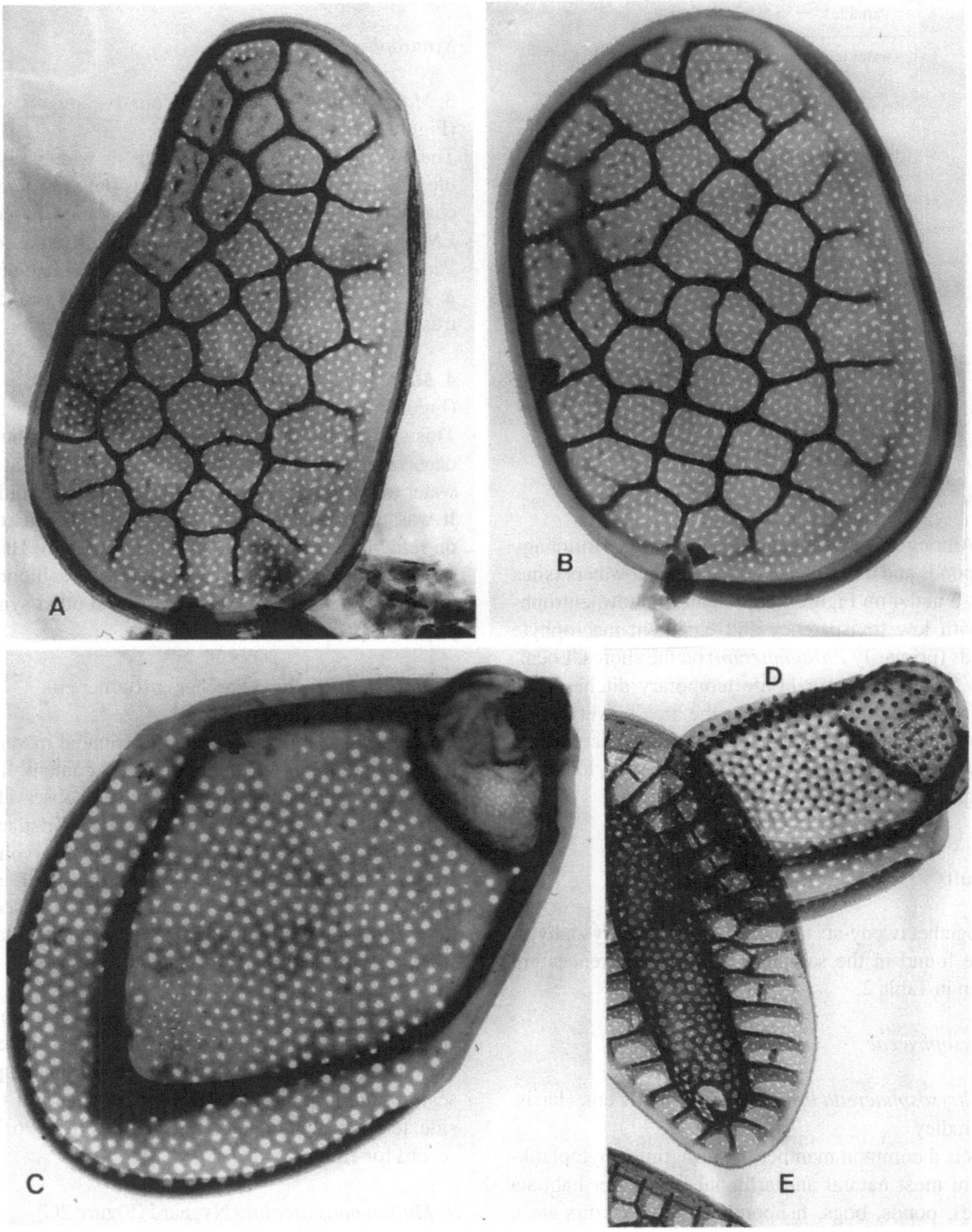

Figure 2. A–B – ***Mallomonas teilingii***. A: apical scale; B: body scale (16 000 ×); C: *M. areolata*, body scale (20 000 ×); D: *M. annulata*, apical scale (20 000 ×); E: ***Synura petersenii*** var. ***petersenii***, body scale (20 000 ×).

Table 2. Silica scaled chrysophytes found in the 6 samples taken in the Hortobágy National Park. +: found in the material collected in spring, 1996; × found in the material collected in summer, 1995. Numbering of localities correspond to that on Figure 1

	1	2	3	4	5	6
Chrysosphaerella brevispina	+	+	+			
Paraphysomonas vestita		+×		+		
Mallomonas acaroides		+		+		
M. akrokomos						+
M. alpina	+×	+	+×	+		
M. annulata						+
M. areolata		+				
M. calceolus						+
M. caudata		+				
M. corymbosa		×				
M. crassisquama			+			
M. cyathellata		×	×			
M. elongata	+	+	+			
M. heterospina						+
M. intermedia					+	
M. paxillata	+					
M. portae-ferreae		+				
M. punctifera		+	+			
M. teilingii	+	+				
M. tonsurata	+×	+×				
Synura curtispina	+		+			
S. petersenii var. *petersenii*	+		+	+	+	+
S. petersenii var. *glabra*	+	+		+	+	
S. spinosa					+	+
S. splendida						+
S. uvella	+	+		+	+	

fishpond of Halastó (Hortobágy) in April 1996. New record for Hungary.

8. *Mallomonas calceolus* Bradley (Figure 6B)
A single scale (3.5 × 1.75 μm) was found in a temporarily flooded roadside *Caricetum*, which mostly resembles the body scales of *M. calceolus*. It has smooth, unperforated anterior and posterior flanges, scattered papillae on the shield, in addition its dome marked with a few papillae. This species is widely distributed, especially in extratropical localities. It has not been recorded previously in Hungary.

9. *Mallomonas caudata* Ivanov em. Krieger
Cosmopolitan, widely distributed in eutrophic waters, mainly during summer. It has been recorded in Hungary several times by both LM and EM (Kristiansen & Padisák, 1992; Kiss & Kristiansen, 1994). With certainty it is a usual member of the spring and summer phytoplankton of most fishponds and reservoires of the Hortobágy.

10. *Mallomonas corymbosa* Asmund et Hilliard (Figures 5A–C)
Widely distributed alkaliphilous to alkalibiontic form, characteristic for water bodies with high nutrient content. Possibly it is one of the most characteristic *Mallomonas* species of the lowland, shallow waters in Hun-

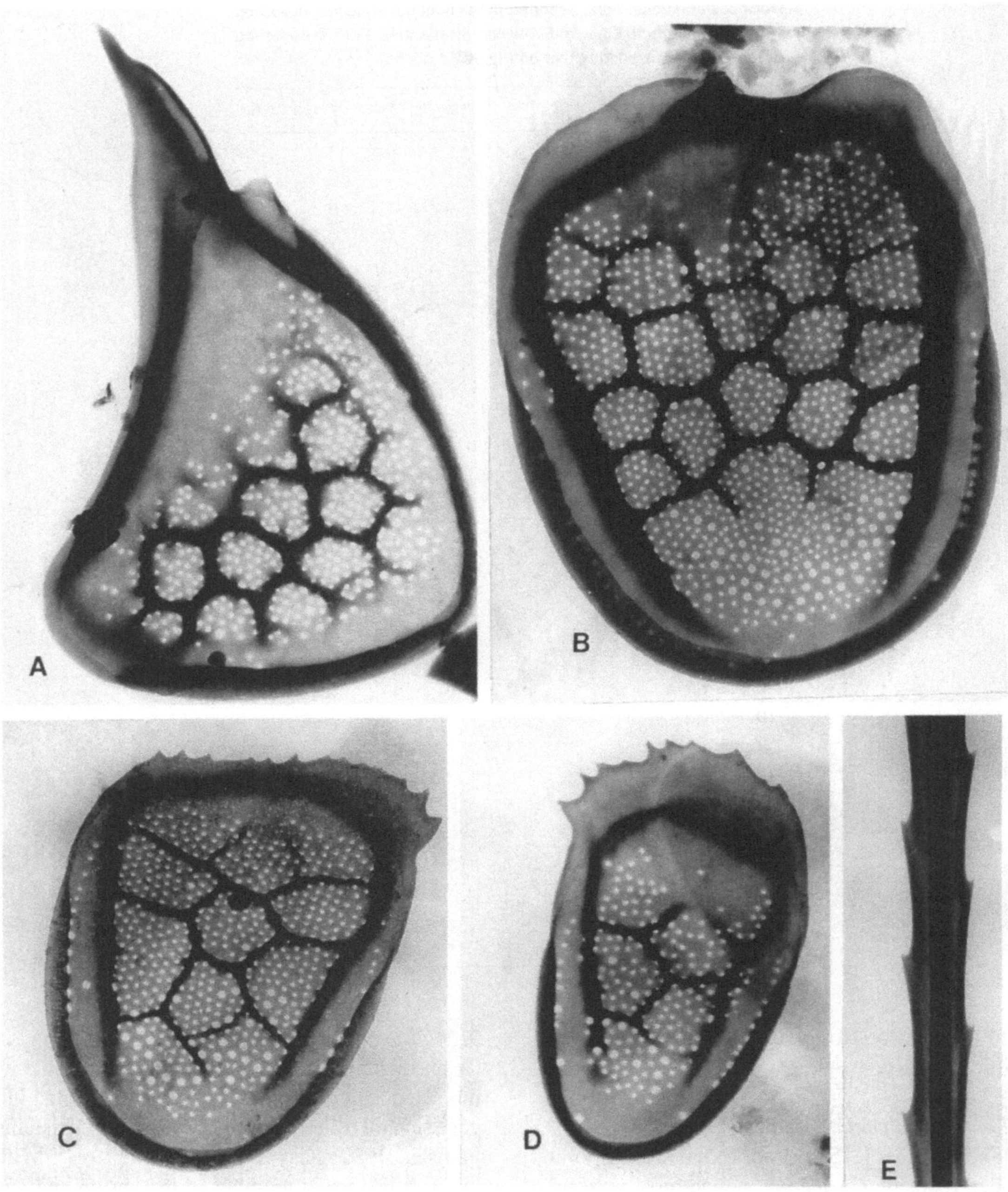

Figure 3. A–E – *Mallomonas punctifera.* A: collar scale; B: body scale; C–D: rear scales; E: part of body bristle (16 000 ×).

gary. New record for Hungary and for Europe, as well. Primarily found in North- and South America and in Eastern Asia.

11. *Mallomonas crassisquama* (Asmund) Fott
Widely distributed mainly in the northern temperate regions showing no special preference in temperature and pH conditions. It appears to be more frequent in eutrophic waters. A single scale was observed in a fish-pond of Halastó (Hortobágy). New record for Hungary.

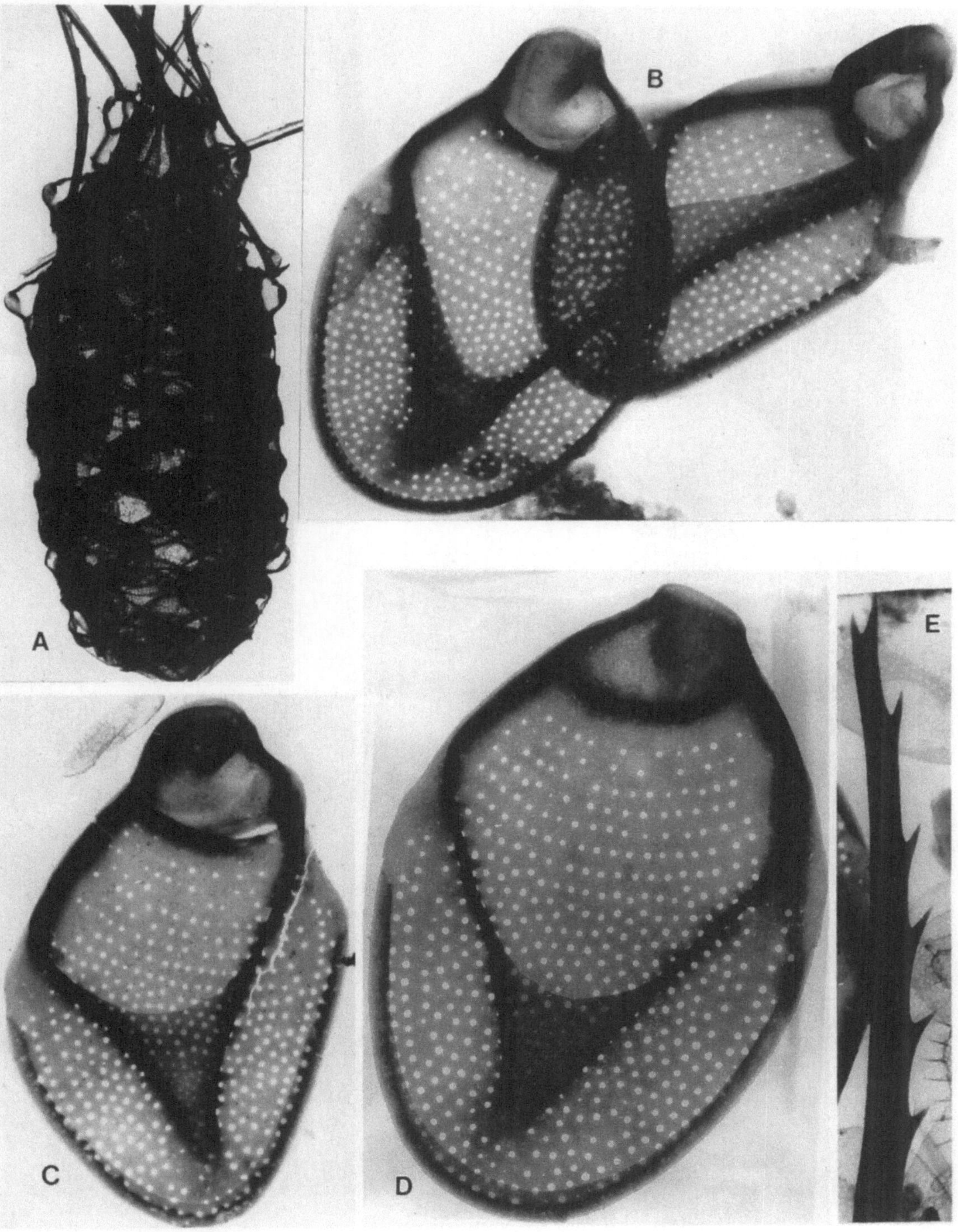

Figure 4. A–E – *Mallomonas elongata.* A: whole cell armour lacking body bristles (5 000 ×); B: body and apical scales (12 000 ×); C: apical scale (12 000 ×); D: body scale (16 000 ×); E: distal part of body bristle (12 000 ×).

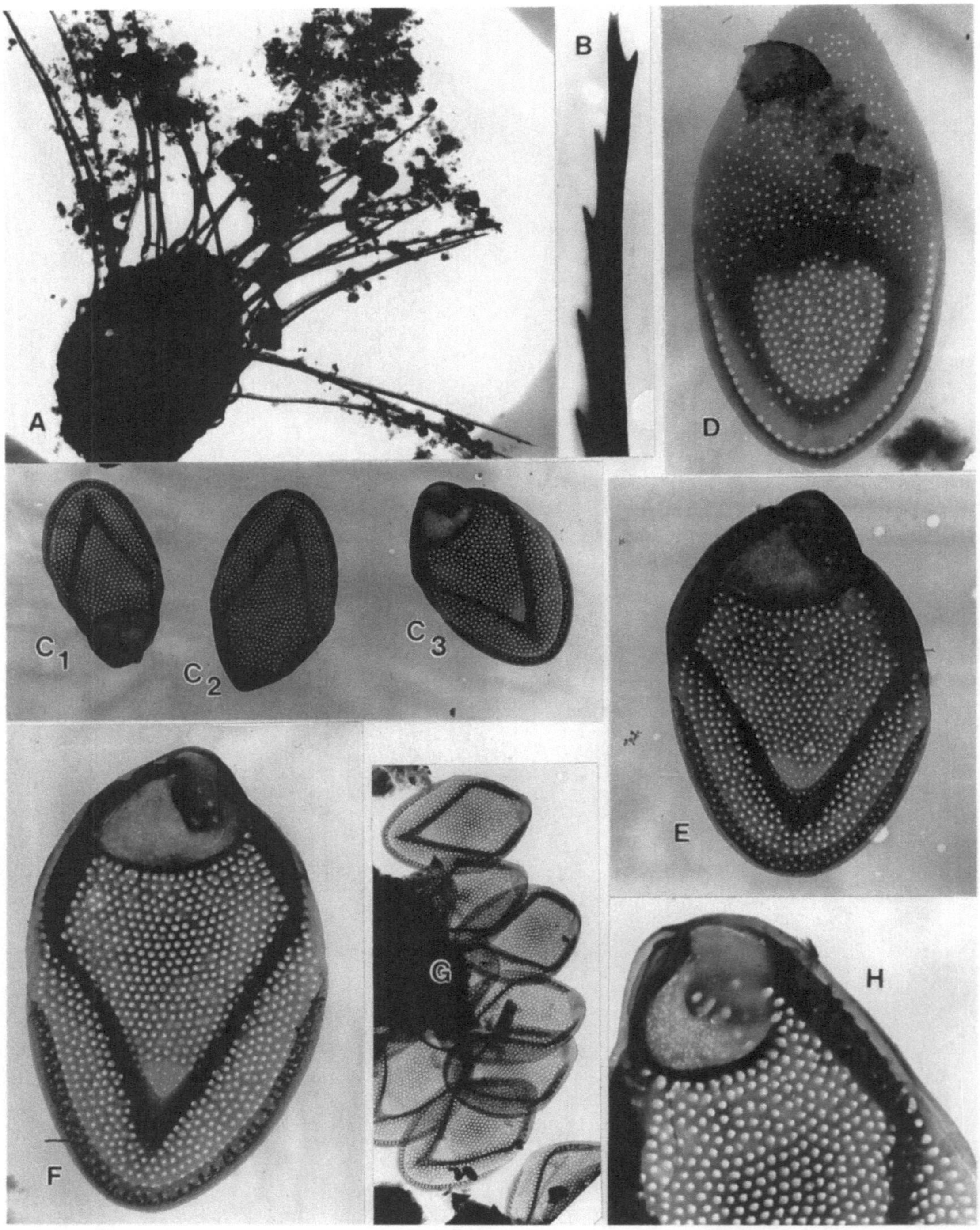

Figure 5. A–C – *Mallomonas corymbosa.* A: whole cell armour lacking posterior body bristles (2 800 ×); B: distal part of body bristle (12 000 ×); C_1: anterior scale; C_2: domeless body scale; C_3: bristle-bearing body scale (5 300 ×); D: *M. akrokomos*, body scale (16 000 ×); E–H – *M. cyathellata.* E: anterior scale (9 400 ×); F: body scale (13 200 ×); G: domeless body scales (4 470 ×); H: distal part of a body scale (19 700 ×).

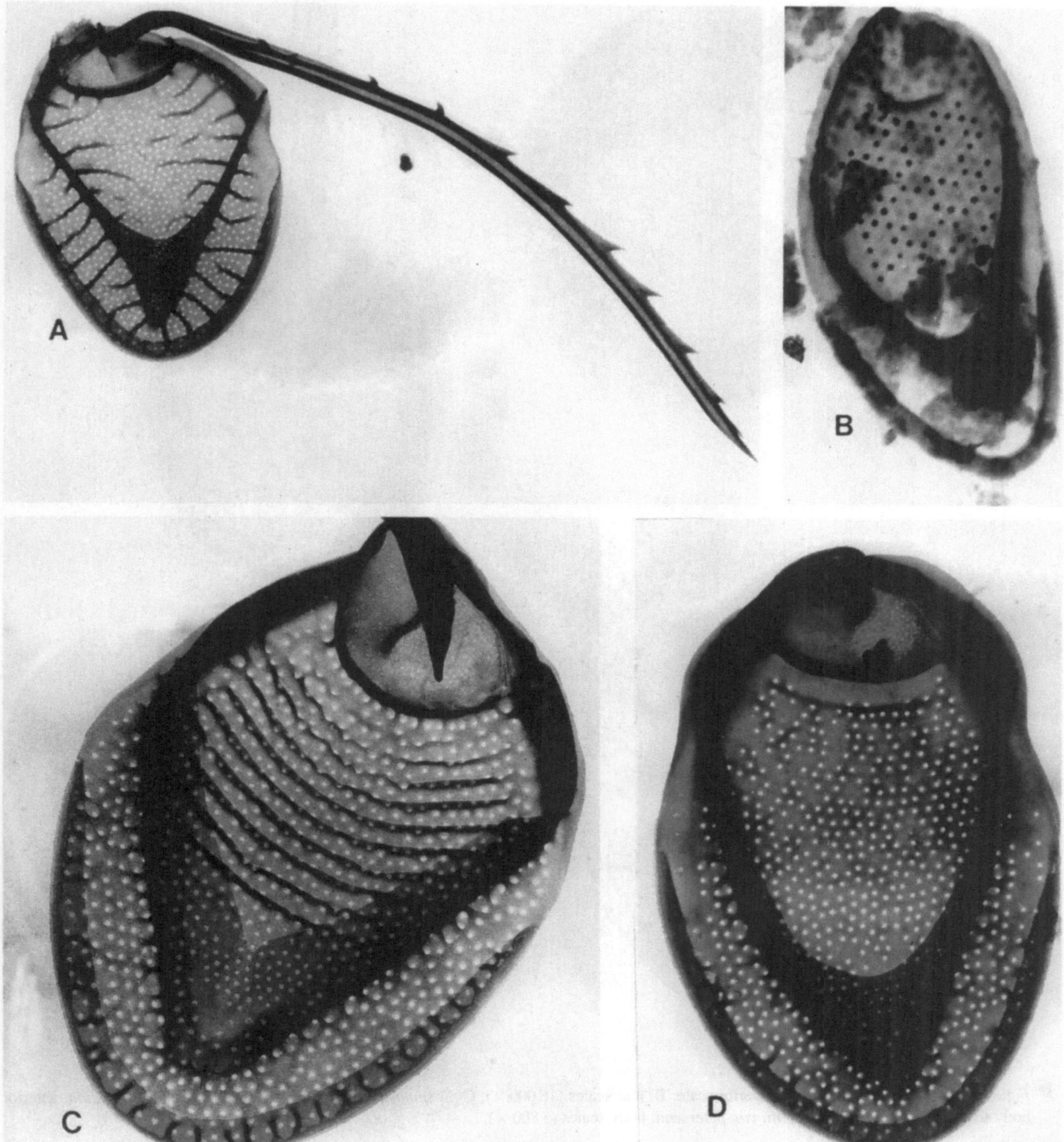

Figure 6. A: *Mallomonas acaroides*, body scale with serrated bristle (8 000 x); B: *M. calceolus*, body scale (20 000 ×); C: *M. portae-ferreae*, body scale (16 000 ×); D: *M. intermedia*, body scale (16 000 ×).

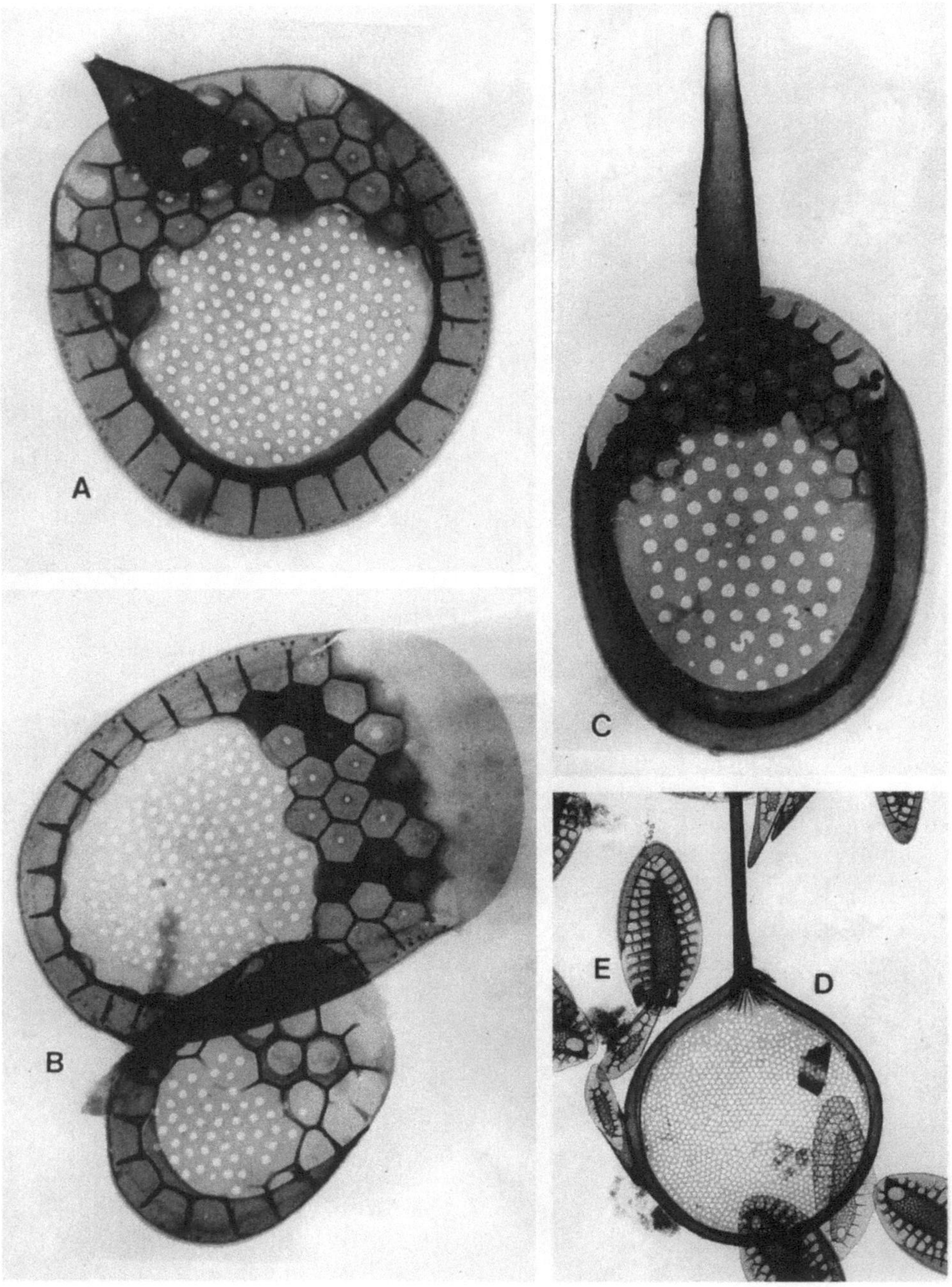

Figure 7. A–B – *Synura uvella.* A: anterior scale; B: rear scales (16 000 ×); C: *S. spinosa*, anterior scale (20 000 ×); D: *S. splendida*, anterior body scale (5 800 ×); E: *S. petersenii var. petersenii*, body scales (5 800 ×).

12. *Mallomonas cyathellata* Wujek et Asmund var. *cyathellata* (Figures 5E–H)
The species has a tropical to mediterranean distribution. Apparently the nominate variety may occur also in neighbourring temperate areas, but in the warmest season of the year (Péterfi & Momeu, 1996).

There are some difficulties in the identification of the species, not only because of its great variability but also due to the forms transitional with related species or groups of species (Siver, 1991). In an attempt to overcome these difficulties in Series Tonsuratae Asmund & Kristiansen (1986) tried to establish some distinctive characters. Unfortunately there are 'cyathellata' populations which do not fit into the proposed scheme. This is the case of our material, too. The cells and bristles agree with the expected features, except that supposedly characteristic goblet-like appendages on the rear scales were not obseved in any of the individuals examined. The continuous secondary layer on the base plate is always present, exhibiting the usual pits and the window, but there are no short ribs on the dome, nor on the distal part of the domeless scales. Instead the anterior submarginal ribs in all scales are marked with evenly spaced struts.

We compared the scale pattern of the Hungarian material with that exhibited by some peculiar 'cyathellata' populations collected from the Romanian Danube Delta, recently investigated by two of the authors (Péterfi & Momeu, 1996). The striking similarities between the Hungarian and Romanian populations became quite evident. Therefore we are certain concerning the taxonomic position of the present Hungarian material, in spite of the absence of cyathi. Maybe the rearmost scales with the goblet-like structures are not always formed, or they can be lost in some successional stage of the population (e.g. old, desintegrating populations). A similar situation could be observed in other species concerning their rearmost, spine-bearing scales. The species was identified for the first time in Hungary.

13. *Mallomonas elongata* Reverdin (Figures 4A–E)
Apparently, this is a widespread, alkaliphilous species occurring mostly in eutrophic habitats during the cold period of the year. Its cells are large, elongate, ellipsoidal to almost cylindrical, covered with large, tripartite scales. The V-rib with very narrow angle, narrowly hooded, as in the micrographs by Siver (1991).The somewhat smaller apical scales at the base of the flagellum are markedly asymmetrical.

This is the first EM record of *M. elongata* from Hungary. Previous records based on LM observations were considered dubious by Kristiansen & Padisák (1992) because it could be easily misidentified (in LM it resembles somewhat to *M. leboimei* Bourrelly).

14. *Mallomonas heterospina* Lund
Very widely distributed, mainly in the northern hemisphere, occurring in various aquatic habitats, mostly under slightly acidic pH conditions. It is more frequent in spring and autumn. Isolated scales were found in a bloom of *Synura petersenii,* together with *S. splendida, Mallomonas annulata* and *M. calceolus* in a small roadside , temporarily flooded *Caricetum. M. heterospina* has not previously been recorded in Hungary.

15. *Mallomonas intermedia* Kisselev em. Péterfi et Momeu var. *intermedia* (Figure 6D)
Apparently acidophilous with scattered distribution in Europe. The present record of the species in Hungary is based on the evidence of a single, though highly characteristic body scale.

16. *Mallomonas paxillata* (Bradley) Péterfi et Momeu
Not very common, with scattered distribution. Few typical body scales seen in samples collected from a water reservoir, Debrecen-Hortobágy road, 96th km section, April 1966. Not previously reported from Hungary.

17. *Mallomonas portae-ferreae* Péterfi et Asmund (Figure 6C)
Alkaliphilous, warm-water form with scattered distribution, mostly in tropical to mediterranean regions. In temperate zones, its occurrence is limited to the warmest period of the year (Péterfi & Momeu, 1996).

Isolated scales and bristles found in April 1996. The species has not previously been found in Hungary.

18. *Mallomonas punctifera* Korshikov (Figures 3A–E)
Widely distributed species, occurring in circumneutral or slightly alkaline habitats. It has been found in the largest fishpond of Halastó (Hortobágy) forming a rich population, but singly also in smaller fishponds in the area. This is the first (EM) record of the species in Hungary.

19. *Mallomonas teilingii* Conrad (Figures 2A–B)
This species has a scattered occurrence in the northern hemisphere. It seems to prefer meso-eutrophic conditions, appears mostly in spring in circumneutral waters.

Our record is the EM confirmation of its occurrence in Hungary (Vízkeleti, 1981, LM).

20. *Mallomonas tonsurata* Teiling em. Krieger
Cosmopolitan, widely distributed in alkaline, slightly to strongly eutrophic water bodies. Earlier LM records of the species from Hungary have been confirmed by the EM observations of Kiss (1987), Ács & Kiss

(1991), Kristiansen & Padisák (1992), Kiss & Kristiansen (1994), as well as by our present findings. It is one of the species which is widely distributed in Hungary during the whole year, in fishponds, reservoirs, canal systems etc.

21. *Synura curtispina* (Petersen et Hansen) Asmund
This species, with a wide but scattered occurrence in eutrophic habitats, has been repeatedly recorded in Hungary: e.g. in the Danube (Ács & Kiss, 1991), Tisza and its side arms (Kiss & Kristiansen, 1994) and Kis-Balaton (Kristiansen & Padisák, 1992).

22. *Synura petersenii* Korshikov var. *petersenii*
(Figures 2E, 7E)
Cosmopolitan, distributed even in strongly eutrophic habitats, often forming water blooms. It was recorded in Hungary by Kristiansen & Padisák (1992) from the Kis-Balaton and by Kiss & Kristiansen (1994) from the Lake Balaton, the rivers Danube and Tisza, their side arms and channels, earlier records by Kiss (1987) and Ács & Kiss (1991).

Synura petersenii is widely distributed in Hungary in all types of aquatic habitats. The present authors identified it in almost all sampling sites. In some cases, heavy blooms were observed (e.g. in April 1996 in flooded roadside areas and on the right side of the reservoir situated at 96th km section of the Debrecen-Hortobágy road).

S. petersenii var. glabra (Korshikov) Petersen et Hansen
It is less frequent than the nominate variety, occurring together with it or forming almost pure populations or blooms. It has not previously been recorded in Hungary.

23. *Synura spinosa* Korshikov (Figure 7C)
The species is widespread, occurring in less eutrophic conditions. It has been found singly in the temporarily flooded roadside *Caricetum* with *Synura petersenii* bloom. The anterior scales of this species has long spines, in the anterior honeycomb area the lack the centrally located pores in the meshes are occluded and the large holes of the basal plate are without thickened rim. It is a new record for Hungary.

24. *Synura splendida* Korshikov (Figure 7D)
This relatively rare species has a scattered distribution, mostly occurring in meso-eutrophic bogs and ponds. It has been found singly in the roadside flooded *Caricetum* with the *Synura petersenii* bloom, together with other silica-scaled chrysophytes. New record for Hungary.

25. *Synura uvella* Stein em. Korshikov
(Figures 7A–B)
Widely distributed in various, mostly eutrophic habitats, blooms are rare. In spite of the fact that it is considered the 'commonest' *Synura* species, and therefore repeatedly recorded by Hungarian phycologists by LM investigations, these records cannot be trusted.

Synura uvella has been found by us in both sampling times in almost all sampling sites, but mostly as single cells or disintegrated colonies. It is new record for Hungary.

Discussion

The chrysophyte flora of the Hortobágy park does not appear to be particularly rich, partly because most of the habitats are culturally influenced. Altogether 26 taxa were found, 14 of which are new for Hungary. The present findings provide EM confirmation of occurrence of some previously, but solely LM recorded species, such as *Mallomonas akrokomos, M. teilingii, M. elongata* and *Synura uvella*. Five of the identified species have already been documented by EM, these are *Paraphysomonas vestita, Mallomonas alpina, M. tonsurata, Synura petersenii* and *S. curtispina* (Hajdu, 1975; Kiss, 1987; Ács & Kiss, 1991; Kristiansen & Padisák, 1992; Kiss & Kristiansen, 1994).

Some of the species are cosmopolitan ubiquists (*Synura petersenii, Chrysosphaerella brevispina, Paraphysomonas vestita*), others exhibit worldwide but rather scattered distribution occurring predominantly in somewhat nutrient rich waters (*Mallomonas alpina, M. tonsurata, M. punctifera, M. caudata, M. corymbosa, M. areolata, M. crassisquama*).

Special attention should be paid to *Mallomonas cyathellata* and *M. portae-ferrae* with predominantly tropical and subtropical distribution. They have recently been found in the Danube Delta (Romania) inhabiting similar, highly eutrophic, fishponds. Their distribution pattern and phytogeographic importance is commented elsewhere (Péterfi & Momeu, 1996).

The occurrence of *Mallomonas teilingii, M. heterospina* and *Synura splendida* is associated by low water temperatures.

The presence of acidophilous species, such as *Mallomonas annulata, M. heterospina* and *Synura splendida,* was determined on the basis of recoveries of single scales, which were, probably, remnants from previous populations.

The most characteristic aquatic habitats of Hortobágy are shallow, eutrophic, alkaline fishponds. *Synura curtispina, S. petersenii, Mallomonas tonsurata* and *M. alpina* were almost invariably found in these fishponds like in other eutrophic Hungarian waters (Kristiansen & Padisák, 1992; Kiss & Kristiansen, 1994). The spring flora (water temperature about 10 °C) is charactrized by *M. elongata, M. punctifera, M. caudata* and *M. teilingii.* The typical species of the summer flora (water temperature about 25 °C) are *M. corymbosa, M. cyathellata* and *M. portaeferreae.*

A quite different silica-scaled chrysophyte flora is supported by the wetlands that are only temporarily flooded during spring. These habitats are also nutrient-rich but their pH falls into the acidic range. The characteristic taxa (*M. annulata, M. heterospina, M. intermedia* and especially *S. splendida*) are commonly found in acidoplilous and mesotrophic habitats of the Carpathian Basin.

The most important conclusion of the present findings is that future investigations must be focused on localities which are not or only slightly influenced culturally. There is a real need for sampling in other types of localities (e.g. acid bogs) especially in winter and early spring.

Acknowledgements

We thank Mrs Zsuzsa Z.-Doma for her careful technical assistance. Water chemical data were made available by the Environmental Protection Authority (TIKÖFE), Debrecen. The authors are also indepted to Dr Constantin Crăcium and Teofil Fărcas who kindly run the EM. This study was supported by the Hungarian National Science Foundation (OTKA No. T-014634).

References

Ács, É. & K. T. Kiss, 1991. Investigation of periphytic algae in the Danube at Göd (1669 river km, Hungary). Arch. Hydrobiol. Algol. Stud. 62: 47–67.

Asmund, B. & J. Kristiansen, 1986. The genus *Mallomonas* (Chrysophyceae). A taxonomic survey based on the ultrastructure of silica scales and bristles. Opera Bot. 85: 1–128.

Hajdu, L., 1975. A *Paraphysomonas vestita* (Stokes) de Saedeleer elöfordulása hazánkban. Botanikai Közlemények 62: 9–12.

Hajdu, L. & G. Zrupkó, 1982. Check-list of algae found in the Hortobágy National Park. In: J. Szujkó-Lacza (ed.), The flora of the Hortobágy National Park. Akadémiai Kiadó, Budapest: 19–48.

Hamar, J., 1976. About the algae of the Kisköre River Barrage and environs. Tiscia (Szeged) 11: 45–58.

Hortobágyi, T., 1954. Les nouveaux micro-organismes de l'établissement piscicole de Hortobágy et du lac Szelid. Acta Bot. Hungarica 1: 89–123.

Hortobágyi, T., 1957. Algák két hortobágyi halastóból. [Algen aus zwei Fischteichen von Hortobágy]. Egri Pedagógiai Föiskola Évkönyve 3: 1–48.

Kiss, K. T., 1987. Phytoplankton species in the Szigetköz section of the Danube during 1981–1982. Arch. Hydrobiol. Algol. Stud. 47: 247–273.

Kiss, K. T. & J. Kristiansen, 1994. Silica-scaled chrysophytes (Synurophyceae) from some rivers and shallow lakes. Hydrobiologia 289: 157–162.

Kristiansen, J. & J. Padisák, 1992. Silica-scaled chrysophytes (Chrysophyceae and Synurophyceae) from the Kis-Balaton Reservoir, Hungary. Ann. hist.-nat. Mus. natn. Hung. 84: 7–16.

Péterfi, I. & L. Momeu, 1996. Three Romanian *Mallomonas* species (Synurophyceae), with special distribution pattern. Hydrobiologia 336: 143–150.

Schmidt, A., 1976. Újabb adatok a Duna magyarországi szakasza algáinak ismeretéhez. In: Környezetvédelem és vízgazdálkodás '76. Vándorgyülés. Sopron: 1–10.

Siver, P. A., 1991. The biology of *Mallomonas.* Morphology, taxonomy and ecology. Dev. Hydrobiol. 63. Kluwer Academic Publ., Dordrecht, 230 pp.

Szujkó-Lacza, J., 1982. The flora of the Hortobágy National Parks and two preserved forests. In: J. Szujkó-Lacza (ed.), The flora of the Hortobágy National Park. Akadémiai Kiadó, Budapest: 7–18.

Vízkeleti, É., 1981. A hegyhátszentjakabi Vadása tó biológiai vízminösége és fitoplankton összetétele 1976–78 között. Savaria (A Vas megyei múzeumok értesítöje) 15: 11–39.

Hydrobiologia **369/370:** 353–361, 1998.
M. Alvarez-Cobelas, C. S. Reynolds, P. Sanchez-Castillo & J. Kristiansen (eds), Phytoplankton and Trophic Gradients.

On the occurrence of *Phacotus lenticularis* (Chlorophyta) in lakes of different trophic state

Ines Schlegel, Rainer Koschel & Lothar Krienitz
Institute of Freshwater Ecology & Inland Fisheries, D-16775 Neuglobsow, Germany

Key words: phytoplankton, trophic state, calcification, autecology, *Phacotus*, Chlorophyta

Abstract

The occurrence of the calcite-loricated chlamydophycean alga *Phacotus lenticularis* was investigated in relation to the trophic state of inland waters. The Phytoflagellate was observed in deep, stratified dimictic lakes, in shallow polymictic lakes, oxbow-lakes and ponds. Habitats of *Phacotus* are rich in lime and vary from oligo- to hypertrophic. Mass developments with cell numbers of 500,000–5.5 × 10^6 individuals l^{-1} were found in lakes with properties in the ranges: TP 34–484 $\mu g\ l^{-1}$, SRP 2–88 $\mu g\ l^{-1}$, NO_2^- 0–32 $\mu g\ l^{-1}$, NO_3^- 0–820 $\mu g\ l^{-1}$ and NH_4^+ 10–740 $\mu g\ l^{-1}$. The most consistently co-incidental site factors were water temperature and pH values. High abundances were observed at temperatures between 15.8 and 24.7 °C and at pH values from 8.3 to 9.6. The results are discussed in the context of numerous records of *Phacotus* published in the literature. The morphometric and ecological data come mainly from the main investigated waters of the Baltic Lake District of northeastern Germany.

Introduction

Phacotus Perty is the type-genus of the chlamydophycean algal family Phacotaceae (Bütschli) Oltmanns. Members of this family are characterized by the development of a lorica, which is often mineralized. *Phacotus lenticularis* (Ehrenberg) Stein represents the type species of the genus *Phacotus*. This unicellular phytoflagellate of 13-17 μm in diameter possesses two flagella. The two-shelled lenticular lorica is the most striking character of this alga. *Phacotus* incorporate $CaCO_3$ in remarkable concentrations in their loricae (Pocratsky, 1982; Steinberg & Klee, 1983; Giering et al., 1990). Therefore, mass developments of *Phacotus* have influence upon the $CaCO_3$ budget of lakes (Koschel et al., 1987). The calcium carbonate budget is closely related to the trophic conditions, especially through the self-purification process of calcite precipitation (Koschel et al., 1990). The calcified shells have polymorphic surfaces (Kamptner, 1950, Hepperle & Krienitz, 1996). This is the reason for the description of several presumably different species of *Phacotus* e.g. *P. lendneri* Chodat (summarized by Giering et al., 1992). Koschel & Raidt (1988) demonstrated that the high variety of crystal forms assumed an influence of different trophic conditions on the structure of the lorica surface. The above-mentioned studies indicate that the alga *Phacotus lenticularis* is of special interest to taxonomists and developmental physiologists as well as to ecologists. Perhaps the ecological role of this calcified phytoflagellate of freshwater habitats is comparable to that of the coccolithophorids, such as *Emiliania huxleyi* (Lohmann) Hay et Mohler, the important primary producer generating calcite in marine habitats (Balch et al., 1992, 1993).

Numerous floristical studies of phytoplankton demonstrate the widespread occurrence of *P. lenticularis* in stagnant inland waters of different morphometric type and trophic degree. *P. lenticularis* is included in many check-lists. However, in most of these records, *Phacotus* is sporadic. Detailed information about the occurrences of *P. lenticularis* concerning both quantitative data, e.g. cell numbers, and ecological data, e.g. nutritional conditions, is scarce. Thus, the objective of this study was to survey the development of *P. lenticularis* in relation to selected ecological variates in several northern German Baltic hardwater lakes of different trophic degree.

Table 1. Morphometric and ecological features of several hardwater lakes of different trophic degree. Maximal observed *Phacotus*-cells in these lakes is shown in Figure 2.

	Stechlin	Gr. Zechliner	Breiter Luzin	Dagow	Tollense	Feldberger Haussee	Goldberger See	Dorfteich Gnetsch
Trophic state	oligotrophic	mesotrophic	mesotrophic	eutrophic	eutrophic	eutrophic	polytrophic	hypertrophic
Mixing frequency	dimictic	dimictic	dimictic	dimictic	dimictic	dimictic	polymictic	polymictic
Area [m^2]	4.3×10^6	2.2×10^6	3.4×10^6	0.24×10^6	17.4×10^6	1.3×10^6	0.18×10^6	3000
Volume [m^2]	96.9×10^6	–	71.8×10^6	1.19×10^6	301×10^6	8.2×10^6	0.16×10^6	1500
Mean depth [m]	22.8	–	21.3	5.0	17.2	6.3	0.9	0.5
Max depth [m]	68.5	38.0	58.5	9.5	33.2	12.0	2.0	1.2
Residence time [a]	30–50	–	23	3–5	3–4	5–6	–	–
Secchi disk [m]**	7.3	3.9	3.5	2.7	2.3	1.5	0.5	0.1
TP [$\mu g \times l^{-1}$]*	14	29	34	49	65	102	490	946
TN [$\mu g \times l^{-1}$]*	493	786	732	1035	1125	1725	–	–
NO_2^- [$\mu g \times l^{-1}$]*	2	3	2	3	7	2	–	–
NO_3^- [$\mu g \times l^{-1}$]*	7	2	12	12	236	19	300–3100	400–5400
NH_4^+ [$\mu g \times l^{-1}$]*	32	80	101	75	87	105	0–840	380–18000
SRP [$\mu g \times l^{-1}$]*	2	7	2	4	24	22	300	3800
Ca^{2+} [$mg \times l^{-1}$]*	47.6	50.4	41.3	65.7	65.5	38.2	–	–
$CaCO_3$ [$\mu g \times l^{-1}$]*	311	–	917	283	464	634	–	–
pH*	8.50	8.46	8.63	8.45	8.55	8.50	8.50	8.00
Alkalinity [$mEq \times l^{-1}$]*	1.94	2.08	2.20	2.47	2.65	1.80	–	–

* Epilimnic summer means 1993–1995. ** Summer means 1995 (with the exception of Goldberger See and Dorfteich Gnetsch which were only sporadically investigated).

Study site and methods

Samples were taken from different depths, generally from the pelagic zones, from more than 60 lakes of the Baltic Lake District of northeastern Germany (Brandenburg, Mecklenburg-Vorpommern). These waters vary from oligo- to eutrophic conditions and are rich in lime (Koschel et al., 1983, 1990; Casper, 1985). From 1987 to 1995, seven lakes were sampled, at least bi-weekly, during the main vegetation periods (April-October) and sporadically during the cold seasons. More details about the ecology of these lakes are given in Koschel et al., 1990, 1993; Koschel & Kasprzak, 1994; Krienitz et al., 1993; Krienitz & Hegewald, 1996 and Padisák et al. (this volume). A data set of main morphological and ecological parameters of some of these lakes is enclosed (Table 1). The other lakes of this area were investigated sporadically during mass developments of *P. lenticularis*. From 1987 to 1989, samples from two shallow waters of the Elbe-Saale-District (Germany, Sachsen-Anhalt) were included for comparison: the polytrophic Goldberger See, an oxbow-lake of the River Elbe in the biosphere reservation 'Mittlere Elbe' and the hypertrophic Dorfteich Gnetsch, an oxidation pond in the village Gnetsch near the town Köthen (Table 1).

Freshwater samples were investigated under the light microscope (LM) NICON 'Optiphot 2' and LEICA 'Diaplan' by means of bright field and differential interference contrast (DIK) methods. Phytoplankton samples were fixed with formaldehyde and counted in sedimentation chambers according to the method of Utermöhl (1958). As far as possible, a minimum of 400 sedimented *Phacotus*-cells were counted in each sample giving a counting accuracy of ± 10%. The biovolumes were calculated after Willén (1976). The volumes were converted into biomass, assuming a specific gravity of 1.00 g cm^{-1}.

Soluble Reactive Phosphorus (SRP), total phosphorus (TP), nitrite (NO_2^-), nitrate (NO_3^-), total nitrogen (TN) and calcium (Ca^{2+}) were measured with the Perstorp Analytical flow-injection analysis system Tecator FIAstar 5010/5030. The $CaCO_3$-concentration was measured using infrared gas analysis (modified according to Proft, 1984). Field pH and water temperatures were recorded by means of a SensoLyt SETA electrode. Transparency was measured using white Secchi disc of 30 cm diameter.

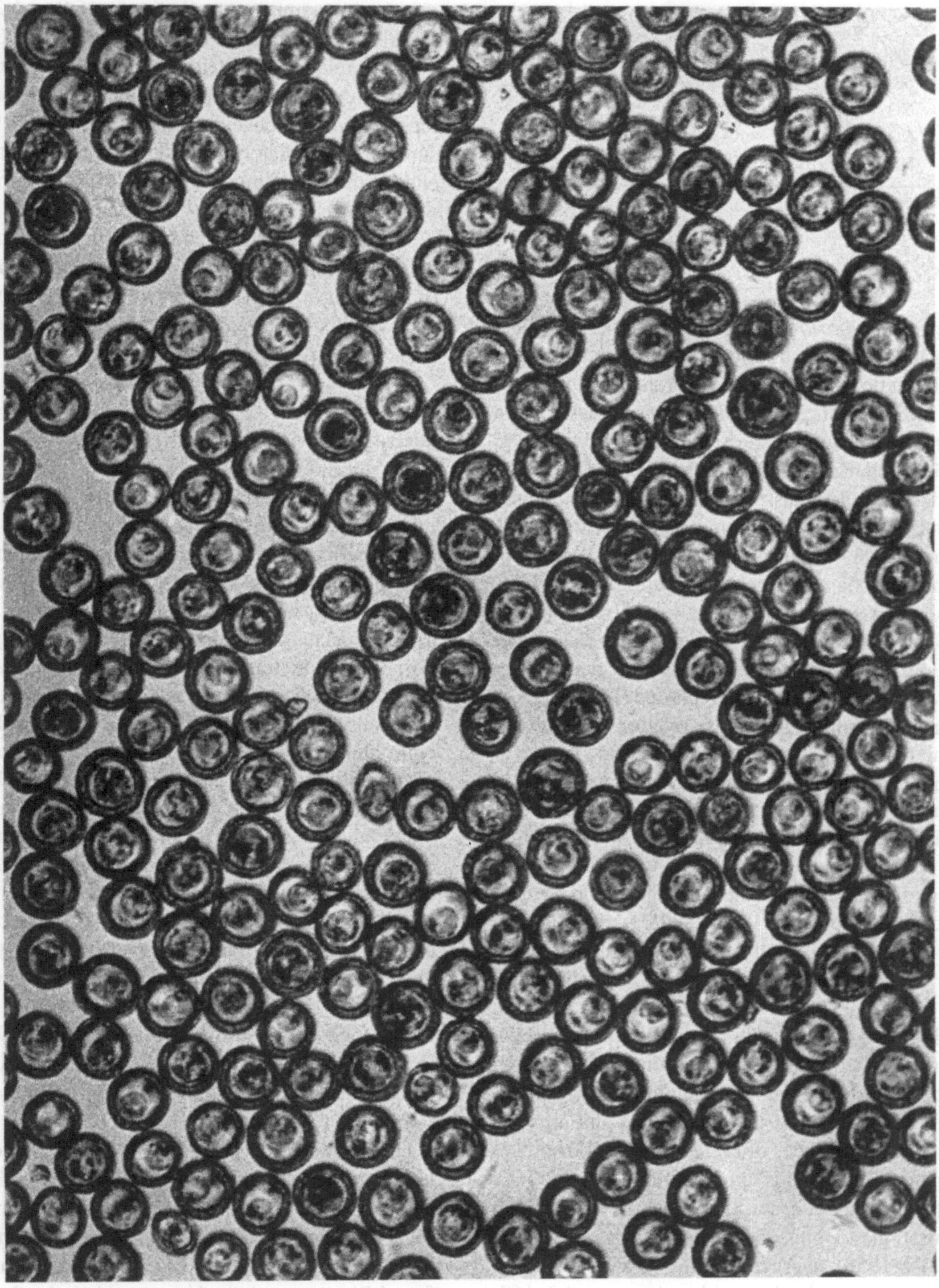

Figure 1. Nearly monospecific *Phacotus lenticularis* population observed in Lake Dagow.

Table 2. Nutritional status of habitats with *Phacotus* cell numbers of more than 500 000 l^{-1}.

	Min–max	Average
TP [$\mu g \times l^{-1}$]	34–484	122
SRP [$\mu g \times l^{-1}$]	2–88	22
NO_2^- [$\mu g \times l^{-1}$]	0–32	4
NO_3^- [$\mu g \times l^{-1}$]	0–820	173
NH_4^+ [$\mu g \times l^{-1}$]	10–740	180

Results

Phacotus lenticularis was found in all types of waters investigated: in deep stratified dimictic lakes, in shallow polymictic lakes, and in dead branches of rivers or ponds, too. In most cases (70% of the samples contain *Phacotus*) cell counts of less than 500 000 cells l^{-1} were observed. Counts of more than 500 000 cells l^{-1} were classified as "mass developments". A population density of about 1 million cells l^{-1} yields 0.2 mg fresh biomass l^{-1}. During periods of mass development the phytoflagellate could account for up to 10%, in extreme cases up to 50%, of the total phytoplankton biomass. Very seldom, especially, after clear water stages in Lake Dagow, near-monospecific populations of *Phacotus lenticularis* were recorded (Figure 1).

Our findings indicate a close relationship between the abundances of *Phacotus* and the trophic degree of the habitats in which they occurred. Maximal observed cell numbers are plotted against the mean summer epilimnetic TP concentration in Figure 2. Mass developments of this flagellate with 2–5.5 × 10^6 individuals l^{-1} were observed only in the lakes that could be categorized as eutrophic (TGL 27 885/01). In oligo- to mesotrophic lakes as well as in poly- to hypertrophic waters the cell counts do not exceed 10^6 individuals l^{-1}.

In Figure 3, the distribution of *Phacotus* concentrations is plotted against concentrations of dissolved phosphorus and dissolved nitrogen in the corresponding samples. Ranges of the phosphorus and nitrogen fraction for occurrences of *Phacotus* at > 500 000 l^{-1} are given in Table 2. It is apparent that we found high cell densities of *Phacotus* under low as well as high concentrations of phosphorus and nitrogen. Solving Redfield's (1958) 'average atomic N/P-ratio' for these lakes we have found values within the range 0.2–157.

Performing the analogous separation of *Phacotus* concentrations against water temperature and pH, we found that the occurrence of *Phacotus* is more profoundly sensitive to these factors (Figure 4). Abundances of more than 500 000 cells l^{-1} were only observed at temperatures between 15.8 and 24.7 °C and always in habitats with a pH value from 8.3 to 9.6.

This relationship between abundance and water temperature leads us to propose that the seasonal pattern of *Phacotus* development normally starts in early summer, immediately after the clear-water phase, and ends at the beginning of autumn. Mass developments were only recorded in July and August.

We have also observed high concentrations up to 260 000 living cells of *Phacotus* per gram fresh weight of sediment of the Lake Dagow during the winter. We suppose this to be typical means of overwintering in *Phacotus* spp.

Discussion

Phacotus lenticularis is a widespread alga recorded in inland waters of the temperate, subtropical and tropical climatic zones of the northern and southern hemispere. This phytoflagellate was first recorded by Ehrenberg (1831, 1838) under the name *Cryptomonas lenticularis* in buckets containing water for fire-extinguishing purposes. Numerous records have been published for European lakes and water bodies of the whole world. For example, Utermöhl (1925) documented the *Phacotus* abundance for numerous northern German lakes of the Baltic type and gave a famous survey about the behavior of this extraordinary flagellate. He supposed that *Phacotus* could exist in mixotrophic nutritional conditions. Skuja (1948, 1956) examined *Phacotus* in Swedish lakes. Nygaard (1949) and Kristiansen (1959) found this alga in various Danish lakes. *Phacotus* was observed as a subdominant member of phytoplankton communities e.g. in Lake Zürich, Switzerland (Thomas, 1950; Messikomer, 1954; Pavoni, 1963), in Lake Mondsee, Austria (Dokulil & Skolaut, 1986) and in Lake Constance (Müller, 1967). Giering (unpubl.) investigated 100 waters of eastern Germany and found *Phacotus* in 70 of them.

In Crystal Lake (U.S.A.), *Phacotus* occurred in all layers of the lake, including the anoxic hypolimnion, and the sediment down to a depth of about 3 cm (Cole, 1955). Munawar & Munawar (1986) checked the phytoplankton of the North American Great Lakes and found *Phacotus* only in the eutrophic Lake Ontario. A detailed comparison between the phytoplankton of the Lake Ontario and the ultraoligotrophic Lake Superi-

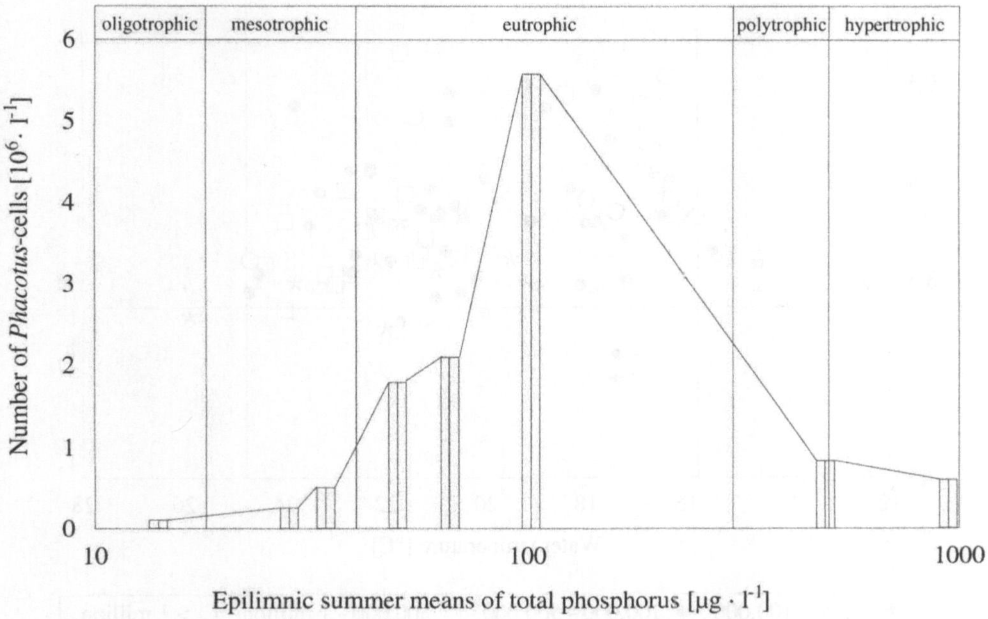

Figure 2. Maximal observed individuals of *Phacotus lenticularis* in northeastern German hardwaters of different trophic state.

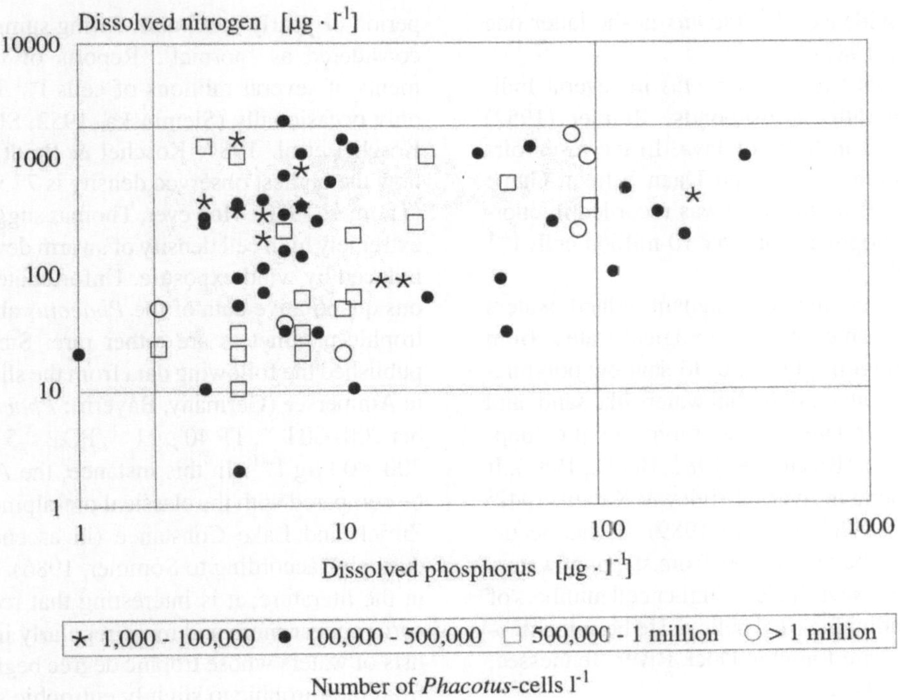

Figure 3. Phacotus-abundances in dependence on dissolved nitrogen and dissolved phosphorus concentrations.

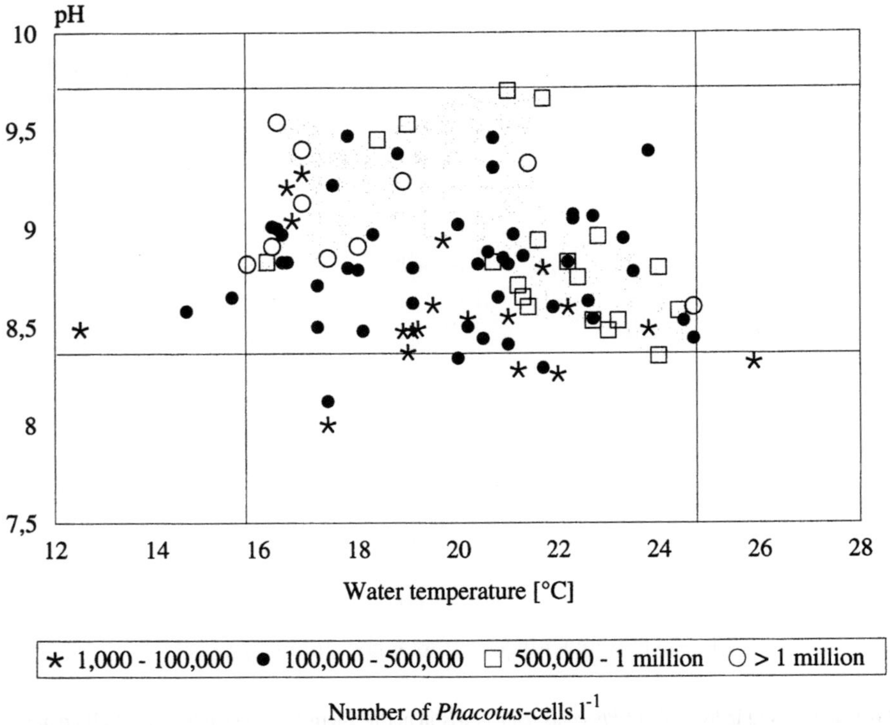

Figure 4. Phacotus-densities in dependence on pH value and water temperature.

or showed no evidence of *Phacotus* in the latter one (Munawar et al., 1987).

Philipose (1958) found *Phacotus* in several Indian waters, generally in fishponds. Ruttner (1952) observed this alga in lakes of Java. In the reservoirs Hardopdam (South Africa) and Daan Viljoen Game Park (Namibia), *P. lenticularis* was recorded in enormous mass developments of over 10 million cells l^{-1} (Krienitz, unpubl.).

Phacotus is an alga of stagnant inland waters of various morphometric and ecological states: from deep stratified oligotrophic lakes to shallow polymictic hypertrophic waters. Artificial waters like sand- and clay pit-lakes, river dams and reservoirs are often populated by *Phacotus* (Blaauboer, 1982; Bucka, 1987). It has also been found in rivers (Brinley & Katzin, 1942; Steinberg et al., 1987; Goldyn 1989). These occurrences may be atributable inflow from stagnant waters like oxbow-lakes. Nevertheless, higher cell numbers of about 240 000 individuals l^{-1} believed to be autochthonously derived, were found in Duck River, Tennessee, U.S.A. (Lackey, 1942).

As a rule, reports of abundance are sporadic. Cell counts from 50 000 to 500 000 individuals l^{-1} in the period of yearly maximum during summer have been considered as 'normal'. Reports of mass developments of several millions of cells l^{-1} have appeared only occasionally (Sieminska, 1952; Sladeček, 1958; Koschel et al. 1987; Koschel & Raidt, 1988). Until now the highest observed density is 77×10^6 cells l^{-1} (Thomas, 1950). However, Thomas suggested that this extremely high cell density of swarm development was induced by wind exposure. Unfortunately, simultaneous quantitative data of the *Phacotus* abundances and trophic parameters are rather rare. Steinberg (1978) published the following data from the slightly eutrophic Ammersee (Germany, Bayern): *Phacotus* cell number 200 000 l^{-1}, TP 40 $\mu g\ l^{-1}$, $PO_4{}^{-3}$ 5 $\mu g\ l^{-1}$, $NO_3{}^{-}$ 300–800 $\mu g\ l^{-1}$. In this instance, the Ammersee can be compared with the classical pre-alpine sites of Lake Zürich and Lake Constance (in ascending order of eutrophy, according to Sommer, 1986). From the data in the literature, it is interesting that records of *Phacotus* appear more and more regularly in the plankton lists of waters whose trophic degree begins to increase from mesotrophic to slightly eutrophic state. This can be illustrated from the data from Lake Zürich and Lake Constance. At the end of the 19th century and the beginning of the 20th no evidence of *Phacotus* had been shown in Lake Zürich (according to Pavoni, 1963). However, its subsequent appearance may be a consequence of eutrophication of the lakes. Fehlmann & Minder (1921) found *Phacotus* in the plankton in summer 1912/1913. Thomas (1950), Messikomer (1954) and Pavoni (1963) had described *Phacotus* developmental patterns from the 1940s to 1960s. Whereas in Lake Constance phytoplankton examinations during 1896 and 1920–1924 provided no evidence of *Phacotus*, this alga was abundant in the lake in the 1960s (Müller, 1967).

Our findings agree with the literature concerning the oligo-, meso- and eutrophic waters investigated. However, there is a large database now available on a spectrum of poly- and hypertrophic shallow lakes and ponds. Uhlmann (1965, 1966) found populations of *Phacotus* of about 100 000 cells l^{-1} in hypertrophic ponds under extreme anthropogenic influence, such as sites subject to high sewage inflow. Sladeček (1958), however, found densities of *Phacotus* of 4.1×10^6 l^{-1} in polytrophic fishponds. Therefore, it can be concluded that *Phacotus* is able to produce similarly high cell concentrations under polytrophic conditions as well as in eutrophic waters.

Neither the concentrations of nitrogen nor those of phosphorus were likely to have had a drastic limiting effect on the growth of *Phacotus*. Populations of this alga were found under conditions describable by N:P ratios extending from 0.2 to 157. One explanation could be the ability of *Phacotus* to use 'microhabitats' of excrements of zooplankton (Bergquist & Carpenter, 1986; Elser et al., 1987) or *Phacotus* could obtain nutients by passing through the digestive tract of zooplankters (Porter, 1973). It may be supposed that ingested *Phacotus* cells leave the intestine of grazers undigested on account of calcite lorica. The marine calcified coccolithophorid *E. huxleyi* is able to grow well in cultures under both P- and N-limited conditions (Paasche & Brubak, 1994). Furthermore, it has been observed that *E. huxleyi* out-competes other planktonic algae at low SRP concentrations (Riegmann et al., 1992). Whether similar arguments may be applied to *Phacotus lenticularis* remains to be investigated.

The increase of autochthonous calcite precipitation (ACP) and increase of population density of *Phacotus* (so-called organismic calcite precipitation [OCP]) quickly follow one another dated or actually coincide (Giering, unpubl., Schlegel, unpubl. data). According to Koschel (1990) and Koschel et al. (1983), the ACP is a natural control mechanism of eutrophication. The occurrence of OCP at the same time as the ACP may be an indication that it is the calcification process of the *Phacotus* lorica which could provide a similar ecological mechanism against eutrophication, by adsorbing N or P on to the calcite crystal of the shells. Otherwise, the OCP could be hypothesized to be in competition with ACP, through incorporation of calcium. This *Phacotus* calcite would be lost for the ecologically effective ACP. Müller & Oti (1981: 898) interpreting the influence of *Phacotus* mass developments in Lake Constance, emphasized that 'these periods of (*Phacotus*) blooms may seriously deplete the calcium budget of lakes'. During recent sediment formation in Lake Zürich, an increase of calcite deposition, driven by ACP as well by OCP, has been observed during the eutrophication period (Fehlmann & Minder, 1920).

Is *Phacotus* an indicator of lake trophic types, sensu Rawson (1956)? The answer is Yes and No. 'Yes' with regard to the quantity: *Phacotus* shows its highest densities in eutrophic waters. 'No' with regard to the quality: However, *Phacotus* does occur practically among lime-rich alkaline stagnant waters, sometimes as ecological 'memory' (Padisák, 1992), and where it sometimes has comprised a dominant or subdominant fraction of the phytoplankton community under appropriate eutrophic conditions.

Acknowledgements

This work was supported by the Deutsche Forschungsgemeinschaft (KR 1262/1–2). We thank M. Coßbau, J. Dalchow, K. Mallok, R. Degebrodt, E. Mach, M. Sachtleben, and A. Scheffler for field sampling and sample treatment. Limnological data were placed to our disposal from the data bank of the Institute of Freshwater Ecology and from the Seenkataster Brandenburg. Our appreciation is also due to Prof. Dr C. S. Reynolds for critical reading of the manuscript.

References

Balch, W. M., P. M. Holligan & K. A. Kilpatrick, 1992. Calcification, photosynthesis and growth of the bloom-forming coccolithophore, *Emiliania huxleyi*. Cont. Shelf Res. 12: 1353–1374.

Balch, W. M., K. A. Kilpatrick, P. M. Holligan & T. L. Cucci, 1993. Coccolith formation and detachment by *Emiliania huxleyi* (Prymnesiophyceae). J. Phycol. 29: 566–575.

Bergquist, A. M. & S. R. Carpenter, 1986. Grazing of phytoplankton: Effects on species growth rates, phosphorus limitation, chlorophyll, and primary production. Ecology 67: 1351–1360.

Blaauboer, M. C. T., 1982. The phytoplankton species composition and the seasonal periodicity in Lake Vechten from 1956–1979. Hydrobiologia 95: 25–36.

Brinley, F. J. & L. J. Katzin, 1942. Distribution of stream plankton in the Ohio River System. Am. Midl. Naturalist 27: 177–191.

Bucka, H., 1987. Ecological aspects of the mass appearance of planktonic algae in dam reservoirs of southern Poland. Acta Hydrobiol. 29: 149–191.

Casper, S. J. (ed), 1985. Lake Stechlin – a temperate oligotrophic lake. Monographiae Biologicae 58: XIV + 550 pp. Dr. W. Junk Publishers, Dordrecht.

Cole, G. A., 1955. An ecological study of the microbenthic fauna of two Minnesota lakes. Am. Midl. Naturalist 53: 213–230.

Dokulil, M. & C. Skolaut, 1986: Succession of phytoplankton in a deep stratifying lake: Mondsee, Austria. Hydrobiologia 138: 9–24.

Ehrenberg, C. G., 1831. Über die Entwicklung und Lebensdauer der Infusionsthiere, nebst ferneren Beiträgen zu einer Vergleichung ihrer organischen Systeme. Abh. Königl. Akad. Wiss. Berlin: 1–154.

Ehrenberg, C. G., 1838. Die Infusionsthiere als vollkommene Organismen – mit Atlas. Berlin und Leipzig, 547 pp.

Elser, J. J, N. C. Goff, N. A. MacKay, A. L. St. Amand, M. N. Elser & S. R. Carpenter, 1987. Species specific algal responses to zooplankton: experimental and field observations in three nutrient-limited lakes. J. Plankton Res. 9: 699–717.

Fehlmann, W. & L. Minder, 1921. Beitrag zum Problem der Sedimentbildung und Besiedlung im Zürichsee. Festschrift Zschokke 11: 3–23.

Giering, B., L. Krienitz, S. J. Casper, T. Peschke & H. Raidt, 1990. LM- and SEM-observations on the asexual reproduction of *Phacotus lendneri* CHODAT (Chlamydophyceae, Phacotaceae). Arch. Protistenkd. 138: 75–88.

Giering, B., L. Krienitz & S. J. Casper, 1992. Zur Taxonomie von *Phacotus lenticularis* (Ehrenberg) Stein (Chlamydophyceae, Phacotaceae). Nova Hedwigia 55: 367–380.

Goldyn, R., 1989. Algae in the seston of the middle section of the River Radunia, its tributaries and dam reservoirs (Northern Poland). Fragm. Florist. Geobot. 34: 201–245.

Hepperle, D. & L. Krienitz, 1996. The extracellular calcification of zoospores of *Phacotus lenticularis* (Chlorophyta, Chlamydomonadales). Eur. J. Phycol. 31: 11–21.

Kamptner, E., 1950. Über den Aufbau des Kalkgehäuses von *Phacotus lendneri* Chod. Österr. Bot. Z. 97: 391–402.

Koschel, R., 1990. Pelagic calcite precipitation and trophic state of hardwater lakes. Arch. Hydrobiol. Beih. Ergebn. Limnol. 33: 713–722.

Koschel, R., J. Benndorf, G. Proft & F. Recknagel, 1983. Calcite precipitation as a natural control mechanism of eutrophication. Arch. Hydrobiol. 98: 380–408.

Koschel, R., G. Proft, & H. Raidt, 1987. *Phacotus*-Massenentwicklungen – eine Quelle autochthonen Kalkeintrages in Seen. Limnologica 18: 457–459.

Koschel, R. & H. Raidt, 1988. Morphologische Merkmale der *Phacotus*-Hüllen in Hartwasserseen der Mecklenburger Seenplatte. Limnologica 19: 13–25.

Koschel, R., B. Giering, P. Kasprzak, G. Proft & H. Raidt, 1990. Changes of calcite precipitation and trophic conditions in two stratified hardwater lakes of the Baltic Lake District. Verh. int. Ver. Limnol. 24: 140–145.

Koschel, R. & P. Kasprzak (Red), 1994. Der Tollensesee. Gewässerökologie – Umweltschutz – Wasser- und Fischereiwirtschaft. Kupijai & Prochnow, Berlin, 62 pp.

Koschel, R., P. Kasprzak & L. Krienitz, 1993. Long term effects by reducing of nutrient loading and following food-web manipulation in a stratified Baltic hardwater lake (Lake Haussee, Germany). Verh. int. Ver. Limnol. 25: 647–651.

Krienitz, L. & E. Hegewald, 1996. Über das Vorkommen von wärmeliebenden Blaualgenarten in einem norddeutschen See. Lauterbornia 26: 55–63.

Krienitz, L., R. Koschel, B. Giering, S. J. Casper. & D. Hepperle, 1993. Phenomenology of organismic calcite precipitation by *Phacotus* in hardwater lakes and ponds of northeastern Germany. Verh. Int. Ver. Limnol. 25: 170–174.

Kristiansen, J., 1959. Flagellates from some Danish lakes and ponds. Dansk. Bot. Ark. 18: 7–81.

Lackey, J. B., 1942. The plankton algae and protozoa of two Tennessee Rivers. Am. Midl. Naturalist 27: 191–202.

Messikomer, E., 1954. Die Algenflora des Zürichsees bei Zürich, Schweiz. Schweiz. Z. Hydrol. 16: 27–63.

Müller, H., 1967. Eine neue qualitative Bestandsaufnahme des Phytoplanktons des Bodensee-Obersee mit besonderer Berücksichtigung der tychoplanktischen Diatomeen. Arch. Hydrobiol. Suppl. 33: 206–236.

Müller, G. & M. Oti, 1981. The occurrence of calcified planktonic green algae in freshwater carbonates. Sedimentology 28: 897–902.

Munawar, M. & J. F. Munawar, 1986. The seasonality of phytoplankton in the North American Great Lakes, a comparativ synthesis. Hydrobiologia 138: 85–115.

Munawar, M., J. F. Munawar & L. H. McCarthy, 1987. Phytoplankton ecology of large eutrophic and oligotrophic lakes of North America: Lakes Ontario and Superior. Arch. Hydrobiol. Beih. Ergeb. Limnol. 25: 51–96.

Nygaard, G., 1949. Hydrobiological studies on some Danish ponds and lakes. Kongel. Danske Vidensk. Selsk. Biol. Skr. 7: 1–293.

Paasche, E. & S. Brubak, 1994. Enhanced calcification in the coccolithophorid *Emiliania huxleyi* (Haptophyceae) under phosphorus limitation. Phycologia 33: 324–330.

Padisák, J., 1992. Seasonal succession of phytoplankton in a large shallow lake (Balaton, Hungary) – a dynamic approach to ecological memory, its possible role and mechanisms. J. Ecol. 80: 217–230.

Padisák, J., L. Krienitz, W. Scheffler, I. Grigorszky & J. Kristiansen, 1997. The phytoplankton of Lake Stechlin (in press, this volume).

Pavoni, M., 1963. Die Bedeutung des Nannoplanktons im Vergleich zum Netzplankton. Schweiz. Z. Hydrol. 25: 219–341.

Phillipose, M. T., 1958. Contributions to our knowledge of Indian algae. II: Volvocales, unicellular forms. Proc. Ind. Acad. Sci. 47b: 287–311.

Pocratsky, L. A., 1982. Nutritional, chemical and ultrastructural characterization of the lorica and extracellular mucilage of *Phacotus lenticularis* (Phacotaceae, Volvocales) Diss. Univ. Tennessee.

Porter, K. G., 1973. Selective grazing and differential digestion of algae by zooplancton. Nature 244: 179–180.

Proft, G., 1984. Die pelagische Kalzitfällung und der Carbonatgehalt von Sedimenten pleistozäner Seen. Acta hydrochim. hydrobiol. 12: 321–326.

Rawson, D. S., 1956. Algal indicators of trophic lake types. Limnol. Oceanogr. 1: 18–25.

Riegman, R., A. A. M. Noordeloos & G. C. Cadée, 1992. *Phaeocystis* blooms and eutrophication of the continental coastal zones of the North Sea. Mar. Biol. 112: 479–484.

Ruttner, F., 1952. Planktonstudien der Deutschen Limnologischen Sunda-Expedition. Arch. Hydrobiol. Suppl. 21: 1–274.

Sieminska, J., 1952. The plankton of the artificial lake at the Roznów dam. Mém. de l'Acad. Pol. Scie. Lettres Sér. B. Scie. Naturelles 18: 109 pp.

Skuja, H., 1948. Taxonomie des Phytoplanktons einiger Seen in Uppland, Schweden. Symb. Bot. Upsal. 9: 1–399.

Skuja, H., 1956. Taxonomische und biologische Studien über das Phytoplankton Schwedischer Binnengewässer. Nov. Act Reg. Soc. Sci. Upsal. Ser. IV. 16: 1–404.

Sladecek, V., 1958. Hydrobiological investigations of a treatment of beet sugar factory's wastes in an experimental lagoon. Sbornik vus. skolu chem. techn. paliv a vodu: 121–150.

Sommer, U., 1986. The periodicity of phytoplankton in Lake Constance (Bodensee) in comparison to other deep lakes of central Europe. Hydrobiologia 138: 1–7.

Steinberg, C., 1978. Limnologische Untersuchungen des Ammersees. Info.-ber. d. Bayer. Landesamtes f. Wasserwirtschaft 6: 1–78.

Steinberg, C. & R. Klee, 1983. Röntgenmikroanalyse der Schale einer einzelligen Grünalge: Chemismus einer *Phacotus lendneri*-Schale. Mikrokosmos 72: 170–173.

Steinberg, C., B. Heindel, R. Tille-Backhaus & R. Klee, 1987. Phytoplanktonstudien an langsam fließenden Gewässern: Donau und Vils. Arch. f. Hydrobiol Suppl. 68: 437–456.

TGL 27 885/01, 1982. Fachbereichsstandard Nutzung und Schutz der Gewässer. Stehende Binnengewässer. Klassifizierung. Berlin: Ministerium für Umweltschutz und Wasserwirtschaft der DDR. 16.

Thomas, E. A., 1950. Auffällige biologische Folgen von Sprungschichtneigungen im Zürichsee. Schweiz. Z. f. Hydrol. 12: 1–24.

Uhlmann, D., 1965. Beitrag zur Limnologie extrem nährstoffreicher Flachgewässer I. Sauerstoff. Nährstoffe. Wiss. Z. Karl-Marx-Universität Leipzig 14: 359–411.

Uhlmann, D., 1966 Beitrag zur Limnologie extrem nährstoffreicher Flachgewässer II. Plankton-Massenwechsel. Wiss. Zeitschrift Karl-Marx-Universität Leipzig 15: 373–423.

Utermöhl, H., 1925. Limnologische Phytoplanktonstudien. Arch. Hydrobiol. Suppl. 5, Stuttgart, 527 pp.

Utermöhl, H., 1958. Zur Vervollkommnung der quantitativen Phytoplanktonmethodik. Mitt. int. Ver. Limnol. 9. 1–38.

Willén, E., 1976. A simplified method of phytoplankton counting. Br. phycol. J. 11. 265–278.

Hydrobiologia **369/370:** 363–372, 1998.
M. Alvarez-Cobelas, C. S. Reynolds, P. Sanchez-Castillo & J. Kristiansen (eds), Phytoplankton and Trophic Gradients.

Euglenophyta found exclusively in South America

(a taxonomic revision)

Guillermo Tell
Depto. Cs. Biológicas, Facultad Cs. Exactas y Naturales, 1428 Buenos Aires, Argentina

Key words: Euglenophyta, South America, geographical distribution, taxonomy

Abstract

A taxonomic revision of the Euglenophyta found exclusively in South America is presented. The taxonomy, geographical distribution and critical analysis of the synonyms are given. More than two hundred species and infraspecific entities from South America are described, many of them recorded several times in different regions. The genus *Trachelomonas* comprises the highest number of new taxa, reaching more than 100 entities. It is followed by *Phacus* (48 taxa), *Strombomonas* (33 taxa), *Euglena* (15 taxa) *Lepocinclis* (14 taxa), and *Distigma, Entosiphon* and *Sphenomonas* with one new taxon each.

Introduction

The taxonomy and the geographical distribution of Euglenophyta from South America had scarcely been studied before 1980. Among the fifty or so papers dealing with the group, only seven were published before this date. The first one dealing with some Euglenoids from Brazil, was published at the begining of the century (da Cunha, 1913).

Up to 1980, the taxonomical knowledge of the Euglenophyta was mainly restricted to Europe and USA. Some of the few studies carried out in South America before 1980 confirmed the presence of many new taxa in the area. Studies carried out during the last 15 years reveal a significant increase in new taxa, including some forms different from those recorded in Europe (i.e.: *Strombomonas diptera* Tell & Zalocar, *Trachelomonas hemisphaerica* Gar. de Emil.). On the other hand, the structures of some South American entities are so complex that their taxonomic positions are still under discussion (i.e.: *Phacus bonettoi* Tell & Zalocar, *Trachelomonas argentinensis* (Gar. de Emil.) Conforti & Nudelman, *T. pyramidata* Couté & Thérézien).

A complete revision of the literature descriptions has been carried out in an attempt to compile the information on new, exclusively South American Euglenophyta. As is usual in this kind of research, the greatest numbers of taxa were found in regions where specialists are available: Argentina, Bolivia, Brazil and Venezuela. More than two hundred species and infraspecific entities have been described, many of them having been recorded several times in the various regions. The geographical distribution, synonyms, bibliography and a new combination are given. The paper by Bourrelly & Manguin (1952), recording new species of fresh-water algae from Guadeloupe, is also included.

The highest number of new euglenoid taxa for South America are ascribed to *Trachelomonas* Ehr. emend. Defl.: there are over 100 entities, with a predominance of new varieties and more than 20 new species. Forty-eight entities are in the genus *Phacus* Duj., with a predominance of new species. Thirty-three new entities of *Strombomonas* Defl. have been described and fourteen new taxa of *Lepocinclis* Perty are included. The genus *Euglena* Ehr. includes fifteen new taxa and the genera *Distigma* Ehr., *Entosiphon* Stein and *Sphenomonas* Stein one new one each.

Even though the taxonomy of South American Euglenophyta has been intensively developed over the last 20 years, little is yet known about the ecology of the group. Only isolated data on the water quality where Euglenophyta live are available and there have been no studies on the ecological requirements of the individual species.

Principal regions studied

ARGENTINA: three main regions of Argentina were studied: the cold south region; the temperate central region; and the warm, north-eastern region.

South Argentina: This area comprises the province of Tierra del Fuego. It is characterized by the presence of great oligotrophic lakes, mountain rivers, peat bogs and some ponds and shallow lakes. Many of the latter water bodies are typically dystrophic. No new taxa have been recorded for this region (Tell & Conforti, 1988).

Central Argentina: The Euglenophyta of the temperate zone of Argentina have been intensively studied by Balech (1944) and Conforti (1983, 1986a, 1986b, 1986c, 1989, 1991). The best-studied aquatic systems are shallow lakes, rivers and ponds surrounding Buenos Aires city and the mouth of the river Paraná. Many new taxa have been found in this region, mainly in the overgrown ponds and polluted rivers near Buenos Aires city.

North-east Argentina: Tell (1980) and Tell & Zalocar (1985) have published two papers on the Euglenophyta of the Mesopotamic warm region, mainly from Corrientes and Chaco provinces. This area is rich in 'esteros' (tropical swamps) and overgrown shallow lakes and ponds, some of them with dystrophic characteristics. Many new taxa have been recorded from this region.

BOLIVIA: Couté & Thérézien (1985, 1994) have studied many samples from the Amazonian Bolivia. Some new taxa for the region have been recorded in the River Ibare which has been most studied. It is interesting to point out that some of the taxa previously thought to be unique to Argentina have since been found in Bolivia.

BRAZIL: Various authors have studied materials from Brazil; the papers of Conforti (1993a, 1993b, 1994) give the highest number of new taxa recorded. Rio Grande do Sul, Manaus and some wetlands near Rio de Janeiro have been the best-studied regions. In most cases, these are waters with high levels of organic matter.

COLOMBIA: Only two papers have been published on the euglenoids of this country. Duque's studies (1995) show special interest because of the number of species recorded. This author enlarged the geographic distribution of some taxa previously only described for Argentina.

VENEZUELA: Yacubson (1980, 1984–85) and Yacubson & Bravo (1982–83) described some new taxa from rivers and ponds situated near Maracaibo lake. These authors also summarized the abiotic properties of the waters concerned.

Methods

The complete literature concerning the taxonomy of Euglenoids recorded in South America was reviewed and listed. Species, authors, references of the publication and geographical distribution were careful checked. Synonyms and combinations proposed by different authors were analysed.

Results

Fam. Astaciaceae

Distigma proteus fo. *minor* Tracana 1985: *Argentina*

Sphenomonas decemlineata Seckt 1922: *Argentina*

Fam. Peranemaceae

Entosiphon ovatum fo. *major* Seckt 1922: *Argentina*

Fam. Euglenaceae

Euglena acus fo. *hyalina* Seckt 1922: *Argentina*

Euglena ehrenbergii var. *brevis* Seckt 1922: *Argentina*

Euglena oxyuris var. *oxyuris* fo. *charkowensis* Bourrelly 1952; *Bolivia* (Thérézien, 1989); *Guadeloupe* (Bourrelly & Manguin, 1952); *Venezuela* (Yacubson, 1980)

Euglena oxyuris var. *oxyuris* fo. *playfairii* Bourrelly 1952: *Bolivia* (Thérézien, 1989); *Guadeloupe* (Bourrelly & Manguin, 1952)

Euglena pringsheimii var. *pringsheimii* Skv. 1967: *Argentina* (Tell & Zalocar, 1985); *Brazil* (Skvortzov, 1967)

Euglena pringsheimii var. *armata* Skv. 1967: *Brazil*

Euglena pringsheimii var. *punctato-striata* Skv. 1967: *Brazil*

Euglena pringsheimii var. *rubra* Skv. 1967: *Brazil*

Euglena pringsheimii var. *subconstricta* Skv. 1967: *Brazil*
Euglena pringsheimii var. *trichita* Skv. 1967: *Brazil*
Euglena pringsheimii var. *vacuolaria* Skv. 1967: *Brazil*
Euglena spirogyra var. *fusca* fo. *minor* Tracana 1985: *Argentina*
Euglena spirogyra var. *fusiformis* fo. *minor* Conforti 1994: *Brazil*
Euglena spirogyra var. *magnifica* Uherk. in Uherkovich & Schmidt, 1974: *Brazil*
Euglena tripteris var. *longa* Conforti 1986 b: *Argentina* (Conforti, 1986 c; Conforti & Tell, 1989)

Lepocinclis acuta var. *major* Tell et Zalocar 1985: *Argentina*
Lepocinclis caudata Da Cunha 1913: *Argentina* (Tell & Zalocar, 1985); *Brazil* (da Cunha, 1913)
Lepocinclis claviformis Conforti 1994 var. *claviformis*: *Brazil*
Lepocinclis claviformis var. *ornata* Conforti 1994: *Brazil*
Lepocinclis glabra var. *raciborskii* fo. *gigas* Conforti 1994: *Brazil*
Lepocinclis ovum var. *ezeizae* Conforti 1991: *Argentina*
Lepocinclis piriformis Da Cunha 1913: *Argentina* (Conforti, 1979); *Brazil* (da Cunha, 1913)
Lepocinclis playfairiana var. *striata* Conforti 1991: *Argentina*
Lepocinclis pseudonayalii Tell et Zalocar 1985: *Argentina*
Lepocinclis salina var. *caudata* Conforti 1994: *Brazil*
Lepocinclis salina fo. *pachyderma* (Defl.) Hüb.-Pest. 1955: *Argentina* (Tell & Zalocar, 1985); *Venezuela* (Deflandre, 1926 b)
Bas.: *Lepocinclis texta* var. *pachyderma* Defl. 1926 b
Lepocinclis salina var. *vallicauda* fo. *minor* Conforti 1994: *Brazil*
Lepocinclis subcylindrica Conforti 1991: *Argentina*
Lepocinclis truncata Da Cunha 1913: *Argentina* (Tell & Zalocar, 1985); *Bolivia* (Thérézien, 1989); *Brazil* (da Cunha, 1913); *Colombia* (Duque, 1995)
Lepocinclis turbiniformis Defl. 1926 b: *Argentina* (Tell & Zalocar, 1985); *Brazil* (Menezes & Fernandes, 1989; Menezes, 1990; Alves-da-Silva & Torres, 1992); *Venezuela* (Deflandre, 1926 b)

Phacus acuminatus var. *cordiformis* Tell et Zalocar 1985: *Argentina*
Phacus acuminatus var. *brasiliensis* Conforti 1994: *Brazil*
Phacus agilis var. *inversa* Bourrelly 1952: *Argentina* (Tell & Zalocar, 1985); *Guadeloupe* (Bourrelly & Manguin, 1952)
Phacus angustatus Conforti 1991: *Argentina*
Phacus argentinensis Conforti 1991: *Argentina*
Phacus bacillifer da Cunha 1913: *Argentina* (Conforti, 1981 a, 1983); *Brazil* (da Cunha, 1913; Menezes & Fernandes, 1989)
Phacus balatonicus var. *minor* Conforti 1994: *Brazil*
Phacus balechii Conforti 1991: *Argentina*
Phacus boliviensis Thérézien 1989: *Bolivia*
Phacus bonaerensis Conforti 1989: *Argentina*
Phacus bonettoi Tell et Zalocar 1985: *Argentina* (Tell & Zalocar, 1985); *Brazil* (Menezes & Fernandes, 1989); *Bolivia* (Thérézien, 1989; Couté & Thérézien, 1994)
Syn.: *Euglena bonettoi* (Tell et Zalocar) Couté et Thérézien 1994
Phacus contortus Bourrelly et Manguin 1952 var. *contortus*: *Argentina* (Tell, 1980; Conforti, 1981 b; Tell & Zalocar, 1985); *Bolivia* (Thérézien, 1989); *Brazil* (Menezes & Fernandes, 1987; Alves-da-Silva et al., 1991; Franceschini, 1992; Alves-da-Silva & Souza Laitano, 1994; Conforti, 1994; Menezes et al., 1995); *Guadeloupe* (Bourrelly & Manguin, 1952); *Venezuela* (Yacubson, 1980)
Phacus contortus var. *complicatus* Bourrelly et Manguin 1952: *Argentina* (Conforti, 1981 b, 1983; Tell & Zalocar, 1985); *Bolivia* (Thérézien, 1989); *Brazil* (Alves-da-Silva & Torres, 1994); *Guadeloupe* (Bourrelly & Manguin, 1952)
Phacus curvicauda fo. *fissurata* Tell et Zalocar 1985: *Argentina*
Phacus elongatus Conforti 1991: *Argentina*
Phacus fenrii Conforti 1991: *Argentina*
Phacus florii Conforti 1991: *Argentina*
Phacus gigas da Cunha 1913: *Argentina* (Tell, 1980); *Bolivia* (Thérézien, 1989); *Brazil* (da Cunha, 1913; Cécy, 1990; Alves-da-Silva & Torres, 1994); *Guadeloupe* (Bourrelly & Manguin, 1952)
Phacus horridus var. *major* Conforti 1994: *Brazil*
Phacus inconspicus Defl. 1926 b: *Venezuela*
Phacus incrassatus (Defl.) Pochm. 1942: *Argentina* (Conforti, 1983); *Venezuela* (Deflandre, 1926 b)
Bas.: *Phacus alata* var. *incrassata* Defl. 1926 b
Phacus inflexus var. *minor* Bourrelly et Manguin 1952: *Guadeloupe*
Phacus lefevrei Bourrelly 1952: *Argentina* (Conforti, 1983, 1986 c; Tell & Zalocar, 1985); *Bolivia*

(Thérézien, 1989); *Brazil* (Huszar et al., 1989; Cécy, 1990; Alves-da-Silva, 1994); *Guadeloupe* (Bourrelly & Manguin, 1952); *Colombia* (Duque, 1995)
Phacus lismorensis var. *insecta* Conforti 1991: *Argentina* (Conforti, 1991); *Brazil* (Alves-da-Silva et al., 1991; Conforti, 1994)
Phacus longicauda var. *tortus* fo. *pseudoroecklii* Thérézien 1989: *Bolivia*
Phacus mangini var. *insecta* Conforti 1991: *Argentina*
Phacus manginii var. *sinuosus* Thérézien 1989: *Bolivia*
Phacus marconii Conforti 1991: *Argentina*
Phacus mariae Defl. 1926 b: *Brazil* (Alves-da-Silva, 1994); *Venezuela* (Deflandre, 1926 b)
Phacus matanzae Conforti 1991: *Argentina*
Phacus meson var. *minor* Conforti 1994: *Brazil*
Phacus multifidus Conforti 1991: *Argentina* (Conforti, 1991); *Colombia* (Duque, 1995)
Phacus myersi var. *myersi* fo. *maior* Uherk. in Uherkovich & Schmidt, 1974: *Brazil*
Phacus myersi var. *myersi* fo. *minor* Uherk. in Uherkovich & Schmidt, 1974: *Brazil*
Phacus novus Conforti 1991: *Argentina*
Phacus onyx var. *symetrica* Tell et Zalocar 1985: *Argentina* (Tell & Zalocar, 1985); *Colombia* (Duque, 1995); *Brazil* (Franceschini, 1992)
Phacus orbicularis var. *minor* Seckt 1922: *Argentina*
Phacus paulae Conforti 1991: *Argentina*
Phacus pentacaudatus Conforti 1991: *Argentina*
Phacus pygmaeus Conforti 1994: *Brazil*
Phacus pleuronectes var. *ungulatus* Tell et Zalocar 1985: *Argentina* (Tell & Zalocar, 1985); *Colombia* (Duque, 1995)
Phacus raciborskii var. *longus* Conforti 1989: *Argentina* (Conforti, 1989); *Brazil* (Conforti, 1994)
Phacus ranula var. *minor* Conforti 1991: *Argentina*
Phacus rodriguesiae Conforti 1994: *Brazil*
Phacus segreti var. *major* Conforti 1994: *Brazil*
Phacus sesquitortus var. *multiannulatus* Conforti 1994: *Brazil*
Phacus suecicus fo. *major* Yacubson 1980: *Venezuela*
Phacus tellii Conforti 1989: *Argentina*
Phacus tropicalis Conforti 1994: *Brazil*

Strombomonas brasiliensis Conforti 1993 b: *Brazil*
Strombomonas bulla var. *cordiformis* Tell et Conforti 1988: *Argentina*
Strombomonas bulla var. *minor* Conforti 1991: *Argentina* (Conforti, 1991); *Brazil* (Alves-da-Silva, 1994)
Strombomonas confortii Zalocar 1991: *Argentina*
Strombomonas diptera Zalocar et Tell in Tell, 1980: *Argentina* (Tell, 1980; Tell & Conforti, 1988); *Bolivia* (Thérézien, 1989; Couté & Thérézien, 1994)
Strombomonas ensifera var. *brasiliensis* Uherk. in Uherkovich & Schmidt, 1974: *Brazil*
Strombomonas ensifera var. *inflata* Tell et Zalocar 1985: *Argentina*
Strombomonas fluviatilis var. *major* Conforti 1993 b: *Brazil*
Strombomonas fluviatilis var. *rugosa* fo. *major* Yacubson 1980: *Brazil* (Alves-da-Silva et al., 1991); *Venezuela* (Yacubson, 1980)
Strombomonas gibberosa var. *grisoli* (Defl.) Hüb.-Pest. 1955: *Venezuela* (Deflandre, 1926 a)
Bas.: *Trachelomonas ensifera* var. *grisoli* Defl. 1926 a.
Strombomonas gibberosa var. *major* Conforti 1993 b: *Brazil*
Strombomonas girardiana var. *girardiana* fo. *boliviensis* Thérézien 1989: *Bolivia*
Strombomonas girardiana var. *triundulata* Tell et Zalocar 1985: *Argentina* (Tell & Zalocar, 1985); *Bolivia* (Thérézien, 1989)
Strombomonas gonitrachela Balech et Dastugue in Balech, 1944: *Argentina* (Balech, 1944; Tell & Conforti, 1988)
Strombomonas helicaudata Balech 1944: *Argentina*
Strombomonas limonensis Yacubson 1980: *Brazil* (Franceschini, 1992); *Venezuela* (Yacubson, 1980)
Strombomonas pizarroi Zalocar 1991: *Argentina*
Strombomonas praeliaris var. *praeliaris* fo. *major* Thérézien 1989: *Bolivia*
Strombomonas prismatica var. *triangulata* Conforti 1986 c: *Argentina*
Strombomonas quadrangularis Menezes 1996: *Brazil*
Strombomonas scabra var. *coberensis* (Defl.) Tell et Conforti 1988: *Argentina* (Conforti, 1983; Tell & Conforti, 1988); *Brazil* (Dewes & Huszar, 1986); *Venezuela* (Deflandre, 1926 a; Yacubson, 1980)
Bas.: *Trachelomonas scabra* var. *coberensis* Defl. 1926 a
Strombomonas scabra var. *hiperintermedia* Conforti 1989: *Argentina* (Tell & Conforti, 1988; Conforti, 1989; Zalocar, 1991)
Strombomonas scabra var. *intermedia* (Yacubson) Tell et Conforti 1988: *Argentina* (Tell & Conforti, 1988); *Brazil* (Alves-da-Silva et al., 1991; Conforti, 1994; *Venezuela* (Yacubson & Bravo, 1982–83)

Bas.: *Trachelomonas scabra* var. *intermedia* Yacubson in Yacubson & Bravo, 1982–83

Strombomonas scabra var. *ovata* fo. *caudata* (Tell et Zalocar) Tell et Conforti 1986: *Argentina* (Tell & Zalocar, 1985; Tell & Conforti, 1986); *Brazil* (Conforti, 1993 b)

Syn.: *Trachelomonas scabra* var. *ovata* fo. *caudata* Tell et Zalocar 1985

Strombomonas scabra var. *ovata* fo. *minor* (Defl.) Tell et Conforti 1988: *Argentina* (Conforti, 1983; Tell & Conforti, 1988); *Venezuela* (Deflandre, 1926 a)

Syn.: *Trachelomonas scabra* var. *ovata* fo. *minor* Defl. 1926 a

Strombomonas scabra var. *saculiformis* Conforti 1989: *Argentina*

Strombomonas tambowika var. *magnifica* Zalocar 1991: *Argentina*

Strombomonas tambowika var. *rhomboidea* Zalocar 1991: *Argentina*

Strombomonas tellii Zalocar 1991: *Argentina*

Strombomonas tetraptera Balech et Dastugue in Balech, 1944: *Argentina* (Balech, 1944; Tell & Conforti, 1988); *Brazil* (Menezes & Fernandes, 1987)

Strombomonas verrucosa var. *argentinensis* Tell et Zalocar 1985: *Argentina*

Strombomonas zuliana Yacubson 1980: *Venezuela*

Trachelomonas abrupta var. *arcuata* Defl. 1926 a: *Venezuela*

Trachelomonas abrupta var. *obesa* fo. *minor* Conforti 1993 a: *Brazil*

Trachelomonas acanthophora var. *minor* Balech et Dastugue in Balech, 1944: *Argentina* (Tell & Zalocar, 1985); *Brazil* (Uherkovich & Schmidt, 1974); *Colombia* (Duque, 1995)

Trachelomonas acanthophora var. *rotunda* Balech 1944: *Argentina*

Trachelomonas acanthophora var. *speciosa* (Defl.) Balech 1944 fo. *speciosa*: *Argentina* (Balech, 1944; Conforti, 1983; Tell & Zalocar, 1985); *Bolivia* (Thérézien, 1989); *Brazil* (Alves-da-Silva & Avila, 1995; Conforti, 1993 a); *Venezuela* (Deflandre, 1926 a)

Syn.: *Trachelomonas speciosa* Deflandre 1926 a

Trachelomonas acanthophora var. *speciosa* fo. *minor* Conforti 1989: *Argentina* (Conforti, 1989; Conforti & Tell, 1989)

Trachelomonas acanthostoma var. *bacterifera* Couté et Thérézien 1994: *Bolivia*

Trachelomonas acanthostoma fo. *tridentata* Tell 1980: *Argentina*

Trachelomonas alata Seckt 1922: *Argentina*

Trachelomonas allia var. *allia* fo. *maxima* Uherk. in Uherkovich & Schmidt, 1974: *Brazil*

Trachelomonas allia var. *obesa* Balech 1944: *Argentina* (Balech, 1944; Conforti, 1983)

Trachelomonas allorgei var. *molesta* (Defl.) Balech 1944: *Argentina* (Balech, 1944); *Brazil* (Menezes & Fernandes, 1989); *Venezuela* (Deflandre, 1926 a).

Bas.: *Trachelomonas fusiformis* Deflandre 1926 a

Trachelomonas allorgei var. *sparsispina* Conforti 1993 a: *Brazil*

Trachelomonas amazonensis Conforti 1993 a: *Brazil*

Trachelomonas amphoriformis var. *spinosa* Conforti 1993 a: *Brazil* (Conforti, 1993 a); *Colombia* (Duque, 1995)

Trachelomonas angustispina var. *unicoronata* Conforti 1993 a: *Brazil*

Trachelomonas argentinensis (Gar. de Emil.) Conforti et Nudelman (in press, a): *Argentina* (García de Emiliani, 1983; Tell & Conforti, 1985); *Bolivia* (Thérézien, 1989); *Brazil* (Conforti, 1993 a); *Colombia* (Conforti & Nudelman, in press, a)

Bas.: *Strombomonas argentinensis* Gar. de Emil. 1983

Syn.: *Strombomonas bourrellyi* Tell et Zalocar in Tell & Conforti, 1985

Trachelomonas armata var. *heterospina* fo. *minor* Conforti et Nudelman (in press, b): *Colombia*

Trachelomonas armata var. *longa* fo. *pseudolongispina* Defl. 1926 a: *Brazil* (C. Bicudo & De-Lamonica-Freire, 1993); *Venezuela* (Deflandre, 1926 a)

Syn.: *Trachelomonas armata* var. *steinii* Lemm. (p.p., sec. C. Bicudo & De-Lamonica-Freire, 1993)

Trachelomonas armata var. *nana* Balech 1944: *Argentina* (Balech 1944; Conforti, 1983); *Venezuela* (Yacubson & Bravo, 1982–83)

Trachelomonas armata var. *nana* fo. *spinosa* Conforti et Nudelman (in press, b): *Colombia*

Trachelomonas armata var. *steinii* fo. *duplex* Couté et Thérézien 1985: *Bolivia*

Trachelomonas bacillifera var. *bacillicollis* Uherk. in Uherkovich & Schmidt, 1974: *Brazil*

Trachelomonas bacillifera var. *victoriae* Tell et Zalocar 1985: *Argentina*

Trachelomonas balechii Tell 1980: *Argentina*

Trachelomonas bernardinensis var. *acanthophora* Thérézien 1989: *Bolivia*

Trachelomonas bernardinensis fo. *major* Yacubson 1980: *Venezuela*

Trachelomonas bolivensis Thérézien 1989: *Bolivia*

Trachelomonas caudata var. *intermedia* Yacubson 1980: *Bolivia* (Couté & Thérézien, 1985); *Colombia* (Duque, 1995); *Venezuela* (Yacubson, 1980; Yacubson & Bravo, 1982–83)

Trachelomonas caudata var. *sparsispina* Yacubson 1980: *Venezuela*

Trachelomonas charkoviensis var. *spinicollis* Conforti 1993 a: *Brazil*

Trachelomonas compressa Yacubson in Yacubson & Bravo, 1982–83: *Venezuela* (Yacubson & Bravo, 1982–83)

Trachelomonas cordata fo. *minor* Defl. 1926 a: *Venezuela*

Trachelomonas crebea var. *obesa* Balech 1944: *Argentina* (Balech, 1944; Conforti, 1983)

Trachelomonas crebea var. *spinosa* Balech 1944: *Argentina*

Trachelomonas crispa Balech 1944: *Argentina*

Trachelomonas curta var. *minima* Tell et Zalocar 1985: *Argentina* (Tell & Zalocar, 1985); *Brazil* (Alves-da-Silva & Avila, 1995), *Colombia* (Duque, 1995)

Trachelomonas curta var. *punctata* fo. *minor* Conforti 1993 a: *Brazil*

Trachelomonas curta var. *subbernardii* Bourrelly et Manguin 1952: *Guadeloupe* (Bourrelly & Manguin, 1952)

Trachelomonas curta var. *tubigera* Defl. 1926 a: *Argentina* (Balech, 1944); *Colombia* (Duque, 1995); *Venezuela* (Deflandre, 1926 a; Yacubson & Bravo, 1982–83)

Trachelomonas dastuguei Balech 1944 var. *dastuguei*: *Argentina* (Balech, 1944; Conforti, 1983; Tell & Zalocar, 1985); *Brazil* (Uherkovich & Schmidt, 1974; Menezes & Fernandes, 1987; Alves-da-Silva et al., 1991); *Colombia* (Duque, 1995; Conforti & Nudelman (in press, b)

Trachelomonas dastuguei var. *depauperata* Tell 1980: *Argentina*

Trachelomonas decora Defl. 1926 a: *Argentina* (Balech, 1944; Conforti, 1983; Tell & Zalocar, 1985); *Brazil* (Menezes & Fernandes, 1989; Xavier, 1990); *Venezuela* (Deflandre, 1926 a)

Trachelomonas denisi Defl. 1926 a: *Venezuela*

Trachelomonas duquei Conforti et Nudelman (in press, b): *Colombia*

Trachelomonas flava var. *stella* Balech 1944: *Argentina*

Trachelomonas fragaria Defl. 1927: *Argentina* (Conforti, 1983); *Venezuela:* (Deflandre, 1927)

Trachelomonas gemmata Yacubson 1980: *Venezuela*

Trachelomonas gracillima Balech et Dast. in Balech, 1944: *Argentina* (Balech, 1944; Tell & Zalocar, 1985)

Trachelomonas hemisphaerica Gar. de Emil. 1983: *Argentina* (García de Emiliani, 1983; Tell & Conforti, 1984); *Brazil* (Menezes & Fernandes, 1989; Alves-da-Silva & Avila, 1995; Menezes et al., 1995)

Syn.: *Trachelomonas ovni* Tell et Zalocar in Tell & Conforti, 1984

Trachelomonas hirsuta var. *coronata* fo. *minor* Conforti 1989: *Argentina*

Trachelomonas hirta da Cunha 1913 var. *hirta*: *Argentina* (Balech, 1944; Tell, 1980; Tell & Zalocar, 1985); *Brazil* (da Cunha, 1913; Menezes & Fernandes, 1989); *Colombia:* (Conforti & Nudelman (in press, b); *Venezuela* (Yacubson & Bravo, 1982–83)

Trachelomonas hispida var. *hispida* fo. *minor* Bourrelly et Manguin 1952: *Brazil* (Alves-da-Silva et al., 1991; Alves-da-Silva & Souza Laitano, 1994; Alves-da-Silva & Avila, 1995); *Guadeloupe* (Bourrelly & Manguin, 1952)

Trachelomonas hispida var. *coronata* fo. *minor* Conforti 1989: *Argentina*

Trachelomonas hispida var. *multispinosa* (Tracana) Tell et Conforti 1986: *Argentina* (Tracana, 1985; Tell & Conforti, 1986); *Colombia* (Duque, 1995)

Bas.: *Trachelomonas hispida* var. *duplex* fo. *multispinosus* Tracana 1985

Trachelomonas hispida var. *rugosa* Couté et Thérézien 1994: *Bolivia*

Trachelomonas hispida var. *sudamericana* Tell 1980: *Argentina*

Trachelomonas horrida var. *parvicollis* Yacubson 1980: *Venezuela*

Trachelomonas horrida var. *spinicolis* Conforti 1993 a: *Brazil* (Conforti, 1993 a); *Colombia* (Conforti & Nudelman (in press, b)

Trachelomonas intermedia fo. *crassus* Tracana 1985: *Argentina*

Trachelomonas intermedia var. *minor* Tell 1980: *Argentina* (Tell, 1980); *Brazil* (Franceschini, 1992)

Trachelomonas kelloggii var. *kelloggii* fo. *crenulata* Couté et Thérézien 1994: *Bolivia*

Trachelomonas kelloggii var. *effigurata* fo. *acuminata* Conforti 1993 a: *Brazil*

Trachelomonas kellogii var. *nana* Balech 1944: *Argentina* (Balech, 1944; Conforti, 1983; Tell & Zalocar, 1985); *Brazil* (Uherkovich & Schmidt, 1974); *Colombia* (Duque, 1995)

Trachelomonas lemmermannii var. *spinosa* Conforti 1983: *Argentina*

Trachelomonas magdaleniana Defl. 1926 a: *Argentina* (Tell & Zalocar, 1985); *Bolivia* (Couté & Thérézien, 1985); *Brazil* (Menezes & Fernandes, 1989; Conforti, 1993 a); *Venezuela* (Deflandre, 1926 a)

Trachelomonas mamillosa var. *ovalis* Yacubson 1980: *Venezuela*

Trachelomonas megalacantha da Cunha 1913 var. *megalacantha*: *Argentina* (Balech, 1944; Tell & Zalocar, 1985); *Brazil* (da Cunha, 1914; Uherkovich & Schmidt, 1974; Alves-da-Silva & Souza Laitano, 1994))

Trachelomonas megalacantha var. *crenulatocollis* Bourrelly et Manguin, 1952: *Argentina* (Tell & Zalocar, 1985); *Bolivia* (Couté & Thérézien, 1985); *Brazil* (Uherkovich & Schmidt, 1974); *Colombia:* Conforti & Nudelman (in press, b); *Guadeloupe* (Bourrelly & Manguin, 1952); *Venezuela* (Yacubson, 1980; Yacubson & Bravo, 1982–83)

Trachelomonas megalacantha var. *heterocantha* Tell et Zalocar 1985: *Argentina*

Trachelomonas megalacantha var. *longispina* Tell 1980: *Argentina*

Trachelomonas minuscula var. *glans* Balech 1944: *Argentina* (Balech, 1944); Brazil (Menezes, 1991)

Trachelomonas molesta (Defl.) Defl. 1926 a: *Argentina* (Tell & Zalocar, 1985); *Venezuela* (Deflandre, 1926 a)

Syn.: *Trachelomonas fusiformis* Defl. 1926 a

Trachelomonas multifacies Yacubson et Bravo 1982–83: *Venezuela*

Trachelomonas naviculiformis Defl. 1926 a var. *naviculiformis*: *Argentina* (Tell & Zalocar, 1985); *Venezuela* (Deflandre, 1926 a)

Trachelomonas naviculiformis var. *bourrellyi* (Bourrelly) Hub.-Pest. 1955: *Guadeloupe* (Bourrelly & Manguin, 1952)

Bas.: *Trachelomonas africana* var. *pulchella* Bourrelly in Bourrelly & Manguin, 1952

Trachelomonas neotropica Balech 1944: *Argentina* (Balech, 1944; Conforti, 1983)

Trachelomonas ovata fo. *minor* Tell et Zalocar 1985: *Argentina*

Trachelomonas papillata Yacubson 1980: *Venezuela* (Yacubson, 1980; Yacubson & Bravo, 1982–83)

Trachelomonas papillata var. *ornata* Bravo in Yacubson & Bravo, 1982–83: *Venezuela*

Trachelomonas parvicollis Defl. 1926 a: *Argentina* (Balech, 1944; Conforti, 1983); *Venezuela* (Deflandre, 1926 a)

Trachelomonas planctonica var. *flexicolis* Balech 1944: *Argentina* (Balech, 1944; Conforti, 1983; Tell & Zalocar, 1985; Conforti & Tell, 1989); *Brazil* (Cécy, 1990); *Colombia:* Conforti & Nudelman (in press, b)

Trachelomonas planctonica var. *vermiculosa* Balech 1944: *Argentina* (Balech, 1944; Conforti, 1983; Tell, 1980)

Trachelomonas pseudoarmata Couté et Thérézien 1985: *Bolivia*

Trachelomonas pseudobulla var. *bulloides* Balech et Dastugue in Balech, 1944: *Argentina*

Trachelomonas punctata var. *torosa* Balech 1944: *Argentina*

Trachelomonas pyramidata Couté et Thérézien 1985: *Argentina* (Tell & Zalocar, 1985); *Bolivia* (Couté & Thérézien, 1985); *Brazil* (Menezes & Fernandes, 1987; Menezes, 1991; Conforti, 1993 a; C. Bicudo & De-Lamonica-Freire, 1993); *Colombia* (Duque, 1995)

Syn.: *Trachelomonas armata* var. *trapeciformis* Yacubson 1984–85; *Trachelomonas pyramidata* var. *ornata* Couté et Thérézien 1985; *Trachelomonas armata* var. *litoralensis* Tell et Zalocar 1985; *Trachelomonas armata* var. *steinii* (p.p., sec. C. Bicudo & De-Lamonica-Freire, 1993)

Trachelomonas raciborskii var. *boliviense* Couté et Thérézien 1994: *Bolivia*

Trachelomonas raciborskii var. *pseudodenisii* Tell et Zalocar 1985: *Argentina*

Trachelomonas raciborskii var. *simplex* Couté et Thérézien 1994: *Bolivia*

Trachelomonas robusta var. *robusta* fo. *undulatocollis* Couté et Thérézien 1994: *Bolivia*

Trachelomonas robusta var. *multispina* Couté et Thérézien 1994: *Bolivia*

Trachelomonas rugulosa var. *obliqua* Bourrelly et Manguin 1952: *Guadeloupe*

Trachelomonas rugulosa var. *paralella* (Tell et Zalocar) nov. comb.: *Argentina* (Tell & Zalocar, 1985); *Brazil* (Alves-da-Silva & Avila, 1995); *Colombia* (Duque, 1995)

Bas.: *Trachelomonas rugulosa* var. *rugulosa* fo. *paralella* Tell et Zalocar, Nova Hedwigia 41: 369, Pl. VIII, Fig. 19a–c, 1985

Trachelomonas saccata Lemm., sec. Deflandre, 1926 a: *Paraguay*

Trachelomonas sculpta Balech 1944: *Argentina* (Balech, 1944; Conforti, 1983; Tell & Zalocar, 1985); *Brazil* (Conforti, 1993 a; Alves-da-Silva & Avila, 1995)

Trachelomonas selecta var. *megaspina* Conforti 1993 a: *Brazil*

Trachelomonas similis fo. *obesa* Conforti 1993 a: *Brazil*

Trachelomonas speciosa Defl. 1926 a: *Venezuela*

Trachelomonas spectabilis Defl. 1926 a var. *spectabilis*: *Venezuela*

Trachelomonas spectabilis var. *hemicalvata* Tell 1980: *Argentina*

Trachelomonas spinosa var. *hirsuta* Couté et Thérézien 1985: *Bolivia* (Couté & Thérézien, 1985); *Brazil* (Conforti, 1993 a)

Trachelomonas spirogyra Balech 1944: *Argentina* (Balech, 1944; Conforti, 1983)

Trachelomonas suboblonga Balech 1944: *Argentina*

Trachelomonas superba var. *curvispina* Couté et Thérézien 1985: *Bolivia*

Trachelomonas volvocina var. *derephora* fo. *punctata* Conforti 1993 a: *Brazil*

Trachelomonas volvocina fo. *major* Deflandre 1926 a: *Venezuela*

Trachelomonas volvocinopsis var. *punctata* (Skv.) Bourrelly et Manguin 1952: *Argentina* (Tracana, 1985); *Brazil* (Menezes & Fernandes, 1987; Alves-da-Silva & Souza Laitano, 1994; Menezes et al., 1995; Alves-da-Silva & Avila, 1995); *Guadeloupe* (Bourrelly & Manguin, 1952)

Trachelomonas volzii var. *ovum* Tell 1980: *Argentina*

Trachelomonas yacubsonii Bravo in Yacubson & Bravo, 1982–83: *Venezuela*

Species et varietates excludendae (see Balech, 1944)

* *Trachelomonas critata* Freng. 1932 var. *critata*: *Argentina*
* *Trachelomonas critata* var. *parvula* Freng. 1932: *Argentina*
* *Trachelomonas bonaeriensis* Seckt 1922: *Argentina*
* *Trachelomonas hirsuta* Freng. 1932: *Argentina*
* *Trachelomonas obtecta* Freng. 1932: *Argentina*
* *Trachelomonas ovolum* Freng. 1932: *Argentina*
* *Trachelomonas ovolum* var. *deformis* Freng. 1932: *Argentina*
* *Trachelomonas porteri* Freng. 1932: *Argentina*
* *Trachelomonas proboscidata* Freng. 1932 var. *proboscidata*: *Argentina*
* *Trachelomonas proboscidata* var. *robusta* Freng. 1932: *Argentina*
* *Trachelomonas reticularis* Freng. 1932: *Argentina*
* *Trachelomonas spinigera* Freng. 1932 var. *spinigera*: *Argentina*
* *Trachelomonas spinigera* var. *densispina* Freng. 1932: *Argentina*
* *Trachelomonas spinigera* var. *robusta* Freng. 1932: *Argentina*
* *Trachelomonas vasculum* Freng. 1932: *Argentina*
* *Trachelomonas yberiensis* Freng. 1932 var. *yberiensis*: *Argentina*
* *Trachelomonas yberiensis* var. *sculpta* Freng. 1932: *Argentina*

Acknowledgements

The author thanks Dr V. Conforti and Dr M. Menezes for their constructive comments on the manuscript.

References

Alves-da-Silva, S. A. & I. R. Avila, 1995. O gênero *Trachelomonas* Ehr. emend. Defl. (Euglenaceae) do Parque Zoológico de Sapucaia do Sul e do Jardim Botânico de Porto Alegre, Rio Grande do Sul, Brasil. Iheringia, Sér. Bot. 46: 27–56.

Alves-da-Silva, S. A. & C. De Souza Laitano, 1994. Euglenaceae pigmentadas do Banhado do Jacaré, em um Parque de Proteção ambiental, Triunfo, Rio Grande do Sul, Brasil. Iheringia, Sér. Bot. 45: 89–116.

Alves-da-Silva, S. A., G. C. Ferraz & J. Reis Torres, 1991. Euglenaceae pigmentadas de dois arroios e do rio Jacuí, Regiâo carbonífera do Município de Sâo Jeronimo, Rio Grande do Sul, Brasil. Rev. Brasil. Biol. 51: 813–828.

Alves-da-Silva, S. A. & J. Torres, 1994. Estudo taxonômico do gênero *Phacus* Duj. (Euglenaceae) no Parque Zoológico, Sapucaia do Sul e no Jardim Botânico, Porto Alegre, Rio Grande do Sul, Brasil. Iheringia, sér. Bot. 44: 45–83.

Balech, E., 1944. *Trachelomonas* de la Argentina. An. Mus. Arg. Cs. Nat. Bernardino Rivadavia 41: 221–305.

Bicudo, C. & M. De-Lamonica-Freire, 1993. *Trachelomonas armata* (Euglenophyceae): an evaluation of the diagnostic features in the species. Arch. Hydrobiol. Algol. Stud. 69: 57–66.

Bourrelly, P. & E. Manguin, 1952. Algues d'eau douce de la Guadeloupe et dépendances. Soc. d'Edition d'Enseignements sup., Paris, 235 pp.

Cécy, I. I. T., 1990. A restinga de Pontal do sul. Municipio de Paranaguá, PR-I. Levantamento ficológico (Euglenophyta) e fisico-química. Arq. Biol. Tecnol. 33: 1–79.

Conforti, V., 1979. Contribución al conocimiento de las algas de agua dulce de la Provincia de Buenos Aires (Argentina), V. Physis (Bs. As.), sec. B, 38: 11–19.

Conforti, V., 1981a. Contribución al conocimiento de las algas de agua dulce de la Provincia de Buenos Aires (Argentina), VII. Physis (Bs. As.), sec. B, 39 (97): 47–51.

Conforti, V., 1981b. Contribución al conocimiento de las algas de agua dulce de la Provincia de Buenos Aires (Argentina), VIII. Lilloa 35: 95–101.

Conforti, V., 1983. Morfología, taxonomía y estereoultraestructura de las Euglenophyta del área bonaerense. Ph. D. Thesis, FCEyN (UBA), Buenos Aires, 277 pp.

Conforti, V., 1986a. Contribución al conocimiento de las algas de agua dulce de la Provincia de Buenos Aires (Argentina), X. Physis (Bs. As.), sec. B, 44 (106): 7–12.

Conforti, V., 1986b. Contribución al conocimiento de las algas de agua dulce de la Provincia de Buenos Aires (Argentina), XI. Physis (Bs. As.), sec. B, 44 (106): 13–18.

Conforti, V., 1986c. Contribución al conocimiento de las algas de agua dulce de la Provincia de Buenos Aires (Argentina), XII. Physis (Bs. As.), sec. B, 44 (107): 65–71.

Conforti, V., 1989. Algunos euglenoideos nuevos o interesantes de la Argentina. Cryptog., Algol. 10: 69–79.

Conforti, V., 1991. Taxonomic study of the euglenophyta of a highly polluted river of Argentina. Nova Hedwigia 53: 73–98.

Conforti, V., 1993a. Study of the Euglenophyta from Camaleâo lake (Manaus-Brazil). I. *Trachelomonas* Ehr. Rev. Hydrobiol. Trop. 26: 3–18.

Conforti, V., 1993b. Study of the Euglenophyta from Camaleâo lake (Manaus-Brazil). *Strombomonas* Defl. Rev. Hydrobiol. Trop. 26: 187–197.

Conforti, V., 1994. Study of the Euglenophyta from Camaleâo lake (Manaus-Brazil). III. *Euglena* Ehr., *Lepocinclis* Perty, *Phacus* Duj. Rev. Hydrobiol. Trop. 27: 3–21.

Conforti V. & A. Nudelman (in press, a). *Trachelomonas argentinensis* (Euglenophyta) nov. comb. Bol. Soc. Arg. Bot.

Conforti V. & A. Nudelman (in press, b). Ultrastructure of the lorica of *Trachelomonas* Ehr. from the Colombian Amazonia. Rev. Hydrobiol. Trop.

Conforti, V. & G. Tell, 1988. Les Euglenophytes de la Terre de Feu, Argentine. Nova Hedwigia 46: 305–317.

Conforti, V. & G. Tell, 1989. Ultrastructure of the pellicle and the envelope of some euglenoid flagellates from Argentina by means of S.E.M. Nova Hedwigia 48: 187–206.

Couté, A. & Y. Thérézien, 1985. Première contribution à l'étude des *Trachelomonas* (Algae, Euglenophyta) de l'Amazonie bolivienne. Rev. Hydrobiol. Trop. 18: 111–131.

Couté, A. & Y. Thérézien, 1994. Nouvelle contribution à l'étude des Euglénophytes (Algae) de l'Amazonie bolivienne. Nova Hedwigia 58: 245–272.

Cunha, A. M. da, 1913. Contribuição para o conhecimento da fauna de protozoários do Brasil. Mem. Inst. Oswaldo Cruz, R.J. 5: 101–122.

Deflandre, G., 1926a. Monographie du genre *Trachelomonas* Ehr. Nemours, 162 pp.

Deflandre, G., 1926b. Sur quelques Euglénacées nouvelles du Vénézuéla. Bull. Mus. Nat. Hist. Nat. 6: 421–422.

Deflandre, G., 1927. Remarques sur la systématique du genre *Trachelomonas* Ehr. Bull. Soc. Bot. 44: 285–287, 657–665.

Dewes, J. T. & V. L. M. Huszar, 1986. Influência dos despejos na Sanga da Barbuda sobre as águas do Lago da Barragem Santa Bárbara, Pelotas, RS, Brasil. In C. E. M. Bicudo & J. G. Tundisi (eds), Simpósio Algas: a energia do amanhâ. Inst. Oceanogr. Univ. Sâo Paulo: 141–161.

Duque, S. R., 1995. Euglenofitas pigmentadas de la Amazonia Colombiana. Rev. Acad. Colombiana C.E.F.yN. 19 (75): 651–659.

Franceschini, I. M., 1992. Algues d'eau douce de Porto Alegre, Brésil (les Diatomophycées exclues). Bibl. Phycol., ed. CRAMER, Berlin, Stuttgart, 81 pp.

Frengüelli, J., 1932. *Trachelomonadi* del Plioceno Argentino. Mem. della Soc. Geol. Ital. 1.

García de Emiliani, M. O., 1983. *Strombomonas argentinensis* y *Trachelomonas hemisphaerica*, especies nuevas de Euglenophyta. Rev. Asoc. Cs. Nat. Lit. 14 (2): 239–242.

Hüber-Pestalozzi, G., 1955. Euglenophyta. Die Binnengewässer. Das Phytoplankton des Süsswassers 16(4). E. Schweizerbart'sche Verlagsbuchhandlung, Stuttgart, 1135 pp.

Huszar, V. L. M., M. Menezes & V. O. Fernandes, 1989. Fitoplâncton de rede da lagoa Cabiúnas, Macaé, Rio de Janeiro, Brasil: uma acontribuiçao a seu conhecimento. Bol. Mus. nac. R. de Janeiro, Nova Sér., Bot 78: 1–43.

Menezes, M., 1990. Estudos taxonômicos sobre o gênero *Lepocinclis* Perty no Município do Rio de Janeiro e arredores, Brasil. Revta bras. Biol. 50: 102–113.

Menezes, M., 1991. *Trachelomonas* (Euglenophyceae) do município do Rio de Janeiro e arredores, RJ, Brasil. Hoehnea 18: 57–73.

Menezes, M., 1996. New taxa of pigmented flagellates from southeastern Brazil. Arch. Protistenk. 147: 101–105.

Menezes, M., E. P. do Nascimento & C. G. Fonseca, 1995. Flora dos estados de Goiás e Tocantins. Criptógamos, Goiânia 4: 1–77.

Menezes, M. & V. Oliveira Fernandes, 1987. Euglenaceae pigmentadas do município de Cáceres e arredores, Mato Grosso, Brasil: uma contribuição a seu conhecimento. Rickia 14: 53–71.

Menezes, M. & V. Oliveira Fernandes, 1989. Euglenaceae pigmentadas (Euglenophyceae) do Noroeste do Estado de Mato Grosso, MT, Brasil: municípios de Barra do Bugres, Cáceres, Juína e porto Esperidiâo. Hoehnea 16: 35–55.

Seckt, H., 1922. Estudios hidrobiológicos en la Argentina. Flagellatae. Bol. Acad. Nac. Cs. Córdoba 25: 430–490.

Skvortzov, B. V., 1967. New and interesting species of *Euglena* Ehr. from the subtropics of Brasil. Nova Hedwigia 14: 279–386.

Tell, G., 1980. Les Euglénophytes chlorophylliennes du nord-est de l'Argentine. Bull. Mus. nat. d'Hist. nat., Paris, 4e sér., 2, sec. B, 1: 21–47.

Tell, G. & V. Conforti, 1984 a. Ultraestructura de *Trachelomonas ovni* nov. sp. (Euglenophyta) en MEB. Nova Hedwigia 39: 201–203.

Tell, G. & V. Conforti, 1985. Ultraestructura de *Strombomonas bourrellyi* nov. sp. (Euglenophyta) en MEB. Cryptogamie Algologie 6: 281–286.

Tell, G. & V. Conforti, 1986. Euglenophyta pigmentadas de la Argentina. Bibl. Phycol., ed. CRAMER, Berlin, Stuttgart, 301 pp.

Tell, G. & V. Conforti, 1988. Quelques *Strombomonas* Defl. (Euglenophyta) de l'Argentine au microscope photonique et électronique à balayage. Nova Hedwigia 46: 541–556.

Tell, G. & Y. Zalocar, 1985. Euglenophyta pigmentadas de la provincia del Chaco (Argentina). Nova Hedwigia 41: 353–391.

Thérézien, Y., 1989. Algues d'eau douce de la partie Amazonienne de la Bolivie. Bilbl. Phycologica, CRAMER, Berlin-Stuttgart, 75 pp.

Tracana, B., 1985. Algas del noroeste argentino (Excluyendo las Diatomophyceae). Opera Lilloana 35: 1–136.

Uherkovich, G. & G. W. Schmidt, 1974. Phytoplanktontaxa in dem zentralamazonischen Schwemmlandsee. Amazoniana 5: 243–283.

Xavier, M. B., 1990. O gênero *Trachelomonas* Ehr. de lagos do Parque Estadual das Fontes do Ipiranga, Sâo Paulo, Brasil. Hoehnea 17: 99–112.

Yacubson, S., 1980. The phytoplankton of some freshwater bodies from Zulia State (Venezuela). Nova Hedwigia 23: 279–339.

Yacubson, S. & C. R. Bravo, 1982–83. Especies de *Trachelomonas* (Euglenophyta) de algunos cuerpos de agua de los distritos Urdaneta y Perija (Estado Zulia, Venezuela). Bol. Ctro. Inv. Biol. 15: 17–47.

Yacubson, S., 1984–85. Algas del río Tucuco y ambientes acuáticos de sus alrededores (Estado Zulía, Venezuela). Bol. Ctro. Inv. Biol. 16: 19–95.

Zalocar de Domitrovic, Y., 1991. *Strombomonas* (Euglenophyta) nuevas o raras de la Argentina. Cryptogamie Algologie 12: 271–281.

MIX
Papier aus verantwortungsvollen Quellen
Paper from responsible sources
FSC® C105338

If you have any concerns about our products,
you can contact us on
ProductSafety@springernature.com

In case Publisher is established outside the EU,
the EU authorized representative is:
Springer Nature Customer Service Center GmbH
Europaplatz 3, 69115 Heidelberg, Germany

Printed by Libri Plureos GmbH
in Hamburg, Germany